# Practical DSP Modeling, Techniques, and Programming in C

# Practical DSP Modeling, Techniques, and Programming in C

DON MORGAN

John Wiley & Sons, Inc.
New York • Chichester • Brisbane • Toronto • Singapore

Publisher: Katherine Schowalter
Editor: Diane D. Cerra
Managing Editor: Micheline Frederick
Text Design and Composition: Electric Ink, Ltd.

This text is printed on acid-free paper.



***Library of Congress Cataloging-in-Publication Data***

Morgan, Don, 1948–
Practical DSP modeling, techniques, and programming in C / Don Morgan.
p. cm.
Includes index.
ISBN 0-471-00606-8 (acid-free paper : paper). — ISBN 0-471-00434-0 (paper/disk).— ISBN 0-471-00613-0 (disk)
1. Signal processing—Digital techniques. 2. C (Computer program language) I. Title.
TK5102.9.M67 1995
621.382'2'0285416—dc20 94-32688
CIP

Printed in the United States of America

10 9 8 7 6 5 4 3 2 1

*To Anita, Rachel, and Don*

# Contents

PREFACE xiii

CHAPTER 1

**What Is a Signal? 1**

An Introduction 1

Concepts and Definitions 4

*The Two Domains 4*

*Continuous and Sampled Data 4*

*Orthogonality 5*

*Transfer Function 7*

*Time Invariance 8*

*Causality 8*

*Linearity 8*

*Characterization of Systems: Impulse Response 9*

*Convolution 11*

Mathematical Foundation 12

*Linear Independence 13*

*Orthogonality 14*

*The Norm 15*

*Orthonormality 15*

*Basis Functions* 17
*Approximating Bases* 17

System Response 19
*The Simple Harmonic Oscillator* 19
*Forced Oscillators* 23
*The LRC Harmonic Oscillator* 24
*Poles and Zeros* 29
*Q and Resonance* 31

Harmonic Analysis 33
*Fourier Series* 33
*Odd and Even Functions* 37
*The Gibbs Phenomenon* 38
*Fourier Series in Exponential Form* 39
*The Fourier Integral* 41
*Initial Conditions, Transients, and Steady State* 42
*The Laplace Transform* 44
*Properties of the Laplace Transform* 46
*Network Analysis* 49
*Magnitude Squared Response* 51

**CHAPTER 2**

**The Real World** **53**

Part One: Seven Inches of Copper Wire 54
*Electronic Circuits* 54
*Data Acquisition* 54
*What the Filter Needs to Do* 56
*Filter Requirements* 57
*Oversampling* 58
*Filters* 58
*Choosing a Filter Class for an Application* 62
*A/D Conversion for DSP* 84
*D/A Conversion for DSP* 86

Part Two: Designing Analog Filters 86
*Introduction* 86
*The Frequency-Dependent Voltage Divider* 87
*Resonant Circuits, Q, and Higher-Order Filters* 90

*Scaled Data 94*
*Filter Design 94*
*Operational Amplifiers in Active Filters 116*
*DACQ 120*

CHAPTER 3

**Signal Analysis** **133**

Introduction 133
*Analysis and Synthesis 134*
*Time Domain Response 134*
*Frequency Domain Response 135*

Spectra 136

Fourier Series 137
*Odd and Even 140*

Fourier Transform 143
*Sine and Cosine Transforms 145*

Discrete Time Theory 146
*The Sampling Theorem: An Intuitive Approach 146*
*Sampling and Aliasing: Two Viewpoints 147*
*Sampling Functions 151*
*Decimation or Downsampling 152*
*Interpolation 153*

Introduction to the Fourier Transform 153
*Theorems Relating to the Fourier Series and Transform 156*
*The Discrete Fourier Transform 159*
*Theorems of the Discrete Fourier Transform 161*

Computing the DFT 162
*The Fast Fourier Transform 166*
*Bit Reversal 170*

The Hartley Transform: A Completely Real Transform 171
*Hartley Transform Theorems 172*

Windowing 173
*Window Types 177*

Software Examples 179
*'Classic' FFT in C: Decimation in Time 179*

*A Faster FFT 181*
*Real and Complex FFTs and IFFTs 186*
*The Discrete Hartley Transform 187*
*The Fast Hartley Transform 189*
*A Fast Hartley Transform Algorithm 193*
*A Windowing Function 196*

**CHAPTER 4**

**Discrete Time Filters** **201**

Introduction 201

Basics of a Discrete Time System 203
*Relationship between Laplace and Z Transforms 203*
*The Z Transform and the Fourier Transform 204*
*Region of Convergence of the Z Transform 206*
*Inverse Z Transform 208*
*Properties of the Z Transform 211*
*Computing in Discrete Time: Difference Equations 211*
*Symbols for Block Diagrams 213*

Linear Time Invariant Filters in Discrete Time 213
*Filtering with the Fourier Transform 217*
*FIR 220*
*FIR Filter Design Techniques 224*
*Convolution 231*
*Convolution in the Time Domain 235*
*Discrete Time: Infinite Impulse Response Filters 239*
*Frequency Transformation 246*

Filter Structures 246

FIR Cascade Structures 248
*IIR Structures 249*
*The IIR Lattice Network 253*

Numerical Methods 254

Software for Discrete Time Filters 255
*Convolution in Frequency Domain 255*
*Generating FIR Filter Coefficients with Fourier Series 256*
*Sum of Products 258*
*An FIR Implementation: Convolution in the Time Domain 258*

*A Realtime FIR Filter 260*
*The Direct Form IIR Filter 261*

CHAPTER 5

**Using Digital Signal Processing** **263**
Introduction 263
*Precision and Accuracy 263*
Finite-Length Arithmetic 264
*Fixed-Point or Floating-Point 264*
*Fixed-Point Arithmetic: Creating and Interpreting 1.15 268*
Sources of Error in Finite-Length Arithmetic: Range 270
Quantization of Filter Coefficients and Intermediate Values 270
I/O Quantization 273
Computational Delay 274
Computational Aids 274
*Division 274*
*Pseudo-random Number Generator 276*
*Polynomial Evaluation: Horner's Rule 276*
*Useful Approximations 277*
*Tables in ROM 278*
*Tables: Interpolation 278*
*Tables: Sine and Cosine 279*
*Circular Buffers: Software 281*
*Bit Reversal: Software 282*
*Hardware Support: Generating ROM Tables 283*
Digital Signal Processes on a Non-DSP 288

CHAPTER 6

**DSP Hardware and Software** **299**
Introduction 299
Desirable Features in a Digital Signal Processor 300
Architectures of Specific Chips 302
*320C30 302*
*Analog Devices ADSP-2101 312*

CHAPTER 7

**Two Final DSP Applications** **327**

Introduction 327

Dedicated Spectral Analysis: The Goertzel Algorithm 328

Adaptive Filtering 331

*System Modelling or Identification 335*

*Adaptive Prediction 336*

*Adaptive Channel Equalization 336*

*Echo Cancellation 336*

GLOSSARY **337**

BIBLIOGRAPHY **343**

ABOUT THE SOFTWARE **347**

APPENDICES **351**

**Appendix A: 12-Bit + Sign Data Acquisition System with Self-Calibration** **353**

**Appendix B: Dual FET-Input, Low Distortion Operation Amplifier** **391**

**Appendix C: 8th Order Continuous-Time Active Filter** **405**

**Appendix D: Voltage-Output, 12-Bit DACs with Internal Reference** **425**

INDEX **437**

# Preface

A few years ago, DSP meant high tech, leading edge. It also meant expensive consultants and unfamiliar equipment. To some people, it meant it was time to go back to school to retread such arcane technologies as network analysis and Fourier transforms. Whatever it meant to you, signal processing and digital signal processing technologies have become so prevalent in every industry that no engineer can ignore their presence. From innovative sorts of musical instruments to analyzing the soil on Venus to echo cancellation on the phone lines, applications of signal processing are common. Digital signal processing has certainly touched almost every area of science and technology of interest today, and will soon do the same for the commercial marketplace.

DSP hardware offers tremendous advantages. The processors are optimized for high-speed arithmetic, many of them have an architecture that allows them access to programs and data simultaneously, and most have an instruction set with parallel instructions designed to speed or eliminate the overhead in filter algorithms and transformation processes. These capabilities find use in many applications, from real-time control to audio and video processing. But it is not only the increase in throughput that makes a difference. With the signal processing in code and not in hardware, we can make filters with virtually perpendicular transition bands, and we can lessen the reliance on analog components that vary both in production and with time and temperature. Turn-around time on changes is faster, and time to production is shorter. All of this and more makes this technology attractive and cost-effective.

My intention in writing this book is to provide a systematic and multidisciplinary view of digital signal processing that will benefit the software engineer, hardware engineer, and student. The mathematical, software, and electronic design and development are deliberately focused to the needs of the subject. During the course of this book, we will approach a number of topics common to the different and sometimes disparate disciplines involved in signal processing again and again. This serves a dual purpose. First, each discipline has its own viewpoint on the subjects we will be discussing which needs to be presented in its own setting, and, second, the opportunity to see a subject from another perspective often enhances one's understanding within his own discipline. We begin with a short tour of the mathematical foundation of signal processing, in the chapters that follow we will see how this basis translates into electronics, both analog and digital, and software. Finally, we see how all of these come together in general practice and in certain digital signal processing applications. My aim is to provide real and useful information pertinent to the subject and its application, to provide tools and capability where I can, to speed the work of the engineer and to reduce the time spent fumbling through book after book looking for a reference. Along with the theory for analog and digital filters, there are schematics for the popular analog filter configurations, along with step-by-step instructions to determine the value of the components. I have made a real effort to write and explain the software in terms of the fundamental mathematical theorems involved so that it isn't simply a matter of copying, but of understanding and control. In order to get real experience, I include the schematics for a data acquisition unit, DACQ, that the reader can build, along with the design strategies and supporting mathematics. The reader can test the routines in the book on the DACQ as well as his own. Throughout, I use illustrations and Mathcad documents to clarify the material as much as possible.

A single caveat: although this is a basic book in digital signal processing, it does require knowledge of at least first-year calculus and the understanding of the principles and operations of microcomputers and programming, as well as knowledge of C and some knowledge of 80X86 assembly language. To make full use of the continuous-time filter design data, a good understanding of analog electronics is required. In addition, the reader will need at least a 386-based computer with a math coprocessor. It is not necessary for someone interested only in the software development to fight his way through the analog design material, nor must someone whose only interest is in learning how to design hardware for DSP applications understand all of the material on software. Since this is a multidiscipline approach, the more technical sections are divided into two parts–one for those who wish a more general understanding, and one for those who have a more specialized interest.

In the text and on the optional disk, Mathcad documents provide further information and explication for each of the topics in the book. These documents and Mathcad allow you to play with the concepts and watch the results change in the form of charts, graphs, or numerical vectors. This way, you can try out a new idea or test your understanding of what is in the text.

The first chapter develops the mathematical basis for signal processing and digital signal processing using simple linear algebra as a basis. We cover orthogonal functions, the Fourier series, the Fourier integral, the basics of Laplace transforms, and the mechanics of the simple harmonic oscillator.

The second chapter takes up the concept of the filter, and covers basic analog design data on how they work and what you need to know to choose the proper components for an application. The chapter ends with the specification for the DACQ circuit.

Chapters 3 and 4 cover the Fourier and Hartley transforms, with some discussion of other useful transforms as well. C code and assembly language routines are given for the Fourier transform. These algorithms also run on the data acquisition unit and display their results graphically on the screen. Later, we take up digital filters and develop software that the reader can use to input data, process it and output it again to a disk, the screen or the DACQ data acquisition unit to drive a speaker or oscilloscope.

Chapter 5 is an extension of Chapters 3 and 4, in that it explores the more practical aspects of implementing the subjects covered in these chapters on *real* processors in *real* systems. Here, we deal with the effects of finite-word arithmetic, I/O and coefficient quantization noise, and fixed-point arithmetic. In this chapter, we also cover some techniques for attaining some of the advantages of the DSP chips on more general-purpose processors. Among the topics here are circular buffers, implementing the butterfly, and using ROMs to pre- and post-compute data.

In Chapter 6, the advantages of application-specific hardware, such as the DSP ICs currently available, are discussed in terms of both hardware and software.

Chapter 7 is a discussion of the uses of the DSP chip and the technology.

Finally, the book ends with a glossary and several appendices with tables of useful mathematical and electronic tables pertinent to the text.

Clearly, in a subject as large as this, there are many aspects. Just as clearly, not everything can be covered in a single book. It is the object of this book, through the material and the manner in which it is presented, to give you a *working* understanding of digital signal processing, rather than simply a theoretical one. I have made every effort to provide what I would want in such a text; I sincerely hope that it is useful to you.

*Don Morgan*
November 1994

The DACQ diagram found on pages 128–129 can be ordered in a large 13½ × 20¼ size directly from the publisher by sending your request to

Tammy Boyd
John Wiley and Sons, Inc.
605 Third Avenue
NY, NY 10158

or by sending email requests to tboyd@jwiley.com

*Very special thanks to Qi Zou for all her help*

# 1

# What Is a Signal?

## AN INTRODUCTION

A signal is an abstraction, an indication, a sign. Because it is an abstraction, there is no absolute criterion it needs to fill, no special color or form. Signals are found in the conduction of heat, electrical flow in a circuit, chemical changes in the brain; they are common to every human sense and everything we perceive. Signals wiggle at high frequencies and low, sometimes at regular intervals and sometimes with no perceiveable regularity whatsoever.

The very nature of our communication with the physical universe is with the motion we call signals. They are ubiquitous and of every form, they are the content and perception of everything we know. The brain communicates over a telegraph-like network of nerves with the various parts of the body; we use the telephone, radio, and television to remain in contact with other parts of our world. The sun lights the fields, birds and babies cry, fires warm, and smells remind us of places we have been. But none of this tells us much about the nature of signals or gives us much handle upon them.

Prior to Fourier and Descartes, there was no reference point for the study or analysis of signals—no way to quantify them, measure them, or deal with them in any rigorous mathematical form. In 1822, Fourier published a treatise describing a trigonometric series whose constants, or coefficients, are derived from the integration of sine and cosine functions, which can be used to describe an arbitrary function given in the interval $-\pi$ to $\pi$. What he described was a system he developed during his studies of heat conduction that could be used to represent periodic waveforms with trigonometric functions. Not all of the implications of what he wrote were clear at the time, but others, such as Euler and Laplace, continued the discussion and expanded upon the theory, enhancing its rigor and, at the same time, giving it more meaning. Representing signals with trigonometric functions means that we can use the same arithmetic on conduction of heat or image processing as we use to measure land.

By placing a signal or function on a set of axes, we can relate its motion to at least one independent variable. It provides us with a graphic representation of the changes that occur in one plane as corresponding changes occur in the plane of the independent variable. The axes also become reference points for viewing and measuring the function or signal. This independent variable is often (but not always) time—we define sound as a pressure in space relative to time, but the independent variable in the definition of Poisson Wave is distance. In addition, a signal may have more than one independent variable—the number of independent variables associated with a signal defines its dimension. The image on a computer monitor is an example of this: the brilliance of any point is a function of two independent variables, the X and Y positions. The ability to quantify and measure signals in such a manner was the beginning of the development of a tool that makes both analysis and synthesis of signals and functions possible. Before this development, such measurements and analyses were as unthinkable as weighing the sunset.

Since the time of Fourier, Euler, and Laplace, a great many people have contributed to the study of signals. Unfortunately, the mathematics involved proved quite daunting for general use, leaving this study mostly in the hands of mathematicians and theoreticians. Until this century, calculators had not come too far and were mechanical nightmares. Then, prior to World War II, George Philbrick began creating computing elements for what he called *computors* comprising analog units called *operational amplifiers*. With these amplifiers, he was able to approximate any arithmetic process desirable, from addition, subtraction, multiplication, and division to integration and differentiation. Devotees approached the subject with excitement and enthusiasm. Philbrick Researches, Inc. developed circuits and published technical books and articles that dealt with, among other things, performing mathematical functions with operational amplifiers. Analog computers quickly found their way into universities, military installations, and some businesses during the 1950s and '60s. Digital computers were also being developed, but they were slow and very much larger than the analog units.

The analog computer accomplished its task quite elegantly, but the components that comprised them changed with time making manufacturing, tuning, and maintenance somewhat troubling. Digital computers, on the other hand, changed little with time but their resolution was limited and they were far too slow to approach anything like real time operation. With the rediscovery of the fast Fourier transform, the digital computer was increasingly used, producing results in hours as opposed to days. As it shrank in size and became easier to maintain and configure, the popularity of the digital computer continued to grow. It was not until the advances made in CPU speed and the development of the Digital Signal Processor, however, that the focus really shifted from analog computing to digital. Even so, analog elements remain in filter technology and signal preparation, and their design still possesses the mystique of an art.

In a digital computer, the march of events is controlled by a clock of predictable and finite resolution. Data is stored in a finite element called *memory*, which is organized into groups of flags known as *bytes* and words with unique addresses. Unlike analog computers, there is a very definite limit to the resolution of mathematical results, and that limit, if it is not constrained by the IEEE 754[1] specification, is constrained by the amount of available memory. On the other hand, digital filters do not change with time; they are independent of the components they are made of, and they are infinitely changeable—and in relatively short order.

The DSP is a different sort of processor than those found in common personal computers. It is designed for a narrow range of applications involving arithmetic operations important to signal processing. Most DSPs feature a high level of parallel processing and pipelining, resulting in a very high throughput, making digital filtering and real time Fourier transforms possible.

Today, we fashion signals to please our senses, such as electronic musical instruments or the special effects found in motion pictures. We use signals as an aid to learn more about ourselves and our environment. Scientists throw radio waves against the moon and distant planets to see what effect their atmospheres and solid parts have on them. The ability to analyze the frequency content of sound allows engineers to enhance selected aspects of a recording. This advance has led to sound equipment that can reproduce a recording as it originally sounded by shaping the sound to its current environment. Signal processing techniques allow us to change the characteristics of a given signal, such as impressing multiple and distinct signals on one carrier, or to unwrap the changes that have already occured to signals, such as noise removal or analysis.

This chapter describes some of the basic definitions and concepts necessary for understanding and applying signal processing techniques. There are several fundamental thoughts in this chapter that are an important part of all of signal processing. The first involves the structure that provides definition for the rest of the subject. These concepts are taken from linear algebra; they concern linearity, orthogonality, and basis functions that provide a mathematical framework for the study of signals. The next has to do with the simple harmonic oscillator and how its characteristics can be used to model system response and filters. There are mathematical and graphic keys here that can help in the development and understanding of discrete and continuous-time systems. Finally, we take up the analysis of signals with the technologies of the Fourier series and integral, as well as the Laplace transform. The subjects are broken into these areas so that they may be referred to separately; this allows the reader to see how each touches upon common subjects from a different angle.

Unfortunately, there is not room enough for a complete discussion on any one of these topics. What discussion there is can only be regarded as reference. For a more complete understanding and a glimpse at a fascinating set of interrelationships that underpin signal processing in particular and form a pervasive web throughout our sciences in general, a bibliography is available at the end of the book.

---

[1] IEEE 754 and IEEE 854 were intended as a definition and standardization of floating point representation for personal computers. They have become a standard in computing generally, though no one actually follows all of the specifications completely. These specifications define the number of bits required in the external and internal representations of the *long real* and *short real* bits so commonly used in software packages today. Most popular floating point chips also subscribe (to some degree) to this specification.

This book takes an overview of the subject of digital signal processing. It is not necessary for the reader to understand all the mathematics to be able to use much of what is in this book—although, obviously, the better the understanding, the more facile the thought process to any solution. Not every reader has the same goals or needs, so the text has been divided into two parts. The first part briefly and simply covers those concepts that will be necessary for the understanding and application of the material in the rest of the book—these are the "Concepts and Definitions." The second part presents the mathematics and physics of the subject in more detail. This part is for those readers who wish a brief reminder or simple reference as they proceed through the book; it begins with "Tools."

# CONCEPTS AND DEFINITIONS

## The Two Domains

We will be dealing with analog and digital systems in both the time and frequency domains. In each case, the *domain* we speak about is the range of the independent variable that the system is referred to. In the time domain, the solutions for a function, $f(t)$, depend upon the point in time passed to the function for evaluation. The results may then be displayed or graphed as a waveform. In the frequency domain, the situation is the same, with the solutions as amplitudes and the result as a *spectrum,* which again may be graphed, relating the contributions of a range of frequencies to the function of interest.

## Continuous and Sampled Data

The input, or signals, that these systems will operate upon can be either *continuous* or *discrete*. Continuous signals have a value for every instant of time, no matter how small, and are commonly called *analog* signals (see Figure 1-1A). As you can imagine, tabulating the values for the continuous-time signal would be an impossible job, since there is no measurement interval small enough that it could not be divided again. There isn't any way in a finite environment to deal with such data—limits on memory, speed, and arithmetic word size require finite time to acquire, process, and store. To represent continuous-time data in a finite system, discrete samples of the input, like snapshots, are taken at regular intervals, quantized, and stored as binary data. The number of such samples per second is called the *sample frequency*, and the time from sample to sample is called the *sample interval.* Figure 1-1B illustrates the results of a discrete time data sampling function. Here, the height of the line depicts the magnitude of the function at that particular point.

According to the *sampling theorem*, sampling can provide enough information about the subject signal to completely recover all the information involved, as long as there are at least two samples for each cycle of the highest *bandwidth* of interest, and the signal itself is bandlimited to the bandwidth of interest. This means that we must sample at twice the bandwidth of the signal we wish to evaluate, and we must remove any content of frequencies outside that bandwidth. Under these circumstances, sampling will tell us everything we need to know about the signal.

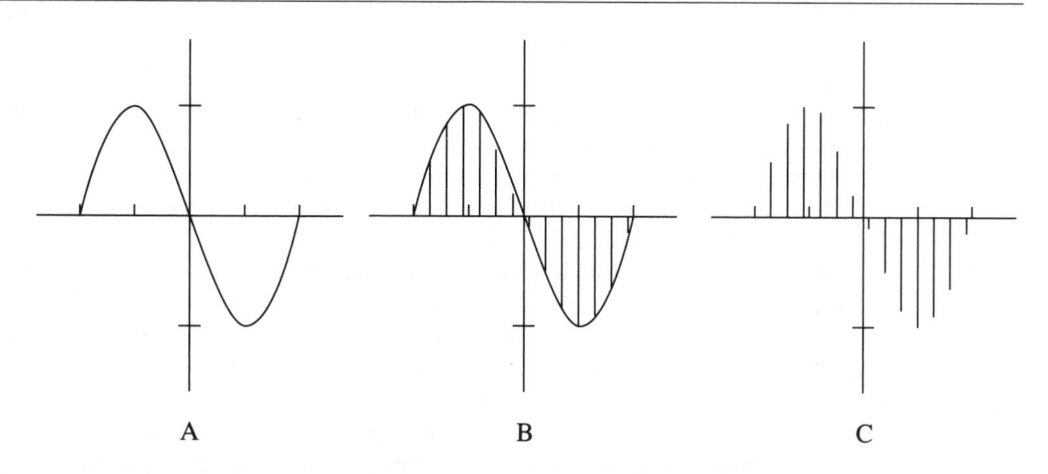


**Figure 1-1.** Analog signal and sampled data. Part A represents an input analog signal, part B is that signal referenced to sample points, and part C shows samples only.

While analog data is often presented in the form of a continuous function that can be evaluated for an arbitrary point in time, discrete time data takes the form of a sum. In Figure 1-1, we see two forms of data. The sinusoid in part A might take the mathematical form

$$h(t) = A\sin\omega t + B\cos\omega t$$

while the sequence of sampled data in part B would take the form

$$x(n) = \sum_{k=-\infty}^{\infty} x[k]\delta[n-k]$$

where $\delta[t]$ is the impulse function.

## Orthogonality

Intuitively, it is not difficult to understand the concept of orthogonality or mutual exclusivity. In order to give someone directions, one needs a reference—"turn left, then right, go, stop" is not guaranteed to get anyone anywhere. It is the same here: we need to be able to describe what we are working with, and to do so we need unique reference points. We do this by defining a system of coordinate systems that we can use to plot data points or vectors. To this end, René Descartes (1596–1650) proposed an axial system for defining space in his 1637 publication of *Discours de la Méthode*. The pair of axes we are so familiar with from geometry and trigonometry are known as the Cartesian axes, though he apparently started with only one axis. Euler (1701–1783) was the first to use two axes, and it was he who contributed a large part of modern notation.

The axes we describe are mutually exclusive—that is, they are independent of one another. That means that they do not mingle, and any point occupies a unique address or specifier relative to each axis. Clearly, if there is any dependency, the definition and results are much more difficult to deal with.

Assuming that each axis is normalized to a unit vector in the direction of increase for each axis, the coordinates for the X axis are (1,0), and the Y axis, (0,1). As you can see, there is no evidence of the X axis in the Y and vice versa.

In Figure 1-2, the point is given meaning within the space prescribed by the axes by its location relative to each axis vector. The coordinates of a point are found by forming a line perpendicular to each axis to the point of interest—the origin of that line is that point's coordinate in terms of that axis. If it is a three-axis system, the point will have one coordinate for each axis, P(x, y, z). A sequence of points, then, would have a sequence of coordinates, each coordinate uniquely describing each point in terms of the reference axis.

On a two-axis system, if we name one axis *time* and the other *amplitude*, we can plot the output of an oscillator upon them. In Figure 1-3, we are using an oscillator producing a tone at exactly 440 cycles per second. Using these axes, this plot represents only one tone and no other. If another tone fits precisely the same set of coordinates, then we say that it is the same tone.

We can say that the sinusoid in the graph is the voltage appearing across the output of the oscillator. The frequency of a periodic function $p=2\pi f$ describes how many times a second

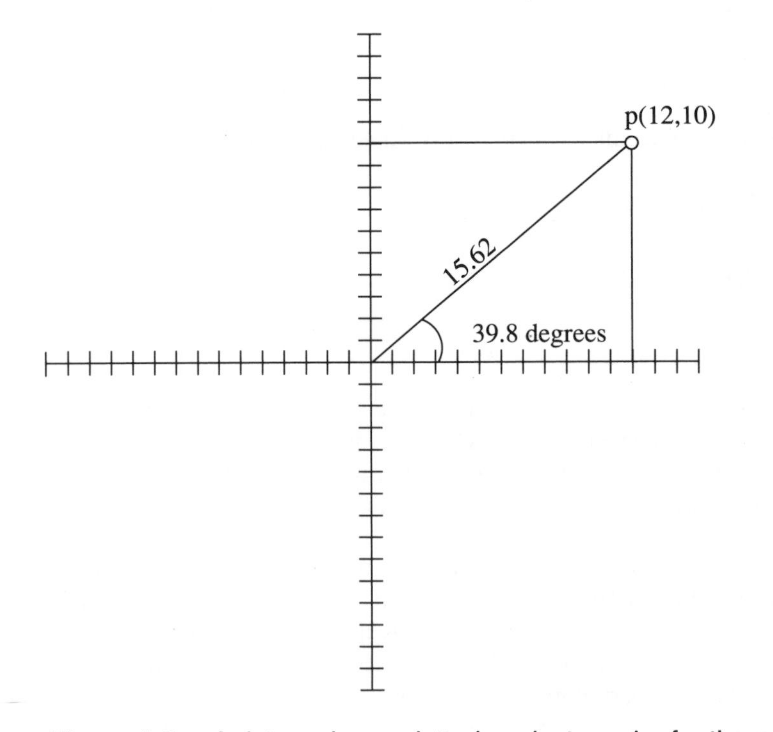


**Figure 1-2.** A datum shown plotted against a pair of orthogonal axes.

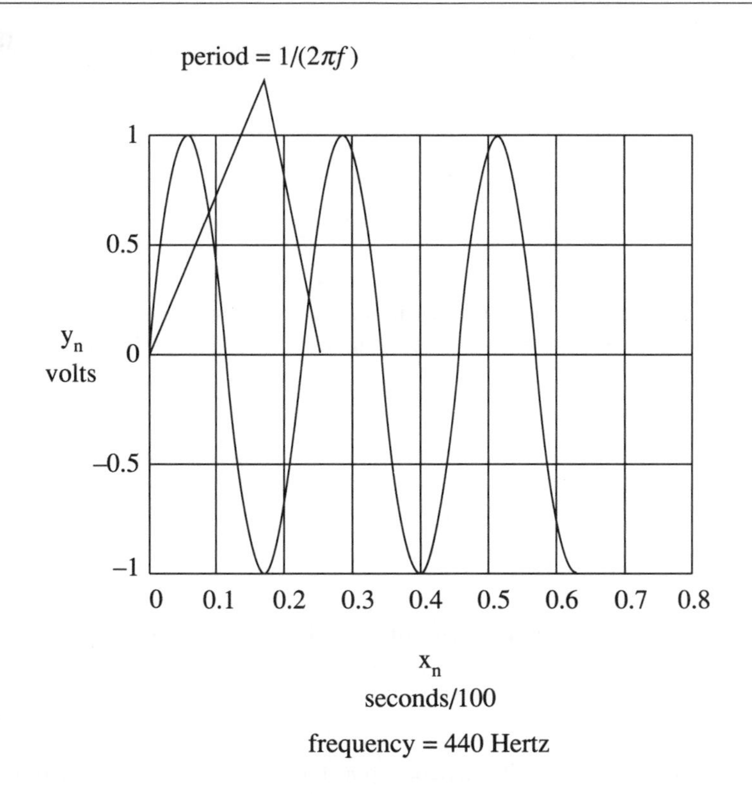


**Figure 1-3.** An analog signal with a frequency of 440 Hertz projected on a coordinate system. The amplitude is in volts, and the period is in seconds.

a complete cycle occurs, in this case 440. At this frequency, one complete cycle takes approximately .002 milliseconds to complete—this is called the *period of the oscillation* and is represented on the graph with the X or time axis. The Y axis is labelled *volts*, and shows the amplitude of the signal, both positive and negative.

This is an illustration involving two axes, but there could be many more. A polynomial of $n$ order will be described by $n$ coefficients relative to $n$ axes. In addition, there are several other coordinate systems of interest to us: the *complex plane* provides a graphical representation of frequency and damping useful in the analysis of networks, and the *z plane* is used for discrete time signal processing. The latter is created by wrapping the imaginary axis of the complex plane around the whole of the left half of the complex plane.

## Transfer Function

These operations are expressed as a ratio of transforms; that is, the transform of the input to the transform of the output. This ratio is called a *transfer function*. For example, assume a filter whose input may be expressed as $A\cos(s)$ and whose output is $B\cos(s)$. We can express the transfer function of such a filter with the equation

$$H(s)=\frac{B\cos(s)}{A\cos(s)}$$

## Time Invariance

Time invariance means that a time shift in an input sequence causes a corresponding shift in the output sequence. We can say that a system is time invariant for all $T$, if the input:

$$x[t]=x[t-T]$$

produces an output:

$$y[t]=y[t-T]$$

## Causality

A system is *causal* if its output at a certain time $t$ is dependent upon its input at time $t$ and its history. Thus:

$$y[t]=H[x[t]] \quad t\leq T$$

A noncausal system does not strictly depend upon the current and previous inputs for its output. One possibility for a noncausal system is that it can anticipate inputs—that is, it can predict the future. As you can imagine, there are many wonderful systems and filters that might be implemented if realtime systems could predict the future, although to this point, they cannot. Most systems cast as noncausal can be rewritten as causal by moving the temporal reference point. For example, the equation for forward difference $y[n]=x[n+1]-x[n]$ will work only if we know the next datum in a sequence, which we would not in a causal system. However, it may be rewritten for a causal system in this fashion: $y[n]=x[n]-x[n-1]$. Depending upon the application, this may serve as well.

## Linearity

Linear algebra describes linearity in rigorous terms. For the purposes of this discussion, two aspects (and their inverses) are important. These two properties are important to all of mathematics. They are necessary for the understanding of finite and infinite sums and for the Fourier series in particular, and they form the basis for Kirchoff's and Thevenin's laws.

1. *Multiplicative*. An input to a system, $x$, produces response $y$. The multiplicative property states that multiplying the input, $x$, by a constant, $a$, will produce the same results as multiplying the output by that same constant. Mathematically, we can express this as $T(ax)=aT(x)$.

The multiplicative property is referred to in some of the literature as *homogeneity* or *scaling*. Multiplicativity also performs division by multiplying by the reciprocal of the constant $a$:

$$T\left(\frac{1}{a}x\right)=\frac{1}{a}T(x)$$

This provides for the possibility of reversing operations of convolution and scaling.

2. *Superposition*. The sum of two (or more) input data produces an output equal to the sum of the outputs produced by the two input data separately, or $H(x+y)=H(x)+H(y)$.

   The principle of superposition, sometimes referred to as *additivity*, states that if we have multiple forces acting on an object, we can solve for the effect of each force individually, and obtain the total effect of all the forces by the sum of the effects of the individual forces. Imagine a small boat crossing a river. In the process, it is affected by cross currents, some undercurrents, and small pools of quiet water. We can predict the direction of travel by solving for the individual vectors resulting from the forces exerted by each of these influences and performing a summation.

   This means that a force considered to be the sum of other forces can be represented as a sum (and its response can also be represented as the sum of the responses of the other forces). In other words, the total force is comprised of its pieces, and its response is comprised of the response of its pieces. At any point in time, then, the state of a signal is the sum of the state of its components. For an electrical network, this sum might be the sum of the currents in its various branches; for a Fourier series, this is the sum of the contributing harmonics at any specific point in time; for a sequence of samples, it can be the sum of the sample history up to a certain point.

## Characterization of Systems: Impulse Response

Clearly, if we are to make or analyze systems, we must characterize them in order to know how the network will respond to or alter an arbitrary input. One way would be to use ultra-precise, broadband sweep generators to input all possible frequencies of given amplitude and phase. We could then attribute any change in the output of the network over the input to the characteristics of the network itself. This process would be difficult and time-consuming, if not impossible.

There is another way. We can input a unit impulse to the network and analyze the results in the frequency domain. This is possible because we find, using the Fourier integral, that the spectrum for a pulse is

$$H(\omega)=\int_{-\infty}^{\infty}f(t)e^{-j\omega t}dt=\tan^{-1}\left(\frac{\sin\left(\frac{\omega T}{2}\right)}{\frac{\omega T}{2}}\right)$$

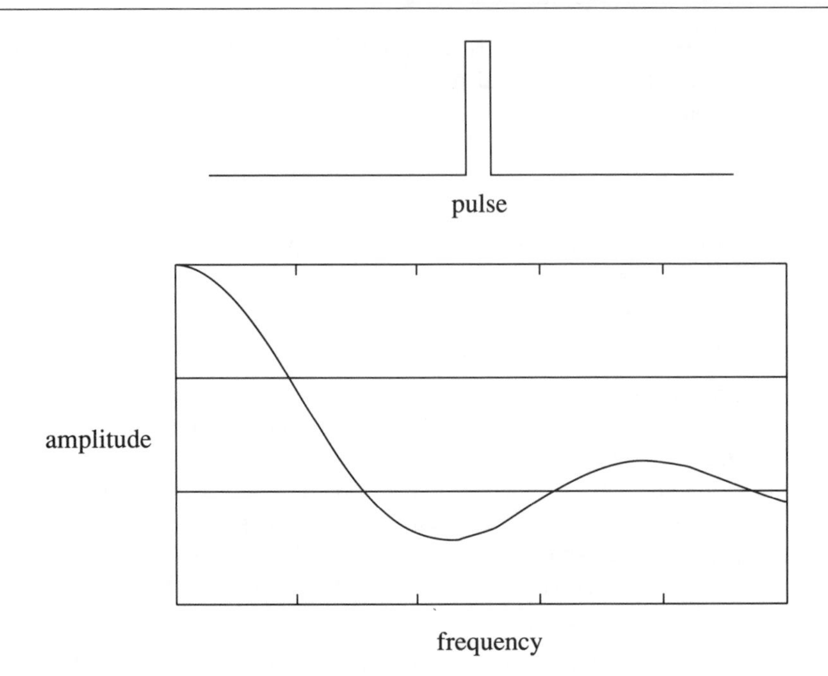


**Figure 1-4.** A unit impulse and its spectrum.

The bandwidth of the spectrum depends upon the width of the pulse. Figure 1-4 is an example of a spectrum of a pulse. The frequency content of the pulse is contained within the area under the curve—the main lobe and side lobes show the magnitude of all the contributing frequencies. As we make the pulse narrower, we find that its bandwidth becomes broader—and since it is of unit area, it becomes taller. This continues until it is narrower than any instrument can measure, resulting in a flat bandwidth containing equal amounts of all frequencies. At this point, it is also taller than any instrument can measure. Applying this pulse to a network is the same as using the sweep generators mentioned earlier.

This impulse, also known as the *Dirac delta function*, symbolized $\delta(t)$, is responsible for a great deal of discussion. There is considerable disagreement over exactly what it is, whether it is an impulse or a distribution, and exactly how to express it. For our purposes, we shall simply say that this impulse is shorter than any equipment we possess can measure and will continue to be so; it is also of unit area, thereby producing a normalized sample. The Dirac delta function is often used in summations to represent the particular sample in a sequence of data, and is therefore also known as the *sample function*. An example is given below:

$$x[n] = \sum_{k=-\infty}^{\infty} x[k]\delta[n-k]$$

where

$$\delta[n]=\begin{cases}0, & n\neq 0\\ 1, & n=0\end{cases}$$

Other functions used in the characterization of systems are the *step function* and the *ramp function.* Both of these functions can be derived from the impulse function by integration. The step function is the first integral of the impulse function, and the ramp function is the first integral of the step function. Figure 1-5 provides examples of the impuse, step, and ramp functions.

## Convolution

Central to the subject of signal processing is the property of convolution. Convolution evolves directly from the concept of superposition described earlier—in fact, the superposition integral and convolution integral can be written so that they appear exactly the same:

$$y(n)=\int_{-\infty}^{\infty} f(k)h(n-u)du$$

or for sampled data systems,

$$y[n]=\sum_{k=-\infty}^{\infty} f[k]h[n-k]$$

Even though the continuous-time form of the convolution is usually expressed a bit differently, they are presented here using the same characters for variables and indices to emphasize the similarity between the expression for discrete time systems and continuous time. In these formulae, we are convolving two functions $f$ and $h$ to produce a third function $y$.

*Convolution* ("folding back"or "winding") describes the process of modifying one function with another function to produce a third. In the time domain, the convolution of two functions is performed using the principle of superposition (see Figure 1-6). In the figure, $g(x)$ is reversed temporally and multiplied pointwise with $f(x)$, and each product summed to yield the next result. As $g(x)$ is moved across $f(x)$, this procedure is repeated at each point. The result is a series of sums representing the convolution of the two functions.

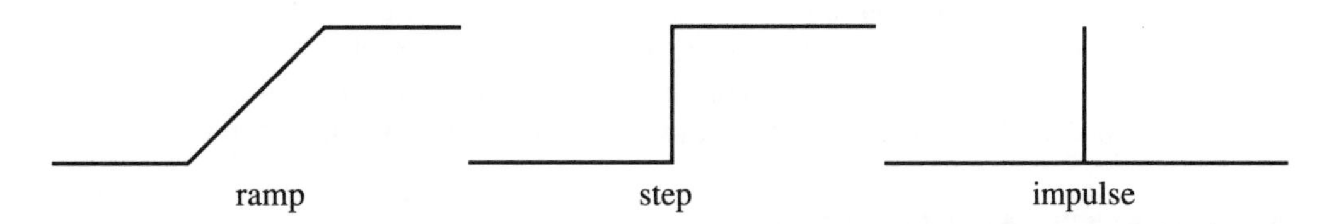


**Figure 1-5.** The unit ramp, step, and impulse are all of unit area and can be derived from each other by either differentiating or integrating. The step is the result of the differentiation of the ramp and an impulse is obtained from the step in the same manner.

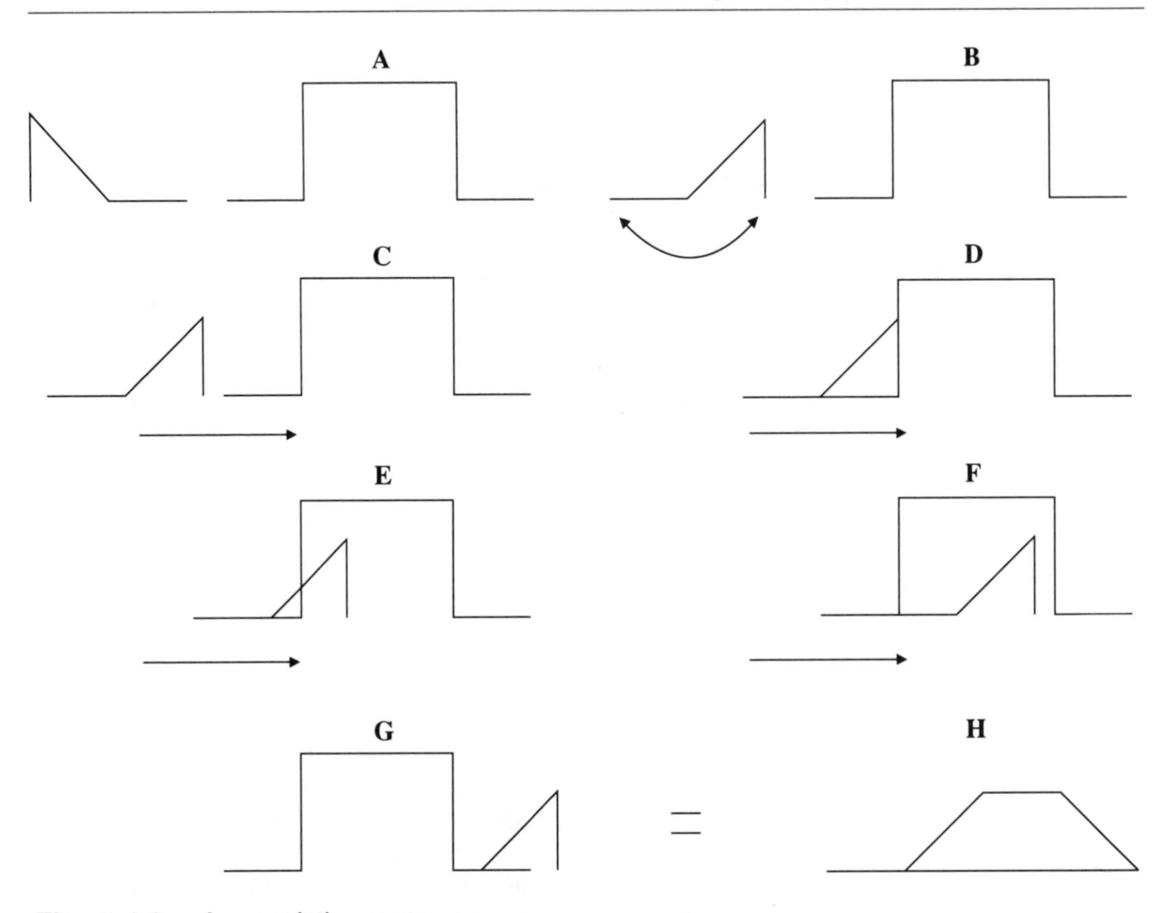


**Figure 1-6.** A convolution sequence.

Filters using this technique can produce very fast results, convolving an input data stream with the transfer function of the desired filter characteristic. This process is the same as polynomial multiplication, which is discussed in greater detail in Chapter 4.

Assuming that all convolution sums exist for a particular sequence, convolutions are commutative, allowing them to be unwound as easily as they are wound.

In the frequency domain, convolution is simply the product of the two spectra of interest: $Y(\omega) = F(\omega) * G(\omega)$. In this expression, $*$ indicates the operation of convolution.

## MATHEMATICAL FOUNDATION

Linear algebra offers an approach to the concept of orthogonality that provides a sound and more rigorous foundation for signal processing and the manipulation of polynomials than does simply the discussion of Cartesian coordinates. It also lays the groundwork for the basic concepts contained in Fourier and Laplacian analyses.

Within linear algebra, the *domain* of a function is defined as the range of values of the independent variable for that particular function. There are three distinct properties associated with the domain of a function:

1. Functions having the same domain are equal if and only if they are equal for every $x$ in the domain, in which case they may be written $f=g$.
2. The sum of functions that have the same domain is given by $(f+g)x=f(x)+g(x)$ for every $x$ in the domain. This is the property of additivity.
3. The product is $(fg)x=[f(x)][g(x)]$, and the scalar multiple $(af)x=af(x)$. This is the property of multiplicativity.

Linear algebra also regards a number, a function, a point, and a vector equally as single objects. A function describes a particular relationship between an independent variable that has a domain or range of values and a dependent variable that is the result of this relationship. As an example, if $f(x)=\sqrt{x}$, for every $x$, the independent variable, there is a function that relates it to $f(x)$, the dependent variable. This means that $f(x)$ refers to a number, not a function as would be denoted by $f$. For example, $f(x)=\sqrt{x}$, and in the case of $x=9$, $f(x)=3$.

## Linear Independence

A very important component in our need to find unique descriptors for our work with signals is the concept of *linear independence*. This is a linearly independent function or sequence that is not parallel to any other, and therefore can provide unique solutions or coordinates.

If we have a sequence of functions, $f_0, f_1, f_2,\ldots, f_k$, that all have the same domain, we can form a *linear combination* by simply mutiplying with another set of scalars, $a_0$, $a_1$, $a_2,\ldots,a_k$, creating the sequence $a_0f_0$, $a_1f_1$, $a_2f_2,\ldots$, $a_kf_k$. Such a sequence, all with the same domain, forms a linear space in which the sum of any two functions is a member, as is any scalar multiple of $f$. A function, such as $f_2$ in the aforementioned sequence, is a linear combination of the other functions in the sequence, if $a_{i\neq 2}=0$ and $a_2=1$.

A zero results if *all* $a_k=0$. This condition is known as *identically zero*, a special linear combination whose sum is zero because *all* of the coefficients are zero. Another combination, such as $f_1+f_2-f_3$ if $f_1(x)=\sin^2 x$, $f_2(x)=\cos^2 x$, $f_3(x)=1$ might also sum to zero, but is not called identically zero because its coefficents are not zero.

A set of *functions* is linearly dependent if any linear combination of elements of the set other than an identically zero combination equals zero. A *sequence* is defined as linearly dependent if at least one of the functions is a linear combination of the others.

If such a sequence of functions forms an axis in a coordinate system, it must be linearly independent of the other axes if it is to be orthogonal to them.

## Orthogonality

We can write an expression for the relationship between vectors as

$$\cos\theta = \frac{(x|y)}{\|x\|\|y\|}$$

which we can rearrange to produce the definition of the *scalar* or *dot product*:

$$(x|y) = \|x\|\|y\|\cos\theta$$

(The double lines around the functions above indicate that they are the *norms* or *magnitudes* of the functions. The single parallel line represents the operation of taking the dot product of the two functions. The dot product may also be described with the use of a black dot.)

This formula serves to illustrate the very special case of $\cos\theta=0$ that will occur when the vectors are perpendicular to one another, or *orthogonal.*

A sequence of functions is orthogonal with another sequence of functions if they do not mingle. This is the same as saying that the dot product of one function with another is zero, in this case identically zero. Figure 1-7 depicts two vectors, $f_0$ with a magnitude of 2 and $f_1$ with a magnitude of 3.

The dot product of these two vectors is equal to the sum of their products multiplied by the angle between them, $\pi/6$. Write this as $2*3*\cos(\pi/6)=5.196$. The vectors in Figure 1-7A are not orthogonal. If they were orthogonal, they would be exactly $\pi/2$ radians apart, so the cosine would be 0, producing a dot product of 0.

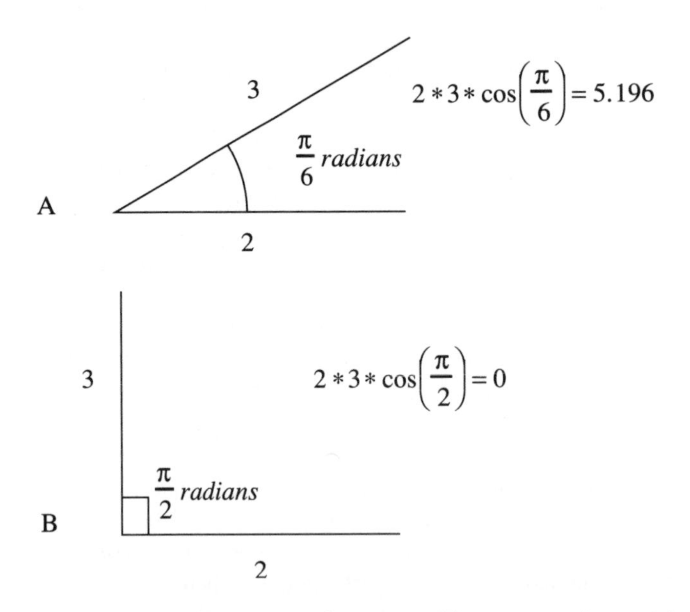


**Figure 1-7.** Two sets of vectors. The vectors in part A are not orthogonal; the vectors in part B are orthogonal.

In the linear space of all continuous functions of period $2\pi$, the dot product is defined as

$$\int_0^{2\pi} f_i(x)\overline{f_j(x)}$$

In that same space, the property of orthogonality is

$$(i \bullet j) = \int_0^{2\pi} f_i(x)\overline{f_j(x)} = 0 \quad i \neq j$$

One set of orthogonal functions exist as the sine and cosine functions. If $f_0(x) = \sin x$, and $f_1(x) = \cos x$, then

$$f(x) = \int_0^{2\pi} f_0(x) f_1(x) = \frac{1}{2}\sin^2 x\Big|_0^{2\pi} = 0$$

## The Norm

The *norm* of a function is a measure of the relative energy of that function. It is the real magnitude of the vector representing a function.

The norm of a vector, written $\|x\|$, is found by taking the square root of the dot product of the vector with itself. In two-dimensional real space, this is expressed as

$$x = (x_0, x_1) \quad y = (y_0, y_1) \quad x|y = x_0 y_0 + x_1 y_1$$

If the $n$-space is complex, the products are taken with the complex conjugate

$$x|y = x_0 \overline{y_0} + x_1 \overline{y_1}$$

which will produce a real result since the product of complex numbers is real.

In the case of the example in Figure 1-8, we would find the norm by first taking the dot product of the vector with itself: $(4,3) \bullet (4,3) = 25$, and then finding the square root of that, $\sqrt{25} = 5$. We may then *normalize* this vector to unity by dividing each of the coordinate sets by this norm:

$$\sqrt{\left(\frac{4}{5}, \frac{3}{5}\right) \bullet \left(\frac{4}{5}, \frac{3}{5}\right)} = 1$$

Normalization is an important feature in simplifying functions and minimizing error in approximations. This is especially true when working with approximating bases, as you will see later.

## Orthonormality

Orthogonal functions may also be *orthonormal*. Since it is possible that we can be dealing with complex variables, we obtain the norm as a square of the function integrated over the prescribed period:

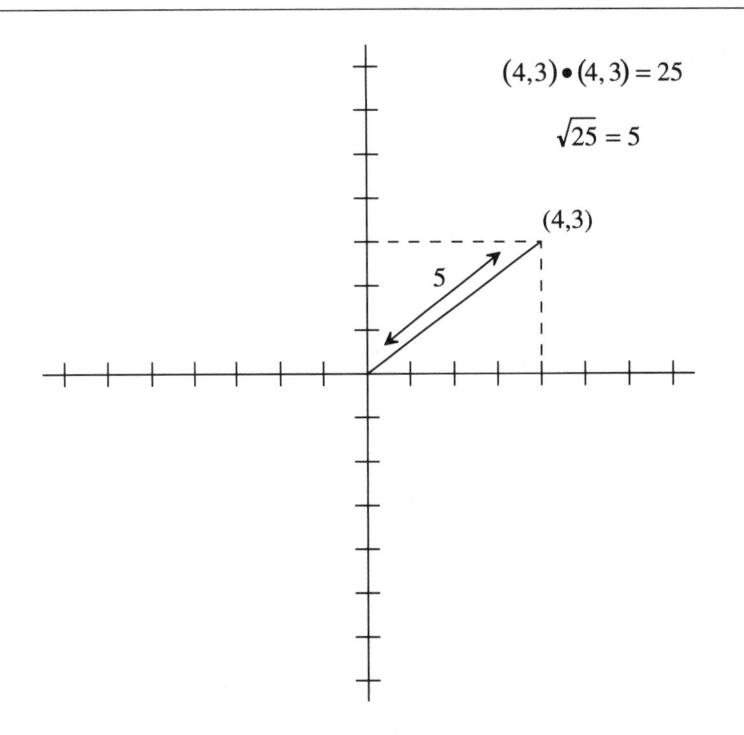


**Figure 1-8.** Finding the Norm.

$$\|f\|^2 = \int_0^T [f(x) \bullet f(x)]dt \text{ if it is real,}$$

$$\text{and } \|f\|^2 = \int_0^T \left[f(x) \bullet \overline{f(x)}\right]dt \text{ if it is complex.}$$

Orthogonal functions are also orthonormal if, over a specified period, the functions are continuous, orthogonal, and have unit energy. Expressed mathematically,

$$\|f\|^2 = \int_0^T [f(x)]^2 dt = 1$$

This means that the radius vector of the function is equal to one over the prescribed period of the function.

To make an orthogonal function orthonormal, it is only necessary to divide the function by its norm. Consider the function $f(x) = \sin \omega t$. Its norm is found with the equation

$$\int_0^{2\pi} [\sin \omega t]^2 dt = \pi$$

and the orthonormalized function is

$$f(x) = \frac{1}{\pi} \sin \omega t$$

## Basis Functions

An important precept of linear algebra is that a group of elements of a linear space can be said to span that space if every element of the linear space is a linear combination of the elements in this group. That same group of elements is also known as a basis for that linear space if it is also linearly independent. Basis functions are very important to our work, because they provide a direct way for the abstract operations in a linear space to become ordinary linear operations with numbers that are the coefficients of the functions or vectors relative to the basis.

Assume an $n$-dimensional linear solution space with $u_0$, $u_1$, $u_2$, ..., $u_n$ forming the basis for this space. If $f$ is an element of that space, then both $f = C_0 u_0(x) + C_1 u_1(x) + C_2 u_2(x) + \ldots + C_n u_n(x)$ and $C_0$, $C_1$, $C_2$, ..., $C_n$ represent $f$, and $C_0$, $C_1$, $C_2$, ..., $C_n$ is the coordinate sequence representing $f$ relative to the basis.

This defines a function or vector *exactly* in terms of a basis, giving it exact coordinates in the same way a point is defined relative to a set of Cartesian axes. This is not to say that there is only one solution set or basis for any sequence—there may be many, but the coefficients do describe each solution uniquely for that particular basis.

The number of basis elements in a linear space also specifies its dimension—that is, a linear space having a basis of $n$ elements is said to be *n-dimensional.* This explains Cartesian two- and three-space, in which the coordinate sequence is given by (x,y) or (x,y,z) for each $f$.

## Approximating Bases

Bases, like functions, can be normalized. This operation creates a new basis that makes it extraordinarily easy to find the coefficients of a sequence or function in that basis:

$$f = \sum_{k=1}^{n} a_k \phi_k$$

where

$$a_k = (f \bullet \phi_k)$$

In words, this means that $f$ is equal to the sum of the product of a set of coefficients $a_k$ and basis functions $\phi_k$ over a specified interval, and that the coefficients are the result of the inner product of the function and the basis functions.

For any function $f$ and a sequence of functions $\phi_0, \phi_1, \phi_2, \ldots, \phi_n$ that satisfies the condition

$$\int_a^b \phi_j(x)\overline{\phi_k(x)}dx = 0 \quad \text{when } j \neq k \text{ and unity when } j = k, \text{ the integral}$$

$$\int_a^b \left|f(x) - a_0\phi_0(x) - a_1\phi_1(x) - a_2\phi_2(x) - \cdots - a_n\phi_n(x)\right|^2 dx$$

assumes its minimum when the coefficients

$$a_k = \int_a^b f(x)\overline{\phi_k(x)}dx$$

(The bar over the functions above indicates that we are working with the complex conjugate.)

In our previous discussions, a basis was defined as a sequence of functions whose linear combinations must exactly equal a given function, but here we are talking about producing a sum that is arbitrarily close—this is an *approximating basis*. This is the same as saying that the norm of the function minus a linear combination of the basis must be less than some positive number $\varepsilon$, or

$$\|f - a_0u_0 - a_1u_1 - a_2u_2 - \cdots - a_nu_n\| < \varepsilon$$

This is true if the basis is both orthogonal and orthonormal.

A remarkable and beautiful example of this is the linearly independent and orthogonal sequence, $1, x, x^2, x^3, x^4, x^5, \ldots$, from which we may obtain

$$e^x = 1 + x + \frac{x^2}{2!} + \frac{x^3}{3!} + \frac{x^4}{4!} + \frac{x^5}{5!} + \cdots$$

from $e^x$ by allowing $x = j\theta$, as in the following:

$$e^{j\theta} = 1 + j\theta + \frac{(j\theta)^2}{2!} + \frac{(j\theta)^3}{3!} + \frac{(j\theta)^4}{4!} + \frac{(j\theta)^5}{5!} + \cdots \quad \text{and collecting the terms}$$

$$e^{j\theta} = \left(1 - \frac{\theta^2}{2!} + \frac{\theta^4}{4!} - \cdots\right) + j\left(\theta - \frac{\theta^3}{3!} + \frac{\theta^5}{5!} - \cdots\right) = \cos\theta + j\sin\theta$$

we derive

$$\sin x = x - \frac{x^3}{3!} + \frac{x^5}{5!} - \frac{x^7}{7!} + \cdots$$

$$\cos x = 1 - \frac{x^2}{2!} + \frac{x^4}{4!} - \frac{x^6}{6!} + \cdots$$

Notice that the powers of the sine function are odd; the powers of the cosine function are even.

With this relationship comes a whole new set of tools:

$$e^{j\theta} = \cos\theta + j\sin\theta$$

$$e^{-j\theta} = \cos\theta - j\sin\theta$$

$$e^{j\theta} + e^{-j\theta} = 2\cos\theta, \quad \frac{e^{j\theta} + e^{-j\theta}}{2} = \cos\theta$$

$$e^{j\theta} - e^{-j\theta} = 2j\sin\theta, \quad \frac{e^{j\theta} - e^{-j\theta}}{2} = \sin\theta$$

$$a + jb = r(\cos\theta + j\sin\theta) = re^{j\theta}, \quad a - jb = r(\cos\theta - j\sin\theta) = re^{-j\theta}$$

These new equalities are especially useful in simplifying both the differential and integral equations found in physics in general and signal processing in particular.

# SYSTEM RESPONSE

## The Simple Harmonic Oscillator

A *harmonic* signal is one that moves in a periodic pattern that is symmetrical about a region of equilibrium, such as a sinusoid. The harmonic oscillator produces such a pattern or signal. It exists throughout nature and is the subject of a great deal of scientific literature. It also provides a very useful model, both physically and mathematically, for the response of a system to stimulus. It is useful in the analysis of networks and harmonics, and in the creation and understanding of filters. The harmonic oscillator can represent mechanical motion, as in the case of a mallet and tuning fork for a musician, or voltage, as for an electrical engineer. Either way, the definitions and mathematical expressions are essentially the same, though the terms and symbolism may vary.

To *oscillate* means to swing back and forth or vary between alternate extremes at regular intervals. The *period* of oscillation is the time it takes to complete one full swing to both extremes and back to the start again; it is the inverse of the *frequency* of oscillation, which is measured in seconds. Assuming that an oscillator produces a pure sinusoid, then a *signal* may be viewed as the sum, composite, or superposition of arbitrary oscillations over an interval. The process of harmonic analysis decomposes signals to these fundamental oscillations or frequencies.

A very important equation in the study of signal processing (as well as many other studies) is the *homogenous linear differential equation,*

$$a_0(x)\frac{d^n y}{dx^n} + a_1(x)\frac{d^{n-1} y}{dx^{n-1}} + a_2(x)\frac{d^{n-2} y}{dx^{n-2}} + \cdots + a_{n-1}(x)\frac{dy}{dx} + a_n(x)y = 0$$

Such equations are described as having an *order* and a *degree*—the order of the equation is the same as the exponent of the highest-ordered derivative in the equation, and the degree of the equation is the power of the highest-ordered derivative after all possible reduction. It is a linear equality, because the sum of two solutions is also a solution, and any scalar multiple of a solution is a solution. There are $n$ solutions to this equation (unless $n=0$), and they form a linearly independent set where each solution is a linear combination: $f=C_0u_0(x)+C_1u_1(x)+C_2u_2(x)+\ldots+C_nu_n(x)$ .

A particular form of this equation is the *linear constant coefficient differential equation* of the following form:

$$m\frac{d^2x}{dt^2}+c\frac{dx}{dt}+kx=0 \qquad \text{(Eq. 1-1)}$$

Please note that the equation is homogenous, that its sum is zero, and that it is linear, meaning that the dependent variable appears only in the zero or first power. This is the equation of a simple harmonic oscillator. It doesn't matter whether the oscillator is mechanical, electrical, or optical, it is defined by this equation. Here $m$, $c$, and $k$ are constant coefficients and may be anything. The first term to the left,

$$m\frac{d^2x}{dt^2}$$

represents acceleration multiplied by mass,

$$c\frac{dx}{dt}$$

is friction, and $kx$ is the displacement that mass has with respect to an equilibrium point multiplied by a constant representing stiffness. We can illustrate a simple and standard analog of the harmonic oscillator with a weighted spring like the one in Figure 1-9.

To begin our discussion, we will imagine a perfect universe and assume that the friction term is equal to zero. This allows us to rewrite the equation in its most basic form:

$$m\frac{d^2x}{dt^2}+kx=0$$

or

$$m\frac{d^2x}{dt^2}=-kx \qquad \text{(Eq. 1-2)}$$

Depending on the tension of the spring and/or the weight of the wrench, the spring stretches under the weight of the iron until it reaches a balance point. At this point, the spring is exerting no force on the hammer at all. If, by chance, the hammer is moved up or down from its equilibrium position to another position, or *displacement*, the spring will attempt to return it to its equilibrium position with a *restoring* force. The magnitude of this force is directly proportional to the displacement experienced by the wrench from its equilibrium position:

$$F=-kx$$

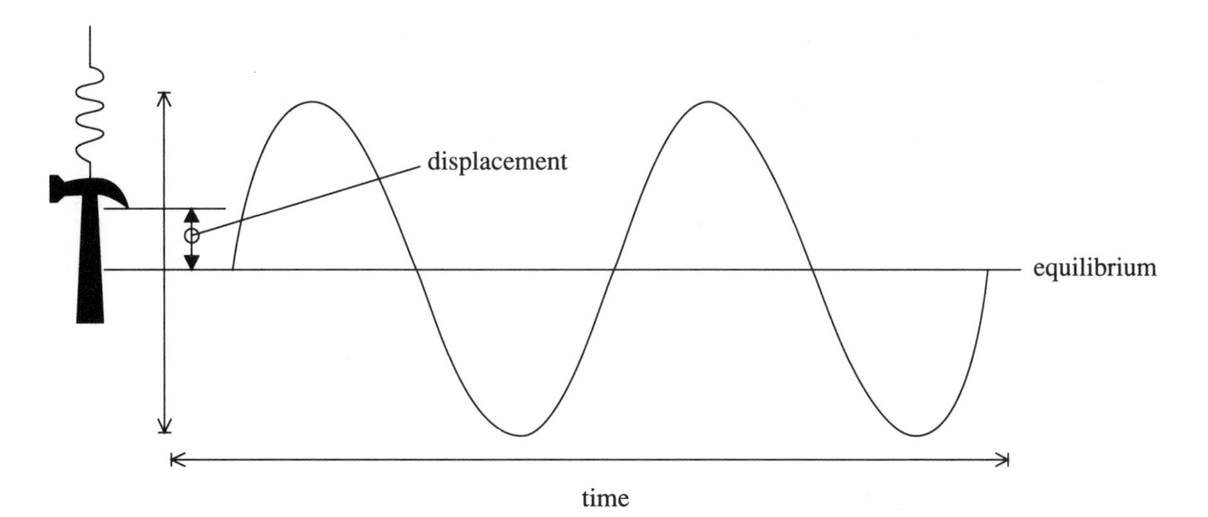


**Figure 1-9.** Hammer on a spring. The illustration shows an initial displacement indicated by the arrows and the sinusoid that would be observed if the oscillating hammer was in horizontal motion as well as vertical.

In this equation, the minus sign indicates that the direction of the force is opposite to the displacement $x$, with $k$, the *spring constant*. When the restoring force is directly proportional to the negative of the displacement, as the equation states, it is a simple harmonic oscillator. In a universe without friction, the wrench might continue to move between its equilibrium position and a displacement on both sides of it forever.

In truth, however, this doesn't happen. The universe isn't perfect—there is another force that must be accounted for in this relationship if it is to work, and that is friction. There are many forms of friction, and many are not at all easy to model. For this work, we are going to assume that friction is proportional to the speed with which the hammer is moving. Examples of what this means are common. A boy in a swimming pool feels no resistance, no force, while he is standing still. However, when he swims, he can feel a resistance that is proportional to how fast he is swimming. In another example, imagine a stone dropping through a barrel of oil: the faster it moves, the faster the oil must crowd past the stone, and the greater the resistance. We have already stated it:

$$F_f = c\left(\frac{dx}{dt}\right)$$

Friction is directly proportional to the speed. Therefore, we have

$$m\left(\frac{d^2x}{dt^2}\right) + c\left(\frac{dx}{dt}\right) + kx = 0$$

which is the form of the linear constant coefficient differential equation we started with.

One of the first things to note about the simple harmonic oscillator is that its natural motion is sinusoidal. If the hammer in Figure 1-9 were set in motion vertically and the entire arrangement were moved before a spectator at a constant horizontal speed, that is what he would see: a sinusoid. We need to know more about the characteristics of this sinusoid.

Starting with the basic equation for the harmonic oscillator and dividing the left-hand member of the equation by $m$, we have

$$\frac{d^2x}{dt^2} = -\frac{k}{m}x$$

In order to simplify the equation further, we choose $-k/m = 1$, which gives us

$$\frac{d^2x}{dt^2} = -x$$

To solve this equation, we must be able to differentiate $x$ twice and arrive at a result equal to the original $x$ with a minus sign. From trigonometry, we recall that $x = \cos\omega_0 t$ satisfies the requirement

$$\frac{dx}{dt} = -\omega \sin\omega_0 t, \text{ and } \frac{d^2x}{dt^2} = -\omega_0^2 x$$

which ultimately means that

$$-\frac{k}{m} = \omega_0^2$$

In these equations, $\omega_0$ represents the *natural* motion of the oscillator, or its *natural frequency*, and is the scaling factor for the equation. Here, $\omega_0 t$ is the phase of the motion, repeating itself every $2\pi$, or period.

What we have defined so far accounts for the period of the motion, but not the amplitude. That depends upon the *initial conditions*. Initial conditions are those things that occurred prior to $t=0$, or time zero, when we started our analysis. In the development above, $x = \cos\omega_0 t$ indicates that at $t=0$, the hammer was displaced from the equilibrium position $[\cos(0) = 1]$. It is possible for the hammer actually to be at the equilibrium position at $t=0$—that is, passing through it—which means that the displacement occurred earlier than that, at some time $t<0$. In such a case, the sine function works well, presenting us with the general solution of this equation:

$$x(t) = A\cos\sqrt{\tfrac{k}{m}}x + B\sin\sqrt{\tfrac{k}{m}}x \qquad \text{(Eq. 1-3)}$$

or

$$x(t) = A\cos\omega_0 t + B\sin\omega_0 t$$

where $A$ and $B$ are coefficients representing amplitudes. In truth, any sinusoidal function can be written in this form, so the concept is very easy to visualize.

Clearly, the harmonic oscillator is periodic, and so must be the differential equation representing it. We can solve Equation 1-2 by setting the following conditions: $x(0)=0$ and $x(\pi)=0$. In order for Equation 1-3 to satisfy these conditions, $A$ must be zero to eliminate the $\cos\omega_0 t$ term, since it is equal to 1 at both boundaries. This will allow solutions for $k=n^2$, where $n=1, 2, 3, 4, \ldots$, making every solution a scalar multiple of a function in the sequence $\sin x$, $\sin 2x$, $\sin 3x$, ... , which is orthogonal relative to the the interval $(0,\pi)$. (When $k$ is negative, the series is complex-valued and usually written with hyperbolic functions.)

To take this another step, it is appropriate to find a solution for $x'(0)=0$ and $x'(\pi)=0$, producing the remarkable sequence:

$$1, \cos x, \cos 2x, \cos 3x, \cdots$$

which is orthogonal relative to the interval $(0,\pi)$.

If the boundary values were 0 to $2\pi$ for either of these two differentials, we would have 1, $\cos x$, $\sin x$, $\cos 2x$, $\sin 2x$, $\cos 3x$, $\sin 3x$, ... , which is orthogonal relative to the interval $(0,2\pi)$. Because $e^{j\omega t}$ is easier to differentiate and integrate, it is often used to simplify equations employing trigonometric functions. (See Approximating Bases for the derivation of this form.) For example, the general form of the linear constant coefficient differential equation

$$m\frac{d^2x}{dt^2} + c\frac{dx}{dt} + kx = 0$$

with $x(t)=Ke^{j\omega t}$, it is

$$-m\omega^2 Ke^{j\omega t} + cj\omega Ke^{j\omega t} + kKe^{j\omega t} = 0 \qquad \text{(Eq. 1-4)}$$

and removing $Ke^{j\omega}$, it is

$$-m\omega^2 + cj\omega + k = 0$$

(Often, we substitute the quantity $s=\sigma+j\omega$ for $j\omega$ in $Ke^{j\omega}$, making it $Ke^{st}$. The $\sigma$ in quantity $s$ is a dimensionless *damping* factor representing rate of decay, which, for the moment, we shall set to zero, leaving the meaning of the Equation 1-4 unchanged.)

None of the elements of this equation can produce motion by itself. As such the sum of an undriven harmonic oscillator of this form will always be zero. This is also called the *natural* or *transient* form of the equation; this is the motion (for a mechanical system) or current (for an electrical system) that can exist without driving force.

Suppose we supply an external source of energy. We would then have a *forced* oscillator that would give us, after a suitable start-up time, a *steady state* response.

## Forced Oscillators

A forced oscillator is one that is supplied with energy from an external source that may or may not be at the $\omega_0$ of the harmonic oscillator. Assume that we are dealing with the same oscillator as before, except that it is now being driven by an outside force:

$$m\frac{d^2x}{dt^2} = -kx + F$$

Here, the external energy, or force, is added to the product of the spring constant with the displacement of the hammer to produce the motion of the oscillator.

Synchronous motion requires very much less work than nonsynchronous motion. Dancing is an example, or pushing a child on a swing. When you synchronize your efforts with the natural motion of the swing, you find that it takes very little effort to keep the child in motion, and with just a little more, you can send him higher and higher. If you are trying to change the motion, slow it down or speed it up, you find yourself fighting the energy in the motion of the swing. So it is with a forced oscillator: if the external energy is at the natural frequency of the oscillator, it is much easier to start and continue the oscillation. The greater the difference, the smaller the amplitude of the oscillation resulting from the external force.

In nature, elements respond differently to different frequencies or rates of change. A certain size and shape of metal or glass may vibrate more freely at one frequency than at another. Forcing it to vibrate at a frequency other than its natural frequency is possible and requires more effort.

The general equation of a forced harmonic oscillator is

$$m\left(\frac{d^2x}{dt^2}\right)+c\left(\frac{dx}{dt}\right)+kx=F$$

In this expression, $x$ is complex-valued, as is $F$, though we are only interested in the real part of each of them. To simplify, we will make $c=m\sigma$ and $k=m\omega_0^2$ (from the previous section) and divide through by $m$:

$$\left(\frac{d^2x}{dt^2}\right)+\sigma\left(\frac{dx}{dt}\right)+\omega_0^2x=\frac{F}{m}$$

Now, to solve using exponential notation,

$$-\omega^2x+j\sigma\omega x+\omega_0^2x=\frac{F}{m}$$

Applying a known force to an oscillator will produce

$$x=\frac{F}{m\left(\omega_0^2+j\sigma\omega-\omega^2\right)}$$

If the friction term $\sigma$ is small as the forcing frequency approaches the natural frequency, the force required to produce oscillation or maintain it drops off greatly. In fact, if there was no friction, almost no force would be required, and the amplitude of oscillation would still approach infinity.

## The LRC Harmonic Oscillator

To this point, the elements of the oscillator have been represented as mechanical, but since we are going to be dealing with electronic signal processing, it is appropriate that we look for electrical analogies. And, not surprisingly, we find them in the *capacitor*, *inductor*, and *resistor*.

The charge $q$ on a capacitor is analogous to the displacement in a mechanical system. Current ($I = dq$/dt) is the velocity, $1/C$ is the spring constant ($C$ is capacitance), $R$ is friction, and the inductor represents mass.

On a capacitor, $V = q/C$, means that the potential difference across a capacitor is proportional to the charge. Since $q$ must accumulate or leave the plates of a capacitor for the voltage to change, the stiffness of this element is directly related to the size of the capacitor. The smaller the $C$, the faster the change and the stiffer it will be.

The voltage drop across a resistor is proportional to a frictional coefficient $R$:

$$V = R\frac{dq}{dt} = RI$$

This is Ohm's law. As you can see, the greater the $I$ through the constant $E$, the greater the voltage drop across the resistor.

Finally, the induced voltage in an inductor is proportional to the rate of change of the current through it:

$$V = L\frac{d^2q}{dt^2} = L\frac{dI}{dt}$$

Arranged as in Figure 1-10, the voltage drop across the entire network must be equal to the sum of the individual voltage drops across each device, according to Kirchoff's law. It is more common in electrical engineering to work in terms of current than charge, and recalling that $I = dq/dt$, we will rewrite the equation in the following general form:

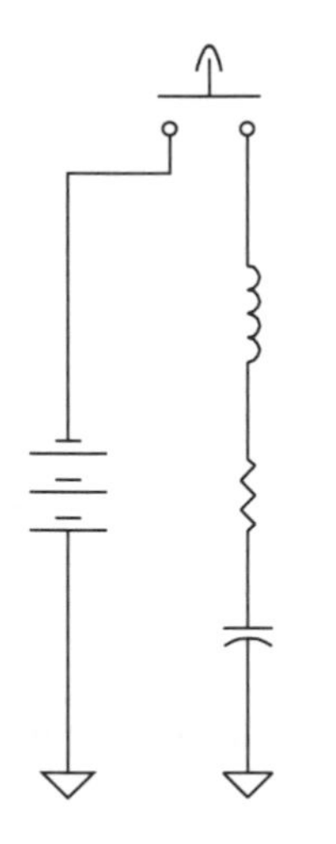

**Figure 1-10.** An LRC circuit. With the switch open, no current flows. When the switch closes, current begins to flow and continues until an equilibrium is reached.

$$L\frac{d^2i}{dt^2}+R\frac{di}{dt}+\frac{1}{C}i=v(t)$$

where $v(t)=0$ for the unforced case.

We will begin with an equation that satisfies Kirchoff's law regarding the sum of voltages in a circuit for an RLC oscillator:

$$L\frac{di}{dt}+Ri+\frac{1}{C}\int i\,dt=v(t)$$

As you can see, the equation is simply the sum of the voltage drops across each of the circuit elements, which would naturally be equal to whatever the driving voltage is. If the driving voltage is a constant, or zero, we can differentiate to rid ourselves of the integral to get the familiar:

$$L\frac{d^2i}{dt^2}+R\frac{di}{dt}+\frac{i}{C}=0 \qquad \text{(Eq. 1-5)}$$

the standard equation for the simple harmonic oscillator.

To get control over this oscillator, we need to go much farther than this. In doing so, it is to our advantage to simplify the notation, so we adopt the operator $D$, which represents *differentiation with respect to time*, and rewrite Equation 1-5:

$$L\frac{d^2i}{dt^2}+R\frac{di}{dt}+\frac{i}{C}=LD^2i+RDi+\frac{i}{C}=0$$

With that simple change, it becomes apparent that all we are dealing with is a simple quadratic:

$$ax+bx^2+c=0$$

whose solutions are:

$$s_1,s_2=\frac{-b\pm\sqrt{b^2-4ac}}{2a}$$

or in this case:

$$D^2i+\frac{R}{L}Di+\frac{1}{LC}i=0 \qquad \text{(Eq. 1-6)}$$

which yields to the quadratic formula:

$$s_1,s_2=-\frac{R}{2L}\pm\sqrt{\left(\frac{R}{2L}\right)^2-\frac{1}{LC}}$$

We know from algebra that the roots of this equation depend upon the contents of the radical

$$\left(\frac{R}{2L}\right)^2-\frac{1}{LC}$$

If this is greater than zero, the roots of the equation are *real and distinct*, if it is zero, the roots are *real and equal*, and if it is less than zero, the roots are *complex*.

The relationship between the real and imaginary components of the oscillator are important—in fact, essential—if we are to analyze and control it. Based upon this, we define a new term for that value of $R$ under the radical that will make the radical disappear, and we will call it the *critical resistance*, $R_{critical}$. This leads us to two more important values: first, *the damping constant*, $\sigma$, which is defined:

$$\sigma = \frac{R_{critical}}{2L}$$

and

$$\left(\frac{R_{critical}}{2L}\right)^2 = \frac{1}{LC}, \text{ or } R_{critical} = 2\sqrt{\frac{L}{C}}$$

We then call the ratio of the actual resistance to the critical resistance the *damping ratio*, and give it the letter $\zeta$:

$$\zeta = \frac{R}{R_{critical}} = \frac{R}{2}\sqrt{\frac{C}{L}}$$

There is another quantity we need, and that is $\omega_0$, the *undamped natural angular frequency*:

$$\omega_0 = \frac{1}{\sqrt{LC}}$$

If we wish to fit these values to the second order differential we are dealing with, we will need to manipulate them somewhat. Doing so, we find that multiplying $\zeta$ by $2\omega_0$ yields

$$2\zeta\omega_0 = 2\frac{R}{2}\sqrt{\frac{C}{L}}\frac{1}{\sqrt{LC}} = \frac{R}{L}$$

and, of course

$$\omega_0^2 = \frac{1}{LC}$$

all of which allows us to rewrite Equation 1-4 in this form:

$$D^2 + 2\zeta\omega_0 D + \omega_0^2 = 0$$

with the roots now equal to

$$s_1, s_2 = -\zeta\omega_0 \pm \omega_0\sqrt{\zeta^2 - 1}$$

This presents us with the interesting consequence that if $\zeta$ is greater than 1, the roots are real; if $\zeta$ is equal to 1, they are real and repeated; and if they are less than 1, the roots are complex conjugates.

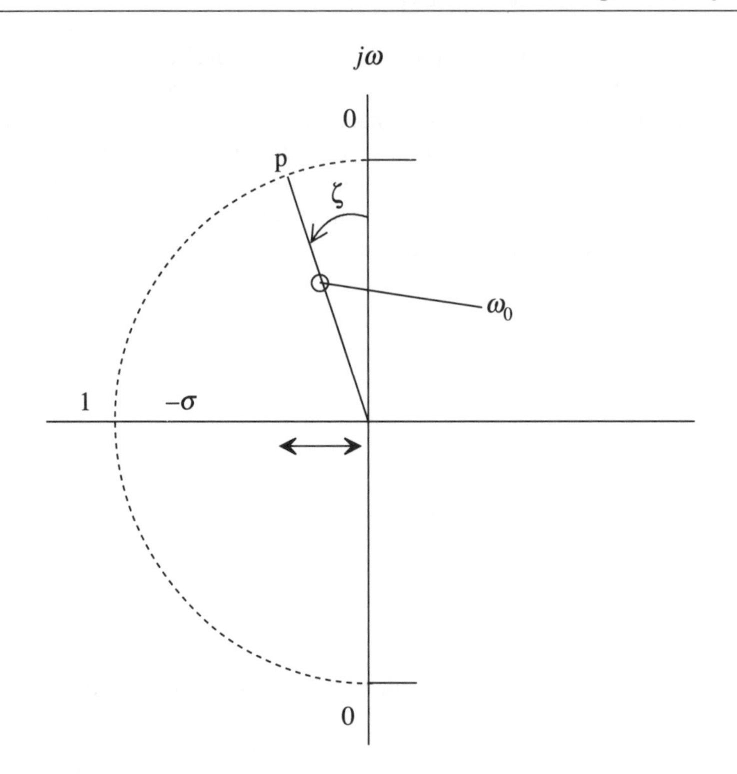


**Figure 1-11.** Argand diagram. The radial arm extending from the origin is of length $\omega_0$, it sweeps through a circle called a root loci.

The Argand diagram illustrates this relationship graphically (see Figure 1-11). A pair of Cartesian axes are drawn: the X axis represents purely *real* values (this is the resistance in the circuit), and the Y axis represents purely *imaginary* values (the frequency of oscillation). As one moves away from the origin along the vertical axis, one is moving into increasing frequencies; as one moves away from the origin along the horizontal axis, one is moving into increasing, nonreactive resistance.

If $\zeta$ is zero or very close to zero,

$$s_1, s_2 = -\zeta\omega_0 \pm \omega_0\sqrt{\zeta^2 - 1}$$

reduces to $s_1, s_2 = \pm j\omega_0$. That is, the roots are purely imaginary and exist only on the Y axis of the Argand diagram. If $\zeta$ is less than one, the roots are complex conjugates

$$s_1, s_2 = -\zeta\omega_0 \pm j\omega_0\sqrt{1 - \zeta^2}$$

For purposes that will become clear later, this complex number is defined as $s = \sigma + j\omega$, where $\sigma$ represents the real part $\sigma = -\zeta\omega_0$, and

$$\omega = \pm\omega_0\sqrt{1-\zeta^2}$$

is the imaginary part.

On the Argand diagram, the locus of the roots of this equation form a circle with its center at the origin and a radius of $\omega_0$, as shown in Figure 1-11, as $\zeta$ is varied from 0 to 1. We find that this is true by taking the real and imaginary part of each root and finding the hypotenuse or radius vector from that point to the origin: $\sigma^2 + \omega^2 = \omega_0^2$.

The angle between the $-\sigma$ axis and the complex roots is the phase angle, and is equal to the arctangent of the imaginary part divided by the real part:

$$\phi = \tan^{-1}\frac{\omega_0\sqrt{1-\zeta^2}}{\omega_0\zeta}$$

It is interesting to note here that because $\omega_0$ is common to both the numerator and denominator, it cancels, which means that regardless of the frequency, the $\zeta$ is constant for a given $\phi_1$ (angle from real axis to $\omega_0$ radial arm).

When $\zeta = 1$, the roots are real, and as $\zeta > 1$ approaches infinity the roots approach $-2\zeta\omega_0$ and 0. There are then three possibilities for the damping ratio. When $\zeta < 1$, the oscillator is called *underdamped*, when $\zeta = 1$, the oscillator is *critically damped*, and when $\zeta > 1$, it is *overdamped*. (See Figure 1-12.)

## Poles and Zeros

Let's examine what happens on the complex plane. When the resistance in the circuit is zero, the roots are purely imaginary, which places them directly on the vertical axis known as the *imaginary axis* as mirror images of one another. As long as the current is oscillatory the roots remain in the left half-plane and $\sigma$ is negative. In Figure 1-13, you can see the complex values of the root-pair (the poles) plotted against the axis of the Argand diagram—$\sigma$ is on the negative horizontal axis, and $j\omega_0$ is plotted relative to the vertical frequency axis.

You will note that if a line is drawn from the origin to either root, we have a right triangle with $\sigma$ and $j\omega_0$ as the length of each side. Solving for the hypotenuse, we have, for $s_1$,

$$\sqrt{(j\omega_0)^2 + \sigma^2} = \omega_0$$

This means that all the roots of this equation are $\omega_0$ from the origin, and since $\omega_0$ remains constant as the resistance varies, the set of all roots must fall on a semicircle around the origin, known as the *root locus*.

As the resistance increases, the two roots will move along the semicircle toward the real axis. At the point where they touch, we have the *critically damped* case—here, $s_1 = s_2$ and $j\omega_0 = 0$, $\sigma = \omega_0$, and the $Q = 1/2$. Any further increase in resistance will cause the roots to move away from one another along the real axis, occupying positions on either side of the semicircle and equidistant from it.

At this point, it is convenient to introduce a notation based upon the contents of the complex plane that is somewhat simpler than the polynomials we have been using.

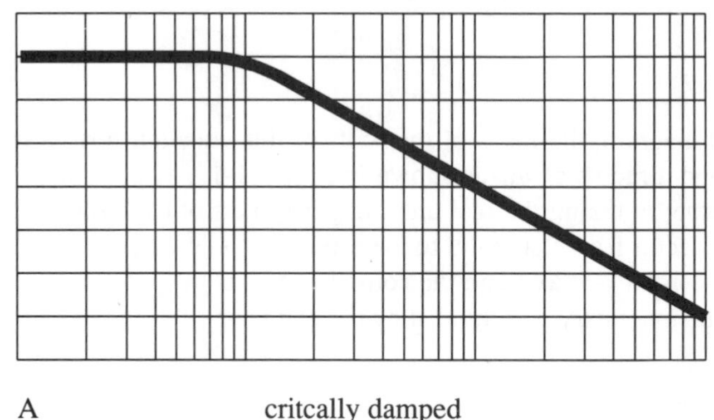

A critcally damped

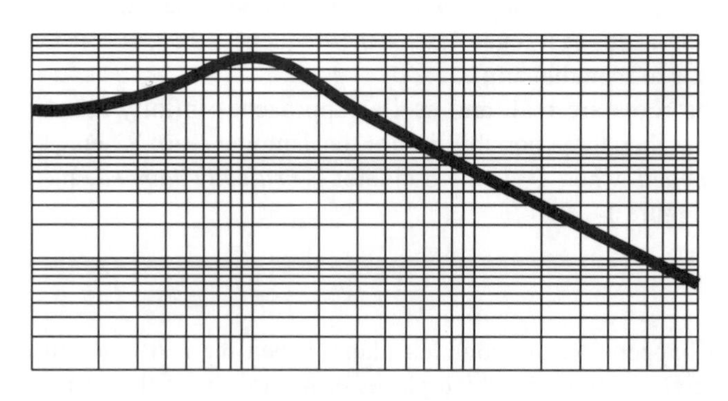

B underdamped

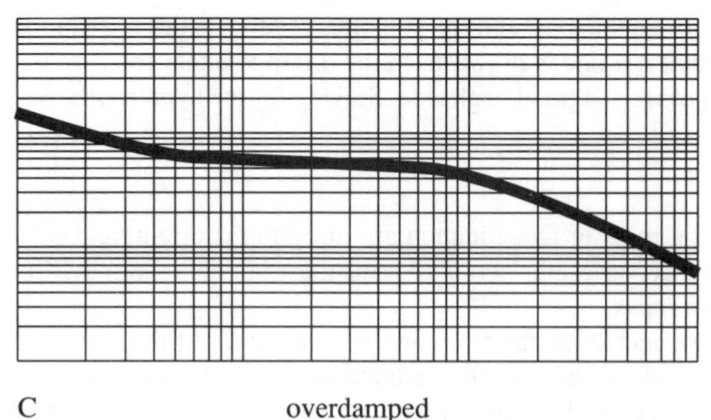

C overdamped

**Figure 1-12.** Three graphs of system responses. A is critically damped, there is no appreciable overshoot or sag, B is underdamped, the peaking near the knee is clear, and C is overdamped, it is sagging and lossy.

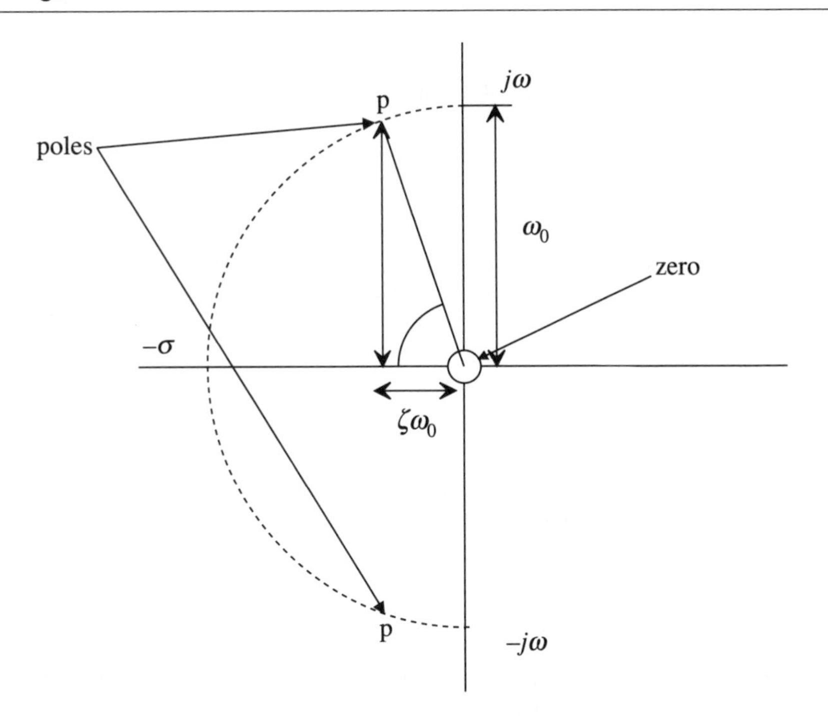

**Figure 1-13.** An argand diagram illustrating a pair of poles and a zero at the origin.

Replacing $j\omega_0$ on the imaginary axis with $s$, and then noting the position of each root of the numerator polynomial with $(s-z_n)$ and the root of each denominator polynomial with $(s-p_n)$, we can completely represent the characteristics of the circuit with a transfer function such as

$$H(s) = H_0 \frac{(s-z_1)(s-z_2)\cdots(s-z_n)}{(s-p_1)(s-p_2)\cdots(s-p_n)}$$

where $H_0$ is a scaling factor. This sort of function is known as a *rational function*, which is by definition a ratio of two polynomials. The magnitude is the length of the line, and the angle is measured relative to the horizontal. Except for the scaling factor, $H$, the *zeros* $z_1$, $z_2$, … , and the *poles* $p_1$, $p_2$, … completely determine this function. Each factor $(s-z_n)$ or $(s-p_n)$ exists as a vector in the complex plane—that is, each has a magnitude and an angle. As you can doubtless see, there is a value for each $z_n$ that can force the entire equality to zero, as well as a value for each $p_n$ that will force it to infinity, hence the names *zeros* and *poles*.

## *Q* and Resonance

Inductive reactances are positive and capacitive reactance is negative. For a series LRC circuit, there will be some frequency for which these reactances are equal and opposite. When this occurs, the total reactance of the circuit is zero, and it is in resonance. This corresponds to a peak in current at the frequency of resonance for the network. We can write magnitude as

$$H(\omega) = \frac{1}{\sqrt{R^2 + \left(\omega L - \frac{1}{\omega C}\right)^2}}$$

which indicates that the magnitude can have a maximum value of $1/R$ when the reactive terms in the equation are equal. This occurs when

$$\omega = \frac{1}{\sqrt{LC}} = \omega_0$$

The circuit is operating at the undamped natural frequency dictated by the reactance involved.

The $Q$ or *quality factor* is the ratio of circuit reactance to resistance. It determines the voltage rise at resonance and, most particularly, frequency selectivity. It is

$$Q = \frac{\omega_0 L}{R}$$

The value of $Q$ increases as the poles of the network approach the $j\omega$ axis; it is inversely proportional to the damping ratio. The less resistance in the circuit, the higher the $Q$.

Circuit $Q$ is commonly specified in terms of the *half-power* points of the circuit. The current at resonance is

$$I = \frac{V}{R}$$

At the half-point powers it has a magnitude equal to

$$I = \frac{V}{\sqrt{2}R}$$

which means that the total reactance of the circuit at that point will be

$$\sqrt{R^2 + \left(\omega L - \frac{1}{\omega C}\right)^2}$$

or written in terms of the roots

$$\pm R = \left(\omega L - \frac{1}{\omega C}\right)$$

Putting the equation into quadratic form

$$\omega^2 \pm \frac{R}{L}\omega - \frac{1}{LC} = 0$$

and solving

$$\omega = \pm\frac{R}{2L} \pm \sqrt{\left(\frac{R}{2L}\right)^2 + \frac{1}{LC}}$$

we then rewrite this using the equalities previously given:

$$\omega = \omega_0\left(\pm\zeta \pm \sqrt{\zeta^2 + 1}\right)$$

The results of this equation are the *half-power frequencies* as they are upon the $j\omega$ axis. They sit above and below $\omega_0$ at a distance prescribed by the damping factor $\zeta$ (refer to Figure 1-13). All of this means that the smaller the damping factor, the higher the $Q$ and the narrower the bandwidth of the system. Highly selective filters need a high $Q$.

# HARMONIC ANALYSIS

## Fourier Series

In terms of harmonic motions in a linear system, it can be said that whether a motion (vibration or signal) is composed of one frequency or more, it can be represented by a superposition of sinusoidal motions. In fact, this general motion can be characterized completely by the sum of the amplitudes and phases of the individual components. Each of these components responds as a harmonic oscillator whose natural frequency is that of the individual sinusoid.

Both of the signals in Figure 1-14 are described by two important formulations: the homogenous differential equation

$$\frac{d^2x}{dt^2} + \gamma x = 0$$

and $f(t) = A\sin\omega t + B\cos\omega t$. Clearly, if they both define the sinusoids in the graphs, they define one another as well.

The function $f(t) = A\sin\omega t + B\cos\omega t$ represents a sinusoid of angular frequency $\omega$. Varying the coefficients produces an arbitrary sinusoid with arbitrary phase (see mathcad document, super.mcd on optional disk). Written in this fashion, we can sum sinusoids of varying $\omega$ by simply adding $A$ coefficients to $A$ coefficients and $B$ coefficients to $B$ coefficients.

In Figure 1-14C we have the result of the summation of 1-14A and 1-14B. That is, if we were to decompose this result, we would have two frequencies, as represented by 1-14A and 1-14B.

Signals, whether simple or complex, can be expressed as a sum of a fundamental frequency $\omega_0$ and integer multiples known as *harmonics*. The second harmonic is twice the natural frequency, $2\omega_0$, the third harmonic is three times the natural frequency, $3\omega_0$, and so on. If the period of $\omega_0$ requires $1/f$ to complete, then $2\omega_0$ will require half that time and so will complete twice, and $3\omega_0$, needing only one-third the time, will complete three cycles in the same time it takes the fundamental to complete one. Of course, all of this repeats with

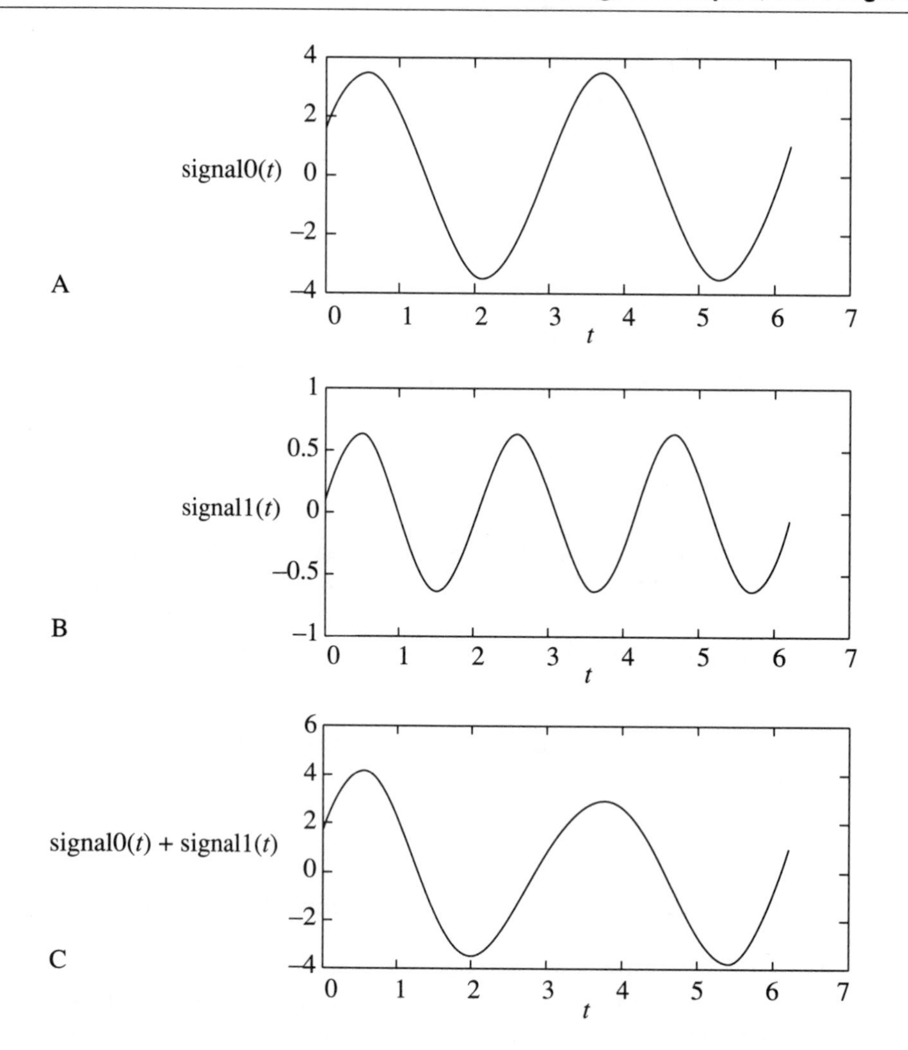

**Figure 1-14.** The sum of two sinuoids. Part C is the algebraic sum of parts A and B.

each period of the fundamental frequency. This relationship is illustrated in the trigonometric series

$$\frac{1}{2}A_0 + \sum_{n=1}^{\infty}\left[A_n \cos(n\omega_0 t) + B_n \sin(n\omega_0 t)\right]$$

for a periodic signal evaluated over an interval $(0, 2\pi)$.

It is a straightforward function that takes the sum of $n$ sinusoids, each sinusoid described by $A\sin\omega t + B\cos\omega t$. To this, we add a DC offset, $(1/2)A_0$, which is merely the average of $f(t)$ over one period. For a trigonometric series to be a Fourier expansion of a

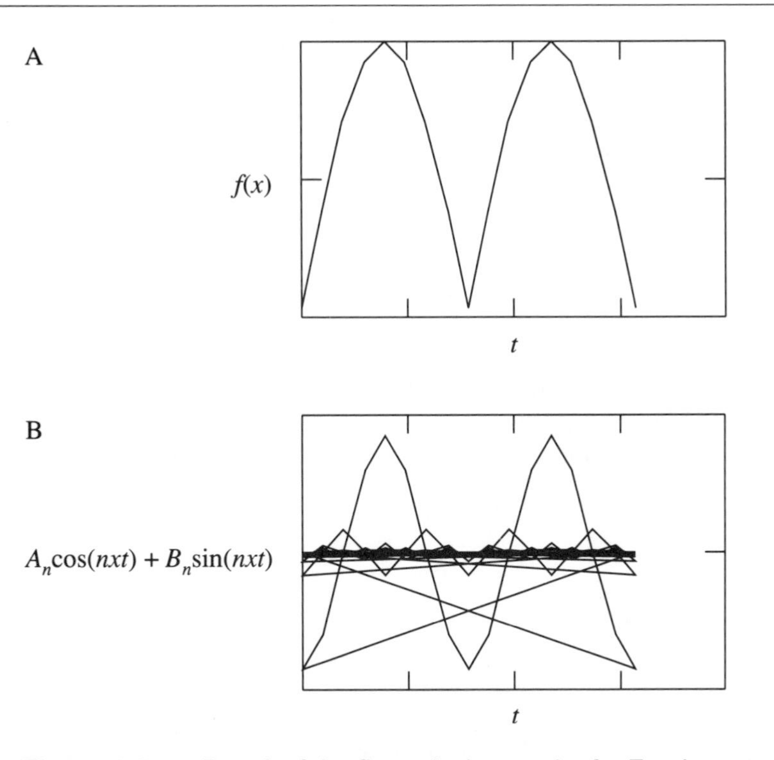


**Figure 1-15.** Part A of the figure is the result of a Fourier expansion of |sin*x*|, part B shows the harmonic components of the sum individually.

function, it must converge to an integrable function $f(t)$ for all except, perhaps, a finite number of values of $t$ in the interval $(0, 2\pi)$.

Simply put, any *periodic* signal can be represented by a sum of sines and cosines that are multiples of the fundamental frequency of that signal. As we have seen, the sine and cosine functions are complete, orthogonal, and independent. This means that we can describe any periodic signal using them, and do so without fear of linear dependency. The equations for approximating the coefficients are found with the help of the orthogonality of the sine and cosine functions themselves. Except for $A_0$, we derive all $A_n$ and $B_n$ by multiplying both sides of the Fourier series by $\cos n\omega t$ for $A_n$ and $\sin n\omega t$ for $B_n$, then integrate over the interval $(0,2\pi)$. $A_0$ is the integral of the function over $(0,2\pi)$. The best approximation is given by the expansion

$$A_0 = \frac{1}{\pi}\int_0^{2\pi} f(t)dt$$

$$A_n = \frac{1}{\pi}\int_0^{2\pi} f(t)\cos n\omega_0 t\, dt \ (n = 0,1,2,\cdots)$$

$$B_n = \frac{1}{\pi}\int_0^{2\pi} f(t)\sin n\omega_0\, dt \ (n = 1,2,3,\cdots)$$

Figure 1-15 is an illustration of this. Figure 1-15A is a graph of the Fourier expansion of the function $f(t) = |t|$ in the period $(0, 2\pi)$. Figure 1-15B is the same expansion without the sum. Here you see the harmonic components that create this function.

The magnitude of the Fourier series may be found from

$$|H| = \sqrt{A_n^2 + B_n^2}$$

and the phase

$$\Phi_n = \tan^{-1} \frac{B_n}{A_n}$$

Though convergence problems do exist for some functions, it is difficult to think of a signal of practical interest that does not have a spectrum. Fourier coefficients exist and the function is convergent if a function satisfies the *Dirichlet conditions*:

1. The function must be absolutely integrable over the period $2\pi$,

   $$\int_0^{2\pi} |f(x)| dx < \infty$$

   or integrable square

   $$\int_0^{2\pi} |f(x)|^2 \, dx < \infty$$

   if convergence to the mean is all that is required.
2. The function must be single-valued.
3. The function has only a finite number of discontinuities within each period.
4. There are only a finite number of maxima and minima within each period.

The utility of the Fourier series is evident. Using this formulation, any convergent periodic function can be approximated and manipulated. Further, the knowledge of this series can help decompose unknown waveforms. Let us take for an example an unknown device, a "black box," that we must characterize for arbitrary inputs (signals). It has two inputs and two outputs, so we attach to the inputs a signal generator that produces pure sinusoids at controlled frequencies, and we sweep a reasonable set of frequencies, recording the response at the outputs as we do. Knowing how the black box reacts to these pure sinusoids, we can predict how it will react to an arbitrary signal input, because we can express that arbitrary signal as a sum of the pure sinusoids we do have information about.

Harmonic analysis makes this Fourier series so useful that it is to our benefit to find ways to use it on as broad a range of functions as possible. The next few paragraphs illustrate some of the tactics that can be employed to extend the range of the Fourier series.

Nonperiodic functions, such as $f(t)$, may be expanded by making the function of interest equal to an identical function that *is* periodic in the interval we are working in. We can deal with the discontinuities at the endpoints by making them equal to the average of the limits approached by each end of the function.

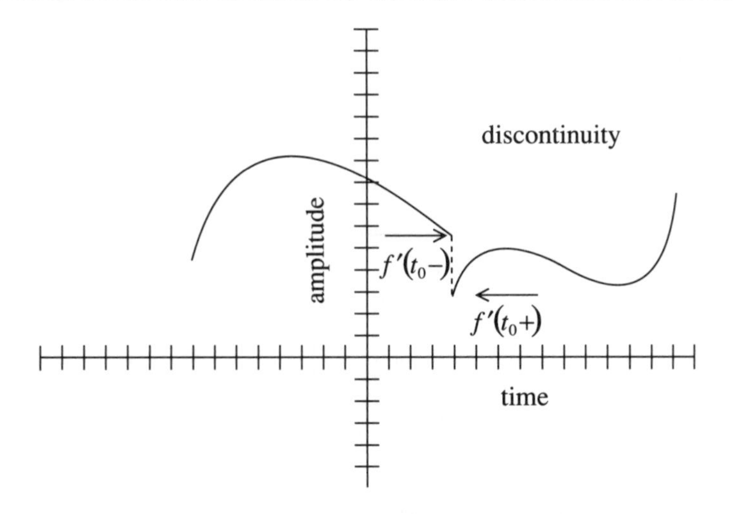


**Figure 1-16.** Function with a discontinuity.

This follows from the theorem, *if $f(t)$ is integrable and has a period of $2\pi$, or if $f(t)$ is integrable in the interval $(0,2\pi)$, its Fourier series will converge to $f(t)$ wherever $f$ is differentiable.* But what if there is a point where a derivative does not exist?

Functions with discontinuities are likewise capable of expansion. If they are integrable in an interval that consists of a finite number of intervals in which it is continuous, it is called *piecewise continuous*, and it has one-sided limits at the endpoints of each such interval.

In Figure 1-16, you can see that $f(t)$ is tending toward a limit as $t$ approaches the discontinuity $t_0$. If $f'(t)$ exists as we approach from the left and tends to a limit $f'(t_0)$, we take the limit as the derivative and write $f'(t_0-)$. In the same manner, we take the derivative as it approaches from the right, $f'(t_0+)$. The Fourier series is then said to converge to $[f(x+)+f(x-)]/2$ at that point, and indeed at every point where the function has a left-hand and a right-hand derivative. Examples of a discontinuous function can be found in the ramp and the square wave.

## Odd and Even Functions

You will quickly realize, after doing a few expansions, that some expansions require only a sine term and some only a cosine term. This *symmetry* is an important aspect in a Fourier evaluation, as it is in much of signal processing.

A function is *even* if $f(\omega t)=f(-\omega t)$, as in the square wave of Figure 1-17. An *odd* function is one for which $f(\omega t)=-f(-\omega t)$, as with the sinusoidal functions depicted in Figure 1-17. For an even expansion, all $B_n$ will be zero, leaving only cosine terms; for an odd expansion, all $A_n$ (including $A_0$) will be zero, leaving only sine terms.

The concept of odd and even functions is an important one to remember. It provides a useful means of reducing the work mathematically and algorithmically in a number of areas. If

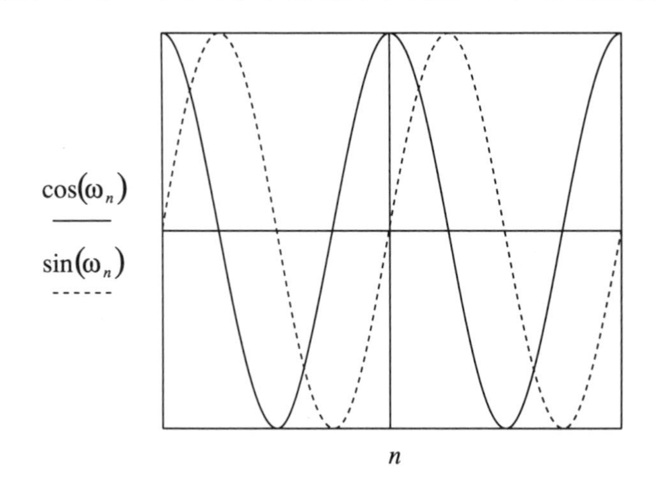


**Figure 1-17.** An example of even and odd functions. In the graph, the cosine wave A is even, the sine wave B is odd.

a function can be expressed as either even or odd, the series becomes that much easier to write, since an even series will have no sine terms and an odd series will have no cosine terms.

## The Gibbs Phenomenon

The Gibbs phenomenon describes the tendency of a partial sum of a Fourier series to overshoot and oscillate about the correct values at points of discontinuity. For an example, note the graph of the square wave (Figure 1-18) whose series was just presented as an odd series. There are "ears" at the rising and falling edges of each square. A greater number of harmonics in the series can improve the situation, but it will not fix it. Actually, as $n \rightarrow \infty$ the overshoot approaches 9 percent.

A significant improvement can be realized by introducing *Lanczos's $\sigma$ factors* (see R.W. Hamming's *Numerical Methods for Scientists and Engineers* for the derivation). These factors represent an average of one cycle of oscillation of the derivative near the discontinuity. The coefficients of the series are computed as usual, but each term of the series is evaluated as multiplied by the following smoothing factor:

$$\sigma_{2n-1} = \frac{\sin(2n-1)\left(\dfrac{\pi}{2m}\right)}{(2n-1)\left(\dfrac{\pi}{2m}\right)}$$

The series becomes

$$\frac{1}{2}A_0 + \sum_{n=1}^{m}\left[\frac{\sin(n\pi/2m)}{n\pi/2m}\right]\left[A_n\cos(n\omega_0 t) + B_n\sin(n\omega_0 t)\right]$$

Figure 1-18 illustrates the effect of using Lanczos's $\sigma$ factors as part of the partial sum of a square wave.

## Fourier Series in Exponential Form

Although the trigonometric representation of the series offers a graphic expression of what is occurring when the series is manipulated, it is cumbersome and difficult to evaluate. The complex exponential form is preferred for calculations. To derive this form, we start with the trigonometric form of the Fourier expansion:

$$\frac{1}{2}A_0 + \sum_{n=1}^{\infty}\left[A_n \cos(n\omega_0 t) + B_n \sin(n\omega_0 t)\right]$$

From Euler's equation

$$e^{\pm j\omega t} = \cos\omega t \pm j\sin\omega t$$

we can write:

$$\frac{1}{2}A_0 + \sum_{n=1}^{\infty}\left(A_n \frac{e^{jn\omega t} + e^{-jn\omega t}}{2} + B_n \frac{e^{jn\omega t} - e^{-jn\omega t}}{2}\right),$$

$$\frac{1}{2}A_0 + \sum_{n=1}^{\infty}\left[\left(\frac{A_n - jB_n}{2}\right)e^{jn\omega t} + \left(\frac{A_n + jB_n}{2}\right)e^{-jn\omega t}\right] \qquad (\text{note}: 1/j = -j)$$

substituting

$$c_n = \frac{A_n - jB_n}{2} \text{ and } c_{-n} = \frac{A_n + jB_n}{2} \text{ and } c_0 = A_0$$

we have

$$c_0 + \sum_{n=1}^{\infty}\left(c_n e^{jn\omega t} + c_{-n} e^{-jn\omega t}\right)$$

which is equivalent to

$$\sum_{n=-\infty}^{\infty}\left(c_n e^{jn\omega t}\right)$$

And we calculate the coefficients with

$$\begin{aligned} c_n &= \frac{1}{2\pi}\int_0^{2\pi} f(t)\cos n\omega t\, dt - \frac{j}{2\pi}\int_0^{2\pi} f(t)\sin n\omega t\, dt \\ &= \frac{1}{2\pi}\int_0^{2\pi} f(t)(\cos n\omega t - j\sin n\omega t)\, dt \\ &= \frac{1}{2\pi}\int_0^{2\pi} f(t)e^{-jn\omega t}\, dt \end{aligned}$$

$$f(t)=\frac{4\cdot A}{\pi}\left(\sin(t)\cdot\sin\left(1\cdot\frac{\pi}{7}\right)+\frac{1}{3}\cdot\sin(3\cdot t)\cdot\sin\left(1\cdot\frac{\pi}{7}\right)+\frac{1}{5}\cdot\sin(5\cdot t)\cdot\sin\left(5\cdot\frac{\pi}{7}\right)+\frac{1}{7}\cdot\sin(7\cdot t)\cdot\sin\left(7\cdot\frac{\pi}{7}\right)\right)$$

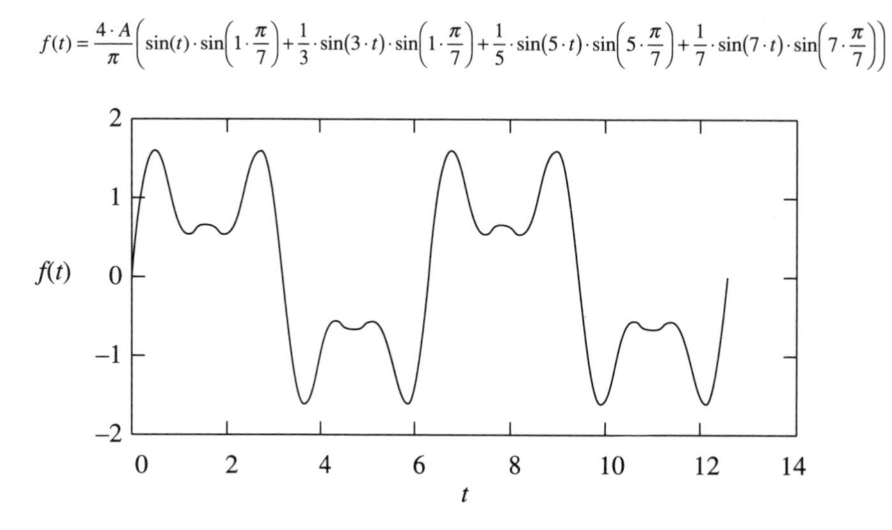


Fourier Expansion of Square Wave

$$f(t)=\frac{4\cdot A}{\pi}\left(\sin(t)\cdot\frac{\sin\left(1\cdot\frac{\pi}{7}\right)}{1\cdot\frac{\pi}{7}}+\frac{1}{3}\cdot\sin(3\cdot t)\cdot\frac{\sin\left(3\cdot\frac{\pi}{7}\right)}{3\cdot\frac{\pi}{7}}+\frac{1}{5}\cdot\sin(5\cdot t)\cdot\frac{\sin\left(5\cdot\frac{\pi}{7}\right)}{5\cdot\frac{\pi}{7}}+\frac{1}{7}\cdot\sin(7\cdot t)\cdot\frac{\sin\left(7\cdot\frac{\pi}{7}\right)}{7\cdot\frac{\pi}{7}}\right)$$

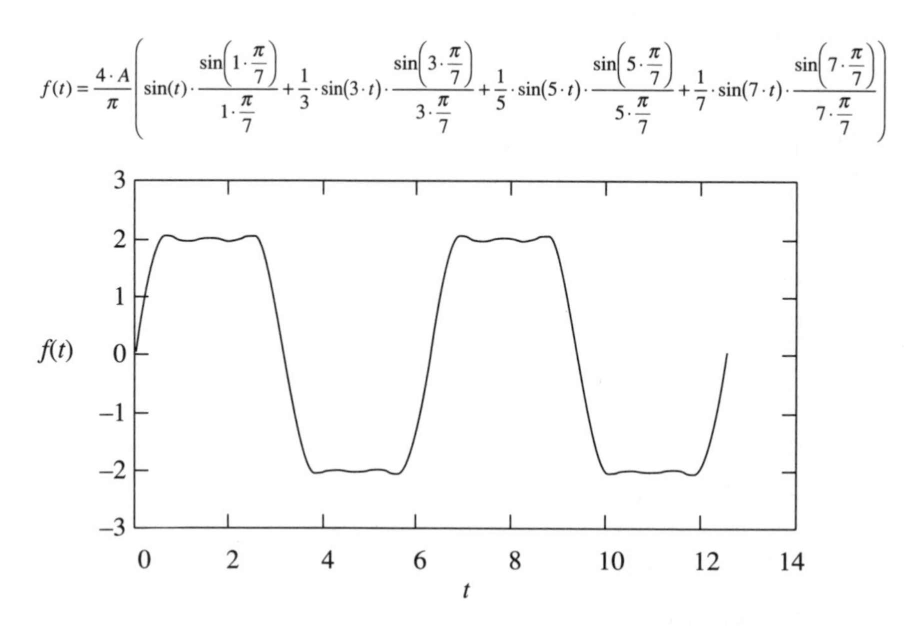


Introducing Lanczo's Sigma Factors

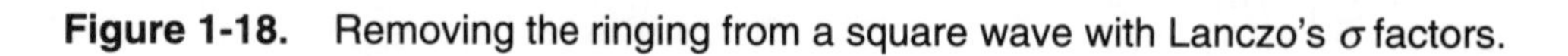

**Figure 1-18.** Removing the ringing from a square wave with Lanczo's $\sigma$ factors.

If the coefficients for a trigonometric Fourier series already exist, it is a simple matter to convert them to an exponential series using the substitutions created for the derivation of the exponential series shown earlier:

$$c_n = \frac{A_n - jB_n}{2}$$

$$c_{-n} = \frac{A_n + jB_n}{2}$$

$$c_0 = A_0$$

## The Fourier Integral

In the last section, we used the Fourier series to approximate periodic functions with sines and cosines. In addition, we have used linear combinations of that decidedly nonperiodic sequence 1, $x$, $x^2$, $x^3$, $x^4$,... to approximate periodic functions. Why can't we approximate nonperiodic functions with periodic functions?

In describing continous and discrete functions, an analogy is often drawn to light and a prism. A prism bends light as a function of its wavelength, resulting in a spread of colors in a regular arrangement from violet through red. If a light source with a random distribution of frequencies is shown through the prism, the result is a continuous spread of colors, one merging seamlessly into the next. However, if a light source that has strong characteristic frequencies is used, such as a mercury lamp, the result is a series of very strong spectral lines. In this second instance, not all colors will be represented. If functions of continuous and of discrete spectra are added together, a function results with both a *continuous* and a *line* spectrum.

The Fourier series represents a sum of periodic functions, sine and cosine, whose nonzero coefficients generate a *discrete spectrum* of the function we are approximating. It is a discrete spectrum because not all frequencies are involved, only integer multiples of the fundamental frequency that we shall call $\omega_0$. In order to approximate a nonperiodic function, we need a *continuous spectrum*, one in which all frequencies contribute. Recall that the Fourier series was a sum of integer multiples of the fundamental frequency called harmonics—except that by the methods discussed in that section, this makes it unsuitable for nonperiodic functions. But if we say that the "period" of our nonperiodic function is actually periodic with its limits at $+\infty$ and $-\infty$, and then allow all frequencies to contribute through integration, it is possible.

We start with the exponential Fourier series:

$$f(\omega_0 t) = \sum_{n=-\infty}^{\infty} \left( c_n e^{jn\omega t} \right)$$

$$c_n = \frac{1}{2\pi} \int_0^{2\pi} f(t) e^{-jn\omega t}\, dt$$

Begin by replacing the $c_n$ in the definition of the series with the actual integral for computing the coefficients. The equation becomes

$$f(\omega_0 t)=\sum_{n=-\infty}^{\infty}\left[\left(\frac{1}{2\pi}\int_{-\pi}^{\pi}f(\omega_0 t)e^{-jn\omega_0 t}\,dt\right)e^{jn\omega_0 t}\right]$$

where $\omega_0$ is the fundamental frequency. As the period in this equation increases from $(-\pi, \pi)$ to $\infty$, the function ceases to be periodic, since it now occupies the entire axis. As the period $T \to \infty$, the summation becomes integration, and we have

$$f(t)=\int_{-\infty}^{\infty}\left(\frac{1}{2\pi}\int_{-\pi}^{\pi}f(t)e^{-j\omega t}\,dt\right)e^{j\omega t}\,d\omega$$

Use the Fourier series for functions that have a discrete or line spectrum, and the Fourier integral for those with continuous spectra.

## Initial Conditions, Transients, and Steady State

Thus far we have been examining the behavior of signals once they have gotten started. We have assumed that the system has always been going and will go on forever. But this ignores the start-up point and what must have existed before that, as well as the turn-off point. *Initial conditions* describe the environment at the moment when the signal is started—these are often assumed to be zero, since nothing was happening before that. This isn't unreasonable—in an LRC oscillator, all voltage drops are zero before current is applied, since none of the elements themselves source current naturally. As the switch is thrown ($T_0+$), there is no current flowing, since an inductor will first store energy in a magnetic field, resulting in an exponential rise of current flow through it. The rate of current flow through an inductor is illustrated in Figure 1-19.

In a harmonic oscillator, the inductor is the mass that must be gotten into motion—it does not determine the displacement from equilibrium or the stiffness. So one would expect that the period of oscillation would be about the same, but multiplied by the exponential character of the rate of rise of current in the inductor. This is true, and it is called *the turn-on transient* (see Figure 1-19A).

Once things have settled, the amplitude of oscillation remains constant, and we have what is known as *steady state response* (see Figure 1-19B).

What happens when we remove current from the oscillator? Something very similar, except that here the important element is the damping of the oscillator. It can be shown that no power is actually dissipated by any of the elements of the oscillator except the purely resistive ones. If there were no damping, the oscillator would go on forever. But even if we remove the $R$ from the LRC circuit, we cannot remove the parasitic resistances from the $L$ and $C$, or the wire and trace that connect them all together. So the oscillator eventually runs out of energy.

The energy that is sapped by the damping in the circuit comes from the same magnetic field the inductor took the time to charge in the first place. This period is known as the *turn-off transient* (see Figure 1-19C).

The important thing to understand is that the character of the system response does not change during turn-on and turn-off transients, merely the driving force—it is the same impulse response as always.

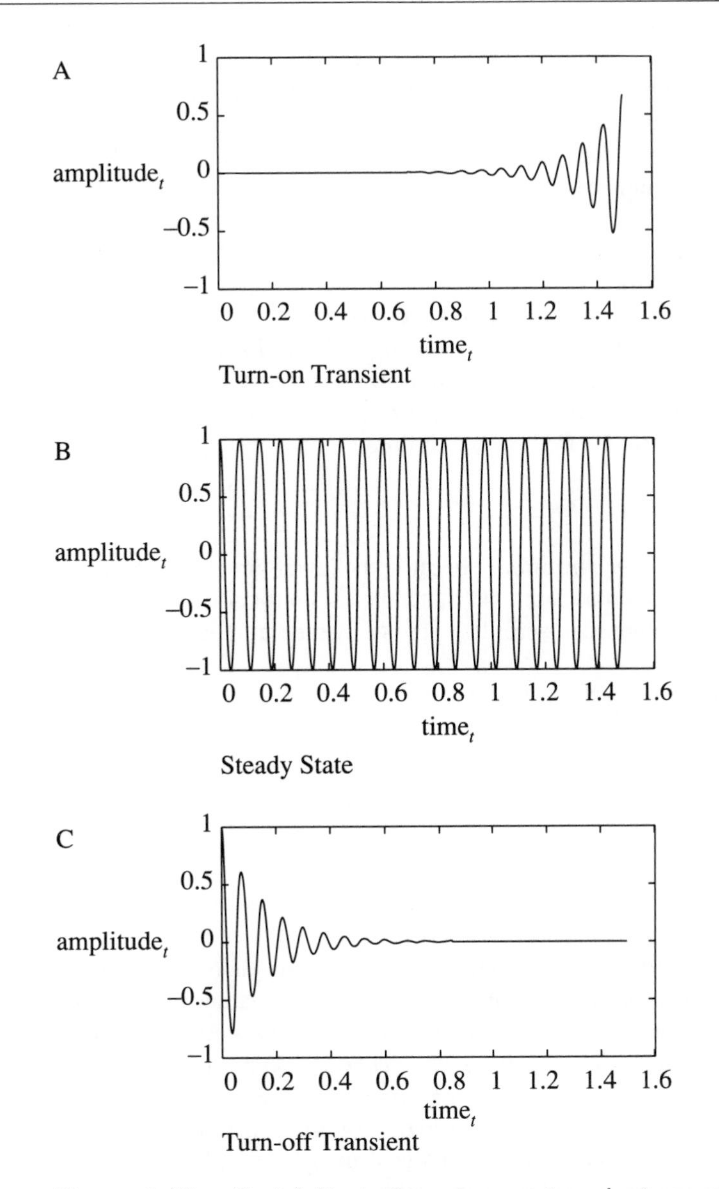


**Figure 1-19.** Part A illustrates a turn on transient occuring just after the application of voltage. Part B is the steady state condition the circuit theoretically reaches after a period of time, and Part C is the turnoff transient the circuit experiences when power is removed.

## The Laplace Transform

The Fourier series handles steady-state sinusoidal functions, but does not have the ability to converge for the decaying factor found in turn-on and turn-off transients, or in transients in general. The *Laplace transform*, however, encompasses the entire complex plane, and can converge for a greater number of functions than the Fourier. In fact, because it can produce the same results as the Fourier transform under the proper circumstances, it is often thought of as a superset of the Fourier. Still, the Laplace transform is useful for much more than that.

The Laplace transform is useful for the solution of differential equations, much the way logarithms simplify multiplication, division, powers, and root extraction. To use logarithms, the numbers would have to be *transformed* into logarithms, the arithmetic performed, and then transformed back into numbers we need. It is the same with the Laplace transform—first the given function of time, $f(t)$, is multiplied by $e^{-st}$, then the product is integrated with respect to time from zero to infinity:

$$L[f(t)]=F(s)=\int_0^{\infty} f(t)e^{-st}\,dt$$

This is the *one-sided Laplace transform*, because the lower limit of integration is 0.

There is a two-sided transform, but it is more useful for situations in which initial conditions are something other than zero. It is safe to assume in most engineering applications that for $t<0$, $t=0$.

With the inverse Laplace transform,

$$L^{-1}\{L[f(t)]\}=L^{-1}[F(s)]=f(t)$$

given by

$$f(t)=\frac{1}{2\pi j}\int_{c-j\infty}^{c+j\infty} F(s)e^{st}\,ds$$

[$c$ is chosen to include all singularities of $F(s)$.]

An important part of these equations in Laplacian mathematics is the quantity $s$, which is a complex number and root of the characteristic equation $s=\sigma+j\omega$. $\omega$ is imaginary and is called the *radian* or *angular frequency*—it finds expression as $\sin\omega t$, or as $\cos\omega t$ in time domain equations. The $\sigma$ is known as the *neper* and represents a nondimensional logarithmic unit. Together, they are defined as the *complex frequency*.

This quantity is associated with the exponential, $e$, to produce both frequency and time varying values. Its effect can be visualized using the $s$-plane itself. Please refer to Figure 1-20 for illustration of the following points:

1. If $\omega=0$ in the equality $s=\sigma+j\omega$, then the $s$ is entirely dependent upon the value of $\sigma$ in terms of the exponential. If $\sigma>0$, then the exponential function will be constantly increasing as in the figure, while if $\sigma<0$, it will be constantly decreasing, and if $\sigma=0$, the exponential function witl have the value 0 forever.

2. If $s=0\pm j\omega$, $e^{\pm j\omega t}$ may be thought of as a unit rotating vector with its direction dictated by the sign. A positive sign means a counterclockwise direction, and a nega-

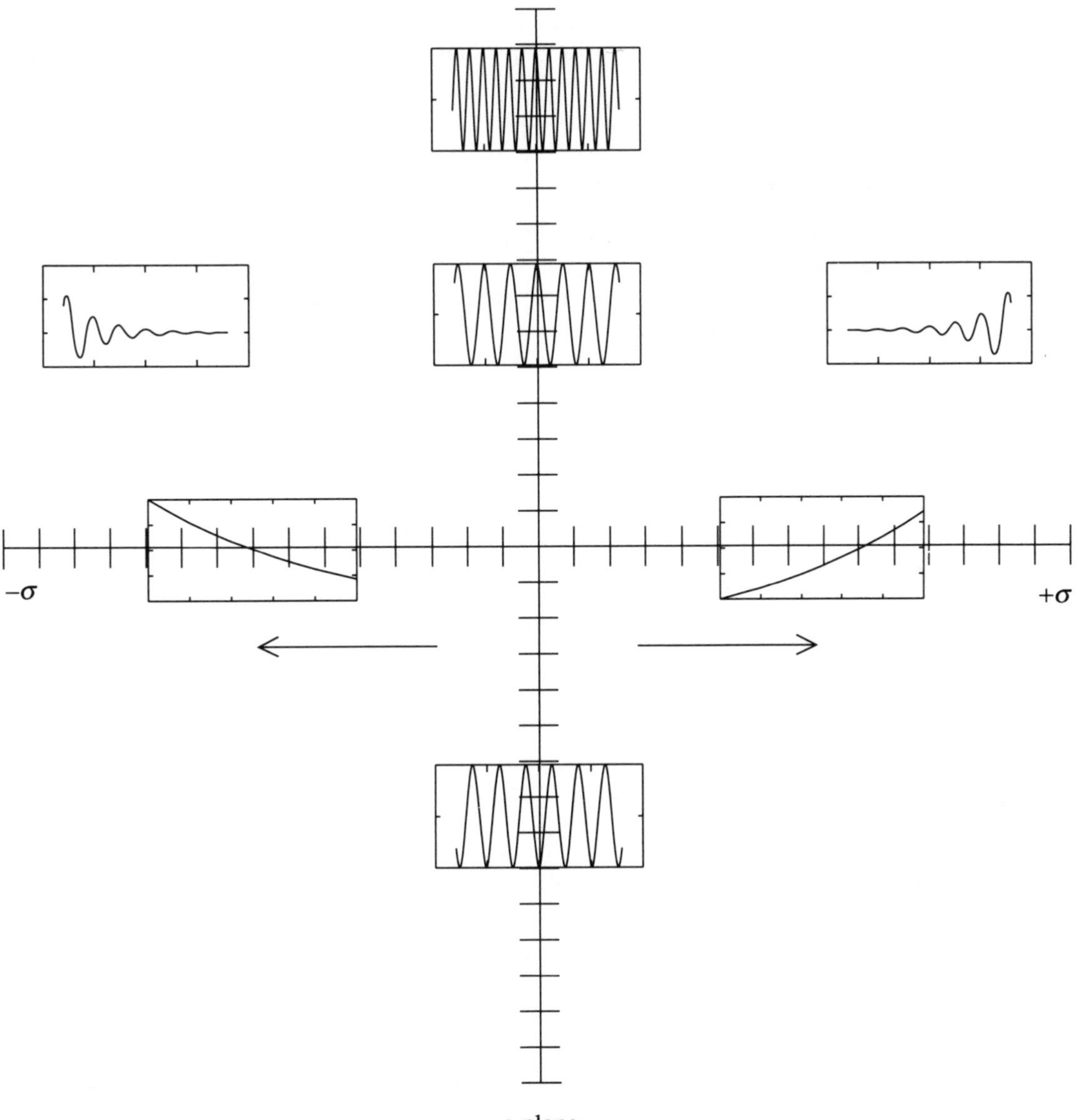

**Figure 1-20.** The S-Plane.

tive sign means a clockwise direction. A positive direction causes a pattern that varies as the cosine of $\omega t$, while a negative direction varies with the sine of $\omega_t$. It is in this form that the Laplace transform will produce the same results as the Fourier transform.

3. When $s=\sigma+j\omega$, the exponential function generates a sequence that is the product of the sinusoid and time varying relationships above and is known as a *damped sinusoid.*

## Properties of the Laplace Transform

Even though the calculations are difficult, there is usually little trouble in finding the transform of a particular $f(t)$. A great deal of work has gone into this subject, and tables exist in abundance listing *transform pairs*. These are common functions of time and their Laplace transforms. Some examples can be found below:

| *Time domain f(t)* | *Frequency domain F(s)* |
|---|---|
| $1$ | $\frac{1}{s}$ |
| $u(t)$ | $\frac{1}{s}$ |
| $\delta(t)$ | $1$ |
| $t$ | $\frac{1}{s^2}$ |
| $t^n$ | $\frac{n!}{s^{n+1}}$ |
| $\frac{t^{n-1}}{(n-1)!}$ | $\frac{1}{s^n}$ |
| $e^{at}$ | $\frac{1}{s-a}$ |
| $e^{-at}$ | $\frac{1}{s+a}$ |
| $te^{at}$ | $\frac{1}{(s-a)^2}$ |
| $1-e^{at}$ | $\frac{-a}{s(s-a)}$ |
| $\frac{1}{\omega}\sin\omega t$ | $\frac{1}{s^2+\omega^2}$ |
| $\sin\omega t$ | $\frac{\omega}{s^2+\omega^2}$ |

| *Time domain f(t)* | *Frequency domain F(s)* |
| --- | --- |
| $\cos \omega t$ | $\frac{s}{s^2+\omega^2}$ |
| $1-\cos \omega t$ | $\frac{\omega^2}{s\left(s^2+\omega^2\right)}$ |
| $\sin(\omega t+\theta)$ | $\frac{s \sin \theta+\omega \cos \theta}{s^2+\omega^2}$ |
| $\cos(\omega t+\theta)$ | $\frac{s \cos \theta+\omega \sin \theta}{s^2+\omega^2}$ |
| $e^{-at} \sin \omega t$ | $\frac{\omega}{(s+a)^2+\omega^2}$ |
| $e^{-at} \cos \omega t$ | $\frac{s+a}{(s+a)^2+\omega^2}$ |

Some important properties of the Laplace transform for our study are:

1. Linearity

$$L\left[af_1(t)+bf_2(t)\right]=aF_1(s)+bF_2(s)$$

2. The Laplace transforms for derivatives:

$$L\left[\frac{df(t)}{dt}\right]=sF(s)-f(0)\text{, and}$$

$$L\left[\frac{d^2 f(t)}{dt^2}\right]=sF(s)-sf(0)-\frac{df(0)}{dt}\text{, with the k}^{\text{th}}\text{ derivative:}$$

$$L\left[\frac{d^{(k)} f(t)}{dt^k}\right]=s^k F(s)=\sum_{n=0}^{k-1} s^{k-1-n} f^{(n)}(0)$$

3. The Laplace Transform transforms for integrals:

$$L\left[\int_0^t f(t)\,dt\right]=\frac{F(s)}{s}$$

4. Frequency shift

$$L\left[e^{-at} f(t)\right]=X(s+a)$$

5. Convolution

$$\text{If } y(t)=\int_0^t h(t-\tau)x(\tau)\,d\tau, \text{ then } Y(s)=H(s)X(s)$$

Let us take the unit step function, described by Oliver Heaviside (1850–1925), as a simple example of the evaluation of a function by the Laplace transform. If $f(t)=1$, then

$$L[f(t)]=\int_0^\infty e^{-st}\,dt=-\frac{e^{-st}}{s}=\frac{1}{s}$$

If you glance at the chart, you will notice that this is the Laplace transform for the *unit step function, $u(t)$*, or 1. But that isn't all: any of the singular functions can occur at any time relative to $t=0$. Perhaps at $t=0$ it is already there, or perhaps at $t=0$ someone throws a switch connecting a battery to our oscillator, or perhaps at $t=0$ nothing happens for a while, then the switch is thrown.

Time relative to $t=0$ is indicated with an offset, say $a$, that denotes the position of the step function relative to zero. Figure 1-21 demonstrates several possibilities. First, there is the possibility that the switch is simply on for all time (part A). Second, the switch may be thrown some measure after $t=0$ indicated by $u(t-a)$, making $t$ greater than $t=0$ by $a$ units (part B)—the impulse occurs when the result of the operation in the parenthesis is zero. Third, the switch may be thrown before $t=0$, represented by $u(t+a)$ (part C). There are still other possibilities relating to when the switch is opened. In fact, one can synthesize a square wave with a sum representing the amplitude and time value of each opening and closing of the switch.

For example, let us say that we have a set of Cartesian axes divided into intervals of time, as in Figure 1-21. Each interval is of equal length, so each can be denoted by a multiple of that length. On this graph we draw a periodic square wave. The formula $v(t)=u(t-a)-u(t-2a)+u(t-3a)-u(t-4a)+\ldots$ is the infinite series time domain representation of that square wave, and

$$V(s)=\frac{e^{-as}}{s}-\frac{e^{-2as}}{s}+\frac{e^{-3as}}{s}-\frac{e^{-4as}}{s}+\cdots, \text{ or } V(s)=\frac{1}{s}\tanh as$$

is the Laplace transform.

Any *periodic* function can be written in closed form like this, using the Laplace transform, because of the ability of the transform to do time shifting by a simple multiplication. This is based upon the almost magical property of the geometric series

$$1+x+x^2+x^3+\cdots$$

which according to Taylor/Maclaurin data will converge to $1/(1-x)$ for $|x|<1$.

With this in mind, we can write

$$F(s)=\frac{1}{1-e^{Ts}}\int_0^T e^{-st}f(t)\,dt$$

to compute the transform of any periodic waveform.

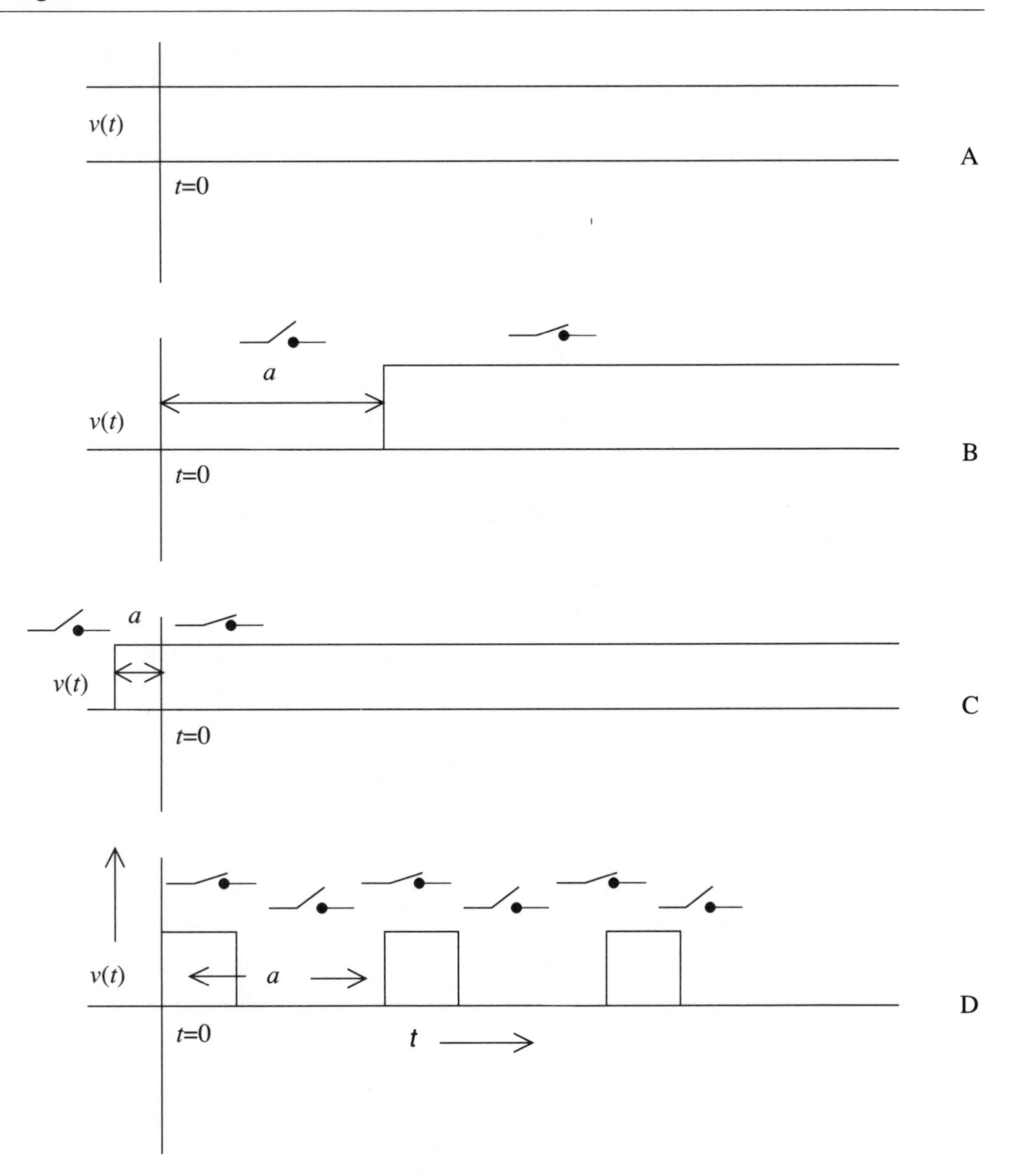

**Figure 1-21.** Square wave generated by pressing and releasing a switch.

## Network Analysis

The predominant passive elements of a network—that is, a filter, oscillator, and attenuator or amplifier—are resistors, capacitors, and inductors. Each of these exhibits markedly different characteristics relative to the others and, in the case of the capacitor and inductor, relative to the input. The resistor is the most straightforward, obeying Ohm's law for both voltage and current: $V=IR$, and $I=V/R$; the voltage drop across the device and the current through it reflect the input in a linear proportionality. The capacitor and resistor, however, change roles depending upon the parameter we are measuring:

$$V_{cap} = \frac{1}{C}\int i\,dt, \text{ and } I_{cap} = C\frac{dv}{dt}$$

while

$$V_l = L\frac{di}{dt}, \text{ and } I_l = \frac{1}{L}\int v\,dt$$

These are the components of a network. Basic to the analysis of a network is its response to various functions, each of unit area. These unit functions are illustrated in Figure 1-5, earlier in this chapter.

The Laplace transform accepts each of these responses as *singular functions*, from which any of the others can be derived. These functions are:

1. The *unit step function*, which is 1 for all $t>0$ and 0 for all $t<0$. This function has the Laplace transform $1/s$.
2. The *unit ramp function*, the integral of the step function, varies linearly with time and has the Laplace transform $1/s^2$.
3. The *unit impulse function*, the derivative of the step function, has a nonzero value only at the beginning of the step function. The unit impulse function has the Laplace transform of 1. From these functions, still others can be generated by simple differentiation.
4. The *unit doublet function*, the derivative of the unit impulse. The unit impulse has a width of zero, the unit doublet is created from the positive and negative going edges of the pulse. The Laplace transform for the unit doublet is $s$.
5. The *unit triplet function*, the derivative of the unit doublet. This is derived in the same manner as the unit doublet and has the Laplace transform of $s^2$. The process can go on forever.

In the case of each of these functions, the area is always the same: one unit. An important thing to remember about these functions is that when you know one, any of the others can be derived simply by the appropriate integration(s) or differentiation(s). If we ignore initial conditions, differentiate with a multiplication by $s$ and integrate with a division by $s$.

Consider the equation that we have been working with for the oscillator:

$$L\frac{d^2 i}{dt^2} + R\frac{di}{dt} + \frac{i}{C} = v(t)$$

If $v(t)$ is a unit impulse, integrating both sides of the equation results in the oscillator's response to a unit step function, while differentiating it gives us the response to a unit doublet function:

$$L\frac{di}{dt}+Ri+\int\frac{i}{C}=\int v(t)$$

a unit step response, and

$$L\frac{d^2}{dt^2}\left(\frac{di}{dt}\right)+R\frac{d}{dt}\left(\frac{di}{dt}\right)+\frac{1}{C}\frac{di}{dt}=\frac{d}{dt}[v(t)]$$

the unit doublet response.

As we mentioned earlier, another name for the impulse function is the *delta function* or *Dirac delta function*, denoted $\delta(t)$.

The Laplace transform has some important implications for our definition of the transfer function. The general expression for a voltage transfer function is

$$\frac{V_{out}(s)}{V_{in}(s)}=H(s)$$

and the Laplace transform of the output divided by the transform of the input is equal to the transfer function $H(s)$. If $V_{in}(t)=\delta(t)$, $V_{in}=1$, which results in $V_{out}(s)=H(s)$. The output from a unit impulse input voltage is completely determined by the transfer function of the sysem. The *impulse response*, a time domain function typically written $h(t)$, is the output produced when a unit impulse is applied to the input. Therefore, we can say the transfer function of a system, $H(s)$, is the Laplace transform of that system's impulse response.

## Magnitude Squared Response

Recalling the general expression for the transfer function, $H(s)=P(s)/Q(s)$, a system's steady state response is that of the transfer function evaluated with $s=j\omega$, $H(j\omega)$. This means that the equation contains both real and imaginary parts, with the real parts in even powers and the odd part in odd powers:

$$H(j\omega)=\frac{A(\omega)+jB(\omega)}{C(\omega)+jD(\omega)}=R(\omega)+jX(\omega)$$

This fact is demonstrated in the *magnitude squared response*, an important quantity containing information valuable for the design and analysis of both continuous-time and discrete time filters. The magnitude squared response is given by

$$|H(j\omega)|^2=\frac{|A(\omega)|^2+|B(\omega)|^2}{|C(\omega)|^2+|D(\omega)|^2}=H(j\omega)H^*(j\omega)=H(j\omega)H(-j\omega)$$

The product of the transfer function with its complex conjugate results in a ratio of even, real polynomials. For example,

$$H(s)=\frac{s+1}{s^2+\sqrt{2}s+1}$$

becomes

$$|H(j\omega)|^2=\frac{j\omega+1}{1-\omega^2+\sqrt{2}j\omega}\frac{-j\omega+1}{1-\omega^2-\sqrt{2}j\omega}=\frac{\omega^2+1}{1+2\omega^2+\omega^4}$$

We can use this network function to derive the phase response of the system. Recalling that the numerator and denominator polynomials of the function comprise real and imaginary parts,

$$H(j\omega)=\frac{\operatorname{Re}A(j\omega)+j\operatorname{Im}A(j\omega)}{\operatorname{Re}B(j\omega)+j\operatorname{Im}B(j\omega)}$$

which we can rationalize by multiplying the numerator and denominator by the complex conjugate of the denominator [Re$B(j\omega)-j$Im$B(j\omega)$] to obtain:

$$H(j\omega)=\operatorname{Re}H(j\omega)+j\operatorname{Im}H(j\omega)$$

yielding the phase as

$$\phi(j\omega)=\tan^{-1}\left(\frac{\operatorname{Im}[H(j\omega)]}{\operatorname{Re}[H(j\omega)]}\right)$$

Alternatively, for electrical systems, one may write these as

$$H(j\omega)=\sqrt{[X(j\omega)]^2+[R(j\omega)]^2}$$

and

$$\phi=\tan^{-1}\frac{X(j\omega)}{R(j\omega)}$$

We have briefly covered some of the basic tools necessary for understanding the origins of many of the functions and inner workings of signal processing from both mathematical and electrical viewpoints. It is a fascinating and involved study with many common parts. It isn't possible to cover everything completely. Much of this information will be amplified in later chapters and tempered with the effects of the real world, which will change things somewhat.

# The Real World

Mathematics, like dreams, can present an unreal idea about the world. We can represent a sinusoid with a few symbols, and depending upon our mood and our domain, we can change it or decompose it as we please without affecting its integrity. Unfortunately, it does not work this way in the real world, where one can not sample, amplify, filter, or even buffer an input signal without changing it in some way. Part of what we learn in applying this data to the real world is that it is impossible *not* to process signals.

The emphasis is on the analog and digital circuitry that we use to interface with the continuous time input. In this chapter, we will look at what we must do to preserve the integrity of signals for our purposes, process them, then reconstruct them. It is important to note that the processes we speak about are essentially the same processes that the physical universe exercises upon all forms of signals all the time. A thorough understanding of these processes, both electronically and theoretically, can produce better applications through an awareness of the effects of the physical universe upon signals. Like the last chapter, this chapter is in two parts. The first part outlines the definitions, foundations, characteristics, and parametric extents of filter technology and how they relate to analog-to-digital conversion, as well as to the reconstruction or creation of analog signals. We will review Butterworth, Chebyshev, and Bessel filters with reference to DSP applications, and provide the mathematics for deriving the poles $\omega_0$ (natural harmonic frequency) and $Q$ (frequency selectivity) for each of these characteristics. This is the data you need to design your own filters, both hardware and software. Most of what is here applies to digital processing as well as analog, and is appropriate for anyone interested in signal processing. The second part of the chapter is devoted to design criteria and procedures in general, and the DACQ design in particular. There are several examples of analog filter designs, along with the algorithms both in the text and in the Mathcad documents on the accompanying disk. There are also examples of popular active filter configurations, including Sallen and Key, multiple feedback, and GIC. A full understanding of this last part will require an electronics background and is not for every reader, though an awareness of these materials can enhance the utility and meaning of the chapters on digital filter implementation.

# PART ONE: SEVEN INCHES OF COPPER WIRE

## Electronic Circuits

In this chapter, we will deal with mathematics, theory, and the real world. Sooner or later, every design must be implemented on actual printed circuit boards, in real-world environments, with parts that are actually available. It can be a rude shock. There are limits to the technology, and it shows up in the parameters of the devices and materials we use and the situations that much of the equipment must function in. It might be instructive in this regard to imagine what it would be like *not* to process a signal. Every electrical signal is in a constant state of convolution with the impulse response of every system it encounters, whether that is a wire, an opamp, a resistor, or a terminal.

Motion takes us, and anything else, through a continuously changing electrical relationship with everything around us. The perfect sinusoid our formulas and symbols describe does not exist. There are other signals invariably embedded within the signal we are concerned with; other phenomena have modified the signal we are interested in. Even the tools we use can sufficiently change the object of our interest to allow us to miss something important about it. Every electrical engineer has had the experience of trying to debug a circuit, only to have the problem disappear as long as his equipment is hooked to it. This in itself is often enough to provide a clue as to the problem, and it is also an indication that we cannot study something without changing it.

These are the phenomena that bend and shape our signals. Wire possesses characteristics that limit and shape the frequency it can carry as well as the power; resistors contribute noise to the current they carry; identical capacitors can be at wide variance in risetime and bandwidth depending upon the material they are made of and when they were made; and an opamp might best be described as a transfer function than an amplifier. Seven inches of copper wire process signals no less than an algorithm or a filter. These are the materials we work with and our tools. It is good to know how they work so that we can use them to produce the best possible products.

## Data Acquisition

A data acquisition system is any arrangement that provides us with the means to translate signals—in this case, continuous-time analog signals—into digital form suitable for interpretation and manipulation by computer. Such a system is illustrated in Figure 2-1.

The signal is initially prepared for conversion by an anti-aliasing filter. Depending upon the application, this may take the form of a specialized IC, or a simple opamp or RC filter. Its purpose is simply to eliminate or reduce unwanted frequency information so that it does not become part of the data that the DSP processes.

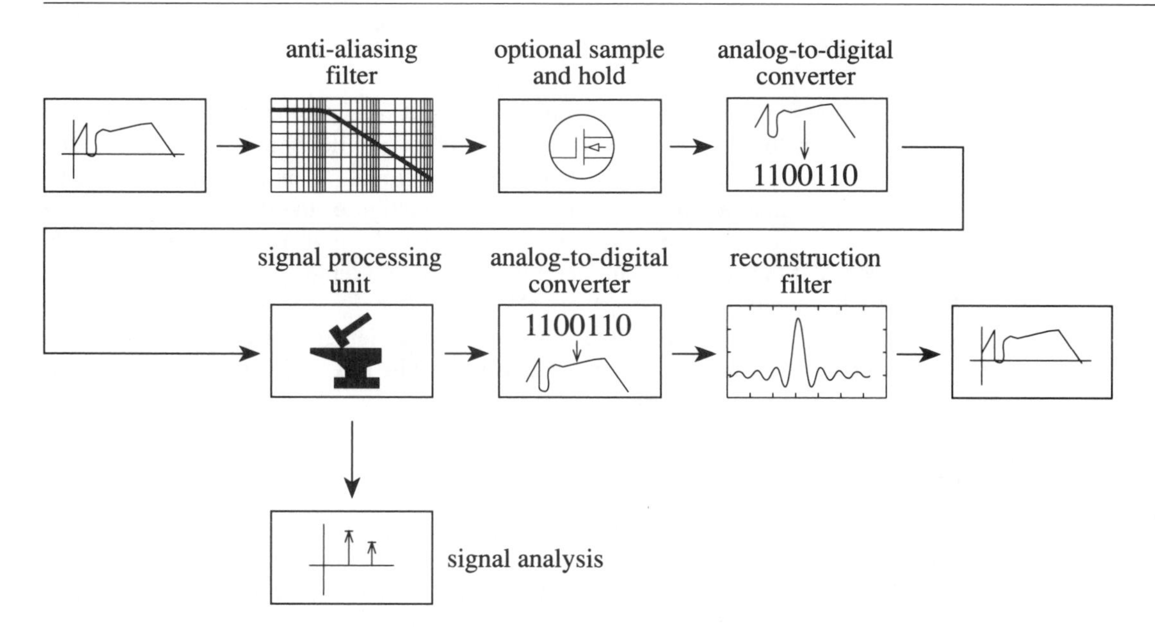


**Figure 2-1.** Block diagram of a data acquisition system.

The next component of the acquisition system is often a sample-and-hold circuit. This device is designed to take a sample of the input signal in an extremely short time and then hold it constant at its output until the next sample is required. The advantages of this circuit are numerous. To begin with, for a reliable conversion, the signal at the input of an A/D must change less than one LSB (of the A/D) during the time it acquires the sample. If the signal is moving faster than that, resolution and accuracy are lost—the signal is smeared, and actually experiences lowpass filtering at the A/D. Using a sample-and-hold means that converters without tracking can be used without reducing the bandwidth of the system. It can have economic benefits, because the cost of the converter goes down if it is not necessary that it be so fast.

The analog-to-digital converter is the heart of the system. This is the device that determines most of the filtering requirements and indicates whether a sample-and-hold is necessary. The more bits, the greater the resolution; the greater the resolution, the more attenuation necessary at and above Nyquist frequencies to make the bits meaningful.[1] It is important that the device actually reflect the needs of the application; otherwise, you may not have the resolution you need, or you may be spending money on bits you don't need. Besides resolution, the A/D is usually characterized for distortion and linearity, characteristics which may or may not be so important depending upon your application. Your needs

[1]The Nyquist frequency is the highest frequency a sampled data system is capable of sampling *and* recovering completely. According to the sampling theorem, the sample frequency must be at least twice any frequency in the bandwidth of interest for this recovery to be possible.

and tolerances are not only an economic concern, they also influence the other components in the system.

After the input is processed, the data may be output for signal construction. For this, we use a digital-to-analog converter. Its function is to take parallel binary data and produce an analog output corresponding to it. Of course, given an input, its output remains constant until its next input, which produces a staircase approximation of the desired output unless filtered. For this, we need another filter tuned to remove the high-frequency components related to the edges of the staircase—in fact, this filter should remove all frequencies above those desired in the output.

In the next few sections, we will discuss each of the elements of the data acquisition system described above. The first subject we approach is the filter, which is one of the most demanding parts of the entire system.

## What the Filter Needs to Do

The ideal lowpass filter has constant amplitude characteristic from its lower cutoff frequency to its upper cutoff frequency, and zero output for all other frequencies. Unfortunately, this ideal does not exist. Try as we might, a perfectly flat passband does not exist, magnitude response does not go instantly to zero dB at cutoff, and we do not get absolute attenuation of all out of band frequencies in the stopband. We may come arbitrarily close in any one of these specifications, but usually at a cost that is not very economical—optimizing one aspect of the filter can skew another.

If we wish a flatter passband (or a more linear phase response), a faster transition time, or more attenuation in the stopband, we can have it with a higher order filter or more preprocessing in the DSP software. The degree to which any of this is required is usually dictated by the dynamic range of the system, the sampling rate at the input, and the software overhead on the DSP.

At first glance, it might seem that all we really need is a filter that rolls off as quickly as possible at the cutoff frequency and then our problems will be solved—but this is not usually the case. Ultimately, all filters roll off at about $6N$ dB per octave ($N$ is the order of the filter), which means that almost any filter of any order will eventually provide the prescribed level of attenuation in the stopband. The truth is, however, that some filter characteristics, such as the Chebyshev, do have a steeper initial transition band than others; they will move from passband to stopband over a shorter range of frequencies than others, and they will do this with a lower order. These characteristics can also contribute passband and stopband ripple, as well as phase distortion in both the passband and stopband.

Few DSP projects can tolerate phase distortion—which, in the time domain, appears as overshoot, undershoot, and distortion of the waveform—yet the filter characteristics with the steepest transition bands also introduce the greatest amount of phase distortion into the passband frequencies. In addition, the steeper the transition band, the greater the distortion. Another consideration is economics. Complex filters and the parts for them can be expensive to purchase and to build, they can also consume larger areas of PCB space. Having high sample rates coupled with digital filtering can provide for lower order anti-aliasing filters, but this will not dismiss the problem completely. As you will see, building a filter is a matter of trade-offs, and really depends upon the needs of your application. Know your application.

## Filter Requirements

We design filters to attenuate frequencies greater than the Nyquist frequency of the application so that they are not detectable by the system A/D converter. Normally, you would make this less than the *rms quantization noise level* defined by the converter. For example, a 12-bit A/D converter has $2^{12}=4096$ levels of resolution. If the reference voltage used for its conversions is 5 volts, each level is equal to 5 volts/4096 = 1 millivolt. Call this value $q$. In order to use all the available bits in your converter for data and not noise conversion, frequencies in the stopband would have to be attenuated to less than $q/2\sqrt{3}$ the *rms quantization noise level*. Therefore, for a 12-bit system, we need

$$Qrms = -20\log_{10}\frac{V_{\text{full scale}}}{V_{\text{q noise level}}} = -20\log_{10}\frac{V}{\left(\frac{q}{2\sqrt{3}}\right)} = -20\log_{10}\frac{5}{\left(\frac{.001}{2\sqrt{3}}\right)} = -84.771\text{dB} \quad \text{(Eq. 2-1)}$$

attenuation in the stopband.

The *dynamic range* of an acquisition system is the ratio of the largest expected signal to the smallest, and describes what the input you want to convert looks like. In this example, we want the stopband attenuation to result in a signal less than the rms noise quantization level, so we make it equal to the absolute value of *Qrms*. Since filter rolloff is approximately 6*N*dB per octave (*N* is the order of the filter), the following formula provides a simplified and conservative estimate of filter order:

$$N = \frac{|Qrms|}{6\log_2\left(\frac{f_s}{2f_{pb}}\right)}$$

*Qrms* is the filter stopband attenuation required (the result of Eq. 2-1 above), $f_{pb}$ is the corner frequency and is assumed to be equal to the desired analog input bandwidth, and $f_s$ is the sampling frequency, the transition period will start at $f_s/2$. Again, assuming a 12-bit A/D requiring −85dB of attenuation in the stopband, a 10 KHz sampling rate, and a 4 KHz cutoff or corner frequency, we have

$$N = \frac{85\text{dB}}{6\log_2\left(\frac{10\text{ KHz}}{8\text{ KHz}}\right)} = 44$$

We need a 44th-order filter to give us full dynamic range without problems from noise. The reason, obviously, is that the cutoff frequency is so close to the Nyquist frequency that it requires a *very* fast transition band.

Be advised, filter design above the 8th order becomes difficult and expensive. For more on this, see Mathcad document Dynamic.mcd.

## Oversampling

One way to decrease the complexity of the filter circuit is with a technique known as *oversampling*. This is very popular in audio equipment, because it allows the designer to simplify his anti-aliasing filters considerably, thereby lowering costs and saving board space. This works by allowing the analog filter a much longer transition period, say $K(f_s/2)$ where $K$ is an integer and $f_s$ is the sampling frequency. A digital filter can then reduce the cutoff frequency to $f_s/2$ in software/firmware. An added advantage here is that the digital filter is now producing more samples than necessary, actually $K$ more samples than necessary, and it is now possible to look at only every $K$th sample, harmlessly neglecting the rest in a process called *decimation*.

Let us assume four times oversampling, a value popular in audio circles because it reduces most of their filter requirement to a 6th order instead of 44th. Using the example from above, we can now make $f_s$ 40 KHz with the transition band starting at 20 KHz, while $f_{pb}$ remains the same at 4 KHz:

$$M = \frac{85\,\text{dB}}{6\log_2\left(\dfrac{40\ \text{KHz}}{8\ \text{KHz}}\right)} = 5.96$$

This represents one of many ways a designer can affect the signal to noise ratio; the filter cannot do it all. Other things intrinsic to the A/D limit the achieveable SNR, such as thermal noise, aperture jitter, nonlinearities, comparator uncertainties, missing codes, slew rate problems, and so on. In actual practice, the values for SNR (signal to noise ratio, as described above) and ENOB (effective number of bits relative to the A/D converter) are determined by applying pure sinusoid signals to the A/D, then performing a Fourier transform on the resulting data. The spectrum of this transform is used to calculate SNR and harmonics of the input frequency. The RMS value of the signal is computed, as well as the RMS values of all the other frequency components in the passband (noise and harmonic distortion). SNR is the ratio of these two quantities.

For examples of the results of such testing, please refer to the LM12H458 datasheet in the Appendix.

## Filters

### Ripple and Phase Shift in the Passband

It is important to remember that the instantaneous value of any signal is the algebraic sum of all the contributing frequencies in its spectrum at a particular phase angle. If there is magnitude ripple in the passband, some frequencies are being attenuated more than others, which will produce an incorrect sum in analysis or reconstruction. By the same token, if certain frequencies have been phase shifted more than others, their sum at any point will be distorted too. Phase content is essential to analysis. Image processing, for instance, depends upon the linearity of phase for edge detection and correlation techniques.

The choice and design of the appropriate filter for a system always involves trade-offs and sacrifices. In the example from the last section, a higher sampling rate permitted the use of a lower order anti-aliasing filter at the cost of phase lag in software. This is acceptable in non-realtime systems, but may be completely unacceptable in realtime systems. It is important to identify the actual needs of the system to determine the kind of characteristic that is required and the amount of phase delay that may be tolerated before deciding on a special filter, high order filter, or any filter at all.

In addition to filter designs employing opamps—usually called VCVSs—there are numerous chips available that can reduce the complexity of the higher orders and higher chip count, involving both continuous time technologies and switched capacitor techniques. Many of these will be examined later in this chapter. Look closely, however, they are not without fault; they can involve overlong group delays and unwanted signals in your design.

### Basic Definitions

Before launching into a discussion of the characteristics and construction of filters, there are some basic concepts you will need to know:

**Monotonic** A filter demonstrates *monotonicity* if the derivative of the magnitude does not change sign over the given range of frequencies. That is, there is no ripple.

**Passband** The passband is the region of frequencies the filter attenuates the least. This usually extends to the $-3$dB point. The passband may not be flat; some classes of filters have *ripples*—that is, variations in attenuation. This means that some passband frequency elements are being attenuated more than others.

**Cutoff Frequency** The cutoff frequency, $\omega_c$, is typically the $-3$dB or half-power point. It is the end of the passband and the start of the *transition region.*

**Stopband** The stopband is that region of the greatest attenuation for the filter; it usually is defined to have some minimum.

**Phase Shift** The phase shift is related to the time delay of the various frequencies going through the passband. If it is not linear (different frequencies experience different delays), the signal that emerges from the passband may be too distorted to use. *Constant time delay* means that the phase shift increases linearly with frequency.

**Group Delay** Filters insert time delays. For some types of filters, there is a nonlinear phase response in the passband. Some frequencies are delayed more than others; this will manifest itself as phase distortion in the output. *Group delay* is a filter characteristic describing the phase response of the filter to different frequencies in the passband and transition band. If the group delay is constant, the filter is linear phase, and the entire pass-band is delayed by a certain time. If the delay is not constant, especially phase-sensitive signals such as digital and video will experience distortion in the form of ringing and attenuation of various harmonics. Thomson (sometimes called Bessel) filters are designed for maximally linear phase and are recommended for applications sensitive to inconstant phase delay.

**Rise Time** This is the time it takes for the filter output to reach 90 percent of its ultimate output after application of a step function.

**Settling Time** This is the time required for the output of the filter to settle to within some amount of the final output, and of course it must stay there.

**Rolloff** Rolloff is the degree of attenuation of an input signal.

**Half-Power Points** This is another way of specifying the $Q$ of a circuit. The current at resonance is $I_{resonance} = V/R$. Since power is proportional to $I^2$, when the current $I = V/\sqrt{2}R$ power is half that of resonance. Referring to the section on "Q and Resonance" in the last chapter, this results in *half-power frequencies* that are $\zeta\omega_0$ either side of $\omega_0$, as indicted in Figure 2-2, approximately equal to $\omega = \omega_0 \pm \zeta\omega_0$. The highest half-power frequency is $\omega_h = \omega_0 + \zeta\omega_0$, and the lowest is $\omega_l = \omega_0 - \zeta\omega_0$. This is demonstrated graphically in Figure 2-2.

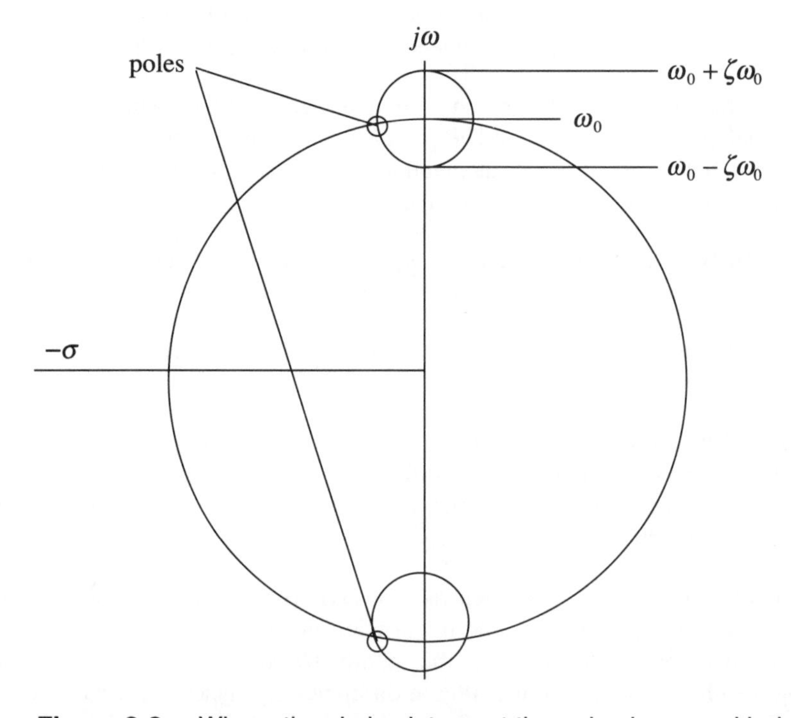


**Figure 2-2.** Where the circles intersect the axis above and below, the system is at halfpower, this is also the 3dB down point.

A line drawn parallel to the real axis from the imaginary axis to the pole $s_1$ forms the radius ($\zeta\omega_0$) of a half circle. Rotate this radius such that it touches the $j\omega$ axis above and below *its own* origin, not the origin of the complex plane. The point above its origin on the $j\omega$ axis is the upper frequency, and the point below is the lower frequency. The *admittance*—the inverse of impedence—of the circuit has dropped to $0.707(1/R)$ at this point. The range for frequencies between the half-power points is called the *bandwidth*, and varies inversely with $Q$. That is, a very high bandwidth is equivalent to a very low $Q$. A half-power point is also the point at which network response is down 3dB.

**LowPass Filter** A lowpass filter consists of a passband that passes frequencies from DC to a specific frequency called the *cutoff* frequency, where the frequency is attenuated to a greater and greater degree until it reaches the stopband and maximum attenuation. It is also characterized by a maximum ripple within the passband. Figure 2-3 illustrates the response of an idealized lowpass filter.

**Bandpass Filter** A bandpass filter passes only a specific frequency band, while attenuating those above and below it. The difference between the upper cutoff and lower cutoff frequencies defines the bandwidth of the bandpass filter. The *center frequency* is a geometric mean, *not an average*, of these two cutoff frequencies:

$$frequency_{center} = \sqrt{frequency_{upper} \bullet frequency_{lower}}$$

Figure 2-4 represents an idealized bandpass filter. Bandpass filters are usually of an even order.

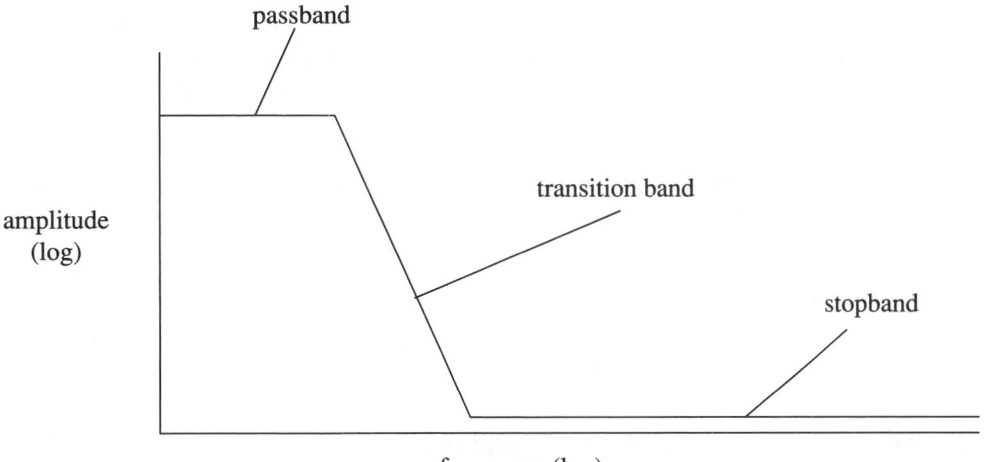


**Figure 2-3.** Idealized lowpass filter.

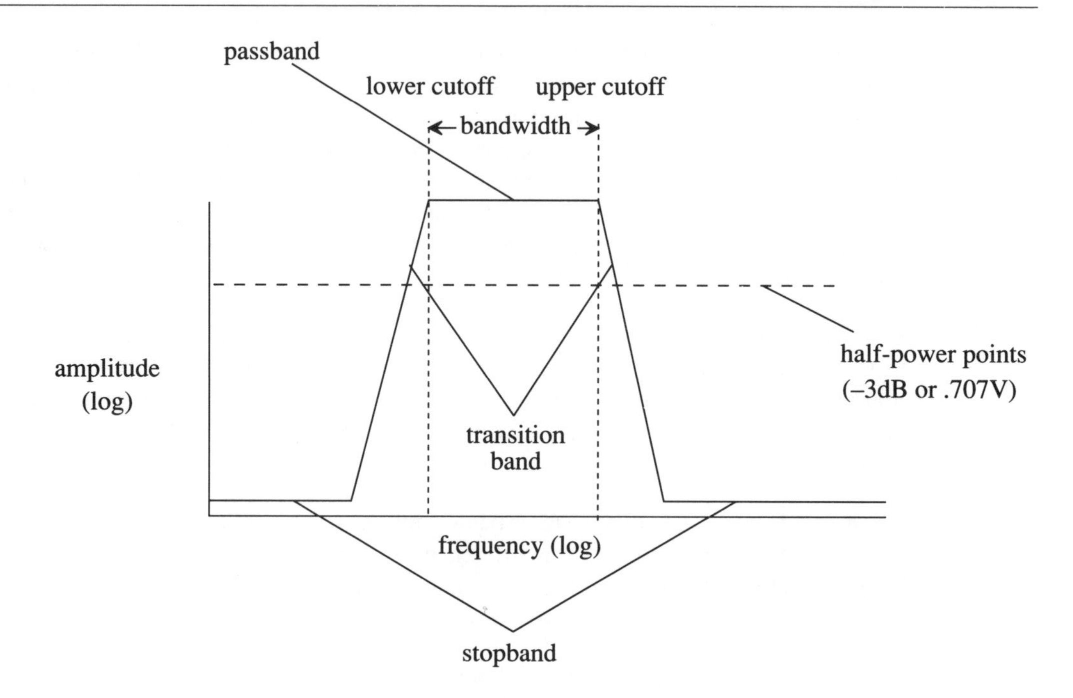


**Figure 2-4.** Idealized bandpass filter.

**Notch Filter** This is a variant of the bandpass filter in which frequencies above and below a specific frequency are passed, while that particular frequency is maximally attenuated. This might also be viewed as a combination of the lowpass filter and highpass filter. Figure 2-5 represents a notch filter.

**Highpass Filter** As the name suggests, the highpass filter rejects frequencies below a specific frequency and passes all those above. In actual practice, of course, this is not possible. Figure 2-6 represents an idealized highpass filter.

**Allpass Filter** The purpose of the allpass filter is to modify the delay characteristics of a signal—that is, to alter the phase shift. It passes frequencies of every order and phase shifts of all or only a specific band of frequencies.

## Choosing a Filter Class for an Application

### Filter Characteristics

There are three basic characteristics of a filter that we might wish to manipulate to obtain the best response for an application. We can optimize a filter for

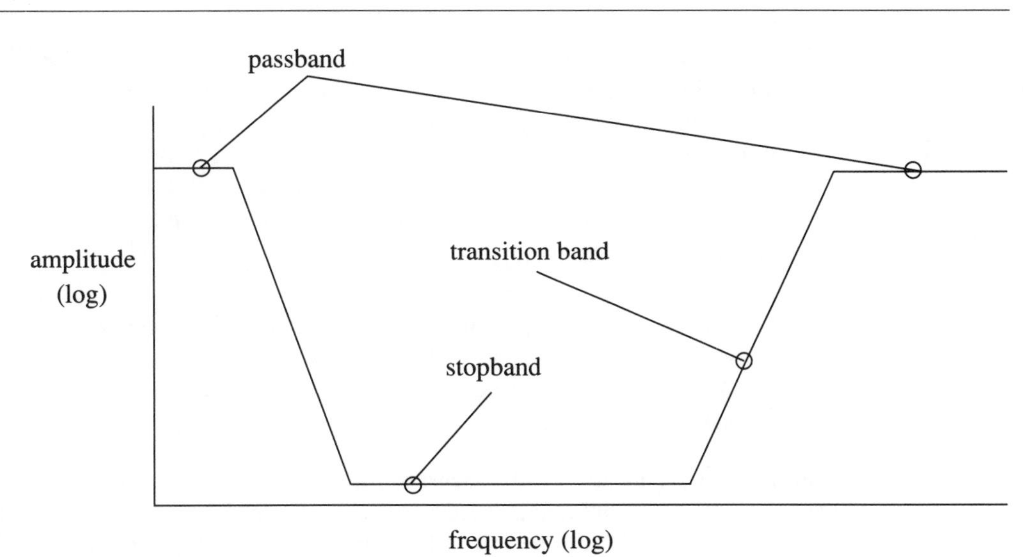


**Figure 2-5.** Notch or bandstop filter.

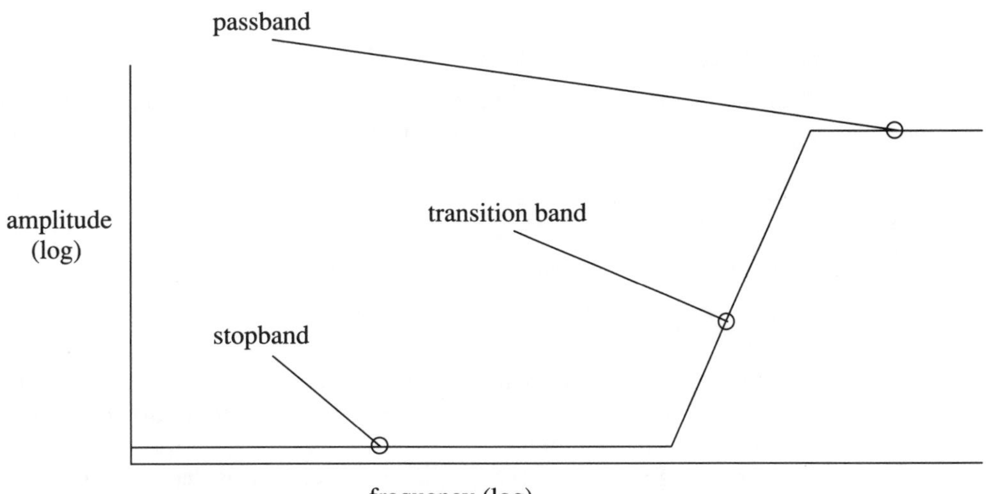

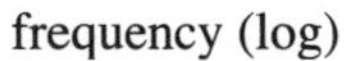


**Figure 2-6.** Idealized highpass filter.

1. *Flatness of passband response.* Since any signal is the sum of its various frequency components, any uneven attenuation will result in signal distortion. Such distortion will be directly evident in the Fourier transform of the signal. The price you are likely to pay for a flatter pass band is a rounder knee at the cutoff point.

2. *Steepness of the transition band.* Even though the ultimate rate of rolloff is approximately given by 6.02*N*dB per octave (*N* is the order of the filter) as one goes deep into the stopband, it is possible to create filters with steeper initial transition bands than others. The obvious benefit is tighter control over aliasing caused by out-of-band frequencies. Unfortunately, the steepness is often produced by allowing ripple in the passband and stopband, which causes varying attenuation on signals in the passband and possibly a higher noise floor in the stopband.
3. *Constancy of phase shift.* This is sometimes referred to as *group delay*. Unwanted phase shift in the passband can distort the waveforms, leading to unacceptable distortion. Constant group delay prevents this, but usually means a much longer and rounder transition band.

In the next sections, we will discuss some of the major classes of filters. There are countless filter designs combining aspects of each of these element, some of which are part of the DACQ acquisition unit described later in this chapter. For each of the points listed above, there is a model or class of filter that exemplifies it. The next sections cover the major classifications of filters, and examples of lowpass filters are given for each. In a final section, we will show how to transform a lowpass filter into any of the others.

## A Magnitude Squared Function

In Chapter 1, we found that the magnitude squared function yields both the magnitude and phase of the system function. A transfer function

$$H(s)=\frac{a_0+a_1s+a_2s^2+a_3s^3\cdots}{b_0+b_1s+b_2s^2+b_3s^3\cdots}$$

becomes a magnitude squared function, as follows:

$$|H(j\omega)|^2=H(j\omega)\overline{H(j\omega)}=H(j\omega)H(-j\omega)$$

In this form, it is easy to manipulate the magnitude for a desired response and then return it to transfer function form by factoring. An example of how this might be done for the Butterworth characteristic follows.

If we wish a maximally flat system response, such as that offered by the Butterworth characteristic, we make it maximally flat at $\omega=0$ and set the first $2n-1$ derivatives of the magnitude squared transfer function to zero at $\omega=0$.

Assuming the transfer function in squared magnitude response form

$$|H(j\omega)|^2=H_0^2\,\frac{1+a_1\omega^2+a_2\omega^4\cdots}{1+b_1s^2+b_2s^4\cdots}$$

in which $H_0^2$ is a scaling factor, we can divide through by the denominator to arrive at another form:

$$|H(j\omega)|^2 = H^2\left[1 + (a_1 - b_1)\omega^2 + \left(a_2 - b_2 + b_1^2 - a_1 b_1\right)\omega^4 \cdots\right]$$

Comparing this to a Taylor series with each $F^{(i)}(\omega)$ at $\omega=0$, we find that the odd derivatives are already zero, and that to set each derivative to zero. It is only necessary to set all $a_i = b_i$.

For a more practical example, let us take a transfer function for which we want to have a maximally flat magnitude, such as

$$H(s) = 2\frac{3s-1}{s^2+ks+1}$$

we need to to find a necessary value of $k$. Manipulating the equation to do so:

$$|H(j\omega)|^2 = \frac{|2(3j\omega+1)|^2}{|1-\omega^2+kj\omega|^2} = 4\frac{1+9\omega^2}{1+\left(k^2-2\right)\omega^2+\omega^4}$$

and equating the coefficients of $\omega^2$, we have $9 = k^2 - 2$ and $k = \sqrt{11}$.

The lowpass filters in the next sections are commonly called *all pole* functions, since the numerator of the network function is a constant, placing the zeros at infinity. The form of this function is

$$H(s) = H_0\frac{1}{B(s)}$$

In this expression, $H_0$ acts as a scaling factor and $B(s)$ actually describes the network function.

## Butterworth

The Butterworth filter produces a maximally flat passband, a monotonic transition band, and a continually decreasing stopband to its zeros at infinity. It also has a round knee at the transition band and nonlinear phase characteristics. Increasing the number of poles produces a flatter passband and sharper cutoff.

In order to get the drop-off at higher frequencies, the zeros of the transfer function are set at infinity, making the numerator of the function a constant. The *normalized* Butterworth squared magnitude function is

$$|H(j\omega)|^2 = H^2\frac{1}{1+\omega^{2n}}$$

Indeed, this relationship will produce the necessary poles for a filter with a Butterworth characteristic. The bode plot of the second order function in Figure 2-7 is also a result of this equation.

The poles of a Butterworth filter are spaced at equal angles around the unit circle on the complex plane, with the roots of the denominator polynomials given by

$$s_i = e^{j\pi[(2i+n-1)/2n]} = \cos\left(\pi\frac{2i+n-1}{2n}\right) + j\sin\left(\pi\frac{2i+n-1}{2n}\right)$$

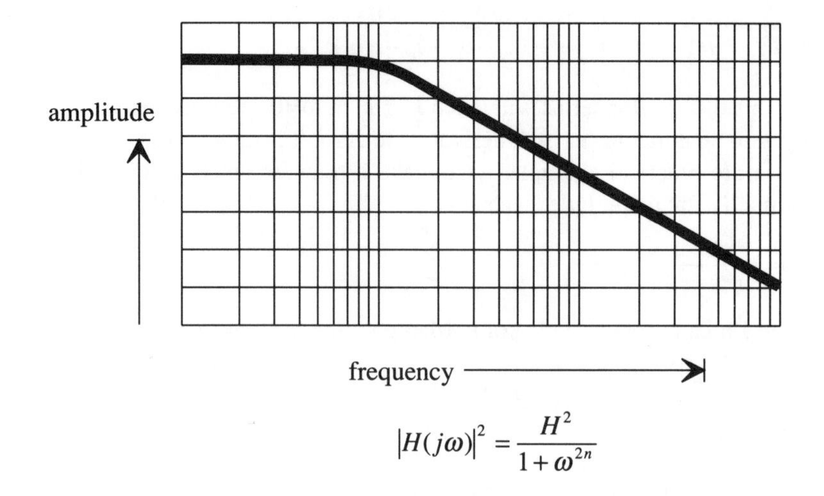


$$|H(j\omega)|^2 = \frac{H^2}{1+\omega^{2n}}$$

**Figure 2-7.** Bode plot of a second order Butterworth filter.

Figure 2-8 is a graph of the pole locations and phase angles for a second-order Butterworth characteristic.

The denominator coefficients for a polynomial normalized for a passband of 0 to 1 rad/s and of the form

$$a_0 + a_1 s + a_2 s^2 + a_3 s^3 \cdots$$

can be calculated with the iterative equation

$$a_k = \frac{\cos[(k-1)\pi/2n]}{\sin(k\pi/2n)} a_{k-1} \quad k=1,\ldots,n \text{ and } a_0 = 1$$

Notice that only the poles on the left half-plane are used; amplifiers or filters with poles in the right half-plane are unstable and unsuitable for most purposes.

The undamped natural frequency $\omega_0$ for the function is computed as the norm of each complex pole with its conjugate:

$$\omega_0 = \sqrt{pole_i * \overline{pole_i}}$$

For the Butterworth, $\omega_0$ is always 1.

The $Q$, then, is equal to the quotient of $\omega_0$ and the difference between each pole and its conjugate:

$$Q = \frac{\omega_0}{pole_i - \overline{pole_i}}$$

The list of poles and $Q$s for the Butterworth characteristic shown in Table 2-1 was computed using the equations above. (Please see Butter.mcd for more examples.)

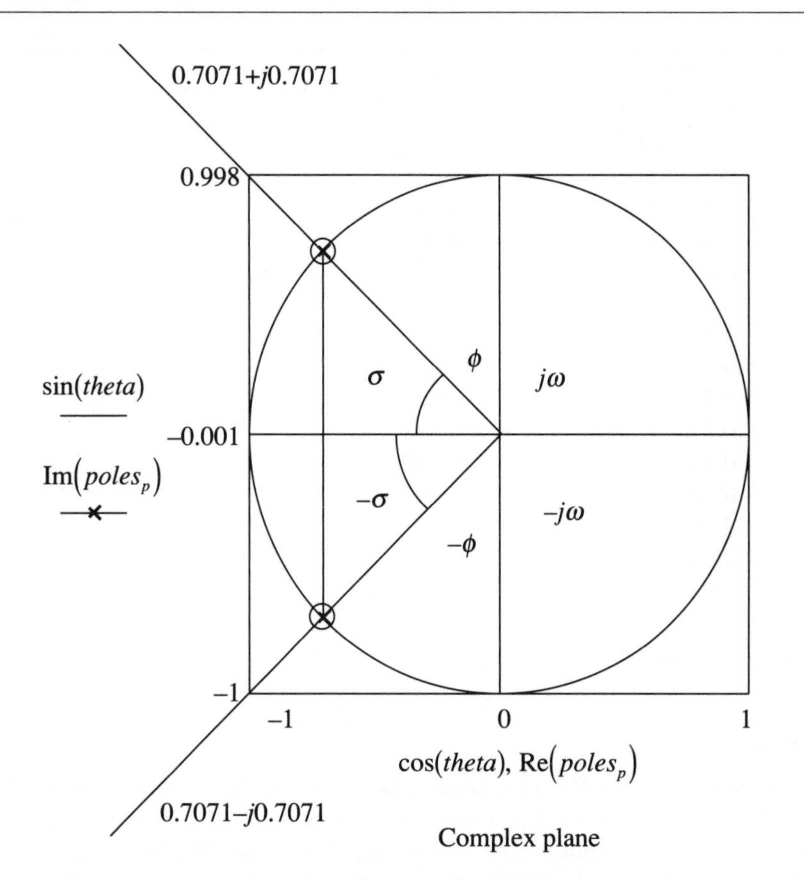


**Figure 2-8.** Argand diagram displaying the placement of poles for a second-order Butterworth filter.

| *order* | *poles* | *Q* |
|---|---|---|
| 2 | $-0.707 \pm j0.707$ | 0.707 |
| 3 | $-0.500 \pm j0.866$ | 1.00 |
| 4 | $-0.383 \pm j0.924$ | 1.307 |
| | $-0.924 \pm j0.383$ | 0.541 |
| 5 | $-0.309 \pm j0.951$ | 1.618 |
| | $-0.809 \pm j0.588$ | 0.618 |
| 6 | $-0.259 \pm j0.966$ | 1.932 |
| | $-0.707 \pm j0.707$ | 0.707 |
| | $-0.966 \pm j0.259$ | 0.518 |

*(continued)*

**Table 2-1.**

| *order* | *poles* | *Q* |
|---|---|---|
| 7 | $-0.223 \pm j0.975$ | 2.247 |
| | $-0.623 \pm j0.782$ | 0.802 |
| | $-0.901 \pm j0.434$ | 0.555 |
| 8 | $-0.195 \pm j0.981$ | 2.563 |
| | $-0.556 \pm j0.831$ | 0.900 |
| | $-0.831 \pm j0.556$ | 0.601 |
| | $-0.981 \pm j0.195$ | 0.510 |
| 9 | $-0.174 \pm j0.985$ | 2.880 |
| | $-0.500 \pm j0.867$ | 1.000 |
| | $-0.766 \pm j0.643$ | 0.653 |
| | $-0.940 \pm j0.342$ | 0.532 |
| 10 | $-0.156 \pm j0.988$ | 3.196 |
| | $-0.454 \pm j0.891$ | 1.101 |
| | $-0.707 \pm j0.707$ | 0.707 |
| | $-0.891 \pm j0.454$ | 0.561 |
| | $-0.988 \pm j0.156$ | 0.506 |

**Table 2-1.** *(continued)*

### Chebyshev

In place of the maximally flat magnitude characteristic of the Butterworth filters, the Chebyshev filter offers a much faster initial transition band by introducing ripple into the passband. This is done through the use of specialized polynomials, often used in mathematics for the implementation of approximations that converge at high speed and introduce a minimum of error into the approximation. These are called Chebyshev polynomials.

These polynomials were named for P.L. Chebyshev, who first used them in his studies of steam engines. They are orthogonal, and actually represent a restatement of the trigonometric identity:

$$\cos(n+1)\theta + \cos(n-1)\theta = 2\cos\theta\cos n\theta$$

Classically, these polynomials are given by the three-term recurrence relation

$$T_{n+1}(x) - 2xT_n(x) + T_{n-1}(x) = 0$$

Substituting $\omega$ for $x$ and rewriting the relation to generate polynomials, we arrive at:

$$T_0(\omega) = 1$$

$$T_1(\omega) = \omega$$

$$T_{n+1}(\omega) = 2xT_n(\omega) - T_{n-1}(\omega)$$

which we can use to derive Chebyshev polynomials of arbitrary order:

$$T_1(\omega)=\omega$$

$$T_2(\omega)=2\omega^2-1$$

$$T_3(\omega)=4\omega^3-3\omega$$

$$\vdots$$

$$T_{n+1}(\omega)=2\omega T_n(\omega)-T_{n-1}(\omega)\ \ (n\geq 1)$$

Figure 2-9 illustrates graphically several orders of the polynomial. The general tendencies of the functions are easy to see.

The filter characteristic for a normalized lowpass equal-ripple approximation is of the form

$$|H(j\omega)|^2=H^2\frac{1}{1+\varepsilon^2 T_n^2(\omega)}$$

$$\varepsilon=\sqrt{10^{r/10}-1}$$

In this equation, $\varepsilon$ represents the magnitude of the deviation allowed in the passband, $r$ represents the passband ripple in decibels, the expression $T_n^2(\omega)$ is the Chebyshev polynomial, and $n$ is the order.

Calculation of the poles of a Chebyshev function will be shown here without proof. More examples are available in the Mathcad document Cheby.mcd.

1. Define the allowable passband ripple, $r$, in dB.
2. Select the order of the filter, *order*.
3. Let $k=1, 2, \ldots,$ *order* and calculate:

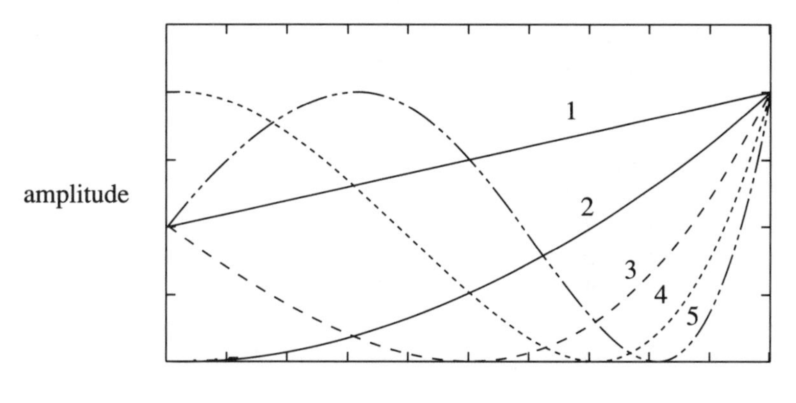


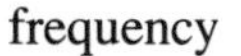


**Figure 2-9.** Plot of the first five Chebyshev polynomials.

$$\varepsilon = \sqrt{10^{r/10} - 1}$$

$$u_k = \frac{2k-1}{2k}\pi$$

$$v = \frac{1}{k}\sinh^{-1}\left(\frac{1}{\varepsilon}\right)$$

$$p_k = \sinh(v)\sin(u_k) \pm \cosh(v)\cos(u_k)$$

Table 2-2 was computed from the poles derived from this algorithm. Each pair of complex poles represents a quadratic polynomial in factored form. The poles and *Q*s are calculated from this quadratic.

| | | *0.5dB Equal-Ripple* | |
|---|---|---|---|
| *order* | *poles* | $\omega_0$ | *Q* |
| 2 | $-0.713 \pm j1.004$ | 1.231 | 0.864 |
| 3 | $-0.313 \pm j1.022$ | 1.069 | 1.706 |
| | $-0.626$ | | |
| 4 | $-0.175 \pm j1.016$ | 1.031 | 2.941 |
| | $-0.423 \pm j0.421$ | 0.597 | 0.705 |
| 5 | $-0.112 \pm j1.012$ | 1.018 | 4.545 |
| | $-0.293 \pm j0.625$ | 0.690 | 1.178 |
| | $-0.362$ | | |
| 6 | $-0.078 \pm j1.008$ | 1.011 | 6.513 |
| | $-0.212 \pm j0.738$ | 0.768 | 1.810 |
| | $-0.290 \pm j0.270$ | 0.396 | 0.684 |
| 7 | $-0.057 \pm j1.006$ | 1.008 | 8.842 |
| | $-0.160 \pm j0.807$ | 0.823 | 2.576 |
| | $-0.231 \pm j0.448$ | 0.504 | 1.092 |
| | $-0.256$ | | |
| 8 | $-0.044 \pm j1.005$ | 1.006 | 11.531 |
| | $-0.124 \pm j0.852$ | 0.861 | 3.466 |
| | $-0.186 \pm j0.569$ | 0.599 | 1.611 |
| | $-0.219 \pm j0.200$ | 0.297 | 0.677 |
| 9 | $-0.034 \pm j1.004$ | 1.005 | 14.579 |
| | $-0.099 \pm j0.883$ | 0.888 | 4.478 |
| | $-0.152 \pm j0.655$ | 0.673 | 2.213 |
| | $-0.186 \pm j0.349$ | 0.395 | 1.060 |
| | $-0.198$ | | |

*(continued)*

**Table 2-2.**

| | | *0.5dB Equal-Ripple* | |
|---|---|---|---|
| *order* | *poles* | $\omega_0$ | *Q* |
| 10 | $-0.028 \pm j1.003$ | 1.004 | 17.987 |
| | $-0.081 \pm j0.905$ | 0.909 | 5.611 |
| | $-0.126 \pm j0.718$ | 0.729 | 2.891 |
| | $-0.159 \pm j0.461$ | 0.488 | 1.534 |
| | $-0.176 \pm j0.159$ | 0.237 | 0.673 |

| | | *1.0dB Equal-Ripple* | |
|---|---|---|---|
| *order* | *poles* | $\omega_0$ | *Q* |
| 2 | $-0.549 \pm j0.895$ | 1.050 | 0.957 |
| 3 | $-0.247 \pm j0.966$ | 0.997 | 2.018 |
| | $-494$ | | |
| 4 | $-0.140 \pm j0.983$ | 0.993 | 3.559 |
| | $-0.337 \pm j0.407$ | 0.529 | 0.785 |
| 5 | $-0.089 \pm j0.990$ | 0.994 | 5.556 |
| | $-0.234 \pm j0.612$ | 0.655 | 1.399 |
| | $-0.289$ | | |
| 6 | $-0.622 \pm j0.993$ | 0.995 | 8.004 |
| | $-0.170 \pm j0.727$ | 0.747 | 2.198 |
| | $-0.232 \pm j0.266$ | 0.353 | 0.760 |
| 7 | $-0.046 \pm j0.995$ | 0.996 | 10.899 |
| | $-0.128 \pm j0.798$ | 0.808 | 3.156 |
| | $-0.185 \pm j0.443$ | 0.480 | 1.297 |
| | $-0.205$ | | |
| 8 | $-0.035 \pm j0.996$ | 0.997 | 14.240 |
| | $-0.100 \pm j0.845$ | 0.851 | 4.266 |
| | $-0.149 \pm j0.564$ | 0.584 | 1.956 |
| | $-0.176 \pm j0.198$ | 0.265 | 0.753 |
| 9 | $-0.028 \pm j0.997$ | 0.998 | 18.029 |
| | $-0.080 \pm j0.877$ | 0.881 | 5.527 |
| | $-0.122 \pm j0.651$ | 0.662 | 2.713 |
| | $-0.150 \pm j0.346$ | 0.377 | 1.260 |
| | $-0.159$ | | |
| 10 | $-0.022 \pm j1.000$ | 0.998 | 22.263 |
| | $-0.065 \pm j0.900$ | 0.902 | 6.937 |
| | $-0.101 \pm j0.714$ | 0.721 | 3.561 |
| | $-0.128 \pm j0.459$ | 0.476 | 1.864 |
| | $-0.142 \pm j0.158$ | 0.212 | 0.750 |

**Table 2-2.** *(continued)*

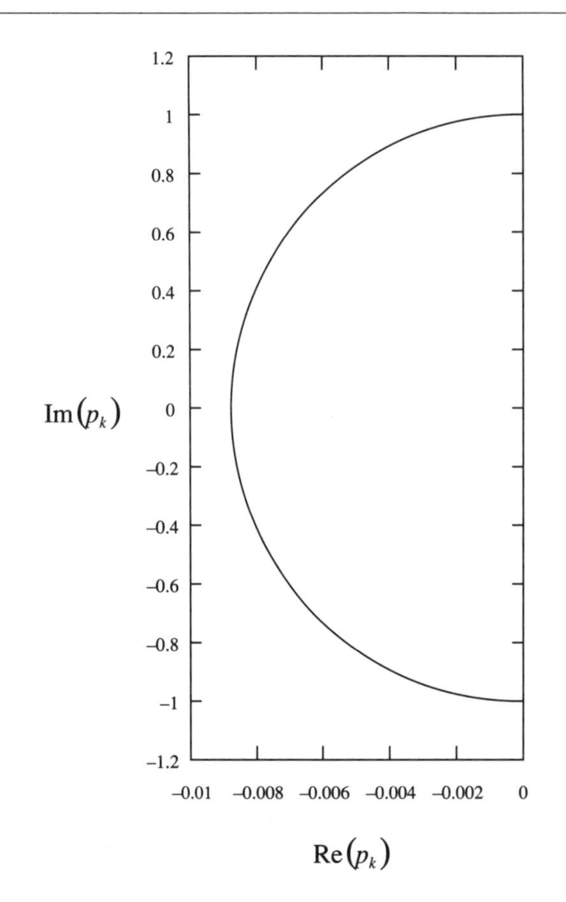


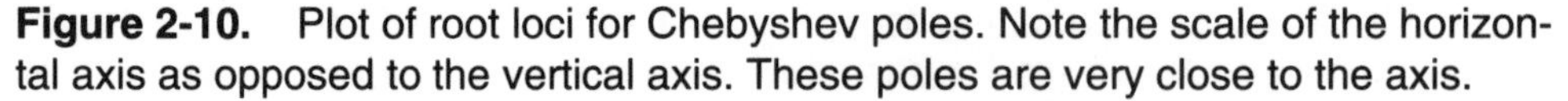

**Figure 2-10.** Plot of root loci for Chebyshev poles. Note the scale of the horizontal axis as opposed to the vertical axis. These poles are very close to the axis.

The poles for an equal ripple filter lie on an ellipse about the origin of the complex plane. Figure 2-10 depicts the pattern of the loci for this function. Please note the scaling in the plot; the poles are very close to the imaginary axis, indicating a very high $Q$.

The bode plots for third- and fifth-order Chebyshev filters are given in Figure 2-11. You will note that the third- and fifth-order Chebyshev polynomials are readily apparent in the passband.

As we stated earlier, this filter has a very fast transition band at the expense of ripple in the passband, and considerable phase distortion as well. In applications for which frequency content is important, this filter can provide a good solution.

There are other versions of the Chebyshev that we will not go into here, such as the inverse Chebyshev, which puts ripple into the stopband instead of the passband. Another filter type that introduces ripple into the passband *and* stopband was developed by Cauer and is called *Elliptic*. It has a transfer function similar to the Equiripple just discussed, but

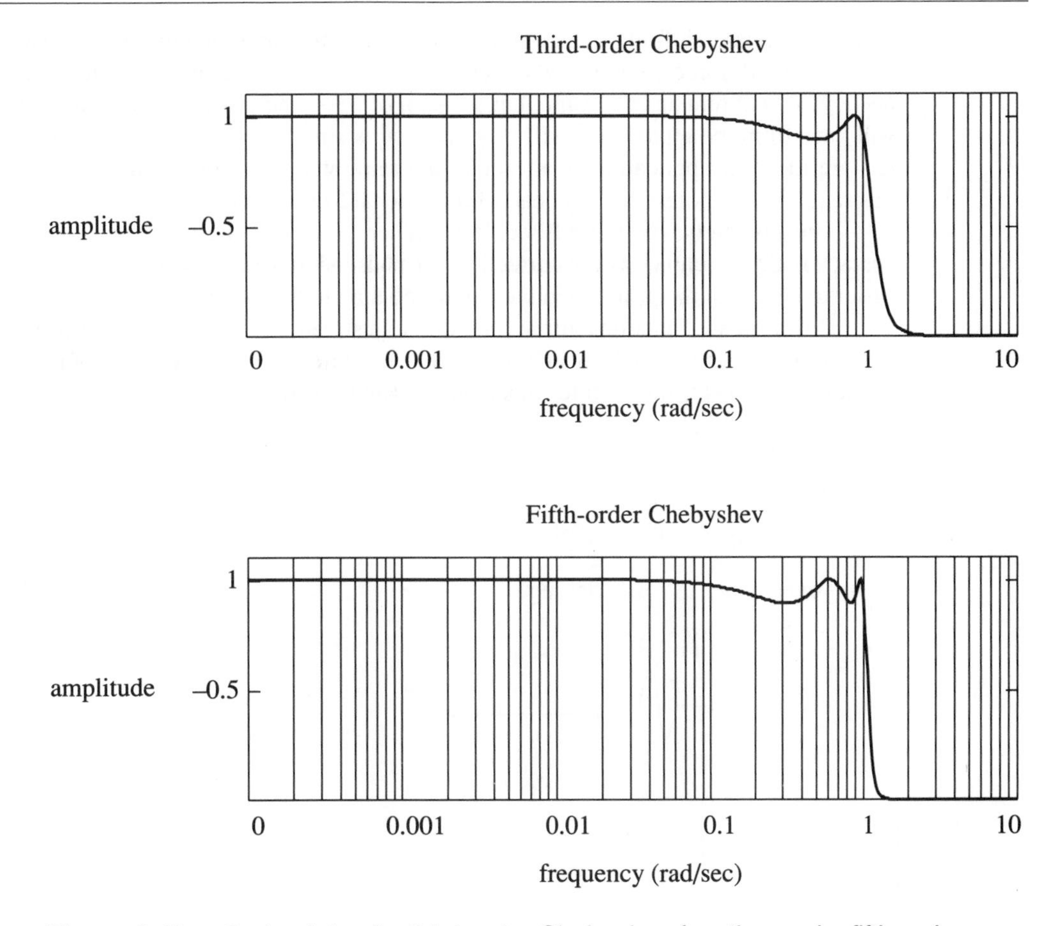


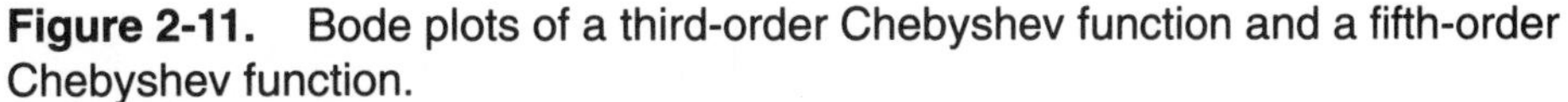

**Figure 2-11.** Bode plots of a third-order Chebyshev function and a fifth-order Chebyshev function.

replaces the Chebyshev polynomial with another rational function whose roots (denominator and numerator) determine the ripple. The discussion of this function is beyond the scope of this book, but may be found in several of the references on filters in the bibliography.

### Thomson or Linear Phase Filters

Passing a signal through an active or passive filter network affects the phase and magnitude of that signal. Butterworth and Chebyshev filters tailor the magnitude response of a network to allow for fast transition bands, but lose valuable phase information.

Instead of optimizing only magnitude response characteristics of the filter, the linear phase filter also provides for linear phase shift in the passband. Lack of this linearity results in overshoot and ringing, as well as unexpected and unwanted distortion of the input because of phase shifting. Continuous-time signals are sums of infinitesimal values of all frequencies. If some frequencies experience more delay than others in the filter, the frequency content remains the same, but the sum of passband frequencies at any point in time will vary from the input. This will produce results similar to Figure 2-12. Linear phase filters are ideal for applications that require the same waveform on the output as the input—this requires a response that is linearly proportional to frequency.

Unfortunately, for most filters, the higher the order, the more non-ideal the phase characteristics. Techniques for correcting for phase shift will be discussed shortly, but it is easier to deal with the problem in the beginning than to try to correct for it later.

Linear phase functions allow for some adjustment in gain and time shifting and provide an approximation of linear phase—that is, time shifting that is proportional to frequency. The transfer function for this kind of ideal transmission is

$$H(j\omega)=Ke^{-j\omega t_{delay}}$$

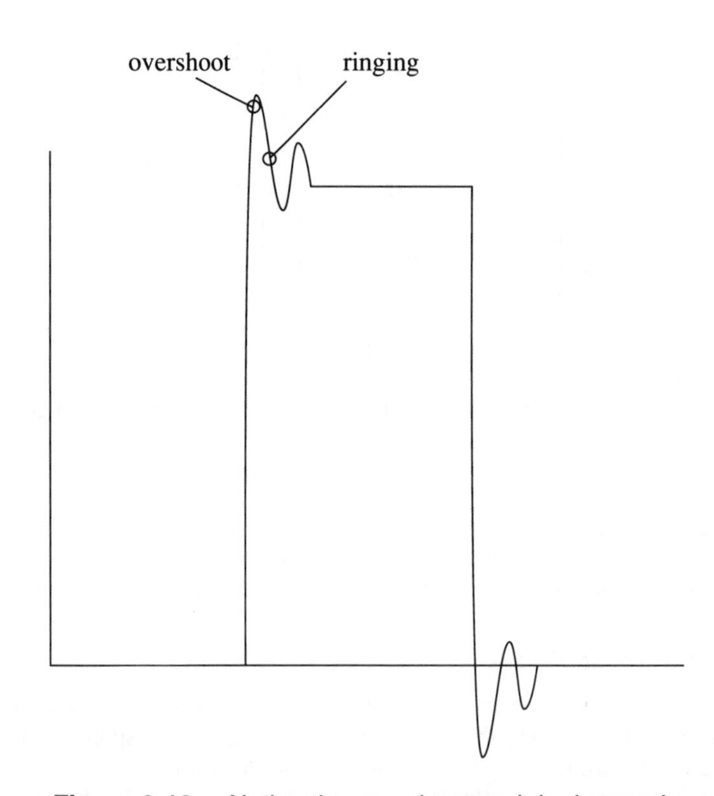


**Figure 2-12.** Notice the overshoot and ringing on the edges of this pulse. This is due to the nonlinear phase response of a filter in which some frequencies are delayed more than others.

If we say that $K$ is the magnitude of the function and is a constant, it is independent of frequency. This results in

$$Arg\,H(j\omega) = -\,\omega t_{delay}$$

that is linearly proportional to frequency, see Figure 2-13.

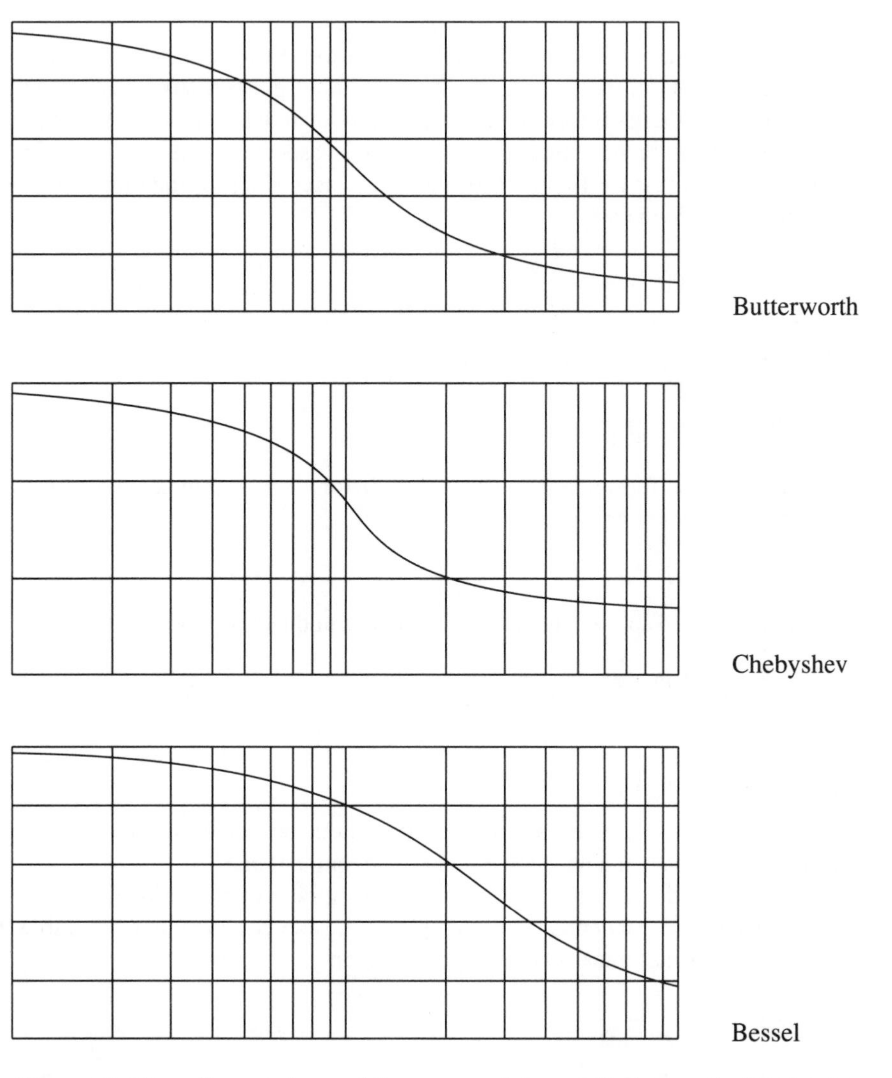


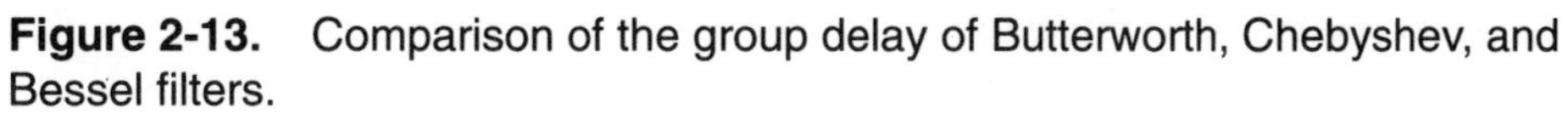

**Figure 2-13.** Comparison of the group delay of Butterworth, Chebyshev, and Bessel filters.

This allows us to define another term, *phase delay*, as the time delay experienced by a sinusoid, $\omega$, passing through the system:

$$\tau_p(\omega) = \frac{-\theta(\omega)}{\omega}$$

And, following that, group delay:

$$\tau_g(\omega) = \frac{-d}{d\omega}\theta(\omega) \qquad \text{(Eq. 2-2)}$$

This is the derivative of radian phase with respect to radian frequency at a given frequency, and results in the time required for an input pulse to arrive at the output. We can approximate this derivative by summing the individual delays associated with the numerator quadratics and subtracting the sum of the delays associated with the denominator quadratics. This requires that we find each of the delays according to Equation 2-2. The delay of a quadratic can be found as

$$\tau_q(\omega) = \frac{-d}{dt}\theta(\omega) = \frac{a_1\omega^2 + a_1 a_0}{\omega^4 + \omega^2\left(a_1^2 - 2a_0\right) + a_0^2}$$

and a first-order function has the form

$$Delay_f(\omega) = -\frac{a}{\omega^2 + a^2}$$

To use these formulae, one can either factor a given transfer function into quadratic sections, such that the polynomial is of the form

$$\omega^2 + a_1\omega + a_2^2$$

or use the poles in Table 2-4 to create the individual quadratics, recalling that

$$a_1 = poles_i + \overline{poles_i}$$

and

$$a_0 = poles_i * \overline{poles_i}$$

Examples of these can be found in the Mathcad documents for each of the filter functions: Butter.mcd, Cheby.mcd, and Bessel.mcd on the optional disk.

For high-order network functions, each quadratic of the rationalized system function may be treated independently. For example, using the coefficients from Table 2-3, we can describe a third-order linear phase system function

$$H(s) = \frac{1}{s^3 + 6s^2 + 15s + 3}$$

finding the phase response

$$\theta(\omega) = -\tan^{-1}\left(\frac{\text{Im}[H(j\omega)]}{\text{Re}[H(j\omega)]}\right) = -\tan^{-1}\left(\frac{-3.677j\omega}{6.459-\omega^2}\right)$$

The group delay is the sum of all the individual delays. Using the formulae above for first- and second-order terms, we would calculate the group delay of the third-order linear phase function in the previous example as follows

$$\tau_g(\omega) = \frac{-d}{dt}\theta(\omega) = \frac{3.677\omega^2 + 23.756}{\omega^4 + .607\omega^2 + 41.724} + \frac{2.322}{\omega^2 + 5.392}$$

The results of this equation are plotted in Figure 2-14.

The following table of poles for linear phase characteristics was developed by finding the roots to the polynomials listed in Table 2-3 using the Laguerre method (see Mathcad document Bessel.mcd).

In non-realtime applications, group delay means little. However, in realtime applications, it can make a great deal of difference how long it takes for a step function to propagate to the output and then to the DSP. Good continuous-time filters usually prove to be little problem, but some of the sampling filters such as the FIR filters and switched capacitor technologies produce a significant delay because of the clock speed.

We derive a Thomson function in a similar manner to the approximation of magnitude functions. First, expand the argument in a Taylor series; this would be the filter function we

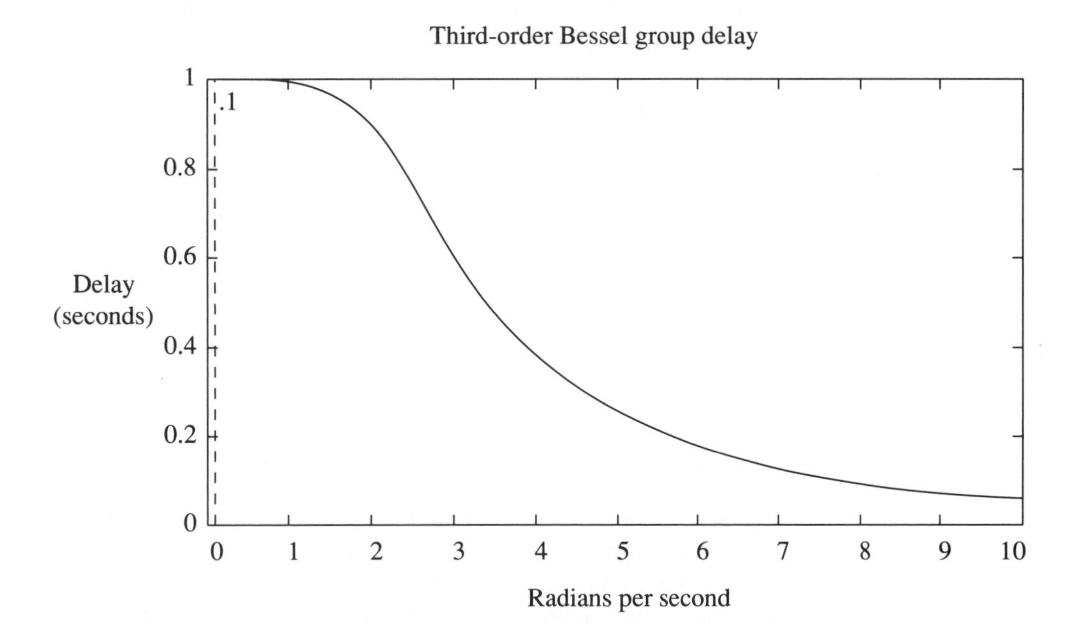


**Figure 2-14.** Graph from grpdly.mcd for third-order system

| $n$ | *polynomial* |
|---|---|
| 1 | $s+1$ |
| 2 | $s^2+3s+3$ |
| 3 | $s^3+6s^2+15s+15$ |
| 4 | $s^4+10s^3+45s^2+105s+105$ |
| 5 | $s^5+15s^4+105s^3+420s^2+945s+945$ |
| 6 | $s^6+21s^5+210s^4+1260s^3+4725s^2+10395s+10395$ |
| 7 | $s^7+28s^6+378s^5+3150s^4+17325s^3+62370s^2+135135s+135135$ |
| 8 | $s^8+36s^7+630s^6+6930s^5+9450s^4+270270s^3+945945s^2+2027025s+2027025$ |
| 9 | $s^9+45s^8+990s^7+13860s^6+135135s^5+945945s^4+4729725s^3+16216200s^2+$ $34459425s+34459425$ |
| 10 | $s^{10}+55s^9+1485s^8+25740s^7+315315s^6+2837835s^5+18918900s^4+91891800s^3+$ $310134825s^2+654729075s+654729075$ |

**Table 2-3.**

select. Then set the first derivative to a nonzero value, providing the delay at origin we require, and solve for the coefficients. This results in some Bessel-like polynomials in the denominators of our transfer functions. Typically, the coefficients of the denominators of a Thomson filter are normalized to a delay of 1 second at direct current, and a slope of $-1$.

There are two standard formulae for developing linear phase functions. First, the denominator coefficients may be found with

$$b_n = \frac{(2n-k)!}{2^{n-k}k!(n-k)!} \quad k=0,\ldots,n-1$$

($n$ is the degree of the denominator). Alternatively, the denominators themselves may be derived with the following recursion formula, using

$$q_1(s)=s+1$$

and

$$q_2(s)=s^2+3s+3$$

to form

$$q_n(s)=(2n-1)q_{n-1}(s)+s^2q_{n-2}(s)$$

Unfortunately, there is no closed form for pole finding with the linear phase filter. The roots must be found separately for the denominator polynomial of each degree. There are a number of popular root finding methods. The Laguerre root finding technique is included as part of the Mathcad document on the optional disk called Bessel.mcd for this purpose. With it, the poles for any degree linear phase filter can be found.

| *order* | *poles* | $\omega_0$ | *Q* |
|---|---|---|---|
| 2 | $-1.500 \pm j0.866$ | 1.7322 | 0.577 |
| 3 | $-1.839 \pm j1.754$ | 2.541 | 0.691 |
| | $-2.322$ | | |
| 4 | $-2.104 \pm j2.657$ | 3.389 | 0.805 |
| | $-2.896 \pm j0.8672$ | 3.023 | 0.522 |
| 5 | $-2.325 \pm j3.571$ | 4.261 | 0.916 |
| | $-3.352 \pm j1.743$ | 3.778 | 0.563 |
| | $-3.647$ | | |
| 6 | $-2.516 \pm j4.493$ | 5.149 | 1.023 |
| | $-3.736 \pm j2.626$ | 4.566 | 0.611 |
| | $-4.248 \pm j0.8675$ | 4.336 | 0.510 |
| 7 | $-2.686 \pm j5.421$ | 6.05 | 1.126 |
| | $-4.070 \pm j3.517$ | 5.379 | 0.661 |
| | $-4.758 \pm j1.739$ | 5.066 | 0.532 |
| | $-4.972$ | | |
| 8 | $-2.839 \pm j6.354$ | 6.959 | 1.23 |
| | $-4.368 \pm j4.414$ | 6.210 | 0.710 |
| | $-5.205 \pm j2.616$ | 5.825 | 0.560 |
| | $-5.588 \pm j0.8676$ | 5.654 | 0.506 |
| 9 | $-2.979 \pm j7.291$ | 7.877 | 1.322 |
| | $-4.638 \pm j5.317$ | 7.056 | 0.761 |
| | $-5.604 \pm j3.498$ | 6.607 | 0.589 |
| | $-6.129 \pm j1.738$ | 6.370 | 0.520 |
| | $-6.297$ | | |
| 10 | $-3.109 \pm j8.233$ | 8.8 | 1.415 |
| | $-4.886 \pm j6.225$ | 7.914 | 0.810 |
| | $-5.968 \pm j4.385$ | 7.405 | 0.620 |
| | $-6.922 \pm j0.8677$ | 6.976 | 0.504 |
| | $-6.615 \pm j2.612$ | 7.112 | 0.538 |

**Table 2-4.**

The greater the order of the linear phase filter function, the more we approach the ideal in magnitude and phase response. These filters are not, however, recommended for sharp transition bands. If linear phase is required, the filtering tasks may have to be shared between the analog and digital environments.

## Delay Equalizers

The *allpass* or *delay equalizer* is a filter that passes all frequencies without discrimination. The general form of the transfer function is

$$H(s)=\frac{(s-\omega_1)\left(s^2-\frac{s\omega_2}{Q_2}+\omega_2\right)\cdots}{(s+\omega_1)\left(s^2+\frac{s\omega_2}{Q_2}+\omega_2\right)\cdots}$$

The voltage transfer function for a first order system is

$$H(s)=-\frac{s-\frac{1}{R1C1}}{s+\frac{1}{R1C1}}$$

The first order allpass filter shown in Figure 2-15A phase shifts the input from 0 to 180 degrees as the frequency of the input moves from DC to infinity. If only phase shift is

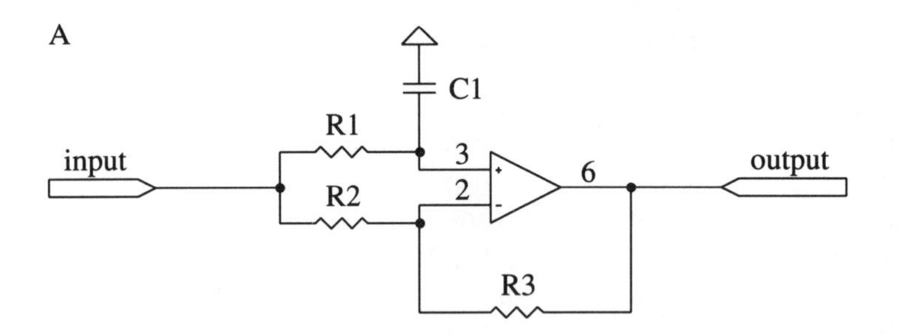


First-order allpass filter

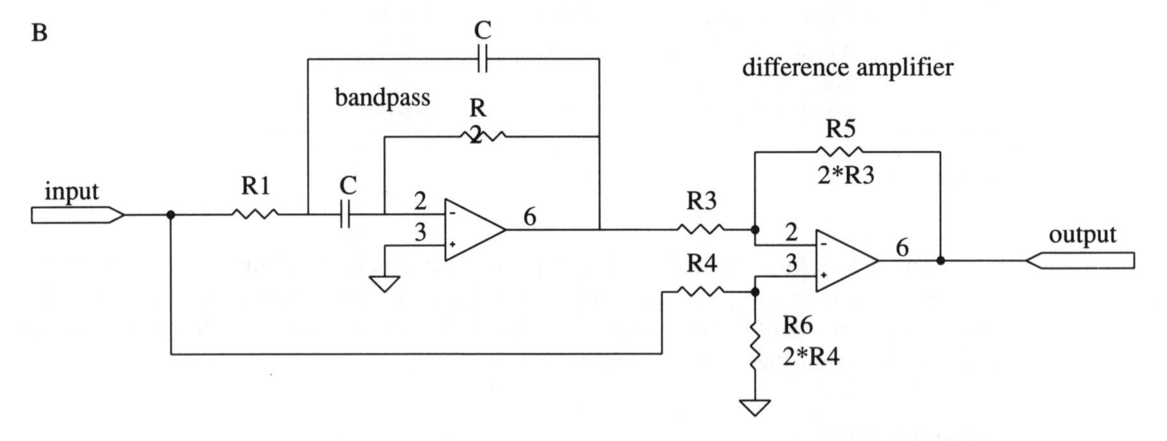


Second-order allpass filter

**Figure 2-15.** Schematics of a first-order and a second-order allpass filter.

to be varied, adjust $R1$ and $C1$ only. Exchanging $R1$ with $C1$ will result in the phase shift range moving to $-180$ to 0 degrees. In addition, the sign of the rational transfer function changes from negative to positive.

It is possible to create a second-order system from a second-order bandpass filter and a difference amplifier. Figure 2-15B is an example of this. The transfer function shows the complex poles and reflected zeros:

$$H(s)=\frac{s^2-\frac{s\omega_0}{Q}+\omega_0}{s^2+\frac{s\omega_0}{Q}+\omega_0}$$

Notice that the difference amplifier has a gain of 2, compensating for the unity gain assumed in the bandpass filter.

The phase for such a filter, by DeMoivre's theorem, is given by

$$\theta(\omega)=-2\tan^{-1}\left(\frac{\text{Im}[G(j\omega)]}{\text{Re}[G(j\omega)]}\right)$$

This filter has all its zeros in the right half of the complex plane, and the poles are mirrored in the left half. It demonstrates unity gain for all frequencies, while allowing phase shift adjustments for specific frequencies.

## Calculating the Order of a Filter

**Butterworth and Chebyshev** It is possible to get a conservative estimate of filter order by using the simple relationship suggested in the section on "What the Filter Needs to Do" later in this chapter. There are additional aspects to the selection of filter order, some of which may be important in your application. In this section, we will approach the determination of filter order from several viewpoints; one may be more appropriate to you than another. First, we will examine filters concerned with magnitude and then those involving linear phase. For filters emphasizing magnitude response, there are several parameters required for the determination of the order:

1. Specification of a passband frequency. This is usually set as the point passband attenuation first falls to the 3dB down point—this is the beginning of the transition band. We will call this $\omega_p$.
2. Maximum deviation allowance. This parameter describes how much the magnitude characteristic is allowed to vary in the passband. Specify this parameter in dB; we will call it $\Delta_p$.
3. Specification of a stopband frequency. We are setting the end of the transition band with this value. It is at this point that we have reached the desired attenuation. Call this value $\omega_s$.
4. Minimum attenuation in the stopband. This will be $\Delta_s$.

To estimate the order we will need, first find the ratio between the stopband and passband frequencies:

$$f_r = \frac{\omega_s}{\omega_p}$$

and another ratio involving the stopband and passband attenuation:

$$A_r = \sqrt{\frac{10^{.1\Delta_s} - 1}{10^{.1\Delta_p} - 1}}$$

The necessary order for a Butterworth filter is then given by

$$O_f = \frac{\ln(A_f)}{\ln(f_r)}$$

and for a Chebyshev

$$O_f = \frac{\cosh^{-1}(A_f)}{\cosh^{-1}(f_r)}$$

Another way to determine the minimum order for a magnitude response filter is to choose a magnitude $A$ (negative and in decibels), again representing the lowest point of magnitude deviation that passband may have for the filter. Choose a frequency $\omega_A$, defining the point at which the magnitude first falls below $A$. Set the 3dB down point, $\omega_c$. Calculate the order with

$$n = \frac{\log\left(10^{-A/10} - 1\right)}{2\log\left(\frac{\omega_A}{\omega_c}\right)}$$

**Linear Phase** With the Bessel filters, we must concern ourselves with linearity of phase as well as magnitude response. This makes the determination of the order of this filter a balance between the two—if both are important, then it will be necessary to build the filter satisfying the specification of the highest order.

The delay error is a ratio between the difference between group delay at a particular frequency and unity and the group delay. Multiply this result by 100 for a percentage:

$$error_{delay} = \frac{1 - \tau_g}{\tau_g}$$

Magnitude error is found by simply solving the transfer function in $s$, substituting the radian frequency of interest for $s$:

$$error_{mag} = 20\log_{10}\frac{H(s)_{ideal}}{H(s)_{actual}}$$

Recall that in a linear phase function, magnitude is independent of frequency, and therefore $H(s)_{ideal}$ is equal to the only constant in the denominator of the function (see Table 2-3). $H(s)_{actual}$ is found as the magnitude response of the function at frequency

$$H(j\omega)=\sqrt{\mathrm{Re}\left[H(j\omega)^2\right]+\mathrm{Im}\left[H(j\omega)^2\right]}$$

Figure 2-16 illustrates the delay for second-order Butterworth, Chebyshev (1dB), and Bessel filters plotted against frequency.

See Mathcad document Berror.mcd for more.

### Filter Transformations

The lowpass filter has become very popular as a *prototype* form for the presentation or development of filters. Tables of quadratics, poles, and zeros—and even component values—exist for all sorts of filters in lowpass form, ready for transformation to other filter types. The technique is fairly straightforward and will be discussed briefly below:

1. To transform a normalized lowpass to a highpass filter, substitute each $s$ in the transfer function with

   $$\frac{1}{p}\left(s=\frac{1}{p}\right)$$

   where p might be any simple or complex conjugate pole. As an example, assume a second-order lowpass filter transfer function

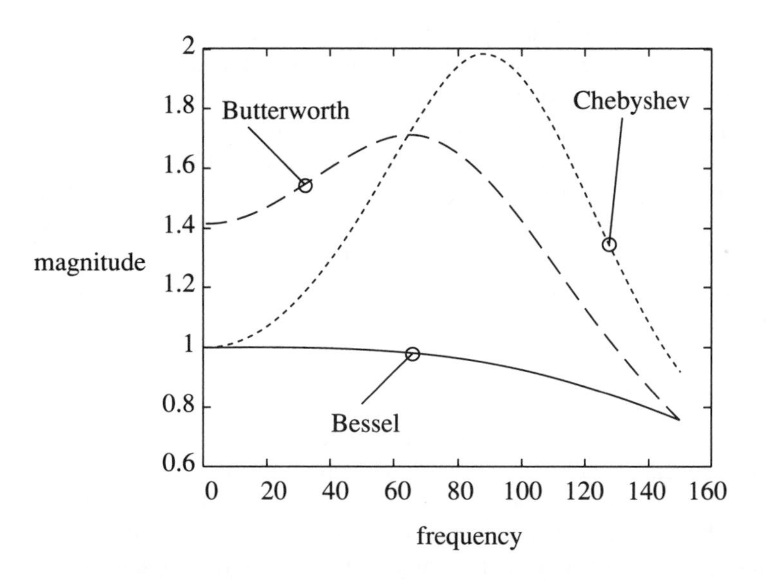


**Figure 2-16.** Comparison of delay error among three filter characteristics.

$$H(s) = \frac{\omega_0^2}{s^2 + s\frac{\omega_0}{Q} + \omega_0^2}$$

which can be transformed into the highpass filter function

$$H(s) := \frac{p^2}{\left(\frac{1}{p}\right)^2 + \left(\frac{1}{p}\right)\frac{\omega_0}{Q} + \omega_0^2}$$

and simplifying further, we have

$$H(s) := \frac{p^2}{p^2 + \frac{p}{Q\omega_0} + \frac{1}{\omega_0^2}}$$

2. To transform a normalized lowpass filter to a *narrowband* bandpass filter, we set $s$ equal to $p+(1/p)$ in the prototype. For bandwidths greater than an octave, use overlapping lowpass and highpass filters.
3. A bandstop or notch filter can be created from a normalized lowpass filter by first transforming the lowpass prototype to a highpass filter, as in step one above, then treat the resulting transfer function as a prototype and transform it to the appropriate notch filter using the equation in step two.

## A/D Conversion for DSP

Follow any necessary filtering and optional sample and hold circuitry is the Analog/Digital Converter. This is where the signal makes the transition to the digital domain and a new set of considerations. A/D conversion provides one of the greatest possibilities for error to the system. Continuous-time signals have infinite resolution, digital electronics does not. It is limited by the word size of the machine you are dealing with, the amount of memory available, and the speed of processing, to list only a few factors. To fully encode a continuous-time signal stream, we would have to sample at infinitesimally small intervals with an infinite dynamic range so that no data was missed. Unfortunately, there is not enough time, room, or bits for that. As a result, we must settle for an approximation of what is going on in the continuous-time world.

### Bandwidth and Aperture Time

One of the initial considerations on any system is its bandwidth. Clearly, for conversion of high-frequency signals, either in rise time or in signal frequency, the sample period must be such that it does not smear the input and lose valuable information. The *aperture time* of the A/D describes the interval during which it acquires the signal. The bandwidth of the device is directly related to this parameter. The longer the aperture time, the lower the bandwidth.

This is a result of the fact that during the time the A/D converter is acquiring the signal, any changes at the input are effectively averaged in the converter.

Many converters are not designed for high-speed sampling, and this lack may result in distortion or loss of high-frequency information. In this case, a sample-and-hold circuit is usually provided with a short enough aperture time to catch and mantain a high bandwidth. Once the sample-and-hold has sampled the input, it holds it constant at the output until the next sample is requested.

The aperture time is the time the A/D converter requires to make its measurement. During the aperture period, the A/D takes a sample of the analog input. This is usually taken on the plates of a capacitor through some kind of impedance. Clearly, changes occurring at the input will affect the charge rate of the capacitor, but they will appear as the result of an integration over the period of the sample. Thus, if there was a square wave at the input of the A/D, and at the moment of the sample, the signal was rising from 0 to 5 volts, the A/D would only approximate what appeared on the plates of the capacitor. This could vary depending upon the aperture period. All of this has the effect of a lowpass filter rolling off the input signal and distorting the resolvable information. Many converters employ a circuit called a *sample-and-hold* to sample the input as quickly as possible, then supply this to the A/D for conversion. This has the effect of increasing the bandwidth without adding substantially to the cost of the converter.

The maximum resolvable change per second detectable by an A/D without a sample-and-hold is

$$\frac{dv}{dt} = \frac{E_{fullscale}}{2^{number\ of\ bits} * T_{convert}}$$

where $2^{number\ of\ bits}$ is the dynamic range in bits of the A/D, $T_{convert}$ is the conversion time in seconds, and $E_{fullscale}$ is the full scale input voltage range of the A/D.

With a sample-and-hold circuit embedded on the chip, we can do better. In that case,

$$\frac{dv}{dt} = \frac{E_{fullscale}}{T_{aperture}}$$

As an example of what this means, let's take an 8-bit A/D with a conversion time of 4 microseconds and a 5-volt analog input range. In this case,

$$\frac{dv}{dt} = \frac{5}{2^8 * .000004} = 4882.8\,\text{v/s}$$

or .0048 volts per microsecond. Given the same A/D using a sample-and-hold with an aperture time of 1 microsecond,

$$\frac{dv}{dt} = \frac{5}{1E-6} = 5E6\,\text{v/s}$$

The aperture is also a source for noise to the system that appears as a random variation in sampling frequency and is known as *aperture jitter*. This becomes more and more critical at higher bandwidth and resolutions. The source of this error can be either external (low-frequency modulation of the clock signal) or internal (from heating).

### Dynamic Range, Resolution, and Quantization Noise

One of the primary criteria for A/D selection is its *dynamic range*, which is usually defined in bits. This tells us how many steps the full range of analog input can be broken up into. If it is an 8-bit converter, then the range is from 0 to 256. That is the full count available in 8 bits. Extending this a little, if the analog input is 5 volts, each bit is equal to just over 19 millivolts. For many applications, this is fine. Of course, it is necessary to see that any offsets from filters and amplifiers in the system do not automatically absorb this bit.

Another factor in how the A/D judges the value of this bit is a source of noise to the system called *quantization noise*. Comparator indecision is the root of this noise, which is directly related to the regeneration time of the internal comparators—the shorter that time, the less likely it is to occur.

## D/A Conversion for DSP

The D/A shares many of the same considerations as the A/D. SNR depends upon accuracy, and is similar to the A/D. The D/A can contribute errors in gain, offset nonlinearity, and nonmonotonicity.

Gain and offset are usually adjustable, but linearity and monotonicity are not. To obtain as linear and monotonic a converter as possible, inspect the integral error—this is the deviation of the output characteristic from a straight line over its range—and the differential linearity, which describes the output change for each change of LSB on the input. The output should be close to a straight line, and should exhibit an equal change at the output for each change of LSB at the input.

# PART TWO: DESIGNING ANALOG FILTERS

## Introduction

Once you have chosen the form of the filter you need from the parameters of your system, you must decide how to implement it. Many DSP applications require something more than a single pole and first-order filter. LRC filters are difficult to implement with consistency; they are expensive and consume printed circuit board real estate very quickly. Active filters are very popular because, in combination with simple, scalable resistor/capacitor components, they can be used to implement poles and zeros in any quadrant of the complex plane. Their uses range from returning gain to lossy LRC and RC circuits to simulating passive elements in *general immitance networks*, such as frequency-dependent negative resistance circuits. These circuits are easy to implement, and have the added advantage that numerous books are available with tables of component values for various filter configurations.

Other new IC approaches involve complex filters in both continuous-time and switched capacitor realizations. The continuous-time filter ICs are usually state-variable or bi-quad circuits; they have the more sensitive components in silicon on the same substrate with the opamps and need only a few off-chip parts to set cutoff frequency and bandwidth.

The switched capacitor filters *integrate* the input signal at sampling frequencies set by internal or external oscillators. These circuits are inexpensive and easily tuned—the cutoff frequency is set by adjusting the switching frequency of the device. Unfortunately, switched capacitor networks suffer the same problem as any sampling system: aliasing. These filters will fold back and output any energy at a frequency that differs from the clocking (sampling) frequency by an amount corresponding to any frequency in the passband. These circuits also manifest clock noise on the output that limits the dynamic range and influences subsequent circuitry. The clock noise, however, is usually easily eliminated with a simple RC network.

Many audio A/D converters are being manufactured with FIR filters implemented in silicon on the same chip. The stopband attenuation can be great with these devices. However, the clocking frequency adds an undesirable group delay to the signal that makes it unusable in a real-time application.

Just as no filter characteristic provides for every need, no implementation is perfect. Depending upon the application, the error budget can be very tight. In the next sections, some of the many variables to consider in constructing a filter for a DSP will be covered.

## The Frequency-Dependent Voltage Divider

Essentially, a filter is a frequency-dependent voltage divider. The order of the filter depends upon the highest power of $\omega$, or frequency, in the transfer function. Figure 2-17A shows a simple network voltage divider consisting of two resistors. Next to this voltage divider is an resistor/capacitor voltage divider, a first-order lowpass filter (the highest power of frequency is 1).

The impedance of a resistor is constant with frequency—that is, it is always the same. A capacitor or inductor has an impedance that varies with frequency, as follows:

$X_C = \frac{-j}{\omega C}$ for the capacitor, and

$Z_L = j\omega L$ for an inductor (see Figure 2-17C).

If we apply an alternating current to both networks, the output of the resistor–resistor network will always be 10 volts, while the output of the resistor–capacitor network will be 99 volts at 10 Hz, 94 volts at 1,000 Hz, and 14 volts at 100,000 Hz. As you can see, this circuit attenuates with an increase in frequency.

The circuit in Figure 2-17B is a lowpass filter with the transfer function

$$\frac{V_{out}}{V_{in}} = \frac{1}{j\omega RC + 1}$$

or

$$H(s) = \frac{1}{s+1}$$

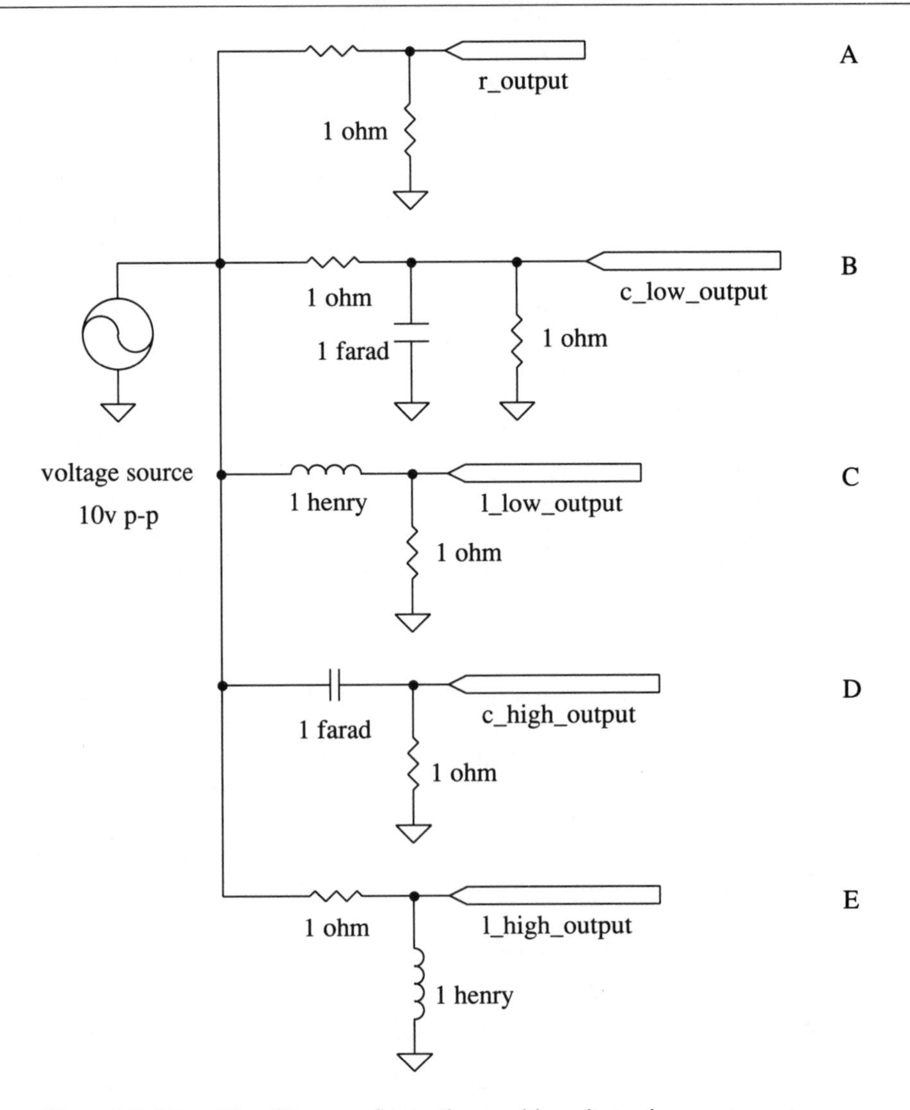

**Figure 2-17.** Five filter configurations with unity values.

Figure 2-17D is an example of a first-order highpass filter with the transfer function

$$\frac{V_{out}}{V_{in}} = \frac{j\omega RC}{j\omega RC + 1}$$

or

$$H(s) = \frac{s}{s+1}$$

Figure 2-18 illustrates the magnitude response for a first order lowpass filter and high-pass filter. Notice the droopy response and long transition band. As simple as the first-order filters are to calculate and construct, they are of limited value as filters, and there are really only two usable forms: the lowpass and highpass filters. The best rolloff or attenuation that we can expect is $1/\omega$; as the input frequency increases, the filter response will halve as the frequency doubles. This equivalent to an attenuation of 6dB per octave.

Second-order filters offer a number of possibilities. The most popular are the lowpass, highpass, and bandpass; others are created from the algebraic sums of these three. And now,

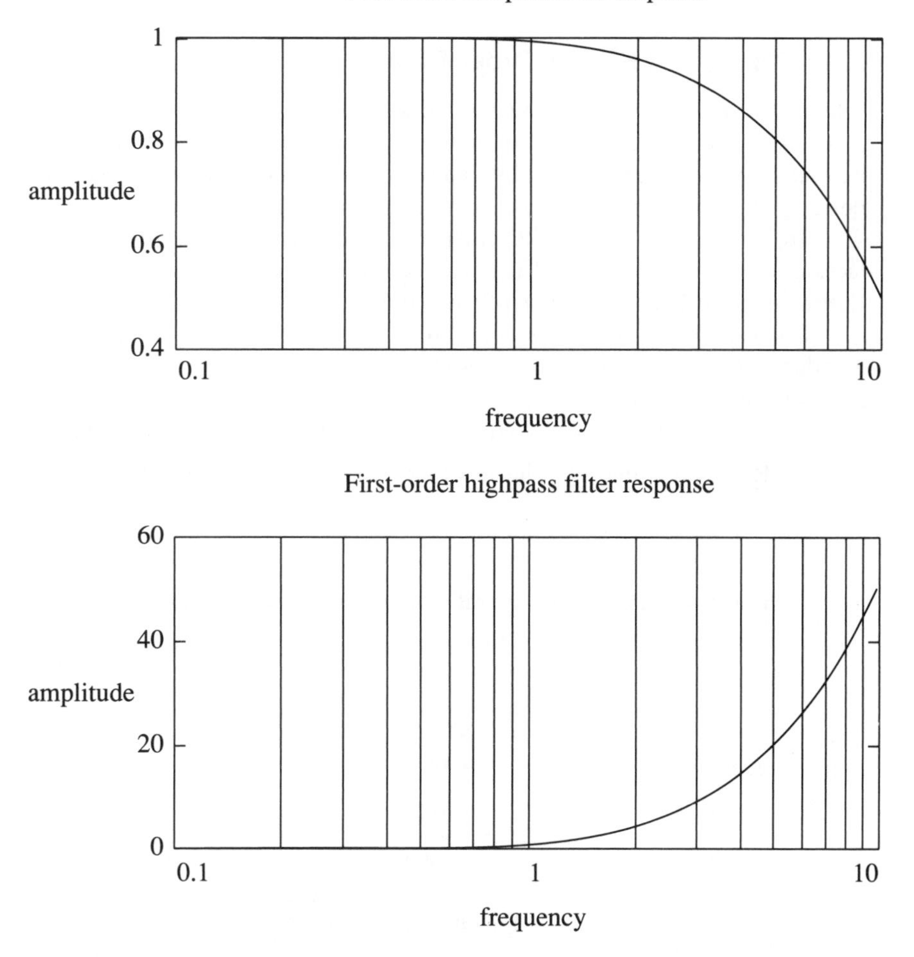


**Figure 2-18.** Bode plots of first order lowpass and highpass filters. Notice the sloping transition bands.

because the second-order filter has the square of frequency (second-order) in its transfer function

$$H(s)=\frac{H_0}{\sqrt{1+\omega^4}}$$

($H_0$ is gain for this second-order Butterworth), the attenuation increases with the square of the frequency—that is, by four with each doubling of frequency, or 12dB per octave.

The higher the order of the filter, the better the rolloff versus frequency. This rate is approximately $6N$ per octave, where $N$ is the order of the filter. It is very often the case that one can construct these higher order filters by simply cascading lower-order filters together, the single pole and the double pole.

## Resonant Circuits, *Q*, and Higher-Order Filters

Active filters have a great deal in common with the simple harmonic oscillator we looked at in the last chapter. In fact, they are forced oscillators, tuned so that they cannot maintain the oscillation by themselves (critically damped). To understand this, we need to take a closer look at $Q$ and resonance as it relates to oscillators and filters.

To construct a two-pole RC filter, we concatenate two single-pole filters. This produces a second-order filter that is very lossy and has a long transition band. Figure 2-19 illustrates just such a construction of a two pole RC filter.

One can also construct a second-order lowpass filter with just three circuit elements: an inductor $L$, a capacitor $C$, and a resistor $R$. As you can see, the response of this filter is different from that of the RC filter.

Two considerations associated with the choice of inductance and capacitance are:

1. Resonant frequency. This is set by the product of the inductance and capacitance:

$$f_c=\frac{1}{2\pi\sqrt{LC}}$$

2. Damping, or $Q$. Set by the ratio of the inductance to capacitance.

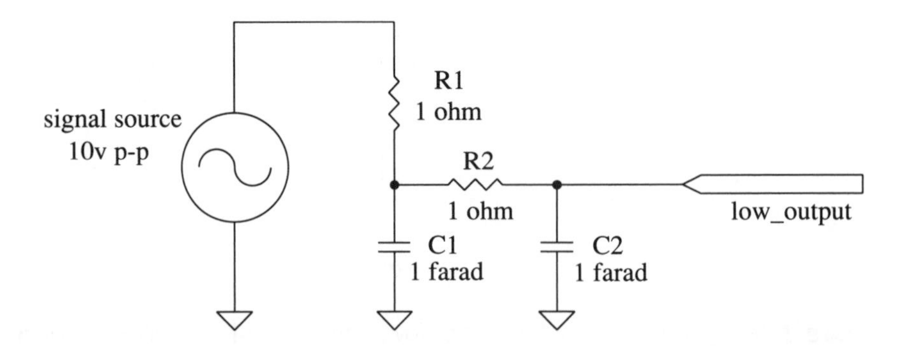

**Figure 2-19.** A two pole RC filter constructed by cascading two single pole RC filters.

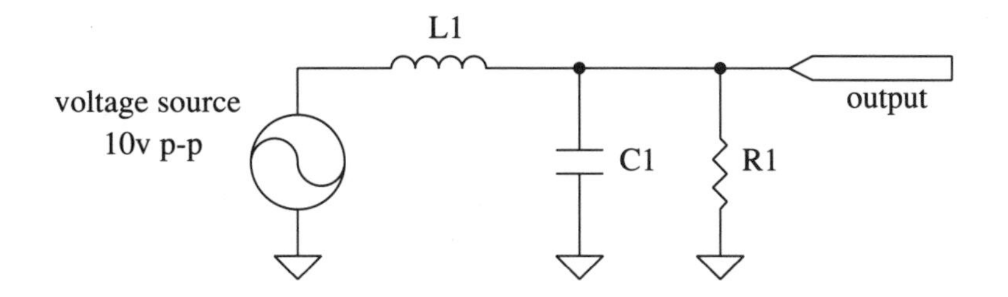


**Figure 2-20.** This second-order LCR filter exhibits a much sharper knee at cutoff and a faster transition band than the two pole RC network in Figure 2-19.

We know that the reactance of a capacitor decreases with increasing frequency. The first-order filter in Figure 2-20 depends upon that fact for its attenuation at higher frequencies. In fact, this attenuation is directly proportional to the reactance of the capacitor. When we create a second-order filter by adding an inductor, we can multiply the qualities of each and produce a far sharper transition period. This is because the reactance of an inductance increases with increasing frequency. In the circuit in Figure 2-20, the cutoff frequency is determined by the product of the inductance and capacitance.

This results from the unique phase relationships that each of the components in this simple circuit have with current and voltage. In order to see how this works, see Figure 2-21 for a schematic of another LRC circuit. This time, the load resistor is removed and another resistor is inserted in series. The inductor and capacitor represent any parasitic or intentional resistance in a network.

In the chart in part B of Figure 2-21, we can see the effect that frequency has on the reactance of the three devices in part A. It is evident from the equations that capacitative reactance is opposite in sign to inductive reactance. At the point at which their magnitudes are equal, they are opposite in sign and thus exactly cancel each other out, leaving the only impedance in the circuit to the resistor and parasitics.

$$Z = R + j\omega L - j\frac{1}{\omega C}$$

This is shown on the chart in Figure 2-21B: the vertical line crosses the X axis at the point where the *sum of reactances* crosses the X axis. This is the point of resonance for the circuit.

Also evident from this relationship is that the effects of this sum will be very much more dramatic for values of $R$ that are small relative to the reactances. The ratio of reactance to resistance is called the $Q$, or quality factor, of the circuit.

First, notice the result in Figure 2-21B: if the capacitor is very large and the inductor small, we have a peak in the magnitude response near resonance—as we would expect—indicating that the circuit is near oscillation. When this happens, the circuit is *underdamped*, $\zeta<1$. If the reactance and resistance in the circuit are made, we will have a flatter response—this circuit is said to be *critically damped*, $\zeta=1$. Finally, when the resistance is the predominant factor, the circuit is *overdamped*, $\zeta>1$. A bode plot of a lowpass filter demonstrating each of these phenomena is shown in Figure 2-22.

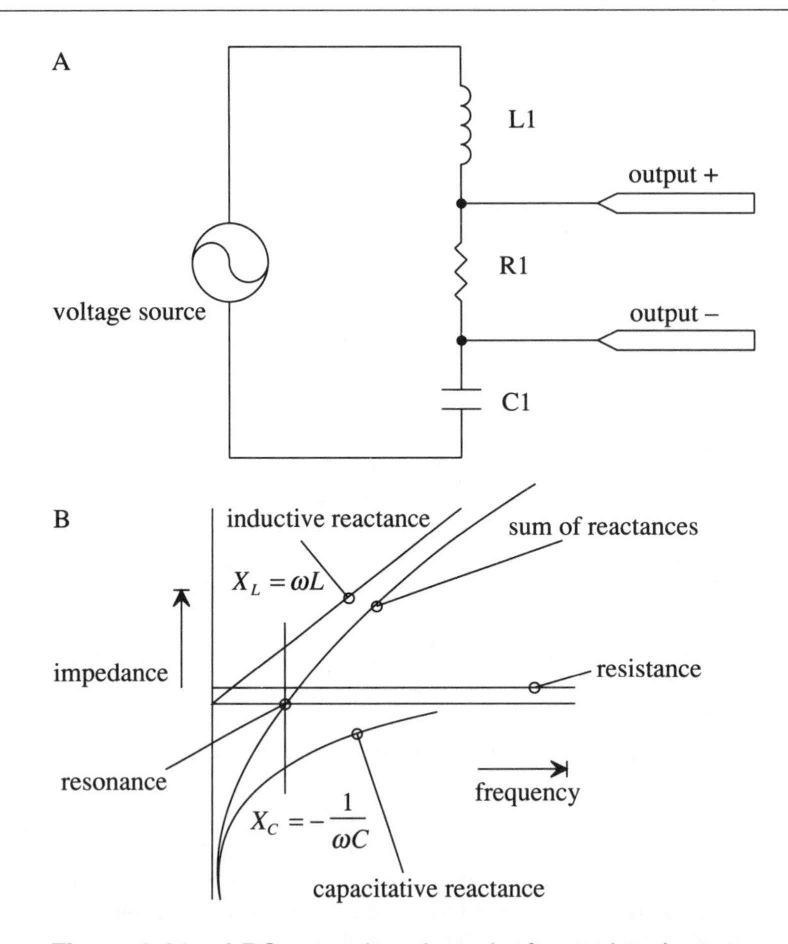


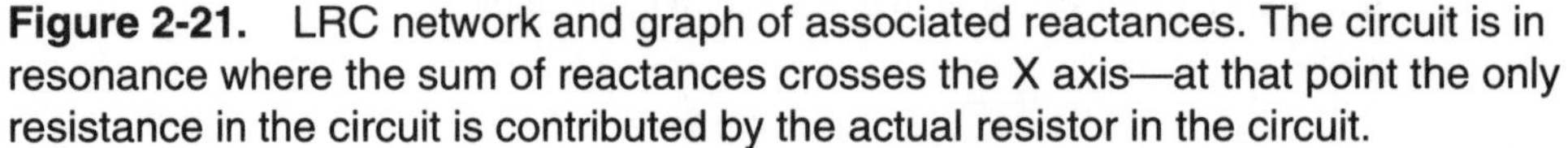

**Figure 2-21.** LRC network and graph of associated reactances. The circuit is in resonance where the sum of reactances crosses the X axis—at that point the only resistance in the circuit is contributed by the actual resistor in the circuit.

Resonance exists when the individual reactances are equal and opposite, resulting in a net circuit reactance of zero. Below resonance, the vector sum of the voltages is in the low right quadrant of the complex plane; current is leading the voltage. At resonance, the sum of the voltages across the inductance and capacitance has become extremely small and is in phase with the current. Above resonance, current now trails, and the voltage vector lies in the upper right quadrant.

At resonance, the energy stored in the storage elements (capacitors or inductors) is constant—the inductance absorbs it as fast as the capacitor gives it up, and then releases it again just as the capacitor needs it. The external source, such as in a forced oscillator, is only required to make up for the losses resulting from the presence of any resistance in the circuit. The terminal voltage and current are in phase, their product is at a minimum, and

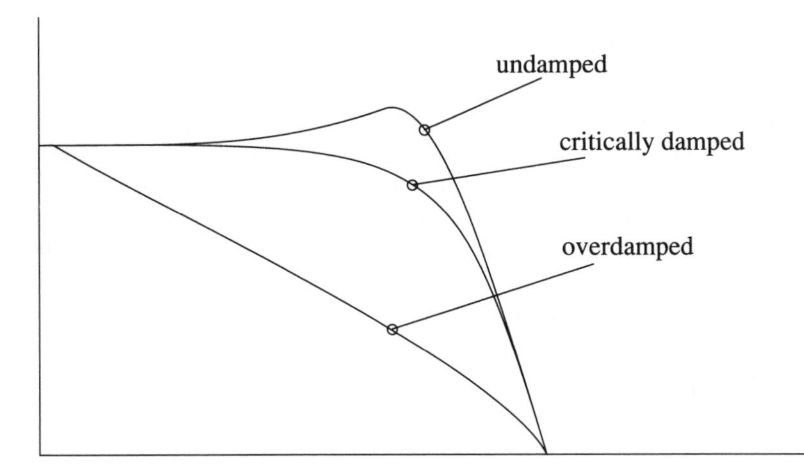


**Figure 2-22.** Bode plot of the effects of resistance in a filter network.

their power factor is unity. Since any external circuit need only supply current to restore the losses from internal resistances, the energy stored may be many times that supplied by the external circuit. The oscillation is self-sustained, damped only by the internal resistances of the circuit. The voltages across the individual reactive elements will be $Q$ times the applied terminal voltage.

With

$$L = \frac{1}{\omega_0^2 C}$$

the total energy in a resonant circuit is given by

$$Energy_{total} = \frac{1}{2} I_m^2 \left[ L(\cos \omega_0 t)^2 + L(\sin \omega_0 t)^2 \right]$$
$$= \frac{1}{2} \frac{I_m^2}{\omega_0^2 C}$$

Here, $I_m$ represents the maximum current and $\omega_0$ is the undamped natural frequency. With this relationship, we will define $Q$ as energy stored divided by energy dissipated in the circuit every cycle:

$$Q = \frac{\frac{1}{2} I_m^2 L}{\frac{1}{2} I_m^2 R / f_0} 2\pi = \frac{\omega_0 L}{R}$$

(Please note the multiplication by $2\pi$ in order to express the value of $Q$ in radians.)

In this equation, dissipated power is

$$\frac{1}{f_0}\left(\frac{I_m}{\sqrt{2}}\right)^2 R = \frac{1}{2f_0} I_m^2 R$$

When the $Q$ is very high, the filter (or oscillator) has fewer internal losses. If it is driven by a signal at or near resonance, it can break into oscillation and sustain that oscillation with very little contribution of energy from the outside. The lower the $Q$, the rounder the cutoff knee will be, and the longer the transition band. Between the two, there is a point at which the knee is as sharp as it can be without risk of oscillation and with maximum rolloff.

This $Q$ and $\omega$ are the same as we find in the system function for a filter, as we see in this lowpass filter transfer function:

$$H(s) = \frac{K\omega_0^1}{s^2 + s\omega_0 / Q + \omega_0^2}$$

## Scaled Data

In the following paragraphs, we will develop data in normalized form for use in constructing filters, along with information necessary for denormalizing them. Whenever possible, the underlying arithmetic is also made available, either in the accompanying text or in the Mathcad document on the optional disk. It is customary to present data normalized to 1 rad/s. This means that we will need to denormalize the results of any calculations in terms of frequency and impedance before we can use them in an actual circuit. If we must divide any capacitance by a target frequency, it will also multiply any resistance in the associated circuit by the same value, as would be expected from the relationship $f = 1/RC$.

In the same vein, if we multiply any resistance by a number to increase or decrease the impedance of a network, we must divide the associated capacitance by the same value. For examples of how this is done, please see the Mathcad documents.

## Filter Design

### What Do We Need to Know to Design a Filter?

We need to know the application. Does it require a steep transition band? Can the CPU share some of the burden by oversampling and doing some of the filtering itself? Is it realtime control or passive post processing? What is the speed of the CPU? Is economics a consideration? Stability? Size? The questions are endless, but the answer is to build to the application.

High speed realtime control systems can tolerate very little analog filtering because of the delays it produces in phase and arithmetic. These situations are usually handled with light, first order filtering and very high speed oversampling. Post processing and non-realtime situations can tolerate greater group delays and may even use filtering supplied by switched capacitor techniques. The answer is to use only what you need.

## Filter Configurations

There are a number of filter formats in common use. We will present some of those with schematics in the next sections. Unfortunately, it is well beyond the scope of this book to present design and development data for all configurations. Only those used in the DACQ unit will be discussed in detail, and a few other popular formats will be presented for information. Please consult the bibliography for references.

Each of the filters (except the first-order) has a lowpass, bandpass, and highpass form, and each (again except the first-order) is able to assume any of the filter types (Butterworth, Chebyshev, and Bessel) we discussed earlier. These filters each have qualities and disadvantages that make them more suitable for some applications than others.

## Cascading Lower-Order Filters

There are two popular techniques for designing filters. Essentially, designing filters by cascading lower-order filters is one of the simplest and most straightforward methods. Circuits for low-order active filters abound in the literature from semiconductor manufacturers and in the library (see the bibliography for some names). All that one needs to do is see that the circuits are suitably isolated from one another so that one circuit does not alter the characteristics of another. High-order circuits of this nature are easier to tune than those designed for a specific order, since each element can be tuned individually without affecting the others.

To design such a filter, one need only select the desired order and factor the transfer function into first- and second-order quadratics. Each section then is built to fit that quadratic (for a second-order filter) or simple pole (for a first-order filter). The transfer function for the whole filter is equal to the product of the transfer functions for the individual sections. Generally, one orders the sections so that the lower $Q$ circuits are first, to avoid gain peaking in the earlier sections, and places any real pole last.

Though it is preferable in most cases to buffer each filter section in a system, it is possible—depending upon your design—to cascade a first-order filter directly with a second-order. It is important, however, to take care for proper impedance matching. Recall that the cutoff frequencies are proportional to $R$ and $C$. Most often, any changes in value will require that the resistor, $R$, be made smaller to reduce losses. *In dividing* R *by any value, always multiply* C *by that same value*. This applies to the first-order section only; all values in the second-order section should remain unaffected.

Of course, it is possible to design a filter of higher order directly in a single circuit without a thought to individual sections. This can be a cost-effective and space-saving approach. It can also be difficult to debug and very sensitive to tolerances and changes in the environment. There will be examples of both types of filters to come, though the concentration here will be on cascading.

## Active Filters

Passive network filters offer problems that make them difficult to use in practice. Passive RC filters, for instance, can produce natural frequencies only on the negative real axis of the complex frequency plane. They are also inherently lossy, so they put a practical limit on the order of filter we can make with them. With an LRC filter, we can place the poles and zeros wherever we choose. However, we must also deal with the inductors, which can be sources of hidden resistance. They can also be large, nonlinear, and expensive, and they can saturate.

The *voltage-controlled voltage source*, usually implemented with an opamp, can make virtually any kind of filter that we choose. With only resistors and capacitors, we can generate complex poles and zeros that give us size, efficiency, and economy gains over LRC filters. These filters are built with RC components and operational amplifiers, and transistors or tubes, and exhibit the same characteristics that we associate with LRC networks—that is, they can oscillate, their poles and zeros are manipulable, and they can act as inductors. They can also supply gain. Single second- and third-order filters can be cascaded to produce still higher-order filters with relative ease, or complete high-order filters can be constructed from a few opamps and passive components. In addition, we can cause the amplifier to gyrate, as you will see in the GIC filters, simulating an inductor with only resistors and capacitors and opamp, allowing the construction of active filters directly from easily attainable tables of LRC values and configurations.

## Second-Order Filter Transfer Functions

The generalized form of the second-order voltage transfer functions as we write them in $s$ do not change from one filter configuration to another, though the electrical transfer functions change with the circuit. The generalized forms are below. The electrical functions follow and are particular to each section:

$$H(s) = H_0 \frac{\omega_0^2}{s^2 + s\frac{\omega_0}{Q} + \omega_0^2} \quad \text{lowpass}$$

$$H(s) = H_0 \frac{\left(\frac{\omega_0}{Q}\right)s}{s^2 + \left(\frac{\omega_0}{Q}\right)s + \omega_0^2} \quad \text{bandpass}$$

$$H(s) = H_0 \frac{s^2}{s^2 + \left(\frac{\omega_0}{Q}\right)s + \omega_0^2} \quad \text{highpass}$$

## Voltage Follower

Using an operational amplifier in voltage-follower mode only means that we are using the amplifier for buffering, perhaps for another stage of filtering, as with a single-order filter used in conjunction with a second- or higher-order filter to add the negative real pole needed. Its purpose is isolation and, perhaps, gain. An example of this is given in Figure 2-23.

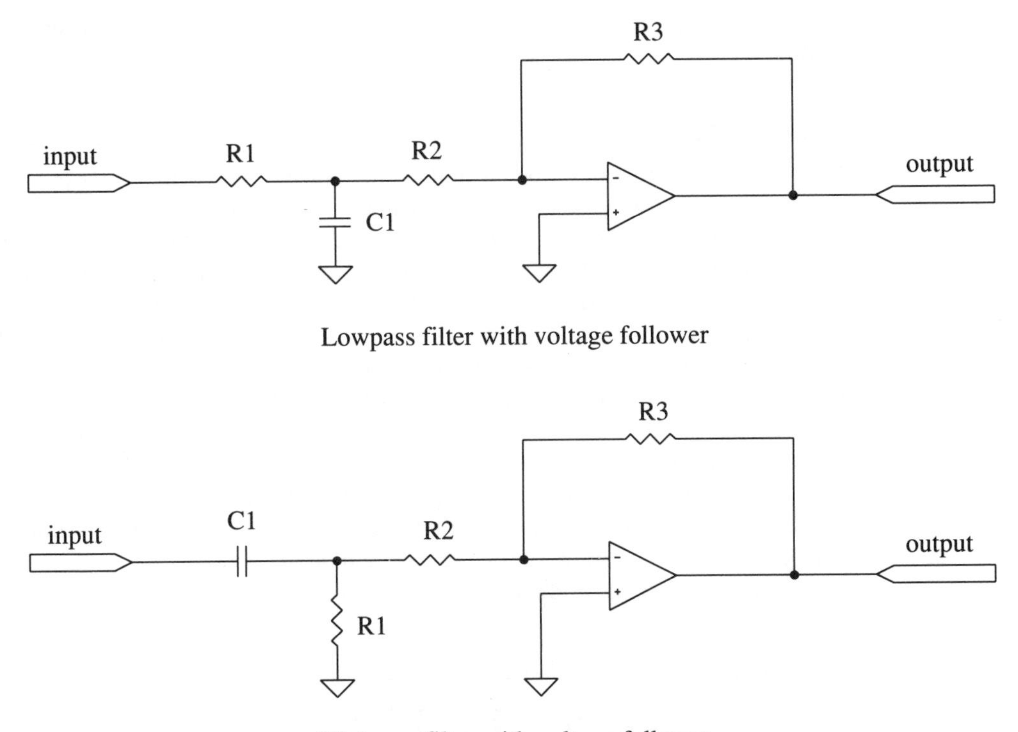


**Figure 2-23.** A lowpass and highpass RC filter buffered by a voltage follower. The amplifiers in this configuration do nothing to enhance the lossy characteristics of the single pole RC filter, they do however buffer the stages for cascading.

The first-order filter is nothing more than a voltage follower buffering an RC network. This could be either a transistor or an opamp in a configuration similar to the one in Figure 2-23.

Due to the nature of the product, there are no good first-order RC bandpass filters, so for this element we have only low- and highpass. The lowpass transfer function is

$$H(s)=\frac{\frac{1}{RC}}{s+\frac{1}{RC}}$$

and the highpass is

$$H(s)=\frac{s}{s+\frac{1}{RC}}$$

This form of a filter does nothing for the droopy and lossy passband of the simple RC first-order filter.

## Sallen and Key

R.P. Sallen and E.L. Key described a simple alternative to buffering a first-order filter. This alternative has come to be known as the Sallen and Key, or sometimes Sallen-Key, filter. This is one of the most popular implementaions of the VCVS filter configurations in use, most probably because of the simple and intuitive approach. What we have is a significant improvement over the RC filter. We gain improvement by feeding back a portion of the signal to the input to sharpen the transition band, as shown in Figure 2-24. The value of the feedback capacitor (C1) is chosen to have a relatively low impedence near cutoff. This prevents some of the droop that normally comes with cascaded RC filters.

This form of the VCVS is simple to implement and requires few components. Tables are commonly available providing precomputed values for the RC elements in the circuit for unity gain, equal resistance equal capacitance, and equal capacitance gain of two versions. High-order filters are achieveable by cascading second- and first- or third-order sections together, or by designing a circuit to a certain order.

The transfer function and related equalities for the circuit in Figure 2-24 are

$$H(s)=\frac{\frac{k}{R1R2C1C2}}{s^2+s\left(\frac{1}{R1C1}+\frac{1}{R2C1}+\frac{K-1}{R2C2}\right)+\frac{1}{R2R1C1C2}}$$

$$k\omega_0^2=\frac{k}{R1R2C1C2}$$

$$\frac{\omega_0}{Q}=\frac{1}{R1C1}+\frac{1}{R2C1}-\frac{k-1}{R2C2}$$

$$\omega_0^2=\frac{1}{R1R2C1C2}$$

In these equations, $k$ is equal to the DC gain, $\omega_0$ is the cutoff frequency, and $Q$ is the quality factor determining the selectivity and bandwidth of the filter.

Once the poles for this filter have been determined, it is a simple matter to solve for the values of the components. To design a filter such as the one in Figure 2-24, we need to know the order, the $Q$, and the cutoff frequency. We will assume, for this example, that we are making a second-order lowpass filter with a cutoff frequency of 10 KHz and we need a Butterworth characteristic.

1. Calculate the poles for a second-order Butterworth. Using the Mathcad document, Butter.mcd, we get

   $$s_n=-0.70711\pm j0.70711$$

2. Fit these into the form of a factored polynomial and assemble the quadratic:

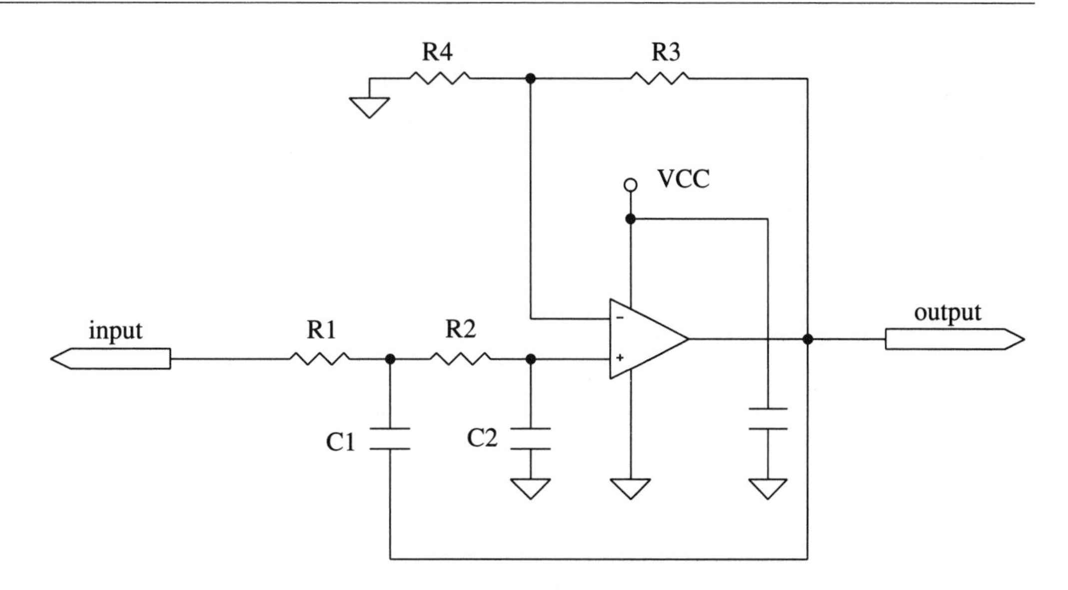


**Figure 2-24.** Classic form of the Sallen and Key lowpass filter.

$$as^2 + bs + c =$$
$$[s_0 - (-0.70711 + j0.70711)][s_0 - (-0.70711 - j0.70711)] =$$
$$s^2 + (0.70711 + j0.70711) + (0.70711 - j0.70711)s +$$
$$(0.70711 + j0.70711)(0.70711 - j0.70711) = s^2 + 1.41421s + 1$$

(Notice that $c$ is 1. This is an earmark of the Butterworth characteristic, where the radius of the root locus is unity.)

3. From the material in the first chapter, we know that $\omega_0$, the cutoff frequency, is

$$\omega_0 = \sqrt{c} = \sqrt{\frac{1}{R1R2C1C2}} = 1$$

$$Q = \frac{\omega_0}{b} = 0.70711$$

4. Now it is time to denormalize the values by multiplying $\omega_0$ by the target cutoff frequency, 10 KHz:

$$\omega_0 * 10\text{Khz} = 10000$$

5. Make $R1 = R2 = R$, and $C1 = C2 = C$.
6. Then we choose a value for $C$ that we can easily get, $C = 0.1$mfd, and solve for $R$:

$$R = \frac{1}{\omega_0 C} = \frac{1}{10000\text{Hz} * .1\text{mfd}} = \frac{1}{.001} = 1000\Omega$$

7. Sometimes it is necessary to adjust the value of $R$. If we want it to be in the 10K ohm range, we simply multiply the value of $R$ by the factor we need, in this case 10, and divide $C$ by the same amount:

   $$10 * R = 10K\Omega$$

   and

   $$\frac{C}{10} = .01\text{mfd}$$

8. We can calculate the gain with the following relationship:

   $$\frac{\omega_0}{Q} = \frac{1}{R1C1} + \frac{1}{R2C1} - \frac{k-1}{R2C2} = \frac{1}{10K\Omega * .01\text{mfd}} + \frac{1}{10K\Omega * .01\text{mfd}} - \frac{k-1}{10K\Omega * .01\text{mfd}} = -k+3$$

   or

   $$k = 3 - \frac{1}{Q} = 1.58579$$

9. Finally, with a value for the gain, we can calculate $R4$ and $R3$. In a noninverting amplifier:

   $$k = 1 + \frac{Ra}{Rb}$$

   or

   $$\frac{R3}{R4} = (k-1) = .58579$$

In other words, for this amount of gain, $R3$ must be 0.58579 times $Ra$. It is good practice to select $R4$ such that the resistance on the inverting and noninverting inputs of the opamp are equal thereby minimizing offsets. This means that the Thevenin equivalent resistance of $R3$ and $R4$ should be equal to the resistance to ground at the noninverting input. That is,

$$R1 + R2 = \frac{R4 * R3}{R4 + R3}$$

Solve these two equations simultaneously to yield an $R4$ of 54K ohms and an $R3$ of 32K ohms. See the Mathcad spreadsheet VCVS.mcd for more examples. Regardless of the method, you will have to adjust these values for the real world.

Actually, any number of approaches might be taken with these design equations. In our example, to maintain the Butterworth characteristic, the gain is 1.58579. A second-order Chebyshev with 0.5dB of ripple approximated with the exact same procedure will produce a gain of 1.84221.

To make a second-order Chebyshev with 0.5dB of ripple and a cutoff frequency of 5 KHz:

1. Calculate the poles:

   $$s_n = -0.71281 \pm j1.00404$$

2. Fit them to a quadratic:

   $$as^2 + bs + c = s^2 + 1.4256s + 1.51620$$

   (Notice that $c$ is not 1.)

3. Calculate the normalized cutoff frequency, $\omega_0$, and $Q$:

   $$\omega_0 = \sqrt{c} = 1.23134$$

   and

   $$Q = \frac{\omega_0}{b} = .863722$$

4. Denormalize the cutoff frequency:

   $$\omega_0 * 5\text{KHz} = 6156.7\text{Hz}$$

5. Make $R1 = R2 = R$ and $C1 = C2 = C$.
6. Choose 0.016mfd for the capacitors, and calculate $R$:

   $$R = \frac{1}{\omega_0 C} = \frac{1}{9.85072^{-5}} = 10151\Omega$$

7. There is no need to adjust the value of $R$.
8. Calculate the gain:

   $$k = 3 - \frac{1}{Q} = 1.8422$$

9. Calculate $R4$ and $R3$:

   $$\frac{R3}{R4} = (k-1) = .8422$$

   $$20302 = \frac{R4 * R3}{R4 + R3}$$

   $$R4 = 44.4k \text{ and } R3 = 37.4k$$

To make a Thomson linear phase filter, second-order, with a cutoff frequency of 9 KHz:

1. Calculate the poles:

   $$s_n = -1.5 \pm j0.86603$$

2. Fit them to a quadratic:

   $$as^2 + bs + c = s^2 + 3s + 3$$

   (Notice that $c$ is not 1.)

3. Calculate the normalized cutoff frequency, $\omega_0$, and $Q$:

   $$\omega_0 = \sqrt{c} = 1.73205$$

   and

   $$Q = \frac{\omega_0}{b} = .57735$$

4. Denormalize the cutoff frequency:

   $$\omega_0 * 9 \text{ KHz} = 15588.45 \text{ Hz}$$

5. Make $R1=R2=R$ and $C1=C2=C$.
6. Choose .006mfd for the capacitors, and calculate $R$:

   $$R = \frac{1}{\omega_0 C} = \frac{1}{9.85072^{-5}} = 10691\Omega$$

7. There is no need to adjust the value of $R$.
8. Calculate the gain:

   $$k = 3 - \frac{1}{Q} = 1.26794$$

9. Calculate $R4$ and $R3$:

   $$\frac{R3}{R4} = (k-1) = .26794$$

   $$21382 = \frac{R4 * R3}{R4 + R3}$$

   $R4 = 101k$ and $R3 = 27.1k$

Again, it is up to you to find the exact values in the real world that fit your needs.

### Third-Order Sallen and Key Filter

Depending upon the ultimate order of the filter you require, it can sometimes be simpler to use a third-order filter with second-order sections to realize it. Figure 2-25 is a schematic of a third-order Sallen and Key filter.

The transfer function for this filter is

$$G(s) = \frac{k * \omega_1 \omega_2}{(s+1)\left(s^2 + s * \frac{\omega_2}{Q} - \omega_2^2\right)} \qquad \text{(Eq. 2-3)}$$

and, in terms of the circuit:

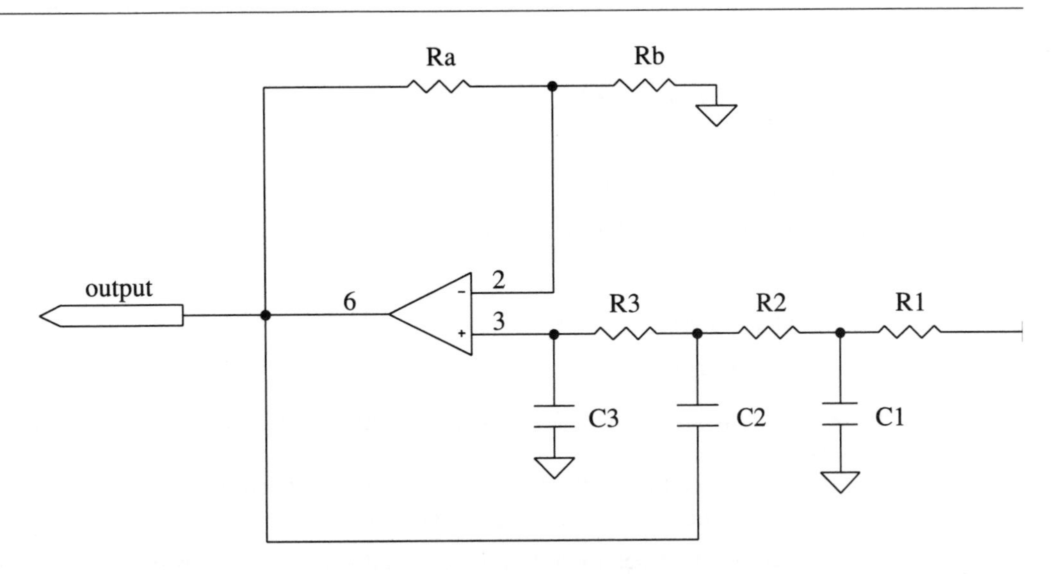


**Figure 2-25.** Third-order Sallen and Key filter. Notice the single real pole ($R1$ and $C1$) cascaded with the second-order Sallen and Key.

$$G(s)=\frac{k}{\begin{array}{l}s^3\cdot R1\cdot R2\cdot R3\cdot C1\cdot C2\cdot C3+\\ s^2\left[R2\cdot R3\cdot C2\cdot C3\left(1+\frac{R1}{R2}\right)+R1\cdot R3\cdot C1\cdot C3\left(1+\frac{R2}{R3}\right)-R1\cdot R2\cdot C1\cdot C2(k-1)\right]+\\ s\left[R1\cdot C1+R3\cdot C3\left(1+\frac{R2}{R3}+\frac{R1}{R3}\right)-R2\cdot C2\left(1+\frac{R1}{R2}\right)(k-1)\right]+1\end{array}}$$

$$R1\cdot R2\cdot R3\cdot C1\cdot C2\cdot C3=\frac{1}{\omega_1\omega_2}$$

$$\left[R2\cdot R3\cdot C2\cdot C3\left(1+\frac{R1}{R2}\right)+R1\cdot R3\cdot C1\cdot C3\left(1+\frac{R2}{R3}\right)-R1\cdot R2\cdot C1\cdot C2(k-1)\right]=\frac{1}{Q\cdot\omega_1\omega_2}+\frac{1}{\omega_2^2}$$

$$\left[R1\cdot C1+R3\cdot C3\left(1+\frac{R2}{R3}+\frac{R1}{R3}\right)-R2\cdot C2\left(1+\frac{R1}{R2}\right)(k-1)\right]=\frac{1}{\omega_1}+\frac{1}{Q\cdot\omega_2}$$

These equations become much more manageable if we make some assumptions about the components before we start. In this case, we are going to set all the capacitors equal to 1. We will also determine that $R1$ is equal to $R2$. We then solve for each component of Equation 2-3:

$$A_1:=\frac{1}{C^3\cdot\omega_1\omega_2}$$

$$A_2 := \frac{1}{C^2}\left(\frac{1}{Q \cdot \omega_1 \omega_2} + \frac{1}{\omega_2^2}\right)$$

$$A_3 := \frac{1}{C}\left(\frac{1}{\omega_1} + \frac{1}{\omega_2 \cdot Q}\right)$$

This results in three simultaneous equations

$$R3 \cdot R1^2 = A_1$$

$$3 \cdot R1 \cdot R3 + \left(R1^2\right)(2-K) = A_2$$

$$R3 + R1(5-2k) = A_3$$

These equations are solvable by numerical methods. (For an example, see the Mathcad worksheet VCVS.mcd.)

If we are to design a third-order filter with a Butterworth characteristic, we need to know the $Q$ and the value we set for the capacitors.

1. Derive the poles for a third-order Butterworth lowpass with a cutoff frequency of 4 KHz:

   $s_0 = -0.5 + j0.86603 \quad s_1 = -0.5 - j0.86603$

   and the negative real pole provided by the single order:

   $s_{odd} = -1$

   Thus, we have

   $(s+1)\left(s^2+s+1\right)$

   and a denominator that looks like this:

   $s^3 + 2s^2 + 2s + 1$

2. Calculate the two *normalized* cutoff frequencies from the second-order quadratic

   $\omega_2 = \sqrt{c} = 1$

   $\omega_1 = 1$

3. Assume all capacitors are equal, and in this case, 1. Make $R1$ equal to $R2$.
4. Solve for $A_1$, $A_2$, and $A_3$:

   $$A_1 := \frac{1}{C^3 \cdot \omega_1 \omega_2} \qquad A_1 = -1$$

   $$A_2 := \frac{1}{C^3}\left(\frac{1}{Q \cdot \omega_1 \omega_2} + \frac{1}{\omega_2^2}\right) \qquad A_2 = 2 + 1.224714774 * 10^{-15} i$$

   $$A_3 := \frac{1}{C}\left(\frac{1}{\omega_1} + \frac{1}{\omega_2 \cdot Q}\right) \qquad A_3 = -2$$

5. Solve the simultaneous equations:

$$R3 \cdot R1^2 = A_1$$

$$3 \cdot R1 \cdot R3 + \left(R1^2\right)(2-k) = A_2$$

$$R3 + R1(5-2k) = A_3$$

Then, since $R1$ equals $R2$,

$$R1 = 1.5906$$

$$R2 = 1.5906$$

$$R3 = .4421$$

6. Denormalize the resistor values relative to our chosen capacitor values. It is more practical to use a 1.2mfd cap in this circuit than a 1 farad cap. To do this, we must divide the 1 farad cap by our new value:

$$C1 = C2 = C3 = 1.2\,\text{mfd} \text{ and } \frac{1\,\text{farad}}{1.2\,\text{mfd}} = 833333$$

Then, we must multiply the resistors by 833333:

$$R1 = R2 = 1.5906 * 833333 = 1325499\Omega \text{ and } R3 = .4421 * 833333 = 36841\Omega$$

These are large values and may not fit in your circuit—it is up to the designer to choose the real-world components that best fit his application.

Sallen and Key configurations for high- and lowpass filters are presented in Figure 2-26. The voltage transfer function for a Sallen and Key bandpass filter in the illustration is

$$H(s) = \frac{sK/R1C1}{s^2 + s\left(\frac{1}{R1C1} + \frac{1}{R2C1} + \frac{1}{R3C1} + \frac{1}{R3C2} - \frac{K}{R2C1}\right) + \left(\frac{1}{R3C2C1}\right)\left(\frac{1}{R1} + \frac{1}{R2}\right)}$$

The Sallen and Key highpass filter transfer function is

$$H(s) = \frac{s^2K}{s^2 + s\left(\frac{1}{R2C1} + \frac{1}{R1C2} + \frac{1}{R1C2} + \frac{1}{R1C1} - \frac{K}{R2C1}\right) + \frac{1}{R2R1C1C2}}$$

As you have seen in previous designs, it is common to find as many equal elements as possible before solving for the rest. In these two designs, making $R1 = R2 = R3$ and $C1 = C2$ is a good starting point.

In the case of the bandpass filter, this leaves us with

$$K = 4 - \frac{\sqrt{2}}{Q}$$

$$RC = \frac{\sqrt{2}}{\omega_0}$$

The highpass filter yields

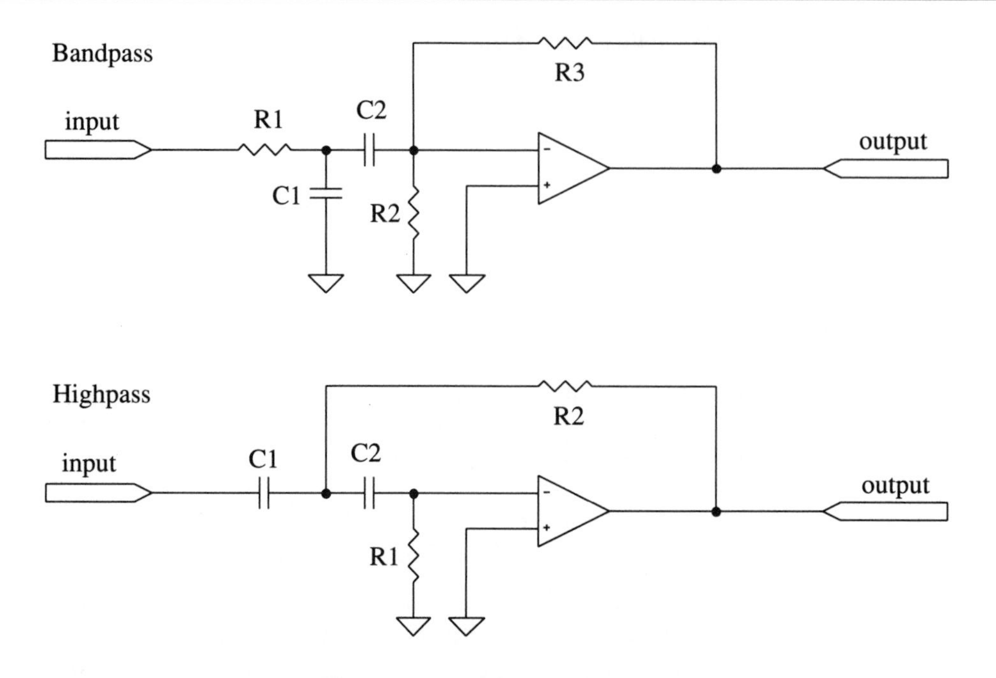


Sallen and Key filter configurations

**Figure 2-26.** Examples of bandpass and highpass filters with the Sallen and Key configuration.

$$\omega_0 = \frac{1}{RC}$$

$$\frac{1}{Q} = 3 - K$$

$$H_0 = K$$

These are the same as those for the lowpass filter we solved for earlier, except for the $1/P$ transformation.

In the next few sections, we will briefly examine some of the other popular forms of filter design employing voltage-controlled voltage sources. Schematics and some design equations will accompany a short description in each case.

### State Variable

The state variable filter provides a lowpass, highpass, and bandpass output simultaneously. It is shown here in a second-order filter function. This design is interesting because all three of the outputs are available, forming the basis for some very interesting topologies. For

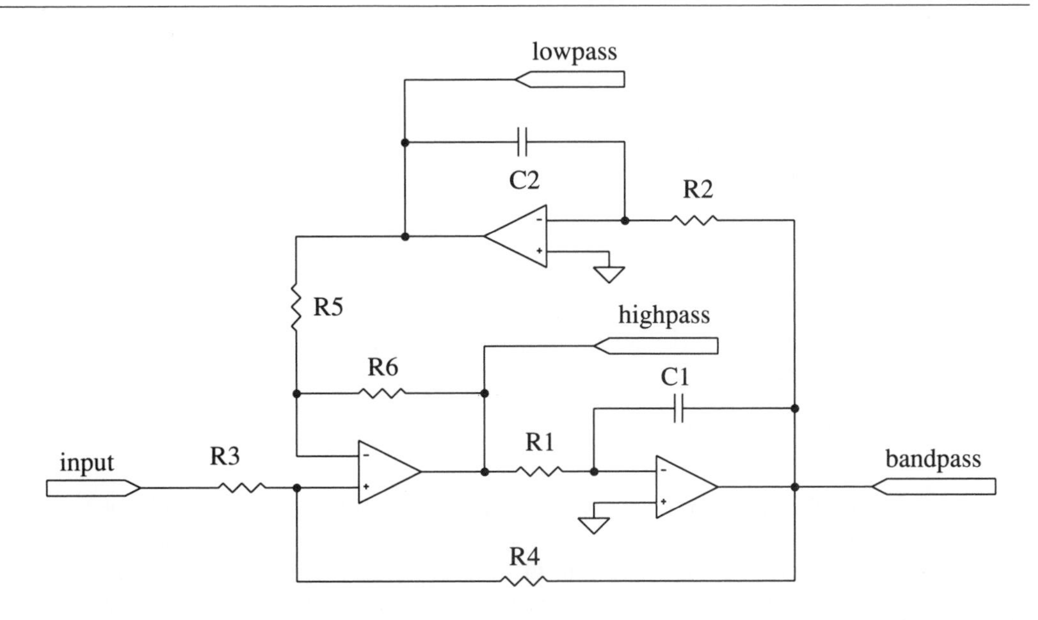


**Figure 2-27.** The state variable provides simultaneous outputs for lowpass, bandpass, and highpass functions. These can be summed in various ways to produce still more configurations.

instance, other outputs are easily attained by summing with one or more of these—a notch filter is a sum of the highpass and lowpass outputs.

As you can see in Figure 2-27, the filter is constructed from two integrators and a summing amplifier. This allows straightforward control of gain, $Q$, and resonance.

Each of the filter outputs conforms to the general expressions for second-order lowpass, highpass, and bandpass filters mentioned in the section entitled *Second Order Filter Transfer Functions*.

The circuit in Figure 2-27 yields the following equations:

$$\omega_0 = \sqrt{\frac{\frac{R6}{R5}}{R1R2C1C2}}$$

$$Q = \frac{1+\frac{R4}{R3}}{1+\frac{R6}{R5}}\sqrt{\frac{R6R1C1}{R5R2C2}}$$

The gains for the separate sections are different:

$$H_{lowpass} = \frac{1+\frac{R5}{R6}}{1+\frac{R3}{R4}}$$

$$H_{bandpass} = \frac{R4}{R3}$$

$$H_{highpass} = \frac{1+\frac{R6}{R5}}{1+\frac{R3}{R4}}$$

## Bi-quad

The bi-quad filter transfer function has a polynomial in the numerator as well as in the denominator:

$$H(s) = H_0 \frac{a+bs+s^2}{c+ds+s^2}$$

Create the filter you want by selecting which coefficients to include and adjusting their magnitudes to fit the needs of the transfer function. Inspection reveals that the numerator of the transfer function is the sum of the three basic transfer functions for high-, low-, and bandpass filters. That is how the circuit is implemented, as well. Please see Figure 2-28 for an example of a biquadratic function created from a state variable filter by summing the high- and lowpass outputs at the noninverting node of an opamp and adding the already inverted bandpass output at the inverting node.

The biquadratic output is found through the superposition of the three inputs to the final amplifier, relative to the gains set by the corresponding resistor network.

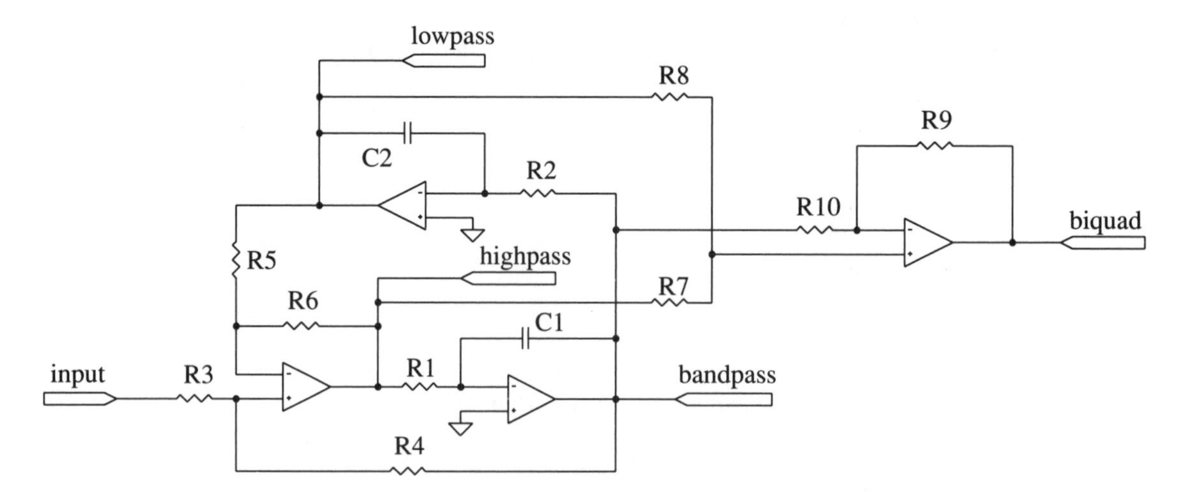


**Figure 2-28.** The biquadratic amplifier as an extension of the state variable filter.

### Multiple Feedback Networks

The *multiple feedback* or *infinite gain* filter has a lowpass transfer function described by

$$H(s) = \frac{\dfrac{-1}{R1R3C1C2}}{s^2 + s\dfrac{1}{C1}\left(\dfrac{1}{R1} + \dfrac{1}{R2} + \dfrac{1}{R3}\right) + \dfrac{1}{R2R3C1C2}} \qquad \text{(Eq. 2-4)}$$

Solving for the components of the generalized lowpass function in Equation 2-4,

$$\omega_0 = \frac{1}{\sqrt{R2R3C1C2}}$$

$$\frac{1}{Q} = \sqrt{\frac{C2}{C1}}\left(\frac{\sqrt{R2R3}}{R1} + \sqrt{\frac{R3}{R2}} + \sqrt{\frac{R2}{R3}}\right)$$

$$|H_0| = \frac{R2}{R1}$$

The bandpass filter is given by

$$H(s) = \frac{\dfrac{-s}{R1C1}}{s^2 + s\left(\dfrac{1}{R2C2} + \dfrac{1}{R3C1}\right) + \dfrac{1}{R3C1C2}\left(\dfrac{1}{R1} + \dfrac{1}{R2}\right)}$$

$$\omega_n = \frac{\sqrt{1 + \dfrac{R2}{R1}}}{\sqrt{R2R3C1C2}}$$

$$\frac{1}{Q} = \frac{\sqrt{\dfrac{R2C1}{R3C2}} + \sqrt{\dfrac{R2C2}{R3C1}}}{\sqrt{1 + \dfrac{R2}{R1}}}$$

$$|H_0| = \frac{\dfrac{R3}{R2}}{1 + \dfrac{C1}{C2}}$$

The highpass implementation is

$$H(s) = \frac{-s^2\dfrac{C1}{C2}}{s^2 + s\dfrac{1}{R2}\left(\dfrac{C1}{C2C3} + \dfrac{1}{C2} + \dfrac{1}{C3}\right) + \dfrac{1}{R1R2C2C3}}$$

$$\omega_n = \frac{1}{\sqrt{R1R2C2C3}}$$

$$\frac{1}{Q} = \sqrt{\frac{R1}{R2}}\left(\frac{C1}{\sqrt{C2C3}} + \sqrt{\frac{C3}{C2}} + \sqrt{\frac{C2}{C3}}\right)$$

$$|H_0| = \frac{C1}{C2}$$

Figure 2-29 shows the three basic implementations of the infinite gain circuit.

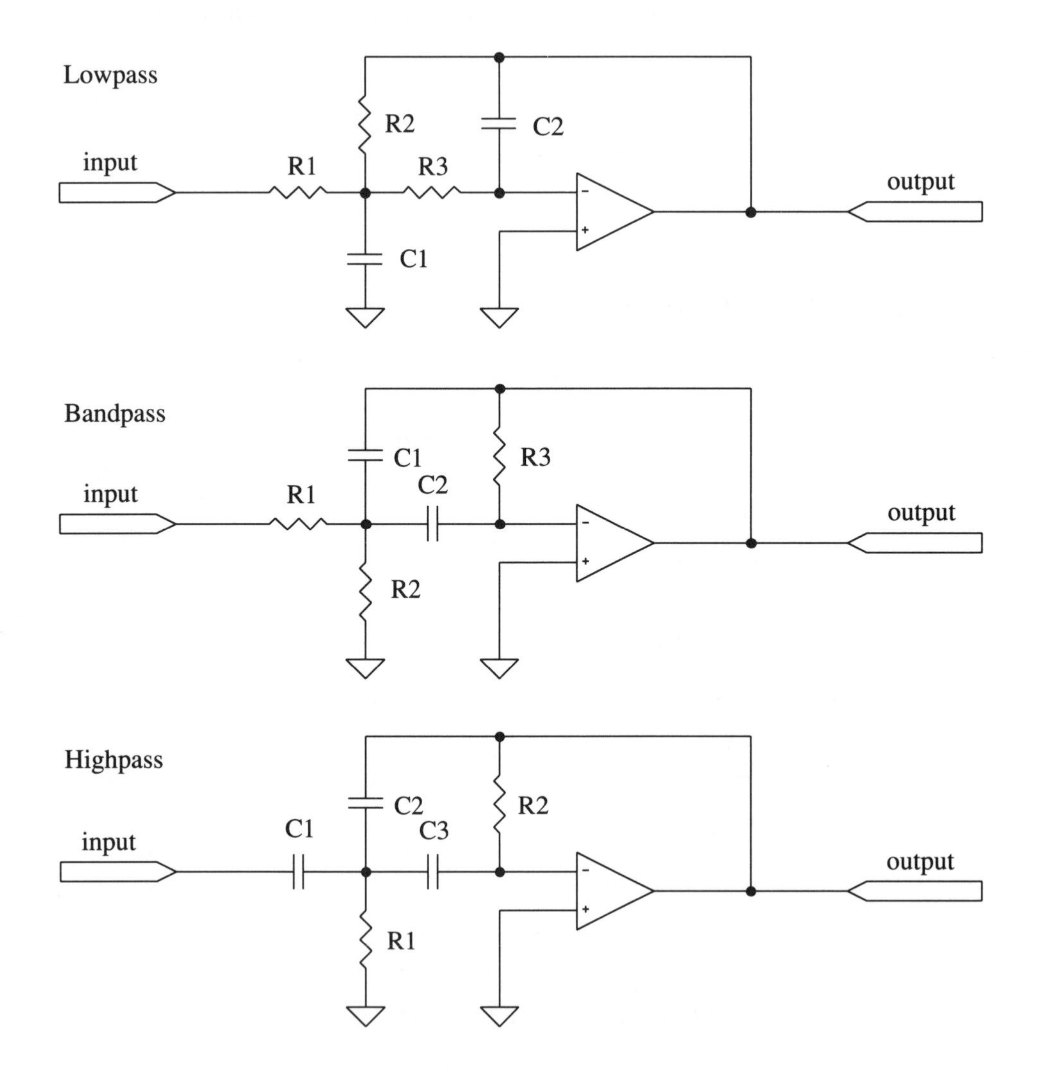


**Figure 2-29.** The Multiple feedback or infinite gain filter.

### Voltage-Controlled Current Sources

There are voltage-controlled current sources, often abreviated *VCIS*, that use an operational amplifier called an *operational transconductance amplifier.* This amplifier can be used to create floating resistors and very large values of resistance in silicon. The OTA produces an output current from an input voltage. Among the advantages of these devices is the fact that they have a higher bandwidth and can be electrically tuned to an application.

The basic characteristic of an OTA is its *transconductance,* which is the ratio of the amplifier output current to the input voltage:

$$g_m = \frac{i_{out}}{v_{in}}$$

This is a high impedance device intended for low-power, low-current operation. All of the basic parameters of the device (such as input offset voltage, input bias current, peak output current, and peak output voltage) are determined by a *bias current*, set with a bias resistor connected to a terminal of the amplifier.

### Passive Network Simulations

**The GIC** Probably one of the simplest and most convenient methods of filter synthesis is passive network simulation. There is already so much information available to designers in the form of tables and prototypes, that this is an attractive possibility. In addition, filters designed in this fashion feature lower cost than their passive counterparts, and generally lower noise than the popular Sallen and Key filter. If we can realize an active filter directly from these tables and avoid the drawbacks inherent in the passive components, then we are ahead. This is possible by defining a two-port active network called a *generalized immitance converter*. This configuration is very useful in synthesizing *grounded* inductors in lowpass filters. Figure 2-30 provides a schematic of a synthetic inductor.

Examining the schematic, you will see that the circuit consists of two opamps connected back to back, with two immitances or impedances between the inputs. Since the voltage across these inputs is ideally zero, then these immitances must be equal, and the voltages $V_1(s) = V_2(s)$ must also be equal. If another immitance is allowed to be connected from the output $V_2(s)$ to ground, we have the following immitance/impedance relationship:

$$Z(s) = \frac{Z_1(s)Z_3(s)Z_5(s)}{Z_2(s)Z_4(s)}, \text{ and}$$

$$Y(s) = \frac{Y_1(s)Y_3(s)Y_5(s)}{Y_2(s)Y_4(s)}$$

Because of this relationship, the GIC can be used to synthesize a great number of elements. It provides us with a starting point for implementing passive LRC networks with VCVSs.

**The FDNR** The FDNR, or *frequency-dependent negative resistor*, is a highly useful variation of the GIC configuration. The FDNR is measured in farad-seconds and has a unique symbol, as shown in Figure 2-31.

It is a one-port active network described by its input admittance:

$$Y(s) = \frac{I(s)}{V(s)} = s^2 D$$

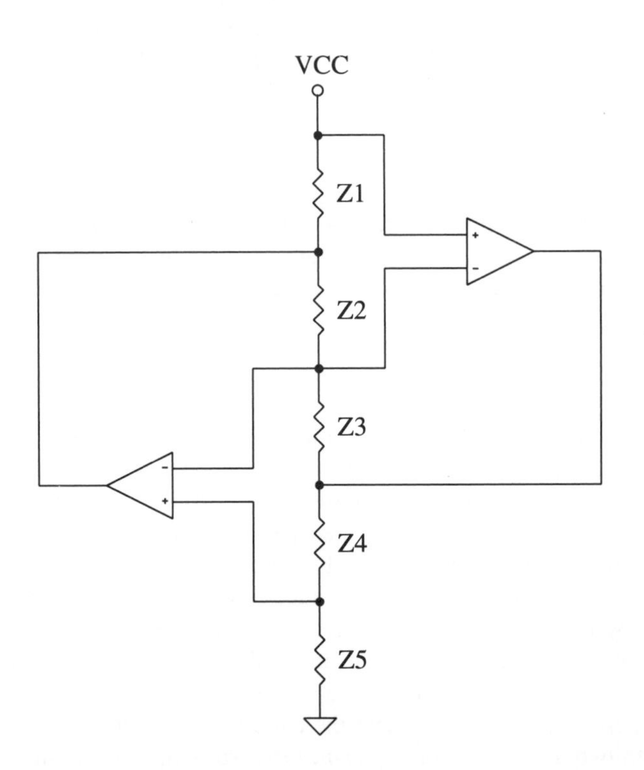


**Figure 2-30.** The GIN allows for the simulation of various passive elements. This is not only very efficient but it means that the designer can use the wealth of passive design tables in existence.

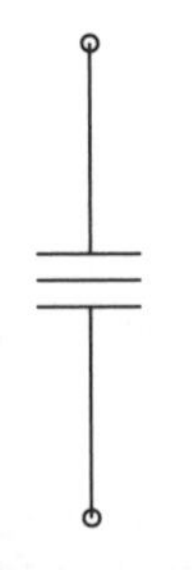

**Figure 2-31.** The symbol for the FDNR.

where $D$ is positive and real. This device can be called a negative resistor, because $s^2 = -\omega^2$, so the entire quantity is a function of negative frequency.

To convert a passive LRC into an FDNR, multiply the passive components by $1/s$. Inductors with an impedance of $sL$ become resistors with resistance $L$, in the same way, resistors of value $R$ become capacitors of value $R/s$, and capacitors with impedance $1/sC$ transform into frequency-dependent variable resistors $D$, with impedance $1/s^2C$. Creating highpass filters from lowpass prototypes simply means scaling the transfer function by $1/s$, meaning that capacitors become resistors with a value $1/C$ and resistors become capacitors with a value $1/R$. Transformation to a bandpass can be more complex, especially if it is narrowband. As is often true of other implementations, when the cutoff frequencies become separated by two or more octaves, the filter is created from lowpass and highpass sections.

The circuit diagram in Figure 2-32 shows how closely the FDNR is based upon the GIC. In this circuit, impedences $Z_3$ and $Z_5$ are capacitative, and the rest are resistors. The immitance function is the same as for the GIC. The value of the element is determined by $D = R2C3R4C5/R1$. In this equation, the notations were kept the same as those of the impedances to avoid confusion.

If we set all the impedances to 1 except for $R4$, then $R4$ becomes the sole determinant in the equation. We set $R4$ equal to $C2$ in the passive network in the example above—that is, to .8746 farad seconds. Figure 2-32 illustrates its inclusion in an actual schematic of a filter—the one used on the DACQ.

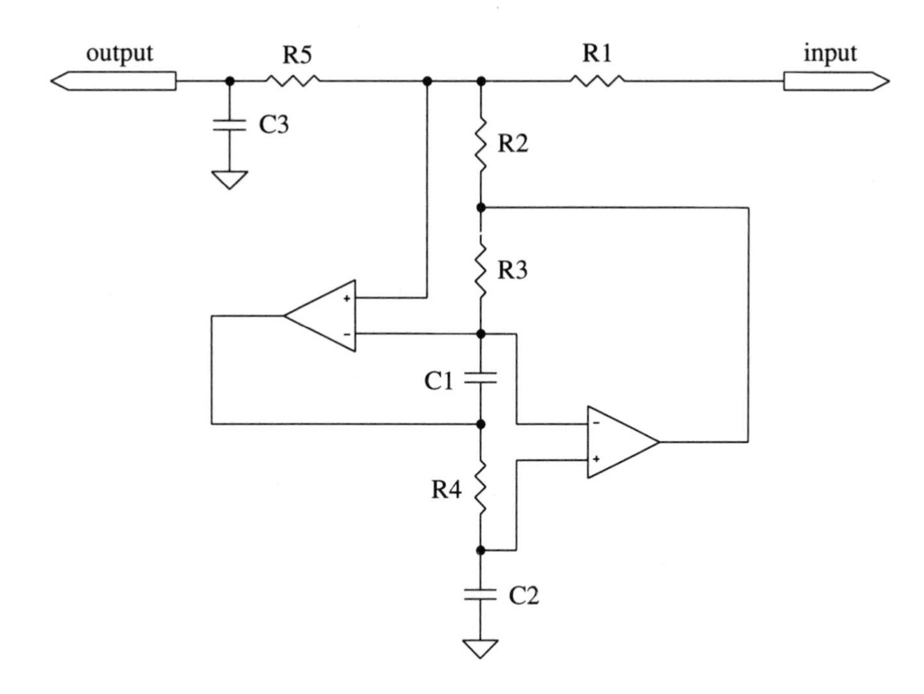


**Figure 2-32.** Schematic of GIC filter element used on DACQ.

You will notice from the schematic that the values of the components are different from those just stated. This is because we were working from values normalized to 1 rad/s. To make these values useful, we have had to scale the frequency determining components and adjust the impedance. As we have seen before, the scaling factor is simply

$$scale = 2\pi f_c$$

All the capacitors in the circuit are divided by this factor.

The impedance is adjusted by picking smaller values for the capacitors (orders of magnitude) and then multiplying the resistors by the ratio of the given value to the target value. To restate:

$$scale_r = \frac{C_{given}}{C_{target}}$$

What we have designed here is an active implementation of a third-order Thomson characteristic. To create a sixth-order, all that is necessary is to cascade two such sections. This is, in fact, what we have done with the DACQ.

### Switched Capacitor

Switched capacitor filters are available from a number of vendors, and they can simplify filter design greatly. It is possible to configure these parts to simulate any of the forms presented in this book and more. They are typically quite compact, requiring only a few external parts to set clock frequencies and filter any clock noise.

For many applications, these can be a valuable resource, as long as their limitations are taken into consideration. There are a few things to keep in mind when using switched capacitor filters:

1. Since switched capacitor filters are inherently sampling systems, they can alias. The clocking frequency must be at least twice the cutoff frequency. This usually isn't a problem, because most chips require a 50:1 or 100:1 ratio.
2. There is clocking noise on the output of the filter. Usually, this is removed quite easily with a simple RC filter that has a pole frequency set at least a decade above the cutoff.
3. Due to a convolution of the internal sampling impulses and the input frequency, energy that is equal to the sampling frequency minus the input frequency is unattenuated, and if that difference corresponds to frequencies in the passband, it will appear. These filters usually require a clock at some multiple of the cutoff or corner frequency—the larger the multiple, the less chance there is that any of that energy will end up back in the passband. For example, suppose we wish a cutoff frequency of 6 KHz and we are clocking at 600KHz. If our input is 3 KHz, frequencies that would be fed back into the passband would be 600 KHz−3 KHz=597 KHz. This is still well out of our range and will not effect the operation of the circuit.
4. There is a distinct group delay associated with switched capacitor filters that does not exist to the same degree with continuous-time filters. This is due to sampling

delays and delays in the integrators. For some applications this can be quite significant, especially if you are interested in real-time processing of signals. Figure 2-33 is an illustration of switched capacitor operation and its analog in continuous-time electronics.

In Figure 2-33, the input signal is switched onto the plates of a capacitor through *SW*1 for a period of time. Then, that switch is opened and *SW*2 closes to communicate that charge to internal integrating circuits. The switching is *nonoverlapping*, and at such a high frequency that the circuit shown in part A of the illustration can be considered a simple resistor, as in part B. Very large values of resistance can be very tightly simulated in this way, and they are programmable.

### Continuous-Time Filter ICs

Continuous-time filter ICs, such as the MAX274 used on the DACQ, offer true second-order filters that can be cascaded to provide up to an eighth-order filter. The filters are con-

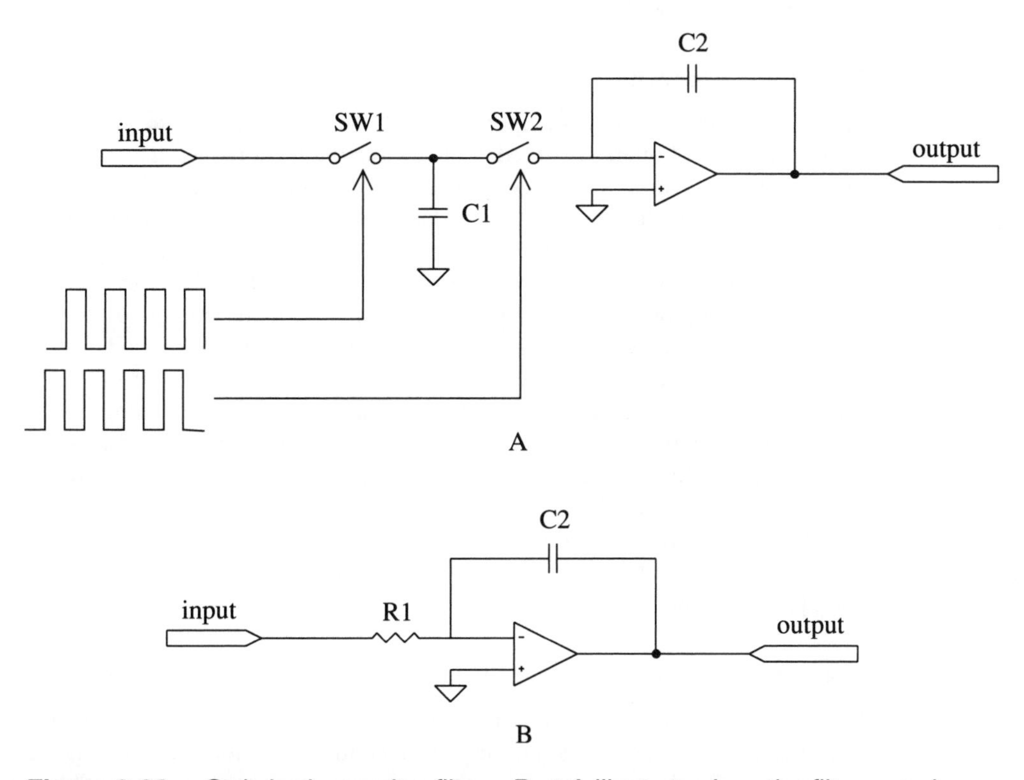


**Figure 2-33.** Switched capacitor filters. Part A illustrates how the filter samples the input and passes the charge to an integrator on alternating clocks. Part B represents the result of this operation.

structed using conventional filter technology—in the case of the MAX 274, Sallen and Key technology—and are quite easily programmed for a number of different filter configurations using the techniques described in this book. Each filter section can be programmed as the user chooses. These sections do not exhibit the problems that sampling introduces to the switched capacitor devices.

The specifications for the MAX274 are included in the appendix for reference. The circuit on the DACQ was designed as two fourth-order filters. The input is Bessel for linear phase and has a cutoff frequency of 5 KHz. The output is Butterworth, and its purpose is to remove the D/A staircasing from the reconstructed signal.

## Operational Amplifiers in Active Filters

For a filter section to work properly, certain aspects of an opamps behavior need to be taken into consideration. The next few paragraphs deal with some parameters of operational amplifiers that are important in the selection of devices for filters and in driving analog-to-digital converters. These are certainly not all—it is assumed that you are familiar with operational amplifiers and the associated circuitry.

### Output Voltage Swing

This is important for circuits employing A/D converters, as data acquisition circuits often do, in that they will often have predetermined analog input ranges. The specifications sheets for an opamp will show what the expected output voltage for a given bias supply will be for a particular part. If the ultimate swing does not cover the entire input range of your A/D converter, you will be losing resolution and dynamic range. If your opamp swings wider than the analog input range of your converter, the output of the converter can saturate at zeros or ones, or provide entirely predictable output, again depriving you of resolution.

There are several ways to solve this problem. For circuit simplicity (but not necessarily low cost), use Cmos amplifiers. These devices can provide a near rail to rail swing that can often be used in circuits that have only one supply voltage for analog signals.

Another alternative is to clamp the input of the A/D so that no excursions outside its input range is allowed, then use gain trimming techniques to see that the output of the amplifiers is always in range.

Finally, separate power supplies can be used—one for the A/D converter and reference, and another for the opamp to ensure that it can swing to both extents of the converters input, but no farther.

### Slew Rate

An opamp's slew rate is important in determining its effective operating frequency. The circuitry and physical construction of an opamp contributes intended and parasitic capacitance in the device. For the output to swing, it must charge those capacitances. This means that an opamp without external filter networks exhibits rolloff at some frequency. Use the slew rate of an opamp as a guide to how quickly it can charge its internal capacitances and make the output change.

$$f_p = \frac{SR}{2\pi e_p}$$

In this equation, $SR$ is the slew rate of the opamp, $e_p$ is the output peak voltage, and $f_p$ is the full power frequency the amplifier is capable of. As an example, the operational amplifiers on the DACQ, the OPA2604, have a slew rate of 25V/microsecond. If we wish to use them in a design with a +/−15V supply, and we expect them to swing +/−12V over the range of the input signal. This formula yields

$$f_p = \frac{25V/\text{microsecond}}{2\pi 24} = \frac{25000000/\text{second}}{150.796} = 165786.4 \text{ Hz}$$

This is the maximum frequency at which we could expect to operate this device at a full +/−12V swing.

## Settling Time

Settling is the time it takes from the application of a unit step input to the time the amplifier output enters and remains symmetrically within an error band about the final value. This can be important for high-speed sampling, or for multiplexing opamps on a single A/D converter. The amplifier must be able to settle to the final value *before* the signal is sampled by the A/D; otherwise, the SNR is degraded because you are sampling bounce and not signal.

Clearly, the settling time indicated on the specification of an opamp is not the whole story. The settling time is also influenced by external capacitances and the magnitude of the driving function. Start with the worst-case figure in the specifications and then examine the application to determine what else could affect the amplifier.

## Bandwidth

Bandwidth is determined by circuitry internal and external to the amplifier. High-frequency operation requires careful consideration to layout, tracewidths, lengths of traces, and positioning of groundplaning. At high frequency, precautions such as guardbanding and groundplaning can do more harm than good by shunting the very signal you wish to amplify to ground.

## Vos

This is an interesting and sometimes difficult subject. If not examined closely, Vos can add an error to your acquisitions that varies from circuit to circuit. This is an offset that appears at the output with no input. It is temperature-sensitive, and though the spec sheets will suggest a value, there is no way to predict its polarity. Whatever that offset is will cause current to flow in the feedback loop, and it will experience gain just like the signals you are trying to amplify.

Just as with system noise, the ultimate voltage offset at the input of your A/D must be less than the LSB of the A/D you are using, or it will cause lost or garbled information to be fed into your calculations. For example, if you are using an A/D converter with a 5 volt input range—this might be described as +/−2.5V or 0 to 5 volts—the value of the LSB is

$$LSB = \frac{input\ range}{2^{number\ of\ bits}}$$

You must use a value less than that to ensure that a bit is not automatically added to or subtracted from every data byte because of the offset.

### Input Currents

Ios is similar to Vos in its ability to skew your data, but it comes from a different source. Ios is generated from the difference in bias currents between the input terminals of the opamp. The absolute value of the difference is the offset current. A small current through a large impedance at the input of an opamp can generate a voltage large enough to influence the output—in fact, this is quite common. This size of the offset can surprise you.

As an example, assume the circuit in Figure 2-34 below. We will say that $R2=90k$ and $R1=10k$. We compute $R3=R1R2(R1+R2)$, giving us $9k$ as a compensation resistor for input bias currents. Since the input transistors on our opamp are not ideal, they draw slightly different bias currents. Let us say that the inverting input draws 350 nanoamps and the noninverting input draws 500 nanoamps. This results in negative potential of 5 millivolts at the noninverting input. Since the potential difference between the inputs of an opamp is zero, the same potential exists at the inverting input, resulting in a current of 450 nanoamps. Subtracting this value from the current drawn at the input, 350 nanoamps, yields −100 nanoamps. The voltage at the output of the opamp, with both inputs grounded through their respective resistors, and neglecting other offsets, is $E_{out}=E_{noninverting}-R2I_{noninverting}=5mv-90k*-100na=-4mv$

As you can see, this encourages the designer to keep resistor values low, when possible. Very often, an attempt is made to reduce this effect by keeping the Thevinin equivalents

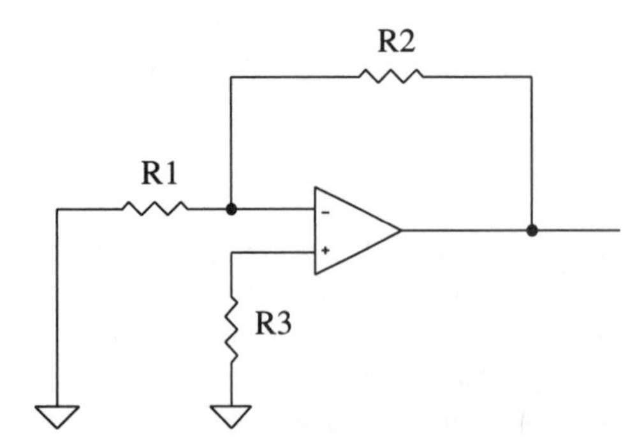

**Figure 2-34.** Simple circuit showing the current paths in an amplifier.

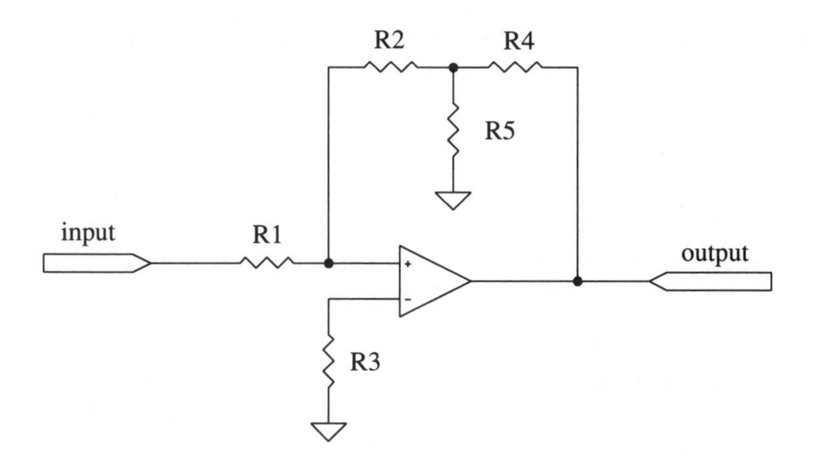


**Figure 2-35.** The Tee circuit can solve some of the problems associated with high resistance paths in an amplifier. This circuit can simulate the impedance necessary without the high resistance.

to the input impedances low at the inputs of the amplifiers and in the feedback loop. Where this is not possible, the circuit in Figure 2-35 offers a solution.

As an aid in determining the equivalent resistance of this Tee-network illustrated in Figure 2-35, use the following formula:

$$R_{feedback} = \frac{R_2R_4 + R_2R_5 + R_4R_5}{R_5}$$

### Drive Capability

The output of the amplifiers, through whatever impedances may be present, must be capable of driving the input to the A/Ds effectively. This is essential if you are using sampling A/Ds. If there is not enough drive capability, the sampling capacitor may not charge sufficiently to represent the input. This can result in having to lengthen sampling time, thereby lowering bandwidth or living with erroneous data.

As an aid to noise immunity, it is sometimes a good idea to put a small LC or RC component right at the input of the A/D. This will have to be scaled to the bandwidth and aperture time of your system.

### Sources of Noise

For amplifiers and A/D converters, noise is a real problem. Besides common impedance coupling along the power and ground planes, and crosstalk between traces, high-resolution A/D converters are finding themselves in close proximity to switching supplies, magnetic contactors, motor starters—noise sources of every description. There is much written on this subject; here are just a few points that can help prevent a noise problem.

1. Place oscillator circuits as far from analog as possible, and isolate any supplies to these circuits with ferrite beads.
2. Resist the urge to mix clocks and switching buffers in the same packages with logic that interfaces to analog circuitry.
3. Use as low-power and slow-speed logic as you can get by with—the faster the signal transition, the greater the amplitude of the resulting harmonics.
4. Keep PC board traces short.
5. If possible, use separate supplies for amplifiers and A/Ds than for those sources of noise. If separate supplies are not possible, see that the returns for the affected circuitry go directly back to the power supply on their own separate traces.
6. Separate the source of noise from the affected circuitry. Magnetic and electrostatic energy diminish as the inverse square of the distance between the source and the affected circuitry. If distance is not possible, separate the circuitry with groundplaning, or even a metal shield. Noise is energy, like light, and travels like light does. This means that a metal shield can take some of the energy to ground and throw a quiet shadow, as well as reflecting that energy.
7. Use separate power and groundplanes for digital and analog circuitry, allowing them to join at only one place in the power supply.
8. Use sufficient decoupling caps of good quality—ceramic for small values and high frequencies, tantalum for larger values. These caps should be placed as close to the power pins as possible. This is to provide as low an impedance to ground as possible for errant signals.
9. Use inductors and beads to clean up power supply traces, especially at the pin. An inductor and a capacitor make a second-order filter that can be very effective at the power pin of an affected IC. Beads work well too, especially on data lines that cannot afford much filtering, but remember that inductors and beads can form resonant circuits that create their own noise. An RC network at the power pins of an IC also filters much of the noise, and lacks the danger of resonance.
10. Feedthrough capacitor filters are very effective at reducing noise on A/D inputs and power supply pins.
11. Decouple completely any power entry or exit to the card. This means that you need at least a 22mfd tantalum cap and a .1mfd ceramic cap very close to the entry or exit pins.

## DACQ

The DACQ is a data acquisition module designed to demonstrate some of the signal processing concepts addressed in this book, and to give you the opportunity to work with these things yourself. The design is based upon the materials presented here as practical demonstration. In the schematics, you will find examples of some of the filter configurations presented in this chapter. In the next few chapters, software will be presented that runs on the card to facility acquisition, analysis, and reconstruction of input signals. The DACQ unit is not necessary for the understanding or utility of the data in this book, but it is very useful for getting a hands-on feel for the subject. Although you are free to build the unit yourself, it is recommended that you purchase the kit to take advantage of a printed circuit board and a layout that is designed to get the most out of the converter. If you're interested, you should familiarize youself with the data sheets included in the appendix of this book.

The DACQ module is designed to reside on the bus of a computer, preferably a fast AT, and communicate over the bus. The core of the unit is the LM12H458—a 12-bit plus sign A/D with eight multiplexed inputs, on-board sample-and-hold, a local 16×32 word FIFO for storing conversions, internal instruction RAM that can control the conversion sequence for up to eight conversions, and watchdog compare capability. The design provides five buffered and filtered inputs, available to the user through a 37-pin D-connector on the rear of the card. Two of the inputs are differential, and three are single ended.

There are three different filter formats on the card. The first is a generalized immitance converter configured as a sixth-order Bessel—it occupies the first two ports and has differential inputs. The next two inputs are single-ended, each with a third-order Sallen and Key filter, and the last is based upon a filter chip using a fourth-order state variable filter.

Finally, a D/A is included for signal creation or reconstruction. It is complete with its own fourth-order filter and an amplifier capable of driving an oscilloscope or speaker.

A full description of the DACQ board itself is enough to fill a good number of pages and really isn't possible within the limitations of this text. The electronics and mathematics involved have been covered in the earlier parts of this chapter, with references to this card. What we offer here is only a brief description of the circuit as represented by the schematics provided. Oscill.exe is available with complete source on the optional disk. It is an oscilloscope program that uses a good number of the techniques discussed in various portions of this book and will not be described in detail here. For those without full source code, the gathering routine and chip select PLD are presented.

### Filters

The DACQ board includes three different filter formats so that you will have the opportunity to see how each works.

### GIC

The schematic in Figure 2-36 is of the one of the two GIC filters on the DACQ. Input is taken from the 37-pin D-sub on the rear of the card. It is differential, so the signal need not share a common ground with the circuit. The input buffer supplies a gain of 2. The filter is in two third-order sections, each comprising a second-order GIC section and a simple pole buffered by an opamp. These two are cascaded as one would two LC sections. The cutoff of this filter is set to 20KHz, making it suitable for the lower sampling rates that a multiplexed circuit might require.

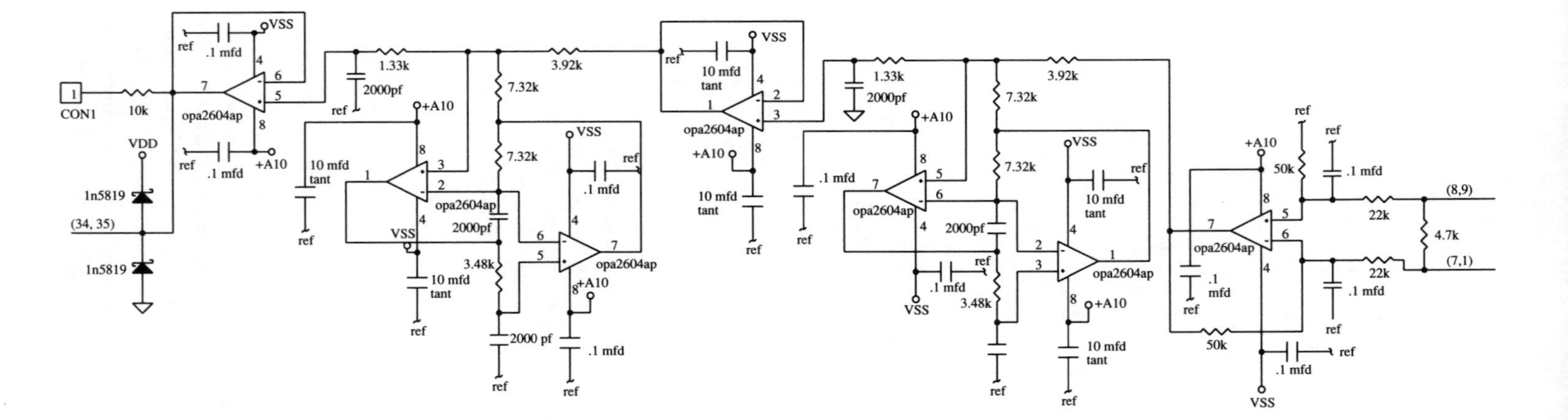


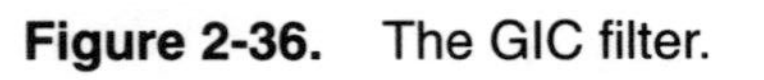
**Figure 2-36.** The GIC filter.

### Sallen and Key

The next two filter sections are classic Sallen and Key, configured as shown in Figure 2-37.

Both of these filters are single-ended, and share a common reference with the GIC filters mentioned previously. Again, the input is taken from the D37 connector through the 10 ohm resistor to the first amplifier, which is only a unity gain buffer. The next opamp supplies any gain the circuit needs and is followed by the third-order Sallen and Key filter. Exercise care when using single-ended inputs, as they are more susceptible to noise. Because the system we sample with this circuit must share a common ground with the card, it will have far less common mode immunity, and will pick up any noise associated with the ground of the system being sampled. The frequency cutoff of this section is 6 KHz.

### Continuous-Time Filter IC

The front end of this filter is a unity gain buffer, which is followed directly by a gain stage. The input comes from the same D37 mentioned before. This actual implementation of the filter section is very similar to the Sallen and Key filter above, except that it is implemented in a single IC. The filter itself is a modified state variable filter with the capacitors embedded on the chip.

This circuit is configured so that the filter chip contributes no offset to the signal from the gain stage. This is accomplished by driving both the A/D and the filter directly from the gain stage, but the output of the filter is AC coupled to the input of the A/D through a lowpass filter set much lower than the cutoff of the filter. Used in this fashion, in band frequencies will be passed to the A/D because the same signal will appear at the output of the filter, as at the gain stage disabling the lowpass filter. The filter will not pass frequencies above cutoff and, consequently, biases the lowpass filter on. In Figure 2-38, the lowpass filter is formed by the 470 ohm resistor coming from the gain stage in conjunction with the 1mfd capacitor coming from the filter. This filter is compact and relatively easy to use. The cutoff frequency for this section is 6KHz.

### LM12H458

The LM12H458 A/D converter is a high-performance system very nicely suited to DSP applications. The specifications on this device are excellent with a high SNR in all modes, low THD, and integral linearity error.

The unit is fully configurable and can be programmed to execute any necessary sequence of conversions, store the results locally, and interrupt the host processor when the stack is full. In addition, the instruction sequence is RAM-based and can be changed on the fly.

### Digital-to-Analog Converter

A 12-bit digital-to-analog converter is also included on the card to allow you to reconstruct signals you have recorded, created, or modified, either to play over a speaker or to display on an oscilloscope.

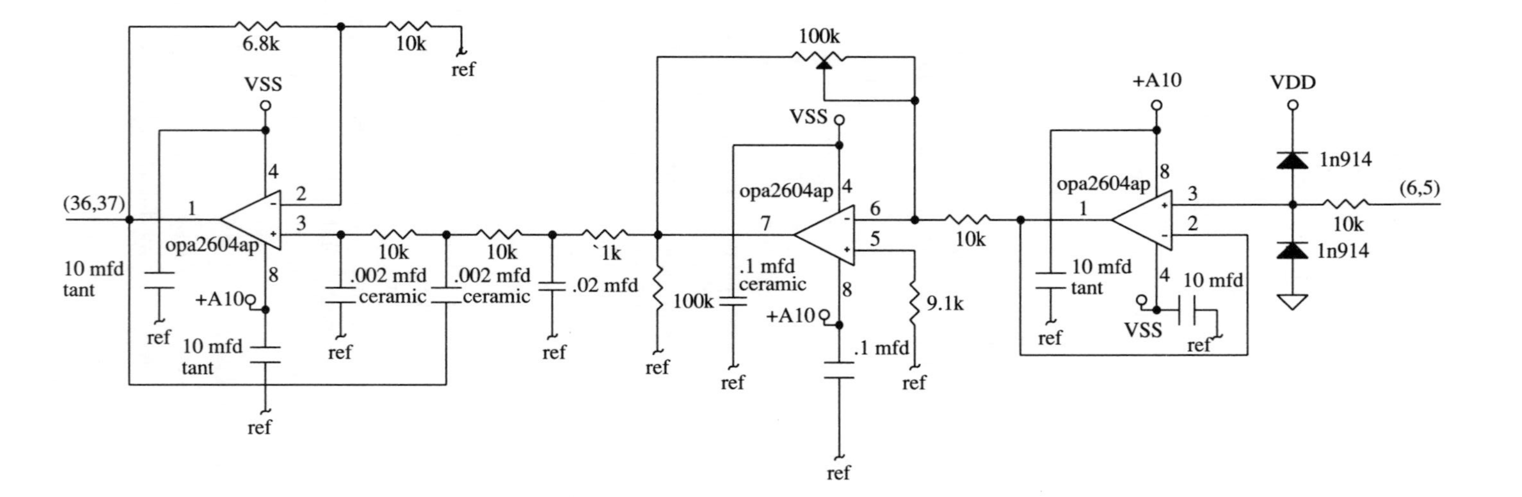

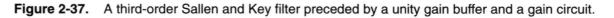

**Figure 2-37.** A third-order Sallen and Key filter preceded by a unity gain buffer and a gain circuit.

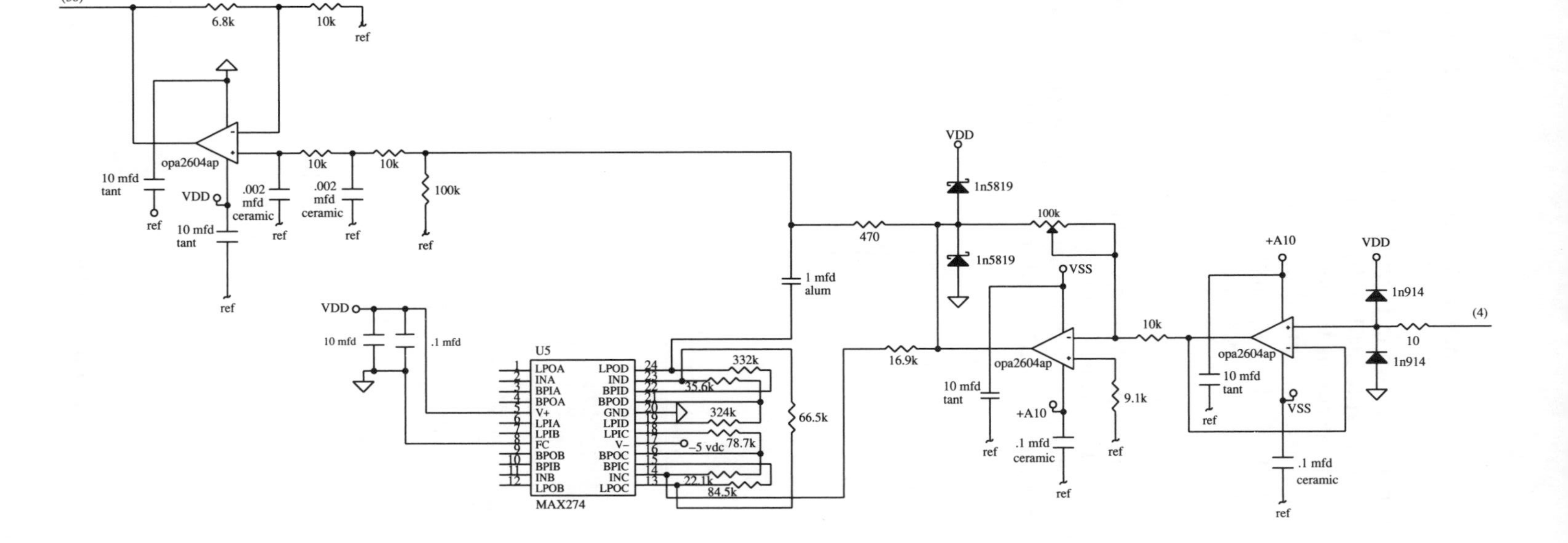


Figure 2-38. Buffer and preamplifier for MAX274.

The converter, as shown in Figure 2-39, is controlled by three lines coming from a PLD. Essentially, you write the low byte and then the high byte, whereupon the device latches the value and performs the conversion. It is followed by one-half of the Max274 continuous-time filter, configured as a fourth-order lowpass in the same manner as described previously. We follow that with a third-order Sallen and Key filter driving an audio amplifier. This output may be used to drive a speaker or an oscilloscope.

## Oscill.exe

Oscill.exe is an oscilloscope program capable of manipulating and displaying data gathered with the DACQ card or created with another program, such as Mathcad. It can read data files from the disk and store on disk what it has gathered or produced. It was designed primarily to interface with the DACQ card and acquire data. You need not have this card to use the program.

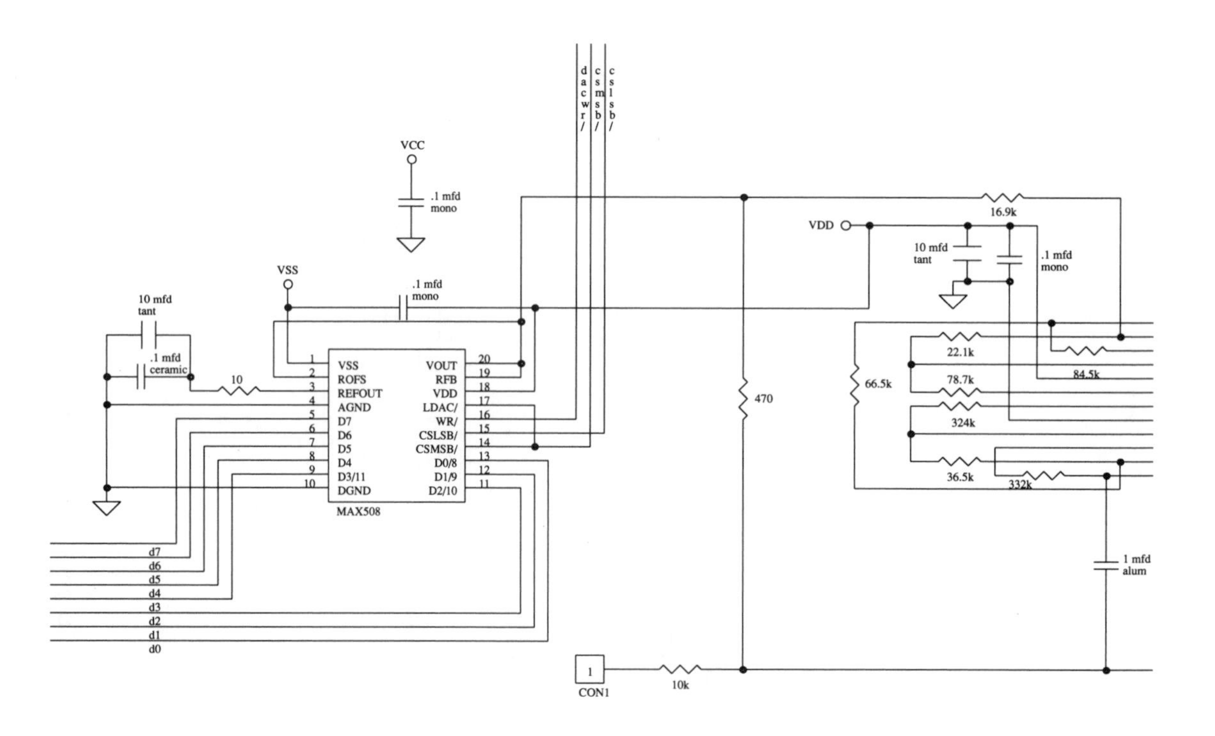

**Figure 2-39.** Reconstruction filter and amplifier.

In addition to data gathering, the program is capable of creating data records of square, sinusoidal, ramp, and exponential waveforms. These may be saved for export to Mathcad or any other program, and be reread at a later time. Arithmetic, such as addition, subtraction, multiplication, and division may be performed upon any two or three waveforms at once. In addition, one waveform may be used to modulate another, and any one of three displayed waveforms may have an FFT performed upon it.

This program is composed of two source files and two header files. These are OSCILL.CPP, DAS.H, GATHER.ASM, GATHER.INC, and they are provided in their entirety on the optional disk. This program contains many examples of the technology discussed in this book. The reader is encouraged to use it, enhance it, and make it his/her own.

### The Gather Routine

Of crucial importance to the software is the data gathering routine. We present it here as reference for those wishing to learn how to use the LM12H458. It must be fast and versatile;

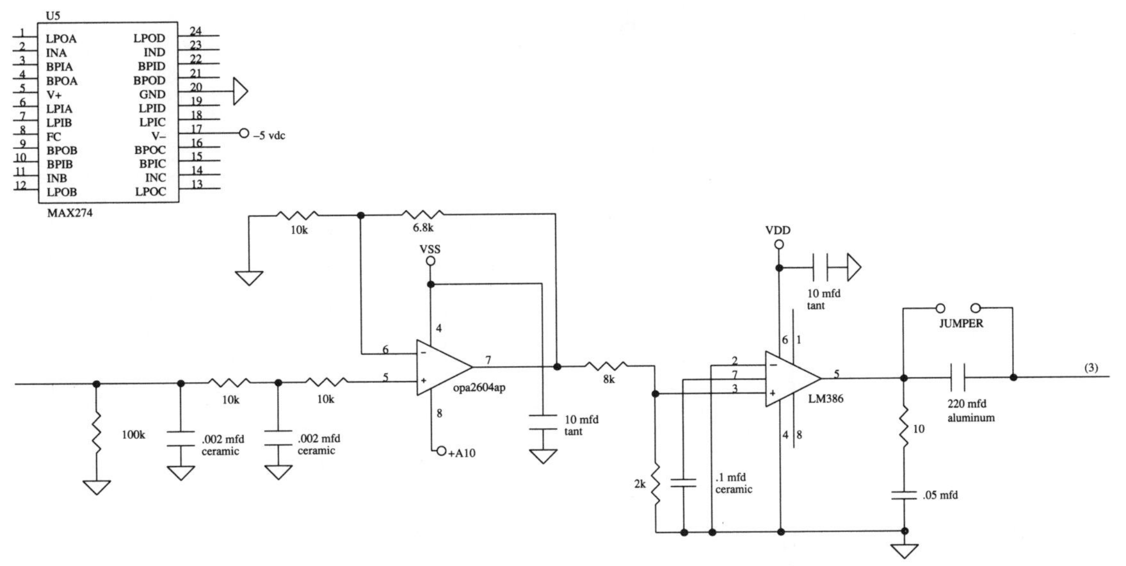

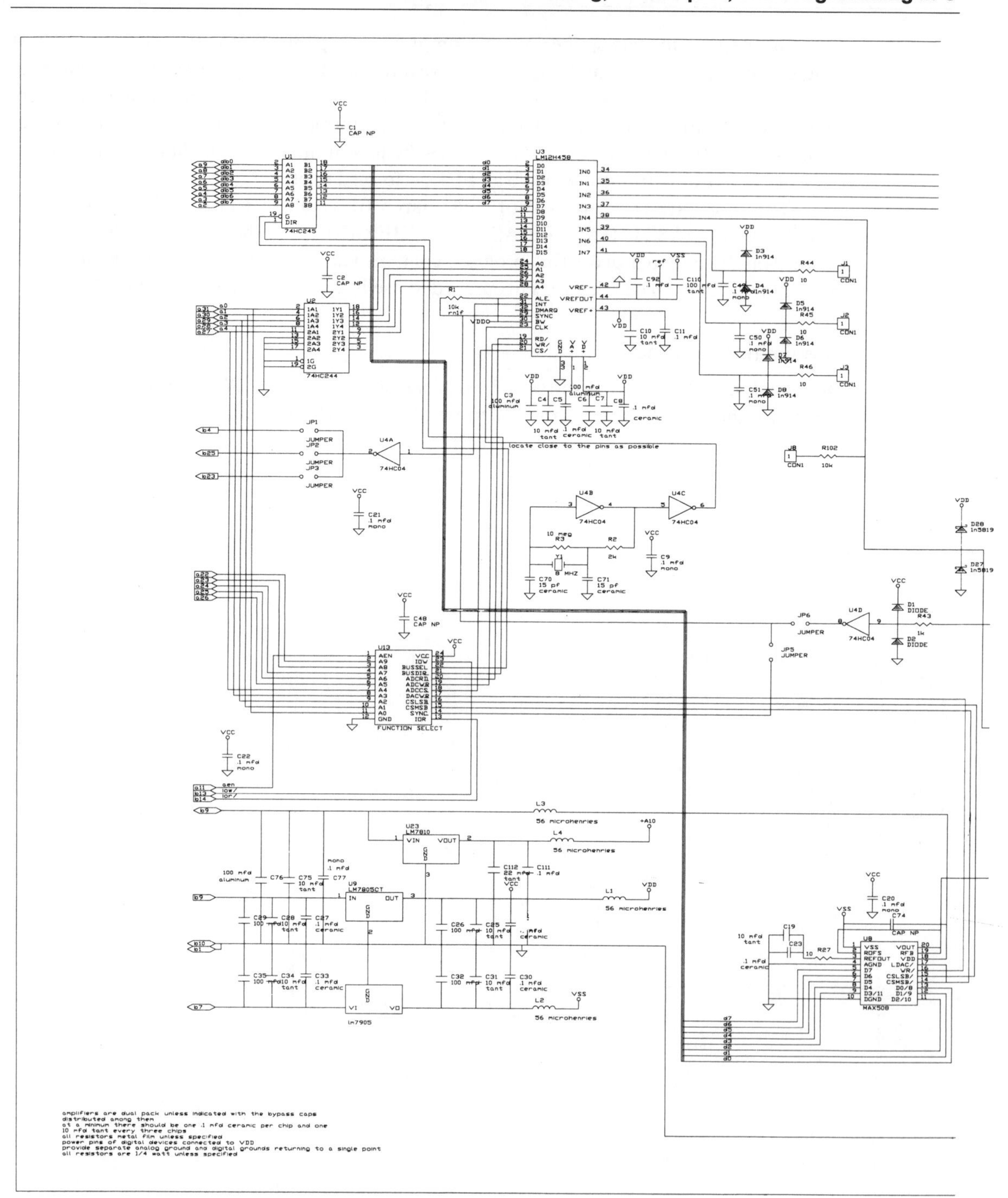


**Figure 2-40.** Complete schematic for DACQ unit.

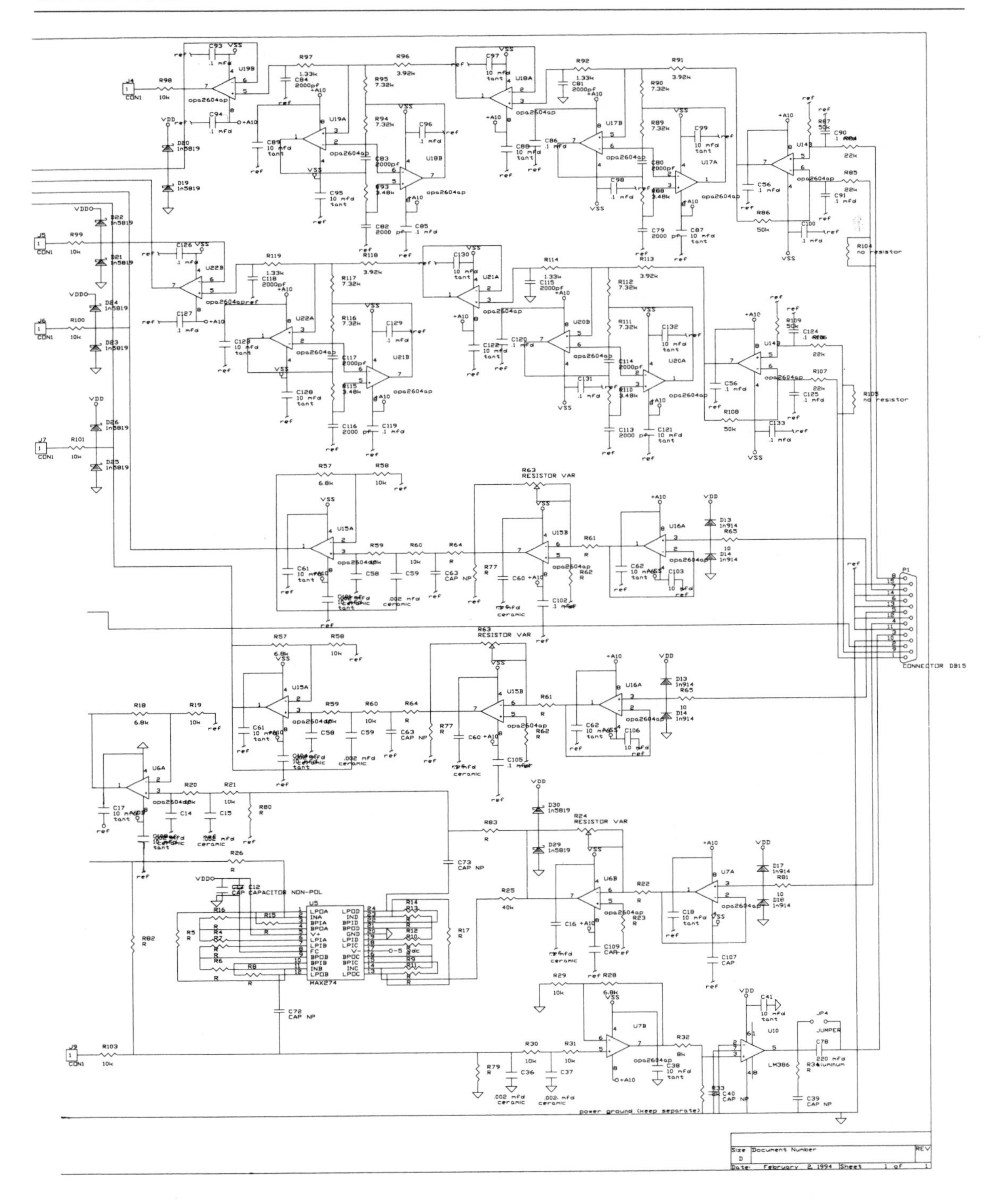

the following routine is designed for that purpose. It is fast enough to serve the data acquisition unit at full speed, and it is capable of triggering on a specific edge and value (although in Oscill.exe it is used only to gather).

```
FIFO            EQU         000000158h
INTSTATUS1      EQU         000000155h

gather proc far c uses ds es eax ebx ecx edx,
        vector:dword, num:word, trigger:word, polarity:word,
        value:word

    local   sample:word, stride:word

    mov     bx, num         ;samples
    mov     stride, bx
    inc     bx              ;n+1
    les     si, vector
    sub     si, 4           ;offset point back for first increment
    mov     sample, 0h

chek_stat:
    mov     dx, INTSTATUS1 ;point at interrupt status register
    in      al, dx
    and     al, 0f8h        ;see if there is anything at all there
    jz      chek_stat

    shr     al, 3           ;get rid of three lower bits
    xor     cx, cx          ;clear cx
    mov     cl, al
    mov     dx, FIFO        ;address of lsbs of data

get_data:
    in      al, dx
    xchg    al, ah
    inc     dx              ;point at upper bits
    in      al, dx          ;get both
    xchg    al, ah
    test    trigger, 1      ;see if trigger is still lit
    jz      went_on

    cmp     polarity, 0     ;positive or negative
    jl      negative
positive:
    cmp     sample, ax      ;got to be greater than the last sample
    jg      failed_trigger
    cmp     ax, value       ;compare current value with test
    jge     go_on
    mov     sample, ax
    jmp     failed_trigger
negative:
    cmp     sample, ax      ;got to be less than the last sample
```

```
    jl      failed_trigger
    cmp     ax, value        ;compare current value with test
    jle     go_on
    mov     sample, ax
    jmp     failed_trigger
go_on:
    mov     trigger, 0       ;clear the trigger
went_on:
    add     si, 4            ;adjust array pointer
    mov     sample, ax
    fild    sample
    fstp    dword ptr es:[si]
    dec     num              ;check count
    jz      got_data
failed_trigger:
    dec     dx               ;point at lower bits
    dec     cx
    jcxz    ck               ;wait for next
    jmp     get_data         ;continue
ck:
    jmp     chek_stat
got_data:
    ret
gather endp
```

## Chip Select

What we offer here is the Cupl source for the chip select PLD used on the DACQ card. Of course, if you are building the card on your own, you are free to choose your own addressing scheme within the constraints of the computer hardware you are using. Nevertheless, it may serve as an illustration and starting point.

```
/*22v10*/

/** Inputs **/

pin [1..10] = [a0..9] ;          /* system addresses a0 - a9      */
pin 11      = ior ;          /* io read strobe            */
pin 13      = aen  ;         /* address enable            */
pin 23      = iow ;      /* io write              */

/** Outputs **/

pin 14      = dir ;          /* 245 direction      */
pin 15      = g ;            /* 245 enable         */
pin 16      = sync ;         /* external sync      */
pin 17      = 458wr ;        /* 458 write strobe   */
pin 18      = 458cs ;        /* 468 chip select    */
pin 19      = 458rd ;        /* 458 read strobe    */
pin 20      = 508wr;         /* 508 write strobe   */
```

```
pin 21     = cslsb;              /* 508 lower byte      */
pin 22     = csmsb;              /* 508 higher byte     */

/** Declarations and Intermediate Variable Definitions      **/

field  ioaddr    = [a0..9] ;
458_eqn          = ioaddr:[120..13f] ;
/*******************************/
508_eqn          = ioaddr:[140..141] ;  /*         I/O Address       */
sync_eqn     = ioaddr:[148] ;      /*        Ranges          */
                    /*******************************/

/** Logic Equations **/

dir     = (!458_eqn # ior) # aen; /*switching polarity of input*/
g       = 458cs & 508wr;
458wr   = (!458_eqn # iow) # aen;
458cs   = !458_eqn # aen;
458rd   = (!458_eqn # ior) # aen;
508wr   = (!508_eqn # iow) # aen;
cslsb   = 508wr # a0 ;
csmsb   = 508wr # !a0 ;
sync    = !sync_eqn ;
```

# 3

# Signal Analysis

## INTRODUCTION

The time and frequency domains complement one another throughout the physical universe. Analysis in the time domain yields certain information, and the same signal or function transformed into the frequency domain reveals even more information—all about the same signal. In radar, for instance, distance to an object is determined from the time it takes for the echo to return to the transmitter from the object, while the Doppler shift (frequency shift) yields the speed of the object. Spectral analysis is used throughout science and engineering for examination, exploration, and heuristics. The uses range from searching for extraterrestrial life to qualifying the bandwidth of operational opamps to generating population statistics. It has both theoretical and practical uses, and ranks among the most important discoveries of science and mathematics.

Applications for the Fourier transform are endless. It is useful in statistics for determining data trends and the validity of the data at hand. Medical researchers are using it to search for correlations between symptoms and diseases. Seismologists have studied earthquake patterns with spectral analysis for many years in an effort to find secrets that will aid them with their predictions. It is used in robot vision for parts recognition and placement; and, of course, the movie industry uses it for special effects. The mechanisms described here may provide some of the most elegant and direct explanations for physical phenomena ever.

In this chapter, we return to the Fourier series and use it to explain the Fourier transform and some of its relatives. We will explore the similarities between the Fourier series and the Fourier transform and use both for signal analysis. There are several Mathcad documents on the optional disk that can be of assistance in understanding the material here by providing additional illustrations and examples. Please refer to those documents while reading the material in this chapter.

Software created for use with the Mathcad documents and the DACQ data acquisition board is also available on the disk. This software and its source are demonstrations of the implementation of much of what is written here. The software on the disk is written in such a fashion that it will not only work with the DACQ card, but will also read data files in the Mathcad format and output them in that same format. This makes it quite easy to create a signal vector in one program for analysis in another. There are also executables for displaying plots of multiple functions, capturing data from the DACQ data acquisiton unit, and computing coefficients for windowing functions. Please refer to the README document included on the disk for more information and instructions on the use of each program.

You should familiarize yourself with the sections on harmonic analysis, decomposition, and synthesis before leaving this chapter. If you are not interested in software design or development, you can skip the section entitled "Software Examples," although some of the concepts pertinent to the Fourier transform are more deeply developed there and may be of interest. Also, the theorems of the Fourier transforms are illustrated in the derivation of the code.

There are examples of some of the code used in the programs on the optional disk. These examples include several different forms of the FFT, the Hartley transform, and a routine for defining variable length windowing functions.

Let us begin with some definitions and clarifications.

## Analysis and Synthesis

Both the Fourier transform and the Fourier series have two forms—a forward and a reverse transform, if you will. One is *analytic*—this is the form that generates the Fourier coefficients for the series and the spectrum of the transform. This form moves us from the time to the frequency domain. The *synthesis* form moves from the frequency domain to the time domain and uses the coefficients or spectrum to generate the function or waveform.

Though we will cover both analysis and synthesis in this chapter, the chief concern is with the development of an analytic set of tools for use in decomposing waveforms and preparing them for further handling.

## Time Domain Response

Generally speaking, we are most experienced in the time domain. The features we study on oscilloscopes, charts, and graphs are often related to time. To understand the complementary nature of the two domains, it is important that we be aware of the particular characteristics of each. There are two time domain properties of practical interest: impulse response and step response.

The response of a system at rest to a unit impulse $\delta(t)$ is called an *impulse response*. If a signal can be understood as a sequence of closely spaced and infinitely narrow impulses that vary only in magnitude, the response of a system to an input signal would be the continuous convolution of the impulse response with the magnitude of each of the input impulses. Carrying this a step further, if we can say that the response of a system to a continuous-time signal is an integral:

$$y(t) = \int_{-\infty}^{\infty} x(t)h(t-\tau)dt$$

where $x(t)$ is the continuous-time signal and $h(t-\tau)$ is the impulse response of the system. This is the superposition integral, which says that $y(t)$ is equal to the continuous convolution of the impulse response and input data.

The *step response* is the output of a system at rest when a unit step is applied. The unit step, as discussed in Chapter 1, is the unit impulse integrated, and so the step response of a system can be obtained by integrating the impulse response. The step response usually is given in terms of the following specifications: steady-state accuracy, rise time, overshoot, and settling time.

*Steady state error* is the deviation from the expected or desired response demonstrated by the system. *Rise time* is defined as the time it takes for the system to reach 90 percent of its final value, it is specified in terms of the damping ratio, $\varsigma$, and the natural frequency of the system, $\omega_0$:

$$t_r = \frac{\pi}{2\omega_n\sqrt{1-\varsigma^2}}$$

*Settling time* is the time it takes for the system to reach and remain within a specified band of the final value:

$$t_r = \frac{4.6}{\varsigma\omega_0}$$

Finally, *overshoot* is the maximum deviation in percentage of the system response from the projected or desire value:

$$overshoot = 100e^{\left(-\frac{\varsigma\pi}{\sqrt{1-\varsigma^2}}\right)}$$

## Frequency Domain Response

In the frequency domain, we have three critical reponses: *magnitude*, *steady state*, and *phase*. The magnitude response is found by solving for $|H(j\omega)|$, the root of the squared magnitude response:

$$|H(j\omega)|^2 = \mathrm{Re}[H(j\omega)]^2 + \mathrm{Im}[H(j\omega)]^2$$

$$|H(j\omega)| = \sqrt{\mathrm{Re}[H(j\omega)]^2 + \mathrm{Im}[H(j\omega)]^2}$$

The phase response, as you will recall, is the ratio of the imaginary to real parts:

$$\theta(\omega) = \tan^{-1}\left(\frac{\text{Im}[H(j\omega)]}{\text{Re}[H(j\omega)]}\right)$$

The steady state response can be obtained by evaluating the transfer function

$$H(j\omega) = |H(j\omega)|e^{j\theta\omega}$$

Also of interest in the frequency domain is the time delay for any particular frequency passing through the system—this is called the phase delay:

$$t_p = -\frac{\theta(\omega)}{\omega}$$

Following from that, we will want to know the group delay of the system—the time it takes for an impulse at the input to arrive at the output:

$$t_{grpdly} = -\frac{d\theta(\omega)}{d\omega}$$

These formulae are of great importance in the design of digital filters and the analysis of digital systems, as you will see.

## SPECTRA

We have been using the expression $H(j\omega)$ for frequency response. When it is used to represent a system function, it stands for the Fourier transform of the impulse response and represents the ratio of frequency response of the output to the frequency response of the input. Ultimately, it tells us about the frequencies involved in producing the impulse response. The frequency response of a time domain function is obtained via the Fourier transform, and results in a spectrum that is a representation of the various contributions of the component frequencies of a waveform. This spectrum can be either *continuous*, as the continuous spread of colors in a rainbow, or discrete, as a *line spectrum* with only specific elements manifest. Sunlight through a prism produces a continuous spectrum, whereas xenon or mercury lamps that have a high concentration of color at certain frequencies produce a line spectrum showing those colors. A signal's spectrum is derived from the inverse of the system's independent variable, which becomes the dependent variable in the domain of the spectrum. Since we are dealing with the time domain in this book, the inverse variable becomes 1/*time*, or frequency. That does not limit the implications of this discussion to these two domains only; the same techniques are used with other independent variables, as well.

All physical phenomena have spectra, or Fourier transforms. It is difficult to imagine a sound without component vibrations or an electrical wave without component frequencies. It is natural for us to think of elements, objects, and even ideas as compositions of other elements, objects and ideas in varying magnitudes. That is the physical universe—if a signal is physically possible, it will have a spectrum.

Theoretically and mathematically, however, it is perfectly possible to create a symbol for a phenomenon that has no spectrum. An oscillator producing a tone at 440 Hertz is producing a harmonic signal *with* a spectrum, but sin$t$ has no spectrum, and neither does the unit impulse or the unit step.

To have a continuous spectrum,

1. The function must be integrable ($-\infty$ to $\infty$).
2. The function must have only a finite number of discontinuities, if any.

Periodic, impulse, and step functions can violate these two criteria, either by not converging for all $s$ or by having infinite discontinuities. Therefore, the spectra they are usually given are line spectra, with coefficients derived from their equivalent Fourier series expansion. Other ways have been devised to include these signals among the ones that converge by multiplying by factors (say, $e^{-ax^2}$) that do converge and then taking the transform as this factor approaches its limit. These functions are called *transforms in the limit.*

Suffice it to say that the signals we will be dealing with in this book are of the physical universe and do have spectra, and though we will also be using mathematical representations of these signals occasionally, we will not concern ourselves with the question of their convergence.

The orthogonality of the sine and cosine functions produces the spectrum. The dot product of a sinusoid with any other sinusoid of a different, nonharmonic frequency will result in zero. To simplify, the spectrum we derive with the Fourier transform is the limit of the sum of the dot product of the input signal with all frequencies. This will result in nonzero magnitudes only when a component of the input signal is identical to the one of the frequencies we are multiplying by. This goes both ways: if we were to run the spectrum we got from a Fourier transform back through the transform, we would again have the input signal.

## FOURIER SERIES

Let us approach the spectrum through the Fourier series and then through the Fourier transform, which is simply the extension of the Fourier series to the infinite interval. In Chapter 1, in the section entitled "Harmonic Analysis," we introduced the general equation for a sinusoid:

$$f(t) = A\sin\omega t + B\cos\omega t \qquad \text{(Eq. 3-1)}$$

or, in exponential notation,

$$f(t) = e^{\pm j\omega t}$$

At that time, we also introduced a trigonometric series known as the Fourier series, which says that a periodic signal is representable by a sum of a fundamental frequency and its harmonics:

$$\frac{1}{2}A_0 + \sum_{n=1}^{\infty}\left[A_n\cos(n\omega_0 t) + B_n\sin(n\omega_0 t)\right] \qquad \text{(Eq. 3-2)}$$

or, in exponential notation,

$$\sum_{n=-\infty}^{\infty}\left(c_n e^{jn\omega x}\right)$$

Let us refer to the trigonometric version of the Fourier series above, at least at first, because it is so illustrative of what we are doing. We can see that the equation for a sinusoid (Eq. 3-1) is embedded in Equation 3-2, indicating that it is a sum of an integral number of sinusoids, with magnitude and phase determined by the coefficients $A_n$ and $B_n$. The formula indicates that it is an infinite sum, in which case we would expect it to be a very close approximation to the actual waveform. In practice, of course, this is not the case—the sum must be truncated at some point, so any waveform synthesized in this manner can only come close. An exact representation would require the contribution of *every* frequency, not just a truncated sum of an integral number.

Since a continuous spectrum is only obtainable by including the contributions of all frequencies, the Fourier series will produce only a line spectrum. This, however, does not prevent efficient analysis and synthesis. A periodic signal can be expanded in a Fourier series to obtain the coefficients which may then be used in analysis or modified and used to compose a new waveform. For example, we could create a simple filter that removed certain frequencies by taking the spectra of an input signal and simply subtracting any component of the frequency we wish to remove. By the same token, you can easily modify a signal by inserting a component or changing the magnitude of an existing component. To illustrate the analysis/synthesis relationship of the Fourier series and a function, begin with the function $f(x)=[1+M\cos(3x)]\cos(9x)+2\cos(13x)$. This equation is a modulated waveform with the addition of noise. It expresses the effect of imposing a waveform $\cos(3x)$ on a carrier $\cos(9x)$ plus noise $\cos(13x)$. In this equation, $x$ is the frequency $\omega_0$, and $M$ is the depth of the modulation of the carrier by the signal $\cos(3x)$. See Figure 3-1 for a plot of this function.

We can immediately derive the Fourier coefficients of the function by selecting a depth for the analysis—in this case, we will take it to 15 harmonics plus the fundamental—and then solving the two integrals for each of the harmonics $n$:

$$A_n=\frac{1}{\pi}\int_0^{2\pi} f(x)\cos(nx)dx$$

and

$$B_n=\frac{1}{\pi}\int_0^{2\pi} f(x)\sin(nx)dx$$

The numerical results are shown in Table 3-1.

You will notice that the $B_n$ coefficients contribute nothing. Values in Table 3-1 that are other than zero are the result of roundoff error in the floating point operations. That is because the function we have chosen to evaluate is based upon the cosine and is therefore *even* (more on this in the next section). To shorten our calculations, we could eliminate the

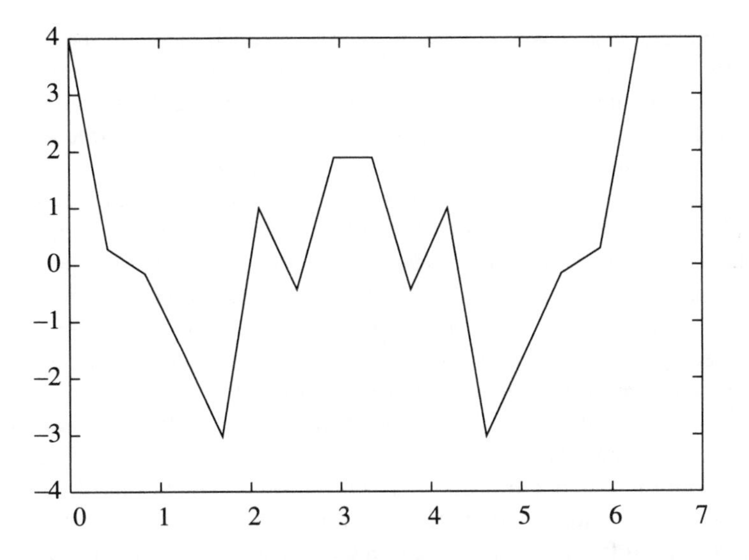


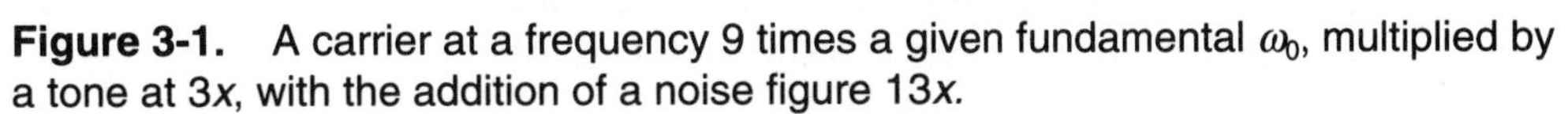
**Figure 3-1.** A carrier at a frequency 9 times a given fundamental $\omega_0$, multiplied by a tone at $3x$, with the addition of a noise figure $13x$.

| $A_n$ | $B_n$ |
|---|---|
| $4.596 \cdot 10^{-13}$ | $0$ |
| $-9.763 \cdot 10^{-10}$ | $-7.771 \cdot 10^{-15}$ |
| $1.012 \cdot 10^{-12}$ | $-1.138 \cdot 10^{-14}$ |
| $-9.802 \cdot 10^{-10}$ | $-1.283 \cdot 10^{-14}$ |
| $9.362 \cdot 10^{-13}$ | $-3.416 \cdot 10^{-15}$ |
| $0.5$ | $-4.87 \cdot 10^{-15}$ |
| $5.327 \cdot 10^{-13}$ | $-5.469 \cdot 10^{-15}$ |
| $4.569 \cdot 10^{-13}$ | $-5.499 \cdot 10^{-15}$ |
| $1$ | $-3.709 \cdot 10^{-15}$ |
| $-9.804 \cdot 10^{-10}$ | $-2.886 \cdot 10^{-15}$ |
| $9.354 \cdot 10^{-13}$ | $-4.597 \cdot 10^{-15}$ |
| $0.5$ | $0$ |
| $2$ | $0$ |
| $-9.763 \cdot 10^{-10}$ | $-1.389 \cdot 10^{-15}$ |
| $4.592 \cdot 10^{-13}$ | $1.958 \cdot 10^{-15}$ |

**Table 3-1.**

$B_n$ coefficients entirely, just as we could eliminate the $A_n$ coefficients in the calculations of an odd function. We know that the coefficients we have derived represent the function, because we can reconstruct it by evaluating the series using the values in Table 3-1.

Now we can plot the coefficients in Table 3-1 relative to the 15 harmonics we used to expand the series in. In this plot, shown in Figure 3-2, the abscissa is an increasing series of integral harmonics of the fundamental frequency, and the ordinate is magnitude. This spectrum makes it very clear what the relative contributions of the different harmonics/frequencies are.

If we choose now to remove the unwanted addition represented by the 13th multiple of the fundamental ($13x$), we could do so by expanding a series based upon that harmonic and subtracting it from the series we created from $f(x)$, as shown in Figure 3-3.

The Fourier series is an infinite sum taken over an interval of 0 to $2\pi$. The Fourier transform is the application of that same technology to an interval spanning $-\infty$ to $\infty$ and taken at infinitesimal increments of the independent variable so that the sum becomes an integral. In the sections that come, you will see that they have a great deal in common with each other and with the Laplace transform. It will become evident that the Fourier series periodic sequence corresponds to the Fourier transform finite-length sequence. We construct periodic sequences from finite-length data by repetition, and we construct finite-length data from periodic functions by a process known as *sampling*.

## Odd and Even

As you may have noticed in the example above, the symmetry of the functions we are dealing with can simplify our work greatly if we know how to use it. Figure 3-4 provides examples of even and odd waveforms and their sum.

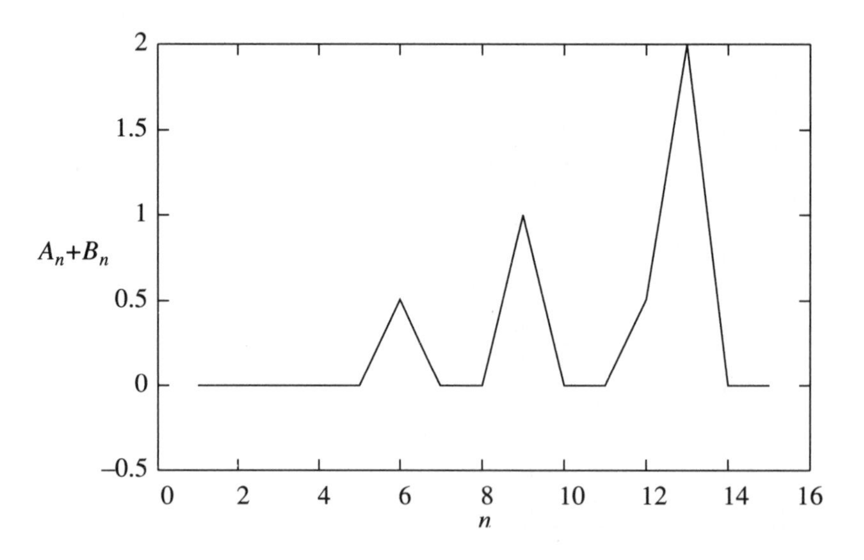


**Figure 3-2.** Here, the magnitudes of the Fourier coefficients are plotted against the harmonics used to expand the series.

$$g(x) := (1 + M\cdot\cos(x\cdot 3))\cdot\cos(9\cdot x) + 2\cdot\cos(x\cdot 13)$$

$$v(x) := 2\cdot\cos(x\cdot 13)$$

A

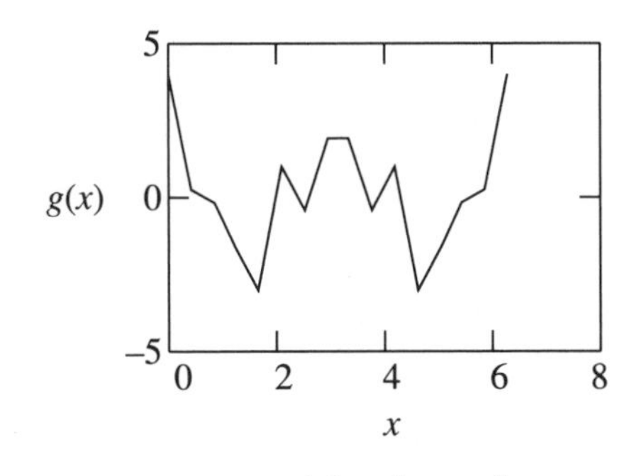


modulated waveform

B

$$\left(\frac{A_0}{2} + \left[\sum_{n=1}^{harmonics} A_n\cdot\cos(n\cdot x) + B_n\cdot\sin(n\cdot x)\right]\right) - \left(\frac{C_0}{2} + \sum_{n=1}^{harmonics} C_n\cdot\cos(n\cdot x)\right)$$

series expression of modulated waveform thirteenth harmonic

**Figure 3-3.** In part A we have the unfiltered waveform *g*(*x*). Part B is a *g*(*x*) with the thirteenth harmonic *v*(*x*) removed.

Basically, a function for which $f(x)=f(-x)$ is true is even, as is the cosine wave in part A of Figure 3-4. A function with the relationship $-f(x)=f(-x)$ is odd; an example of such a function is the sine wave in part B of Figure 3-4. Even functions are also referred to as symmetrical, and odd functions as antisymmetrical. Even and odd functions obey the following rules:

1. $f_{even} + f_{even} = f_{even}$
2. $f_{odd} + f_{odd} = f_{odd}$
3. $f_{odd} \times f_{odd} = f_{even}$
4. $f_{even} \times f_{even} = f_{even}$
5. $f_{odd} \times f_{even} = f_{odd}$

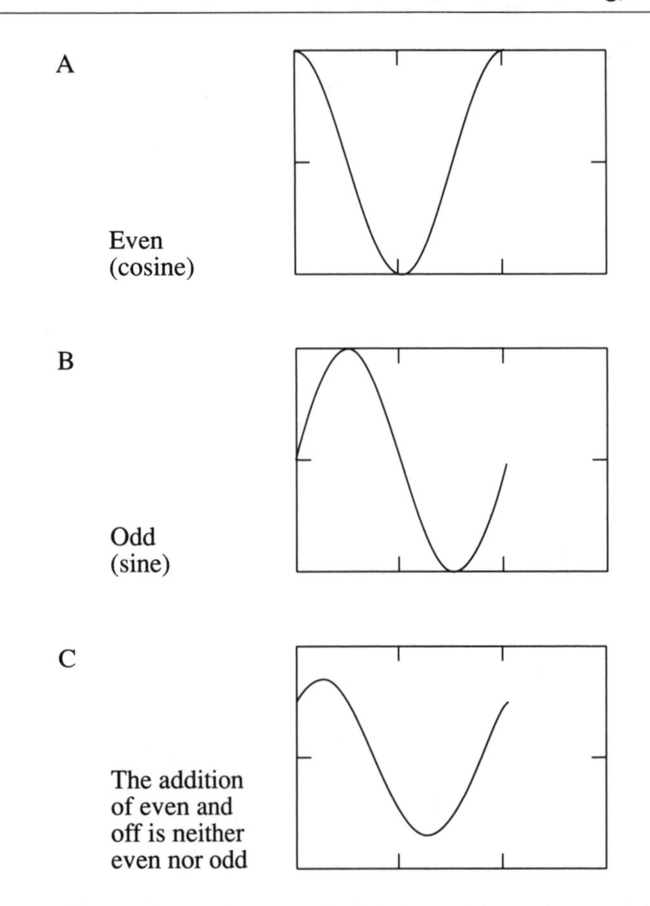


**Figure 3-4.** Even and odd sinusoids and a sum that is neither even or odd. Part A is an example of an even function, and part B is odd. Part C is their sum, which is neither even nor odd

As you can see from Figure 3-4, the sum of even and odd produces neither.

It is a simple and useful fact that any function can be written in terms of its even and odd parts:

$$f(x) = f_{even}(x) + f_{odd}(x)$$

Both even and odd parts represent the mean of the function with reflections in the vertical axis, positive (even) and negative (odd). The even part of the function is

$$f_{even}(x) = \frac{1}{2}[f(x) + f(-x)]$$

and the odd part is

$$f_{odd}(x) = \frac{1}{2}[f(x) - f(-x)]$$

What is even and what is odd can change with a change in the origin. This is most evident with functions, such as sine and cosine, that can change from fully even to fully odd as the origin changes. This quality is directly reflected both in the Fourier transform of the function and in a Fourier series expansion. If the function is even, its transform and expansion are even; if it is odd, so are its transform and expansion. We can rewrite the Fourier transform with even and odd parts:

$$F(\omega) = 2\int_0^\infty f_{even}(t)\cos(\omega t)dt - 2j\int_0^\infty f_{odd}(t)\sin(\omega t)dt \qquad \text{(Eq. 3-3)}$$

just as we show the Fourier series:

$$f(x) = \sum_{n=0}^{\infty} A_{k_n}\cos(kx) + B_{k_n}\sin(kx)$$

If a signal is purely even or purely odd, only half needs to be computed, since the other half will sum to zero anyway. You will find this concept reflected throughout this study.

## FOURIER TRANSFORM

In an effort to understand the relationship between the Fourier series and Fourier transform, let us investigate the impulse function and its spectrum. You will recall from earlier discussions that the number of spectral lines contained within the spectrum of an impulse (a sinc envelope) is proportional to the duration of the impulse and its frequency. A rectangular pulse is represented by the following sum

$$f(t) = \sum_{n=-\infty}^{\infty} \frac{1}{k} \frac{\sin\left(\frac{nt\pi}{T}\right)}{\frac{t\pi}{T}} e^{jn\omega t}$$

You will notice the sin*c* or $\sin(x\pi)/x\pi$ function in the expression. The sinc envelope (see Figure 3-5) is the form of the spectrum assumed by a rectangular pulse train of any frequency or duration.

One could drop a line from each point of this envelope to the horizontal axis to represent the magnitudes of the harmonics, or integer multiples of the fundamental frequency, that comprise the impulse. If you study the series expansion, you will see that the ratio of the width to the period has a direct effect upon the number of spectral lines contained within the envelope. As the duration of the pulse diminishes *or* the period increases, the number of lines within the envelope increases. Each time $nt$ is equal to $T$ or an integer multiple thereof, the harmonic represented by $n$ and all of its multiples vanish. These are the zeros of the sinc function. If we were to increase the period of that impulse function to infinity—as would happen if it were not periodic—we would approach a point at which the

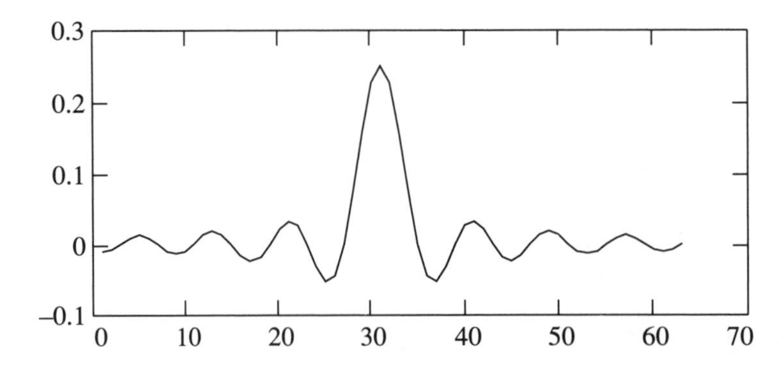


**Figure 3-5.** The magnitude spectrum of an impulse which turns out to be a sinc (sin*x*/*x*) function. The envelope represents the magnitudes of the Fourier coefficients of the the harmonics of the fundamental frequency comprising the impulse.

distance between each line is infinitesimally small and all frequencies are contained within the envelope. As the period of the pulse train lengthens or the pulse becomes narrower—depending upon your viewpoint—more frequency components of a smaller amplitude are added to the line spectrum, but at a smaller amplitude, resulting in a proportionately smaller envelope.

In the end, the measure between each of the harmonics becomes smaller and smaller, finally becoming so small that what we have is no longer a sum but an integral:

$$f(t) = \int_{-\infty}^{\infty} \left[ \frac{1}{2\pi} \int_{-\infty}^{\infty} f(t) e^{-j\omega t} dt \right] e^{j\omega t} d\omega$$

which is nothing more than the equation for the Fourier coefficients substituted into the equation for the Fourier series, with the frequency deltas made vanishingly small. This equation is a form of the Fourier integral. The Fourier transform,

$$F(\omega) = \int_{-\infty}^{\infty} f(t) e^{-j\omega t} dt$$

and the inverse Fourier transform,

$$f(x) = \int_{-\infty}^{\infty} F(\omega) e^{j\omega t} d\omega$$

are a transform pair.[1] The first equation is analytic, producing the spectrum, and the second is a synthesis equation, reconstructing the function.

This is also how the Fourier integral devolves from the Fourier series. Because it is an integral, spanning and including all frequencies, it is possible to use it to analyze aperiodic phenomena.

The Fourier transform will produce a valid spectrum based upon a sample stream in the same way it would with a continuous-time signal, though in this form it is usually written as a sum and called the *Discrete Fourier transform*, with an analytic form given by

$$F[v] = \frac{1}{N}\sum_{n=0}^{N-1} f[t]e^{-2j\pi\left(\frac{vt}{N}\right)}$$

and its synthetic form given by

$$F[t] = \frac{1}{N}\sum_{v=0}^{N-1} f[v]e^{2j\pi\left(\frac{vt}{N}\right)}$$

The higher the bandwidths of interest, the smaller the deltas must be to gather enough information about the signal to analyze and reconstruct it. This brings us to a very special aspect of digital signal processing—how to gather the data. Here again, we meet the sinc, or impulse function.

## Sine and Cosine Transforms

Two obvious consequences of the data on odd and even functions are the sine and cosine transforms. The cosine transform is

$$F_{\text{cosine}}(\omega) = 2\int_{-\infty}^{\infty} f(t)\cos(\omega t)\,dt$$

Comparing this to Equation 3-3, we can see that it is the even part of the Fourier transform. It has, as an inverse transform,

$$f(t) = 2\int_{-\infty}^{\infty} F_{\text{cosine}}(\omega)\cos(\omega t)\,d\omega$$

In the same manner, the sine transform is defined:

$$F_{\text{sine}}(\omega) = 2\int_{-\infty}^{\infty} f(t)\sin(\omega t)\,dt$$

---

[1]There are at least three popular forms of writing the Fourier transform. The first, and perhaps the simplest, demonstrates the reciprocality of the transform by changing only the sign of $j$ between the transform and its inverse:

$$F(\omega) = \int_{-\infty}^{\infty} f(x)e^{-j\omega t}\,dx$$

and

$$f(n) = \int_{-\infty}^{\infty} F(\omega)e^{j\omega t}\,d\omega$$

The other two involve changing scaling factors as well as the sign of $j$:

$$F(\omega) = \int_{-\infty}^{\infty} f(x)e^{-jx\omega}\,dx$$

and

$$f(x) = \int_{-\infty}^{\infty} F(\omega)e^{jx\omega}\,d\omega$$

Except where noted, we will be using the first version in this book.

and its inverse:

$$f(t) = 2\int_{-\infty}^{\infty} F_{\text{sine}}(\omega)\sin(\omega t)\,d\omega$$

# DISCRETE TIME THEORY

We will begin our examination of discrete time theory with several different approaches to the subject of sampling and its results. We will see what must be done to derive adequate information and what must be done to avoid aliasing and distortion. Included here are some techniques for increasing the resolution of data, called *interpolation*, and zooming in on spectral fields, called *decimation*.

## The Sampling Theorem: An Intuitive Approach

A digital computer stores and references data as collections of flags organized as bits, bytes, and words. Continuous-time signals comprise infinitesimal amounts of *every* frequency, but these are only representable with a Fourier or Laplace integral. To analyze and synthesize a signal on the digital domain, we must convert the analog signals into digital by taking samples at regular intervals. These samples can be compared to snapshots of a single instant in the continuous stream of the analog signal. We get the snapshots, or samples, of our signal at discrete intervals, equally spaced and separated by an integer multiple of a sampling interval, the inverse of the sampling frequency. Each sample is created by multiplying a unit impulse by the signal at the instant of each of these intervals:

$$x[t] = \sum_{k=0}^{\infty} x(t)\delta(t-kT)$$

The brackets indicate a discrete time sequence and $\delta(t-KT)$ represents the impulse, or Dirac delta function, interpreted as a series of delayed impulses. $K$ is an index, $T$ is the period of the sampling interval, and $t$ is the origin. Used in this context, $T$ is usually normalized to 1 and the index alone is thought of as the sample time, and the equation above has the meaning:

$$x[t] = x(t)\delta(t) + x(t)\delta(t-1) + x(t)\delta(t-2) + \cdots$$

The result of this multiplication is really a value equal to the signal at the moment of sampling. These samples can be used to analyze or recreate the original analog signal, if care is taken to see we get adequate information about the signal. Unfortunately, there are problems associated with sampling that must be understood before it can be used effectively.

If you have ever watched a movie in which a wheel was turning, such as a wagon wheel or bicycle wheel with spokes, then you have some idea of what sampling can do. A movie is a series of snapshots of something that occurred in continuous time. These snapshots are taken at regular intervals and at a rate high enough for the motion to appear smooth and continous. At first, as the wheel begins moving, it appears to be going in the correct direction for the direction of travel—that is, clockwise for forward motion. But as the bicycle or wagon gets up to speed, the wheel slows, perhaps even stops for a moment before reversing direction and turning counterclockwise—turning the reverse of what it should to continue in the same direction! If each frame is a sample, then we have 30 samples per second of the wheel turning. At low enough frequencies, this is enough to tell us how fast the wheel is going, but as the angular velocity of the wheel approaches half the frame speed, the wheel slows noticeably and stops—this is the Nyquist frequency. As it accelerates past this point, the wheel seems to reverse and go in the opposite direction—this is frequency misidentification.

Therein lies one of the problems with sampling: the sample rate. As we said before, continuous-time signals are composed of varying amounts of frequencies from the entire spectrum. If the sampling frequency is not high enough, and the signal not bandlimited, energy from frequencies outside the bandwidth of our interest can become confused with that from within it.

In Figure 3-6, you will see in part C that several signals can share common peaks and crossover points. If high-frequency signals such as those in the graph are allowed to be included in the input signal, and the sampling rate is low enough—at points indicated by the circles, for instance—then the high-frequency signals will appear as lower-frequency signals in any spectrum of this data. High-frequency signals can mimic lower-frequency signals, and must be removed before sampling so that they do not interfere.

## Sampling and Aliasing: Two Viewpoints

This interference by other frequencies that doesn't originate in the bandwidth of interest is called *aliasing*. The phenomenon of aliasing can be explained in a number of ways. In the next paragraphs, we will approach this concept in two ways; each provides its own insight and understanding.

According to Nyquist and Shannon, in order to recover a signal function $f_c(t)$, we must sample it at more than twice its highest bandwidth. To restate: in order to retain enough data to analyze and reconstruct a signal, we need to sample it at a rate more than twice its bandwidth. An example of this was given in Figure 3-6, part C. Clearly, unless we can get at least two samples of, for instance, a sinusoid, we cannot recreate it, let alone identify what frequency it is.

One does not sample with a single pulse, but with a series of pulses separated by a fixed interval—this is called a *pulse train*. Applying the Fourier integral to a series of unit impulses, we find that such a sequence will contribute equal amplitudes at integral multiples of the sample frequency and zeros elsewhere. This means that a pulse train in the time domain is again a pulse train in the frequency domain, with a period equal to the sampling frequency. To say it still another way, the spectrum of a pulse train is a pulse train.

In part A of Figure 3-7, the period of the sampling pulses is $t_s$ and the sampling frequency is $f_s = 1/t_s$. The spectrum of this signal will be represented by a *line* at intervals equal to the sampling frequency, $f_s$. This line is periodic, appearing again and again at integer multiples throughout the continous spectrum.

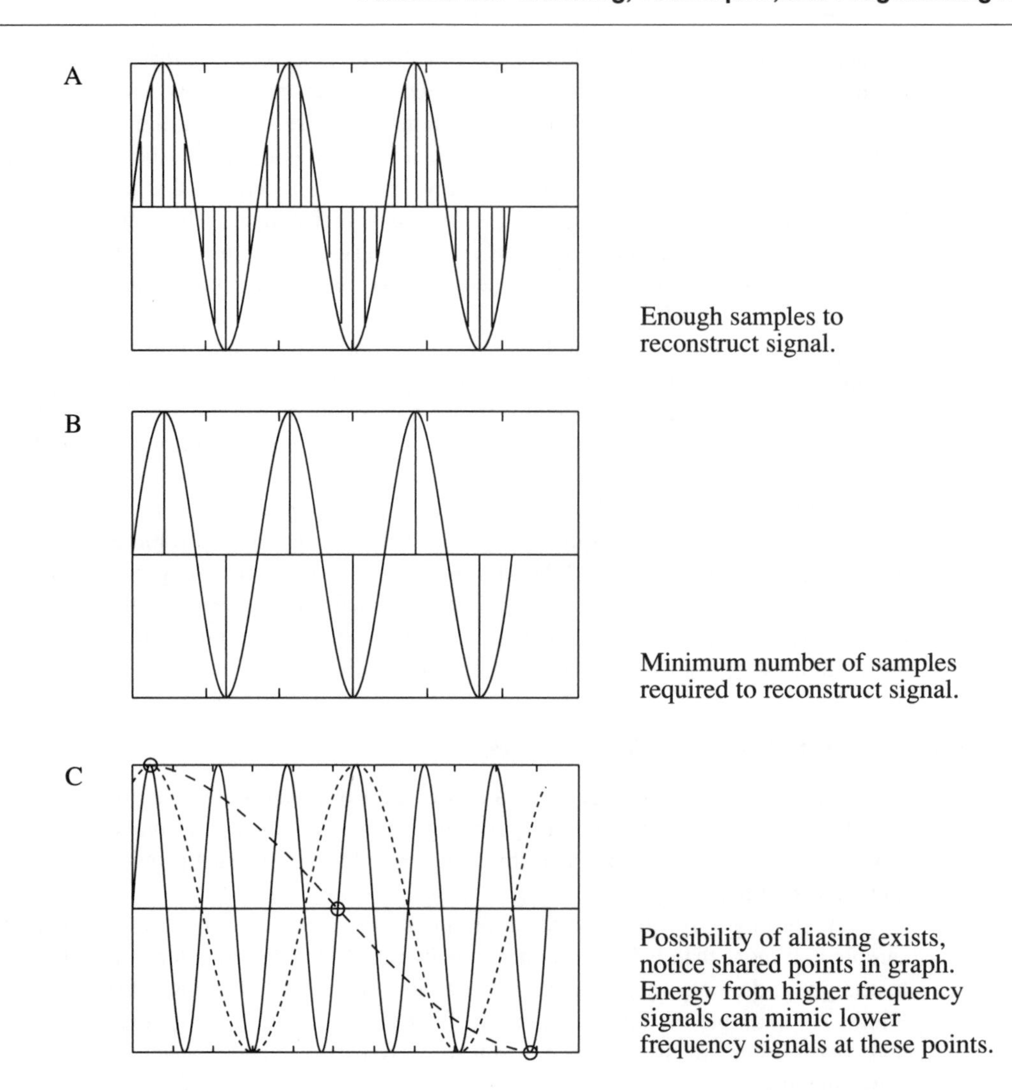

A
Enough samples to
reconstruct signal.
B
Minimum number of samples
required to reconstruct signal.
C
Possibility of aliasing exists,
notice shared points in graph.
Energy from higher frequency
signals can mimic lower
frequency signals at these points.


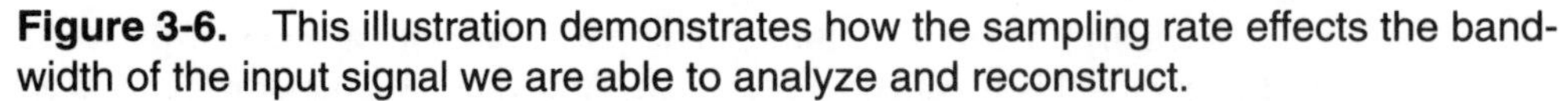
**Figure 3-6.** This illustration demonstrates how the sampling rate effects the bandwidth of the input signal we are able to analyze and reconstruct.

Let us take as an example an application that requires a bandwidth of 5 KHz—that is, 0 Hz to 5 KHz. We will need to sample the input at least twice that rate to get unambiguous data at 5 KHz—we will choose 12 KHz. The period of the samping pulse train in part A of Figure 3-7 is $t_s = 1/12000 = 8.5E{-}5$ seconds and the period of the spectrum is $f_s$ repeated at integer intervals. *A periodic waveform has a periodic spectrum, and the frequency of the sampling pulses determines its period.*

Referring to Figure 3-8, if the input contains signals that are of a bandwidth greater than $f_s/2$, the system will have incomplete information on them, and the data that is present

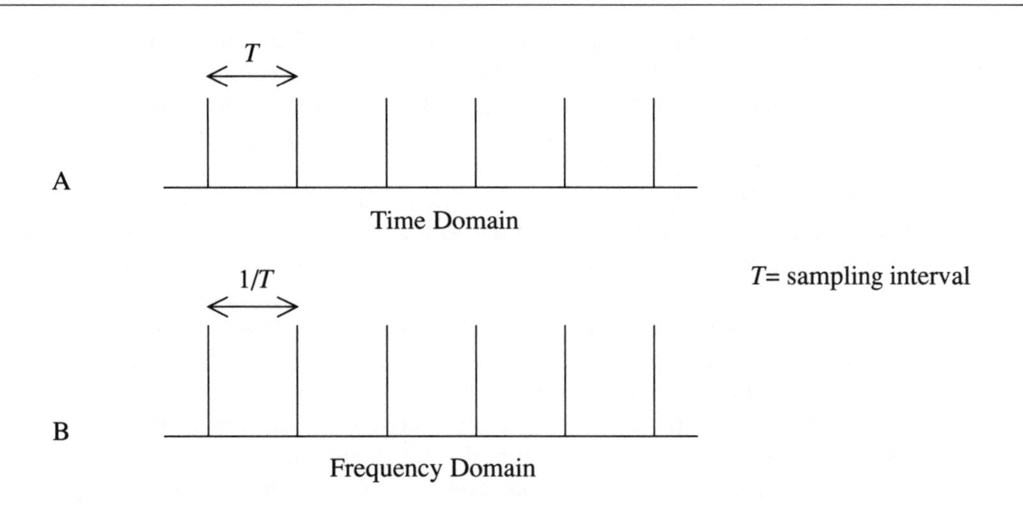


**Figure 3-7.** Part A is a pulse train with the period of the pulses clearly shown; part B is a pulse train with the period clearly marked that is at 1/t to the original.

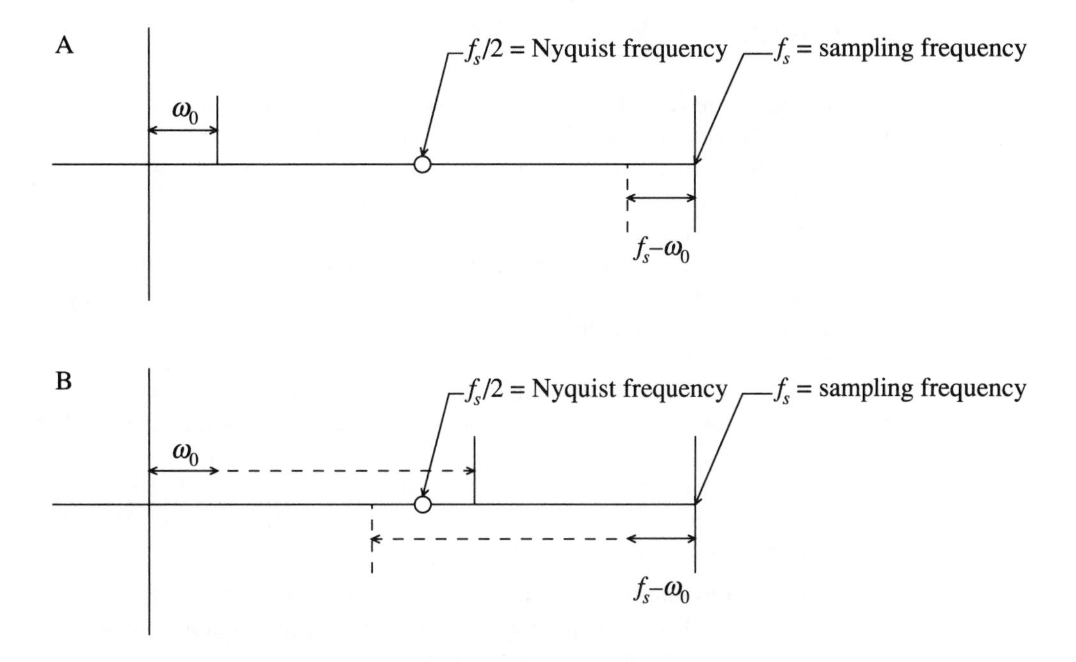

**Figure 3-8.** A frequency at $\pm\omega_0$ will repeat at $f_s \pm \omega_0$, $2f_s \pm \omega_0$, $3f_s \pm \omega_0$,...

will be interpreted as part of the signals we do have adequate data on. This introduces errors into the data, and ultimately into the analysis and reconstruction of the signal. This is because the spectrum of the signal is repeated at intervals of $f_s$, and any element in the spectrum is also represented $f_s \pm f_p$, where $f_p$ is the frequency of a spectral line (distance from the origin of the spectrum $f_0$) and $f_s$ is the sampling frequency. A signal at a frequency less than half the sampling frequency will have a line in the fundamental spectral interval at a distance $f_p$ from the origin of the spectrum, and an alias at $f_s - f_p$. If a signal has a frequency component at exactly one-half the sampling frequency, the line representing it on the spectrum and its alias will coincide: $f_0 + 1/2$ and $f_s - 1/2$ are the same place. A frequency component greater than half the sampling frequency will then appear within the spectral interval, $f_0 \leq \omega < f_0 + 1/2$, mimicking frequencies of the bandwidth of interest and distorting the information by adding information (energy) from unwanted sources.

If there is data or signal content in the samples that is outside that period, it will add itself in, and reconstruction of the original will be impossible.

This explains aliasing from one point of view, but there is another model that also yields a good deal of insight into this subject and others involving the problems of deriving meaningful transforms and windowing.

Just as convolution in the time domain is equal to multiplication in the frequency domain, so multiplication in the time domain is equal to convolution in the frequency domain. Multiplying a periodic pulse train by an analog signal produces a convolution in the frequency domain of the spectrum of the analog signal with the spectrum of the impulse train:

$$F\left( x(t) \sum_{n=-\infty}^{\infty} \delta(t - nT) \right) = X(f) * \left( f_s \sum_{m=-\infty}^{\infty} \delta(f - mf_s) \right)$$

This results in reflections of the original spectrum at periodic intervals along the frequency axis. The convolution will have the form of a sinc function and it will contain within its envelope the signals that impulse function is capable of sampling, in other words, the spectrum of the analog signal. This we know from the Fourier series. Since the result of this convolution is repeated along the frequency axis at intervals equal to the sampling frequency itself, the analog signal must be bandlimited to $f_s/2$ or its spectrum will overlay the next copy of that spectrum on the frequency axis. Thus aliasing represents a cyclical convolution, a subject we will treat more fully later in this chapter. An illustration of this phenomenon is presented in Figure 3-9.

As the pulse width of the sampling impulse function becomes wider, or as the sampling rate itself slows, the $(\sin x)/x$ envelope broadens. If it becomes so broad that it touches or overlaps the next image or copy along the frequency line, aliasing will occur.

You have probably been convinced by now that aliasing should be avoided if it is not actually to be part of the process, as it can be in certain mixing procedures. Here are two simple but important points regarding aliasing:

1. In order to sample and reconstruct a continuous-time signal without alteration, it is necessary that we employ a samping rate, $1/T$, that is at least twice the frequency of the highest frequency component of interest.

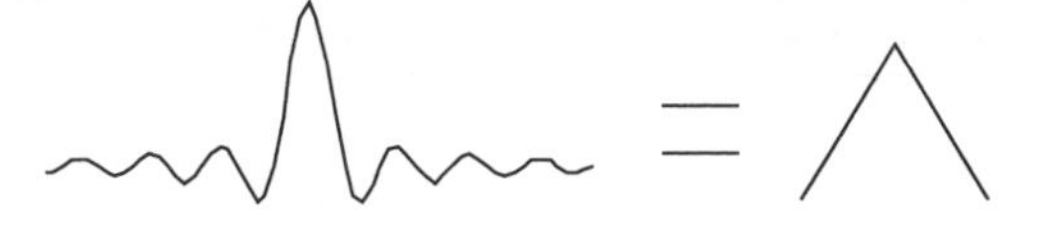

The Sinc envelope contains Fourier coefficients of contributing frequencies.

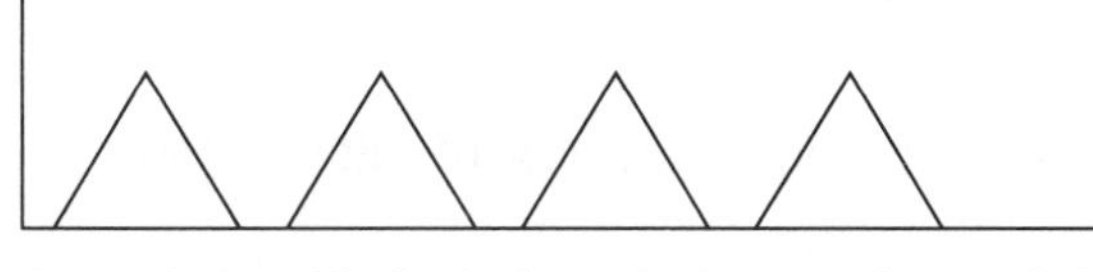

Properly bandlimited, the only frequencies available are less than the Nyquist frequency—the replicas created by the sampling pulses are well separated.

Without bandlimiting, the frequency components of a signal can overlap, resulting in *aliasing*.

**Figure 3-9.** Nonbandlimited signals can alias by allowing sin*x*/*x* overlap.

2. Noise and other frequencies outside the band we are interested in will require an anti-aliasing filter with sufficient attenuation in the stop band to limit any of these other signals to a range less than the dynamic range of our data acquisition unit (A/D).

## Sampling Functions

At this point, we will reintroduce some of the functions we have been talking about with their Fourier transforms. These are listed in Table 3-2.

For sampled signals, each of the functions in Table 3-2 is a sequence represented in terms of the sampling function, the unit impulse.

1. The unit impulse is equal to 1 at time zero and 0 everywhere else:

   $\delta[t]=1,\ t=0$ and $0$ at $t\neq 0$

| | Function of time $x(t)$ | Function of frequency $X(s)$ |
|---|---|---|
| unit impulse function | $\delta(t)$ | $1$ |
| unit step function | $u(t)$ | $\frac{1}{2\pi s}+\frac{1}{2}\delta(s)$ |
| rectangular function | $u(n_0t-n_1t)-u[n_0t-(n_2+1)t]$ | $\sin cx$ |

**Table 3-2.**

2. The sampling function is a continuous sequence of unit impulses:

   $\delta[n]$

3. The unit step sequence is represented in terms of the sampling function:

   $$u[n]=\sum_{k=0}^{\infty}\delta[n-k]$$

4. The unit rectangle is also a sequence:

   $$R_m[n]=\sum_{n=0}^{M-1}\delta[t_0-n]$$

The spectrum of the unit rectangle is the sinc function with a width equal to the inverse of the period of the sampling function times the width of the rectangle. Of course, since this a convolution of the impulse train and the rectangular pulse, we must also consider the transform of the impulse train, which itself is impulse train with a period equal to the inverse of the period of the sample impulse train. Thus, the sinc function is repeated as the transform of the sample impulse sequence.

## Decimation or Downsampling

Decimation is the process of using only certain samples from the input sequence. Usually, this is denoted as every $M$th sample. This has the effect of speeding up the time sequence and broadening the spectral information, the frequency range expanded by $M$. To use this mechanism, it is important to lowpass filter the input to cut off at a frequency $f_s/2M$. Otherwise, frequencies from the decimated spectrum could be aliased into the new interval. Used in this manner, it might be compared to time-lapse photography.

Decimation is used for zoom functions on such instruments as spectrum analyzers. In this context, a sample rate is chosen that will satisfy the Nyquist requirements for the highest resolution of interest (i.e., the narrowest frequency range or greatest magnification). Samples can then be taken from the input data stream at a lower rate by dividing the sample rate by a suitable number, depending upon the bandwidth chosen for examination. For a fixed-length DFT, the higher the resolution, the narrower the analysis frequency range.

Lowering the resolution (dividing the input sample rate by a larger number) broadens the bandwidth and coarsens the analysis.

Other applications for decimation include radio, frequency division multiplexing, data compression, and situations where data only occupies a certain band that can be downsampled into a lower group of frequencies for easier handling.

### Interpolation

Interpolation is a filtering technique used on small data sets to read in intermediate information for the generation of high-resolution data from tables. This method increases the granularity of available information and allows finer approximations for waveform generations and data interpretation. It is based upon the table lookup technique, using the data itself as a pointer into the table for translation, but it also incorporates any fractional residue to estimate an intermediate result.

Interpolation is often used to reduce the memory requirements for various math tables, such as sine and cosine. Although it does increase the processing time because of the additional steps, this is usually easily offset by what is gained in memory. Depending upon the amount of storage available, the table may be as simple as function values at certain intervals for low-level linear interpolation, or as complex as the function values and their derivatives for much higher resolution. More information and examples are available in Chapter 5 and on the optional disk in Oscill.cpp, where interpolation is used to fit a small, single cycle waveform to a much larger vector.

## INTRODUCTION TO THE FOURIER TRANSFORM

In a previous example, we evaluated the function

$$g(x)=\left[1+M\cos(3x)\right]\cos(9x)+2\cos(13x)$$

using the Fourier series. We did this using the trigonometric form of the series, because the notions are so much easier to visualize in that way. It is appropriate now to reevaluate the function, first using the exponential form of the series, then using the Fourier transform to show the similarities and dual nature of each.

To begin with, we find the coefficients of the Fourier series by inserting the function $g(x)$ in the integral

$$c_n=\frac{1}{L}\int_0^{2\pi} g(x)e^{-j\omega nt}\,dt$$

which results in the following values of $c_n$:

$c_n$

| | |
|---|---|
| 1 | $2.298\cdot10^{-13}$ |
| 2 | $-4.882\cdot10^{-10}$ |
| 3 | $5.059\cdot10^{-13}$ |
| 4 | $-4.901\cdot10^{-10}$ |

$c_n$

| | |
|---|---|
| 5 | $4.681 \cdot 10^{-13}$ |
| 6 | 0.25 |
| 7 | $2.664 \cdot 10^{-13}$ |
| 8 | $2.284 \cdot 10^{-13}$ |
| 9 | 0.5 |
| 10 | $-4.902 \cdot 10^{-10}$ |
| 11 | $4.677 \cdot 10^{-13}$ |
| 12 | 0.25 |
| 13 | 1 |
| 14 | $-4.882 \cdot 10^{-10}$ |
| 15 | $2.296 \cdot 10^{-13}$ |

Plotting these against the scale of increasing harmonics of the fundamental, we get the results shown in Figure 3-10.

Of course, we can reconstruct the wave form by using these coefficients in the series expansion. Thus, we can move between the time domain and the frequency domain easily with the Fourier series.

The Fourier integral and transform will not converge for periodic waveforms, but if we consider the waveform generated by this function as aperiodic, existing only in the inter-

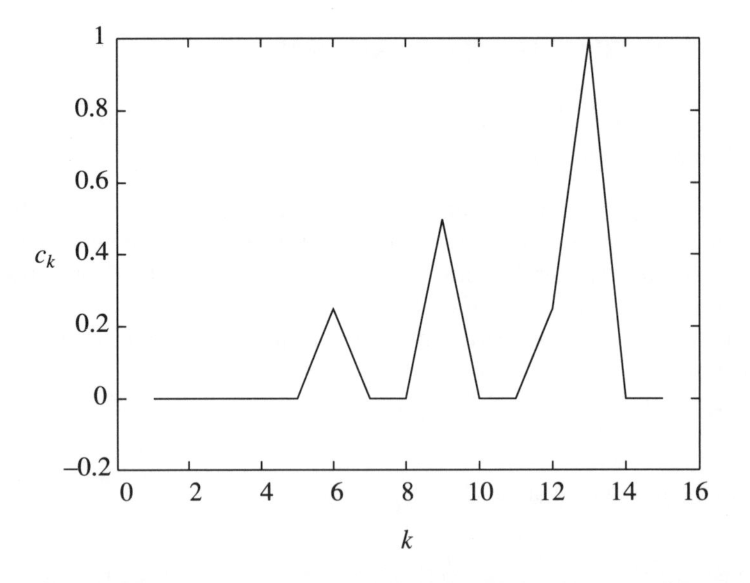


**Figure 3-10.** The coefficients of the exponential Fourier series plotted against a scale of harmonics. Notice the difference in scaling between this spectrum and that in Figure 3-2.

val of inspection, then we can analyze it with the Fourier transform as well. Let us place the function in the Fourier transform:

$$H(j\omega) = \int_{-\infty}^{\infty} g(x)e^{-j\omega x}dx$$

We have, as the result of this transform, the following complex coefficients:

| | $H(j\omega) =$ | |
|---|---|---|
| 0 | 1 | |
| 1 | 1.136 | −0.226$j$ |
| 2 | 4.285 | −1.775$j$ |
| 3 | 0.805 | −0.538$j$ |
| 4 | 0.156 | −0.156$j$ |
| 5 | 0.404 | −0.605$j$ |
| 6 | 0.945 | −2.281$j$ |
| 7 | −0.231 | +1.63$j$ |
| 8 | 0 | |

These may also be plotted against a scale of ranks or indices representing multiples of $dx$ used in the analysis; see Figure 3-11.

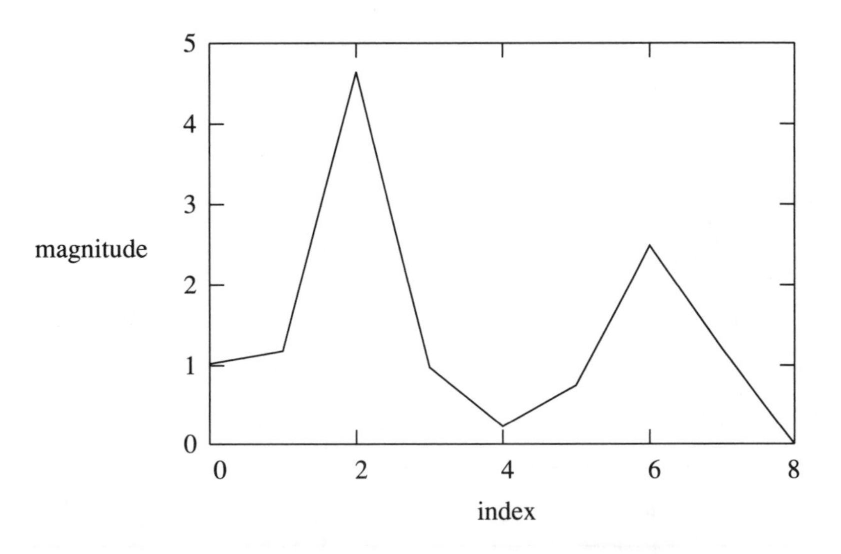


**Figure 3-11.** A plot of the Fourier transform of $g(x)$ showing the relative magnitude of its frequency components. In this case, the frequency of any individual component is determined as (*rank/vector length*) * *sampling frequency.*

Once again, the transform can be used to synthesize the original waveform: use the same transform, but reverse the sign of $j$:

$$h(x) = \int_{-\infty}^{\infty} F(\omega) e^{j\omega t} d\omega$$

## Theorems Relating to the Fourier Series and Transform

There are a number of theorems relating to the Fourier series and the Fourier transform that are very useful in signal processing. Table 3-3 lists some of the more useful theorems.

| *Theorem* | *Time domain* | *Frequency domain (Fourier series)* | *Frequency domain (Fourier transform)* |
|---|---|---|---|
| Linearity (Additivity) | $ax(t) + by(t)$ | $aX(n) + bY(n)$ | $aX(s) + bY(s)$ |
| Multiplication | $x(t)y(t)$ | $\sum_{m=-\infty}^{\infty} X(n-m)Y(m)$ | $\int_{-\infty}^{\infty} X(s_1)Y(s_0 - s_1) ds_1$ |
| Convolution (circular or periodic) | $h(t) = \sum_{t'=0}^{N-1} f(t')g[t - t' + NH(t' - t)]$ <br> $H(\cdot)$ is the unit step function | $X(n)Y(n)$ | $H(s)X(s)$ |
| Time convolution | $\int_{-\infty}^{\infty} h(t_0 - t)x(t) dt$ | $X(n)Y(n)$ | $H(s)X(s)$ <br> (provided the record length is long enough) |
| Time shifting | $x(t_0 - t)$ | $e^{\frac{-j2\pi nt}{T}} X(n)$ | $e^{-j2\pi as} X(s)$ |
| Frequency shifting | $e^{\frac{j2\pi mt}{T}} x(t)$ | $X(n - m)$ | $X(s_0 + s)$ |
| Similarity | $x(ax)$ | $\left\|\frac{1}{a}\right\| X\left(\frac{s}{a}\right)$ | $\left\|\frac{1}{a}\right\| X\left(\frac{s}{a}\right)$ |
| Parseval's theorem | $\frac{1}{T}\int_T \|x(t)\|^2 dt$ | $c_0^2 + \sum_{n=1}^{\infty} \frac{1}{2}\|2c_n\|^2$ | |
| Rayleigh's theorem | $\int_{-\infty}^{\infty} \|x(t)\|^2 dt$ | | $\int_{-\infty}^{\infty} \|X(s)\|^2 ds$ |

*continued*

**Table 3-3.**

| *Theorem* | *Time domain* | *Frequency domain (Fourier series)* | *Frequency domain (Fourier transform)* |
|---|---|---|---|
| Autocorrelation | $x(x)*x*(-x)$ | $\|X(k)\|^2$ | $\|X(s)\|^2$ |
| Duality | $X(t)$ | $x(-k)$ | $x(-s)$ |
| Conjugation | $x*(t)$ | $X*(-k)$ | $X*(-s)$ |
| Real part | $\mathrm{Re}[x(t)]$ | $\frac{1}{2}[X(k)+X*(-k)]$ | $\frac{1}{2}[X(s)+X*(-s)]$ |
| Imaginary part | $\mathrm{Im}[x(t)]$ | $\frac{1}{2j}[X(k)-X*(-k)]$ | $\frac{1}{2j}[X(s)-X*(-s)]$ |

**Table 3-3.** *(continued)*

In addition to the facilities it offers to signal processing, the Fourier transform has a foundation and utility in mathematics generally, because it is based upon the $n$th roots of unity. Thus, two more properties are available that may prove valuable. These are shown in Table 3-4.

You will see from Table 3-3 that both the Fourier series and the Fourier transform possess the basic elements of linearity, as well as additivity and mulitplicativity, and therefore obey the laws of superposition.

We showed in the first chapter that convolution in the time domain was a sliding multiplication of two functions that is summed over time. In the frequency domain, we can say that if function $h(t)$ and $g(t)$ have transforms $H(s)$ and $G(s)$, then the convolution of these two time domain functions is equal to the inverse Fourier transform of the product of their transforms, $H(s)$ and $G(s)$. In other words, the convolution of two functions is the multiplication of their transforms.

Convolutions are commutative

$$f*g=g*f$$

| *Mathematical property* | *Function* | *Fourier transform* |
|---|---|---|
| Differentiation | $\frac{d^n}{dt^n}x(t)$ | $(j2\pi\omega)^n X(\omega)$ |
| Integration | $\int_{-\infty}^{t} x(t)\,dt$ | $\frac{X(\omega)}{j2\pi\omega}+\frac{1}{2}X(0)\delta(\omega)$ |

**Table 3-4.**

associative

$$x*[y*z]=[x*y]*z$$

and distributive

$$x*[y+z]=x*y+x*z$$

This theorem is one of the most useful properties of the Fourier transform. With it, new functions can be formed through the convolution of two other functions. This allows us the ability to filter and shape other functions. Deconvolutions provide us with the means of removing noise and other embedded signals, thereby recovering the original. Please refer to the section entitled Theorems of the Discrete Fourier Transform for more detail on the subject of frequency domain convolution.

Correlation is the process of relating items, objects, or signals through a statistical connection. It is used in the study of earthquakes, in medical science in the search for better diagnostic techniques, in sonar and radar, and in communications where the autocorrelation is used to dig signal information out of a noisy input.

Random noise will have no relationship with itself or the input signal, and so will not have the concentrated energy of sinusoid at any particular frequency. Autocorrelating a signal and then filtering for selected frequencies based upon the results can form the foundation of an adaptive filter. Such a filter is developed in greater detail with source code in Chapter 7. Autocorrelation can also be used as a heuristic method to finding and eliminating correlated noise. In image processing, it is useful in identification and image enhancement.

Another theorem related to the convolution theorem is Parseval's theorem for the Fourier series, or Rayleigh's theorem for the Fourier transform. Simply, these state that the integral of the squared modulus of a function is equal to the squared modulus of its spectrum. The total energy of a signal is related to the total energy in the transform with this theorem. This function is helpful in determining how energy is distributed in a system. Along with the correlation techniques mentioned above, it can be very useful in identifying and removing noise energy from a signal.

The similarity theorem and the time shifting theorem are two properties important in the development of the fast Fourier transform, among other things. The similarity theorem formalizes the relationship between time and frequency scaling. It states that as one of the transform pairs expands horizontally, the other contracts horizontally and grows vertically to maintain a constant area within its bounds. In less formal terms, if you expand a sequence to double its length by inserting a zero between neighboring components of the original sequence, the elements of the transform are repeated.

$$X\{a \ b \ c \ d\}=e \ f \ g \ h$$

and

$$X\{a \ 0 \ b \ 0 \ c \ 0 \ d \ 0\}=e \ f \ g \ h \ e \ f \ g \ h$$

Time shifting does not influence frequency, only phase, and this in proportion to frequency. If the shift is $a$, the phase delay will be

$$\frac{a}{s^{-1}}$$

The time interval that delays the waveform has different effects upon different harmonic components. The higher the frequency, the greater the change, since the shift represented by $a$ occupies a larger proportion of the period of the higher harmonic conponents. This means that, in general, a component of period $s^{-1}$ will suffer no change of amplitude, but will be delayed in phase by

$$2\pi\frac{a}{s^{-1}}$$

This information is useful in the design of linear phase FIR filters, as you will see in Chapter 4.

The modulation theorem is well known in radio and television. Multiplying a continuous-time function by $\cos\omega x$ results in a a modulated signal that has a transform split into two halves, each having half the original strength and shifted about the origin by $\pm\omega/2\pi$.

## The Discrete Fourier Transform

The Laplace transform of a continuous-time signal, $x(t)$, is

$$X(s) = \int_0^\infty x(t)e^{-st}\,dt$$

We use this to solve differential equations involving continuous-time systems. For discrete time systems, we must deal with sequences of data sampled from the continous-time signal at regular intervals.

An analog signal may be represented as such a sequence of sampled data in the following form:

$$x[t] = \sum_{k=-\infty}^{\infty} x(t)\delta(t-kT)$$

This is a convolution sum, in which a continuous time signal, $x(t)$, is sampled at intervals, $T = 1/f_s$, delayed by $k$ multiples of $T$. The result of this sum is a sequence, $x[t]$, indicated by the brackets around the variable. This sum actually looks like this:

$$x(t)\delta(t) + x(t)\delta(t-T) + x(t)\delta(t-2T) + \cdots$$

Since the impulse function is zero everywhere except at $k$multiples of the sample period, $T$, where it represents the value of the function at that point, we could write the Laplace transform of this function as

$$X(s) = \int_0^\infty x[nT]e^{-st}\,dt$$

Because the impulse function serves to punch out a value only at prescribed intervals, we can reduce this to a sum:

$$X(s)=\sum_{n=0}^{\infty}x[nT]e^{-nst}$$

Finally, if we make $\sigma=0$ in the equality $s=\sigma+j\omega$, we have the discrete time Fourier transform:

$$F(v)=\frac{1}{N}\sum_{t=0}^{N-1}f[t]e^{-j2\pi(v/N)t}$$

with its inverse given by[2]

$$f[t]=\sum_{v=0}^{N-1}F(v)e^{j2\pi(v/N)t}$$

This formulation says that the spectrum $F(v)$ is the sum of the products of the input vector, with discrete frequencies given by

$$\frac{v}{n}$$

where $v$ is the data increment and $n$ is the vector stride or length. Unlike the Fourier integral, this spectrum is not the result of all frequencies, but only of certain ones dependent upon the length of the input vector.

The transform will converge if the sequence is absolutely summable, and all finite sequences are absolutely summable. In addition, a stable sequence is absolutely summable, so all stable sequences have transforms. It is important here to notice that this formulation is finite: nothing is known before its start and nothing is known after its finish. A discrete time transform has as its start and origin $t=0$, and as you can see from the formulation, it has a definite size—that is, the input vector must be as large as size $n$ or padded to fit.

All the Fourier transform theorems have sampled data representations. Of the most important is the convolution theorem:

---

[2]As with the Fourier transform, the discrete Fourier transform has more than one notation. The one in the text follows that of Ronald Bracewell from his book, *The Fourier Transform and Its Applications* (New York: McGraw-Hill, 1986). Another very popular version is given by

$$F[k]=\sum_{n=0}^{N-1}f[n]e^{-j2\pi\frac{kn}{N}}$$

and

$$x[n]=\sum_{k=0}^{N-1}f[k]e^{j2\pi\frac{kn}{N}}$$

$$y[n] = \sum_{k=-\infty}^{\infty} x[k]z[n-k]$$

In general, to change a theorem from continuous time to discrete time, it is generally only necessary to replace the integration with a discrete sum.

## Theorems of the Discrete Fourier Transform

Most of the theorems associated with the Fourier transform are represented in the same way with the DFT. There are a few notes, however, that are worth mentioning:

1. The DFT is reciprocal (as is the FT) by alternating the sign of $j$. However—and this depends upon author and notation—there may be a scaling factor also associated with either the forward transform or reverse that must be taken into account.

2. Up to now, we have described the convolution of two sequences as a serial product having a length $M+N-1$, where the length of the first sequence is $M$ and the length of the second is $N$. (See Chapters 1 and 4.) This is known as *noncyclic convolution*, and is the same convolution indicated by the convolution integral.

3. The Discrete Fourier Series possesses a convolution property that is based upon the periodic nature of the signals involved. Aliasing is not a problem because we are dealing with periodic signals. The DFT offers an operation similar to that of the discrete Fourier series, in that it is a point-by-point multiplication on the frequency domain. However, like convolution in the time domain, it requires $M+N-1$ elements in the output vector to produce an unaliased result.

   In order to come to the same result using the frequency domain as you would on the time domain, the spectra of the two sequences to be convolved must be composed mostly of zero elements. If the sequences grow in nonzero elements so that they extend beyond the Nyquist point, the two sequences will overlap with a resulting flat non-zero offset in the region of the overlap. This is aliasing.

   An operation called *cyclic convolution* does exist and is defined as

   $$h(\theta) = \int_0^{2\pi} f(\theta)g(\theta - \theta')d\theta'$$

   This operation represents the multiplication of two DFTs without meeting the requirements for the noncyclic convolution. Figure 3-12 illustrates both forms of convolution.

4. If the forward transform is scaled by $1/N$, then the following two properties hold. First, the sum of the input sequence is equal to the product of the length of the input sequence and the first term of the DFT. Second, the inverse is also true—that is, the first value of the input sequence is equal to the sum of the elements of the DFT of the sequence.

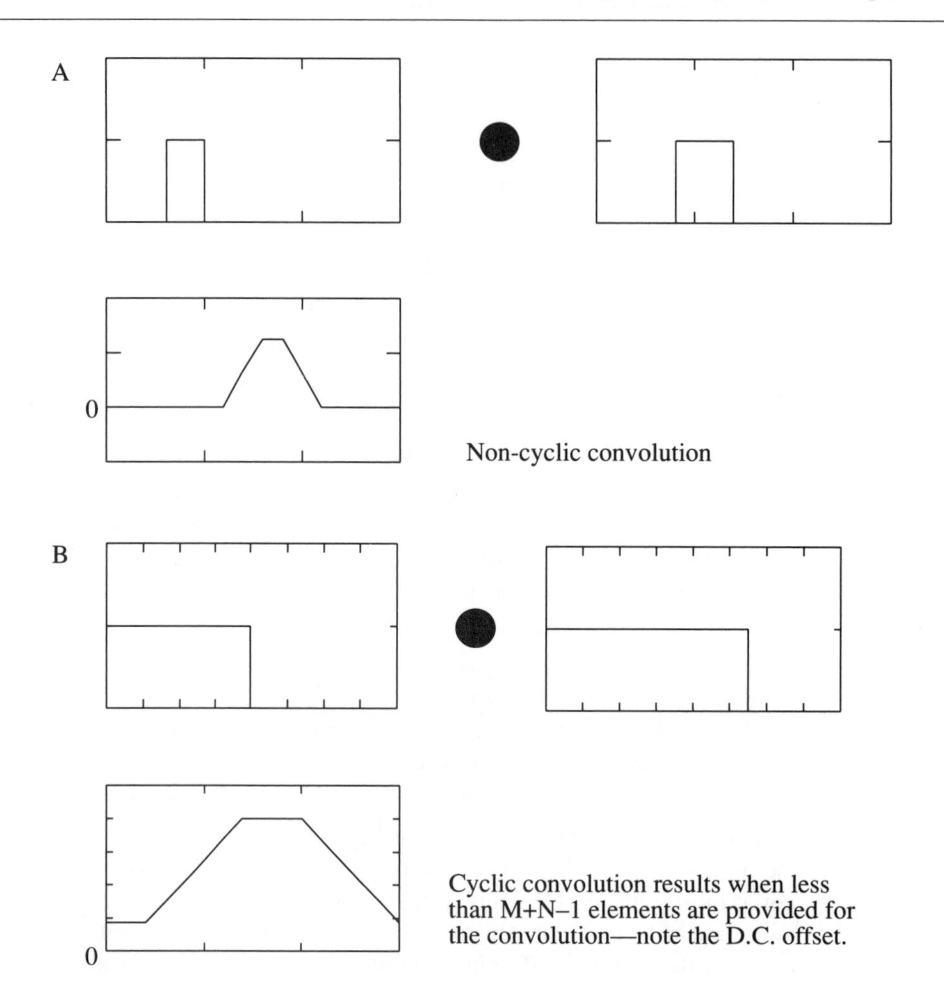


**Figure 3-12.** Circular convolution. Part A is the convolution of two rectangles that occupy only a small part of their individual cycle lengths—that is, most of the length is zero. Part B is the same convolution; however, more than half the length is occupied by the rectangle.

## COMPUTING THE DFT

The discrete Fourier transform is defined as the (scaled) sum of the product of an input sequence with a quantity

$$W^k = e^{-j2\pi v/N}$$

These are called *phase factors* or *twiddle factors*. The twiddle factors represent phase angles relative to the unit circle.

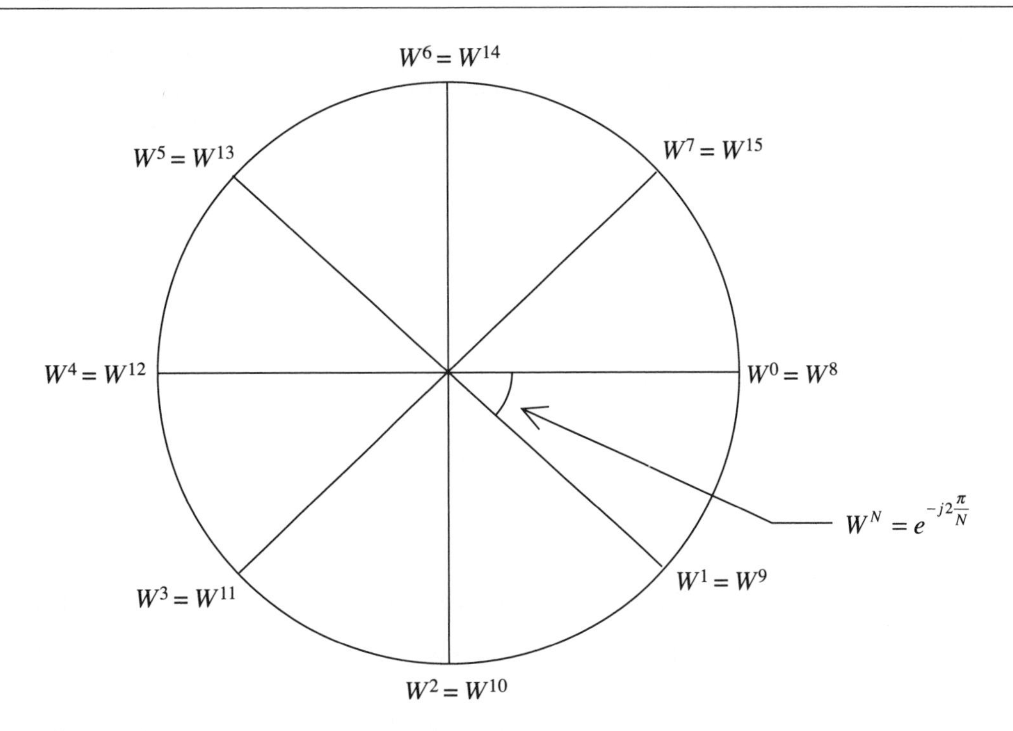


**Figure 3-13.** The phase angle of w is proportional to the power to which it is raised, it will have as many increments as vector elements. This factor is often called an $N$th root of unity, because $W^n = e^{-j2\pi} = 1$. On the unit circle, this would give it a radius of one and a phase equal to $-1/n$ turns.

The length of the input vector determines the number of twiddle factors. Figure 3-13 illustrates the periodicity and symmetry of these factors around the circle. By *periodicity*, we mean that each phase factor is repeated every $N$ samples:

$$W^k = W^{k+N}$$

The symmetry is demonstrated by

$$W^k = W^{k+\frac{N}{2}}$$

The circumference of the circle may be visualized as the *entire sample period* of the DFT—that is, if we are sampling at 10 KHz and our sample length is 1024 data points, the circumference of the circle is equal to 1024/10000 = 0.1024 seconds. It represents 1024 samples, but only 1024/2 = 512 frequency indices, because of the Nyquist point which lies at exactly 1/2 the way around the circle. The entire second half of the circle contains complex conjugates of the first half. Each of the 512 indices point to frequency magnitudes that contribute to the signal from which the input sequence was derived. In this way, the circle is our spectrum. The phase angles are located at exact integer multiples, or powers, of the fundamental angle, and correspond to the integer multiples of the fundamental harmonic in the Fourier series.

In the example that follows, we will produce an eight point DFT of an impulse such as the one in Figure 3-14. For this case, our twiddle factors are:

$$w = e^{-j2\frac{v\pi}{n}}$$

$$w^0 = 1$$

$$w^1 = 0.707 - 0.707j$$

$$w^2 = -j$$

$$w^3 = -0.707 - 0.707j$$

$$w^4 = -1$$

$$w^5 = -0.707 + 0.707j$$

$$w^6 = j$$

$$w^7 = 0.707 + 0.707j$$

According to the sum, the magnitude for each frequency ($v/n$) is calculated as

$$F(v) = f(0) + f(1)W^v + f(2)W^{2v} + f(3)W^{3v} + f(4)W^{4v} + \cdots + f(n-1)W^{(N-1)v}$$

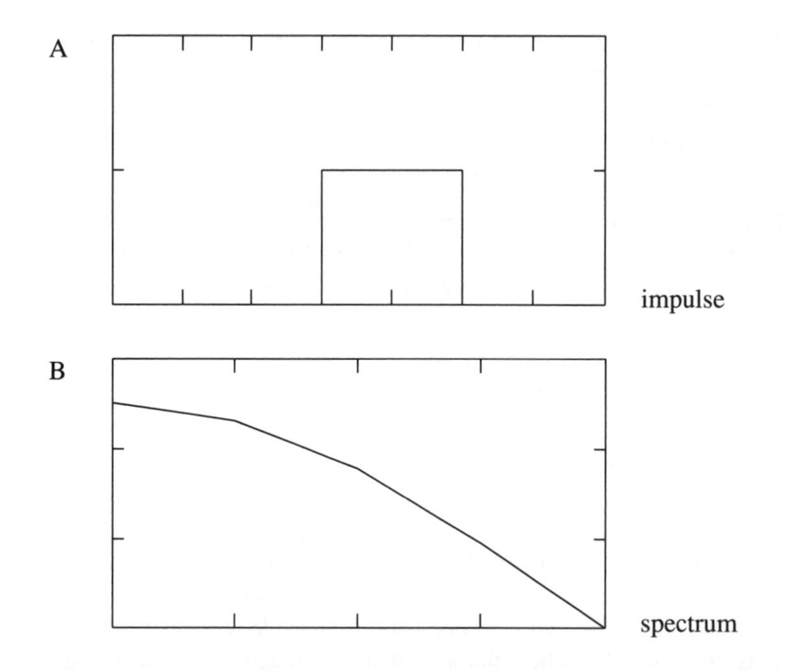


**Figure 3-14.** Part A is a pulse input as a sampled data signal to the DFT; part B is a spectrum based upon this data

For an eight-point DFT, with an input vector formed from a sequence such as that depicted in part A of Figure 3-14,

$$f(0)=0$$
$$f(1)=0$$
$$f(2)=0$$
$$f(3)=1$$
$$f(4)=1$$
$$f(5)=0$$
$$f(6)=0$$
$$f(7)=0$$

The equations would solve as:

$$F(0)=\frac{1}{n}\left(f(0)+f(1)W^0+f(2)W^0+f(3)W^0+f(4)W^0+\cdots+f(7)W^0\right)=0.375$$
$$F(1)=\frac{1}{n}\left(f(0)+f(1)W^v+f(2)W^{2v}+f(3)W^{3v}+f(4)W^{4v}+\cdots+f(7)W^{7v}\right)=-0.213-0.213j$$
$$F(2)=\frac{1}{n}\left(f(0)+f(1)W^v+f(2)W^{2v}+f(3)W^{3v}+f(4)W^{4v}+\cdots+f(7)W^{7v}\right)=0.125j$$
$$F(3)=\frac{1}{n}\left(f(0)+f(1)W^v+f(2)W^{2v}+f(3)W^{3v}+f(4)W^{4v}+\cdots+f(7)W^{7v}\right)=-0.037+0.037j$$
$$F(4)=\frac{1}{n}\left(f(0)+f(1)W^v+f(2)W^{2v}+f(3)W^{3v}+f(4)W^{4v}+\cdots+f(7)W^{7v}\right)=0.125$$
$$F(5)=\frac{1}{n}\left(f(0)+f(1)W^v+f(2)W^{2v}+f(3)W^{3v}+f(4)W^{4v}+\cdots+f(7)W^{7v}\right)=-0.037-0.037j$$
$$F(6)=\frac{1}{n}\left(f(0)+f(1)W^v+f(2)W^{2v}+f(3)W^{3v}+f(4)W^{4v}+\cdots+f(7)W^{7v}\right)=-0.125j$$
$$F(7)=\frac{1}{n}\left(f(0)+f(1)W^v+f(2)W^{2v}+f(3)W^{3v}+f(4)W^{4v}+\cdots+f(7)W^{7v}\right)=-0.213+0.213j$$

This is equivalent to the matrix operation

$$\left[\begin{bmatrix} 1 & 1 & 1 & 1 & 1 & 1 & 1 & 1 \\ 1 & w & w^2 & w^3 & w^4 & w^5 & w^6 & w^7 \\ 1 & w^2 & w^4 & w^6 & w^8 & w^{10} & w^{12} & w^{14} \\ 1 & w^3 & w^6 & w^9 & w^{12} & w^{15} & w^{18} & w^{21} \\ 1 & w^4 & w^8 & w^{12} & w^{16} & w^{20} & w^{24} & w^{28} \\ 1 & w^5 & w^{10} & w^{15} & w^{20} & w^{25} & w^{30} & w^{35} \\ 1 & w^6 & w^{12} & w^{18} & w^{24} & w^{30} & w^{36} & w^{42} \\ 1 & w^7 & w^{14} & w^{21} & w^{28} & w^{35} & w^{42} & w^{49} \end{bmatrix} \bullet \begin{bmatrix} f(0) \\ f(1) \\ f(2) \\ f(3) \\ f(4) \\ f(5) \\ f(6) \\ f(7) \end{bmatrix}\right] \bullet \frac{1}{n} = \begin{bmatrix} 0.375 \\ -0.213-0.213i \\ 0.125i \\ -0.037+0.037i \\ 0.125 \\ -0.037-0.037i \\ -0.125i \\ -0.213+0.213i \end{bmatrix}$$

As you can see, this does produce the expected result, but the multiplications involved in this computation can take a very long time, since it requires $N^2$ complex multiplications and $N^2$ complex additions to produce a result *for each frequency*. For an example, see Bitsnpcs.mcd on the optional disk.

## The Fast Fourier Transform

The theory behind the FFT can be explained in two ways. One is based upon the matrix operation on page 165, in which the matrix is factored to give

$$\left[\begin{bmatrix} 1 & 0 & 0 & 0 & 1 & 0 & 0 & 0 \\ 0 & 1 & 0 & 0 & 0 & w & 0 & 0 \\ 0 & 0 & 1 & 0 & 0 & 0 & w^2 & 0 \\ 0 & 0 & 0 & 1 & 0 & 0 & 0 & w^3 \\ 1 & 0 & 0 & 0 & w^4 & 0 & 0 & 0 \\ 0 & 1 & 0 & 0 & 0 & w^5 & 0 & 0 \\ 0 & 0 & 1 & 0 & 0 & 0 & w^6 & 0 \\ 0 & 0 & 0 & 1 & 0 & 0 & 0 & w^7 \end{bmatrix} \bullet \begin{bmatrix} 1 & 0 & 1 & 0 & 0 & 0 & 0 & 0 \\ 0 & 1 & 0 & w^2 & 0 & 0 & 0 & 0 \\ 1 & 0 & w^4 & 0 & 0 & 0 & 0 & 0 \\ 0 & 1 & 0 & w^6 & 0 & 0 & 0 & 0 \\ 0 & 0 & 0 & 0 & 1 & 0 & 1 & 0 \\ 0 & 0 & 0 & 0 & 0 & 1 & 0 & w^2 \\ 0 & 0 & 0 & 0 & 1 & 0 & w^4 & 0 \\ 0 & 0 & 0 & 0 & 0 & 1 & 0 & w^6 \end{bmatrix}\right.$$

$$\left.\bullet \begin{bmatrix} 1 & 1 & 0 & 0 & 0 & 0 & 0 & 0 \\ 1 & w^4 & 0 & 0 & 0 & 0 & 0 & 0 \\ 0 & 0 & 1 & 1 & 0 & 0 & 0 & 0 \\ 0 & 0 & 1 & w^6 & 0 & 0 & 0 & 0 \\ 0 & 0 & 0 & 0 & 1 & 1 & 0 & 0 \\ 0 & 0 & 0 & 0 & 1 & w^4 & 0 & 0 \\ 0 & 0 & 0 & 0 & 0 & 0 & 1 & 1 \\ 0 & 0 & 0 & 0 & 0 & 0 & 1 & w^4 \end{bmatrix} \bullet \begin{bmatrix} 1 & 0 & 0 & 0 & 0 & 0 & 0 & 0 \\ 0 & 0 & 0 & 0 & 1 & 0 & 0 & 0 \\ 0 & 0 & 1 & 0 & 0 & 0 & 0 & 0 \\ 0 & 0 & 0 & 0 & 0 & 0 & 1 & 0 \\ 0 & 1 & 0 & 0 & 0 & 0 & 0 & 0 \\ 0 & 0 & 0 & 0 & 0 & 1 & 0 & 0 \\ 0 & 0 & 0 & 1 & 0 & 0 & 0 & 0 \\ 0 & 0 & 0 & 0 & 0 & 0 & 0 & 1 \end{bmatrix} \bullet \begin{bmatrix} f(0) \\ f(1) \\ f(2) \\ f(3) \\ f(4) \\ f(5) \\ f(6) \\ f(7) \end{bmatrix}\right] \bullet \frac{1}{n} = \begin{bmatrix} 0.375 \\ -0.213-0.213i \\ 0.125i \\ -0.037+0.037i \\ 0.125 \\ -0.037-0.037i \\ -0.125i \\ -0.213+0.213i \end{bmatrix}$$

With only two nonzero elements a row, the number of multiplications is reduced from $n^2$ to $2n$ per factor. The last factor before the input vector is simply a rearrangement commonly called *bit reversal*.

The second and more popular method of explaining the FFT employs two of its theorems, the shift theorem and similarity theorem. For this explanation, let us suppose an input vector (8 7 6 5 4 3 2 1) for which we wish to find the DFT. We know that a DFT may be represented as a sum of two smaller DFTs:

$$DFT(0\ 7\ 0\ 5\ 0\ 3\ 0\ 1) + DFT(8\ 0\ 6\ 0\ 4\ 0\ 2\ 0) = DFT(8\ 7\ 6\ 5\ 4\ 3\ 2\ 1) =$$
$$\{4.5\ \ 0.5-1.207j\ \ 0.5-0.5j\ \ 0.5-0.207j\}$$

and the similarity theorem provides us with the ability to do this. According to the similarity theorem, a sequence may be expanded by inserting zeros between each component, with the result that if

$$DFT(8\ 6\ 4\ 2) = \{A\ B\ C\ D\}$$

then

$$DFT(8\ 0\ 6\ 0\ 4\ 0\ 2\ 0) = \frac{1}{2}\{A\ B\ C\ D\ A\ B\ C\ D\} =$$
$$\{2.5\ \ 0.5-0.5j\ \ 0.5\ \ 0.5+0.5j\}$$

The shift theorem then allows us to change the phase of the DFT so that it becomes the transform of a sequence with the same values but different phase shift. This makes it possible for us to find the DFT of the sequence

$$DFT(7\ 0\ 5\ 0\ 3\ 0\ 1\ 0) = \frac{1}{2}\{E\ F\ G\ H\ E\ F\ G\ H\} =$$
$$\{2\ \ 0.5-0.5j\ \ 0.5\ \ 0.5+0.5j\}$$

and then, by multiplying each element by a corresponding phase factor to produce a shifted result,

$$DFT(0\ 7\ 0\ 5\ 0\ 3\ 0\ 1) =$$
$$\frac{1}{2}\{2\ \ (0.5-0.5j)W^1\ \ (0.5)W^2\ \ (0.5+0.5j)W^3\ \ 2W^4\ \ (0.5-0.5j)W^5\ \ (0.5)W^6\ \ (0.5+0.5j)W^7\} =$$
$$\{2\ \ -0.707j\ \ -0.5j\ \ -0.707j\}$$

If we then add the two smaller transforms, as we indicated earlier, we do obtain the transform for the whole input sequence. Let us step through the example again in more detail.

Initially, the sequence is broken up into successively smaller sequences. Most fast Fourier transforms employ a radix of 2, meaning that they break the input sequence into pairs. For our example, we will simply break the sequence into two sequences, each containing $n/2$ data points; the even-numbered are in one, and the odd in the other. For the eight-point DFT we have been working with, the smaller sequences would look like this:

$$\{8642\} \text{ and } \{7531\}$$

Each of these produces a DFT, and those DFTs are combined by stretching, shifting, and simple addition. Stretching is accomplished through the similarity theorem:

$$\{80604020\} = \frac{1}{2}\{F(0)\ F(2)\ F(4)\ F(6)\ F(0)\ F(2)\ F(4)\ F(6)\}$$

The second sequence follows suit with

$$\{7531\} = \{F(1)F(3)F(5)F(7)\}$$
$$\{7050301\} = \frac{1}{2}\{F(1)F(3)F(5)F(7)F(1)F(3)F(5)F(7)\}$$

The shifting follows the shifting theorem, which states that a shift in the frequency domain is equivalent to a multiplication by

$$e^{-j2\pi a\left(\frac{v}{n}\right)}$$

or in this case $W$. In this way, the second DFT becomes

$$\{0705\,0301\}=\frac{1}{2}\{F(1)\ WF(3)\ W^2F(5)\ W^3F(7)\ W^4F(1)\ W^5F(3)\ W^6F(5)\ W^7F(7)\}$$

These two DFTs are then added to form the final DFT.

After some examination, we find that the number of the terms in these equations need only be computed once and used as needed. Figure 3-15 is an illustration of the contributions that symmetry and periodicity make to this process known as *The Butterfly*, while Figure 3-16 actually shows how the butterfly fits into the computation of a form of FFT called *decimation in time*.

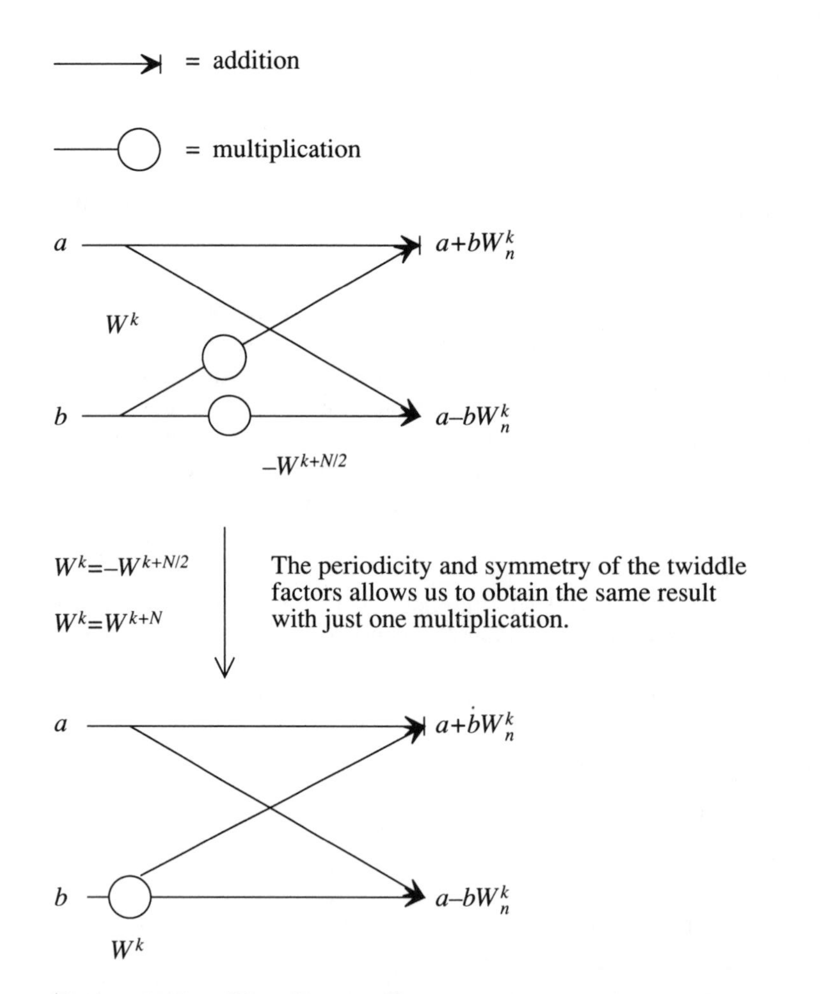


**Figure 3-15.** The diagram illustrates the contributions that the symmetry and periodicity of the twiddle factors make to the butterfly.

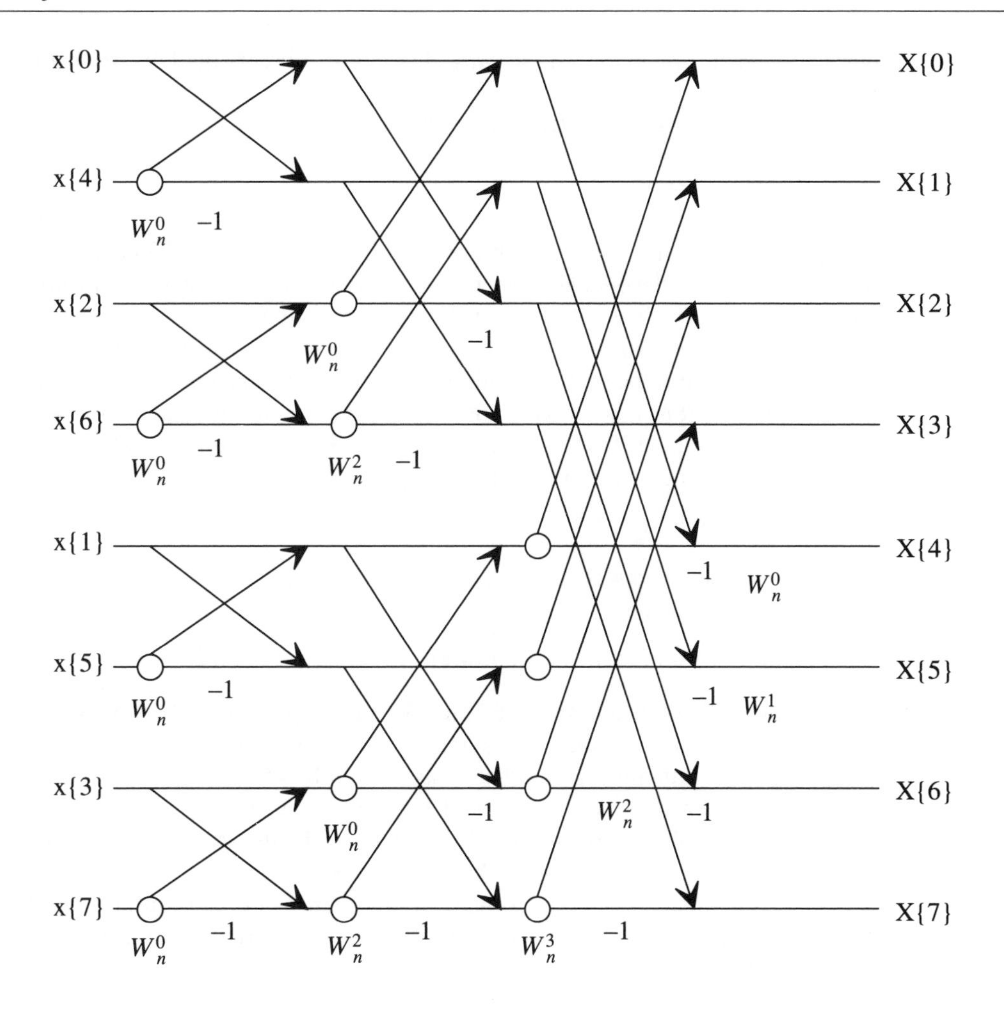


**Figure 3-16.** The figure shows the multiplication and addition steps involved in the FFT. The bit reversal step is not shown and as a result the output is a scrambled version of the input. The arrow in this figure represents multiplication; lines without arrowheads are addition. Perform the multiplication first, then the addition.

This is a process of decomposing the input data stream into smaller and smaller subsequences; the multiplication by the twiddle factor is done before each summation, so the output sequence is in the proper order as it is produced.

There is another manner of doing this, though, and that is *decimation in frequency*, which decomposes the output sequence into smaller sub-sequences. Here, the multiplication comes after each summation. In this way, the routine reorders the DFT so that the data will leave in the proper order.

Code for both is given later in this chapter.

## Bit Reversal

An important part of the FFT is the bit reversal involved either in the beginning, as in decimation in time, or at the end, as in decimation in frequency algorithms. This process has the magical effect of somehow reordering the data into a usable vector.

Though this process may appear mysterious, it isn't. Bit reversal is the process of dividing the data sequence into smaller and smaller sub-sequences, separated into odd and even groupings. Let us take for an example the sequence of eight data points we used in the preceding FFT example If we wish to perform a radix 2 FFT on this data set, we must first separate the data sequence into odd and even parts:

$$(1\ 2\ 3\ 4\ 5\ 6\ 7\ 8) \rightarrow (1\ 3\ 5\ 7\ 2\ 4\ 6\ 8)$$

Now we have two sets of four-point data structures—but we want sets of two, so we must break it down again:

$$(1\ 3\ 5\ 7\ 2\ 4\ 6\ 8) \rightarrow (1\ 5\ 3\ 7\ 2\ 6\ 4\ 8)$$

Now we have pairs of data to find DFTs for. This is precisely what bit reversal does.

If we give each member of our input data stream a binary number that we call an index, we end up with information that looks like Table 3-5.

Then we can reverse the order of the bits in the binary numbers—that is, bit 0 becomes bit 2 and bit 2 becomes bit 0. If we then reorder the sequence based upon these bit-reversed indices, we have the results shown in Table 3-6.

Thus, the input stream is automatically reordered into pairs suitable for transformation. This is the same process used in the Hartley transform, where it is called *permutation*.

| *Index* | *Input data* |
|---|---|
| 000 | 1 |
| 001 | 2 |
| 010 | 3 |
| 011 | 4 |
| 100 | 5 |
| 101 | 6 |
| 110 | 7 |
| 111 | 8 |

**Table 3-5.**

| *Input data sequence* | *Index* | *Bit reversed index* | *Output data sequence* |
|---|---|---|---|
| 1 | 000 | 000 | 1 |
| 2 | 001 | 100 | 5 |
| 3 | 010 | 010 | 3 |
| 4 | 011 | 110 | 7 |
| 5 | 100 | 001 | 2 |
| 6 | 101 | 101 | 6 |
| 7 | 110 | 011 | 4 |
| 8 | 111 | 111 | 8 |

**Table 3-6.**

## THE HARTLEY TRANSFORM: A COMPLETELY REAL TRANSFORM

Before leaving this discussion of the Fourier transform, it seems appropriate to present another transform that is popularly used, especially in video processing. It is interesting to note the similarities in the derivation of the Fourier and Hartley transforms, as well as the similarities in their properties and theorems. There has been a good deal of debate upon which of the two transforms is faster, the fast Hartley or the fast Fourier transform. In the last few years, much of the debate has died down; most decided that it was a dead heat, or that the FFT (or one of its interpretations) had the edge. Even so, the Hartley transform is especially suited for particular applications, especially those involving real values. The Hartley transform also possesses a very useful convolution theorem, which, in certain forms, may be many times faster than that same convolution performed with the FFT.

The Hartley transform is a completely and symmetrically reciprocal real transform that produces a sequence of coefficients that is the same length as the data sequence that was input to it. It can do this because it is a real transform and does not require the redundancy necessary in the DFT to express both the real part and the sign-reversed imaginary part. An additional advantage is the complete symmetry of the transform, which makes any algorithm based upon this transform truly able to go both ways without change. The Hartley transform finds some of its greatest uses in image processing work, where one of the functions is almost always symmetrical, and in applications where phase is not of concern.

The Hartley transform was developed by Ralph V. L. Hartley. He is best known to electrical engineers for the Hartley oscillator, a circuit that has the ability to produce pure sinusoidal output and requires only a handful of parts. In 1942, he presented a set of relationships in "A More Symmetrical Fourier Analysis Applied to Transmission Problems" (Proceedings of the Institute of Radio Engineers, vol. 30, 1942) that function as a truly symmetrical transform—that is, the forward and inverse forms are identical:

$$\psi(\omega) = \frac{1}{\sqrt{2\pi}} \int_{-\infty}^{\infty} V(t) e^{j\omega t} dt$$

$$V(t) = \frac{1}{\sqrt{2\pi}} \int_{-\infty}^{\infty} \psi(\omega) e^{j\omega t} d\omega$$

An interesting and important facet of the Hartley transform is that the Fourier transform of a function can be derived from it as well. In a section earlier in this chapter, we discussed the even and odd qualities of a function as they pertain to its transform; here, we use them by separating $\psi(\omega)$:

$$\psi(\omega) = e(\omega) + o(\omega)$$

$$e(\omega) = \frac{\psi(\omega) + \psi(-\omega)}{2}$$

$$o(\omega) = \frac{\psi(\omega) - \psi(-\omega)}{2}$$

Then, noting that the odd and even parts are the sine and cosine components of the equation,

$$e(\omega) = \frac{1}{\sqrt{2\pi}} \int_{-\infty}^{\infty} V(t) \cos \omega t \, dt$$

$$o(\omega) = \frac{1}{\sqrt{2\pi}} \int_{-\infty}^{\infty} V(t) \sin \omega t \, dt$$

and recalling one of *Euler's identities*,

$$e^{-j\omega t} = \cos \omega t - j \sin \omega t$$

we can form the Fourier transform by recombining the even and odd parts so that

$$s(\omega) = e(\omega) - jo(\omega) = \frac{1}{\sqrt{2\pi}} \int_{-\infty}^{\infty} V(t) e^{-j\omega t} dt$$

To complete this presentation, here is the manner in which the Hartley transform of a function may be obtained from the Fourier transform:

$$\psi(\omega) = \text{Re}[H(\omega)] - \text{Im}[H(\omega)]$$

## Hartley Transform Theorems

Many of the Fourier theorems correspond directly with the Hartley theorems, though they are not all the same. Three of the most important are the convolution, similarity, and shift theorems.

The convolution theorem obeyed by the Hartley transform differs somewhat from that of the Fourier transform, in that it is not a simple point-by-point multiplication. Instead, one of the DHTs is separated into its odd and even parts for the multiplication and sum:

$$H(v) = H(v)G_e(v) + H(-v)G_o(v)$$

Clearly, if one of the functions is purely even or purely odd, the operation is significantly simplified. If the function $G(v)$ is odd, then $G_e(v) = 0$, and the resulting convolution is

$$H(v) = H(-v)G_o(v)$$

In certain cases, this can give a convolution based upon this theorem a real advantage over the same convolution done using Fourier theorems. Two very important theorems for the development of the Hartley transform are the similarity theorem and the shift theorem. The similarity theorem plays much the same role in the fast Hartley transform as it does in the FFT. It states that if you wish to expand a sequence by inserting zeros between neighboring values, you will have a repeating transform—that is,

$$DHT\{a\ 0\ b\ 0\ c\ 0\ d\ 0\} = A\ B\ C\ D\ A\ B\ C\ D$$

The shift theorem also plays as important a part in the construction of the fast Hartley transform as it does in the FFT, though it differs from that of the fast Fourier transform:

$$DHT[f(t+a)] = H(v)\cos\left(\frac{2\pi a v}{N}\right) - H(-v)\sin\left(\frac{2\pi a v}{N}\right)$$

Together, they allow us to accomplish the transform in small pieces, expand the pieces with the similarity theorem, then shift the pieces into position with the shift theorem.

Examples of this theory are given in greater depth in the next section, when we present code based upon the Hartley transform.

## WINDOWING

Now that we have described some methods for transforming sequences from the time domain to the frequency domain, it is appropriate to say a word about the result. The spectrum we have as a result of this transformation may be inaccurate and misleading, because we are using only a small finite number of data points for any particular transformation. The reason for the inaccuracy is that convergence is always more rapid for some values than others. The effect this has on the Fourier transform and possible remedies is the subject of the next few paragraphs. (For the following discussion, please refer to the Mathcad document winfft.mcd.)

Analysis on data sampled in finite intervals called *windows* can exhibit distortion in the spectrum. The Fourier transform is applicable to continuous-time nonperiodic data integrated from $-\infty$ to $+\infty$; in this case, a sinusoid will have a single peak. Taking only a small portion of that range for analysis, however, is equivalent to multiplying the signal by a rectangle which, on the frequency domain, has the effect of convolving the spectrum of the signal with the spectrum of the rectangle. This distortion can manifest itself in a number of problems, including broadening of the spectral peaks and difficulty in separating frequencies. This distortion is often called *leakage*, and is usually lessened with longer record lengths or with *window weighting*, a subject to be discussed shortly.

Before going on, we would like to develop a clear picture of what occurs when we convolve the spectrum of the rectangle, a $\sin\omega/\omega$ envelope, with the spectrum of the signal. You will recall that the spectrum of a pulse train is a pulse train whose period on the frequency axis is equal to the frequency of the pulse train itself. A sampling function is a sequence of sampling pulses at a fixed frequency with an origin and a finite length, such a sampling function is depicted in Figure 3-17.

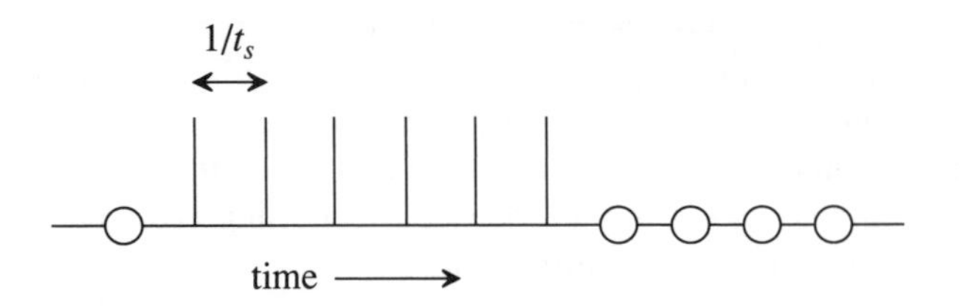


**Figure 3-17.** The sampling function is zero everywhere, except during the sampling sequence.

The spectrum of the sampling function is a $\sin\omega/\omega$ function repeated along the frequency axis at intervals equal to the frequency of the sampling function, $1/t_s$. The width of the main lobe is equal to the inverse of the product of the number of sampling pulses in the sequence, with the frequency of the pulse train:

$$width = \frac{1}{samples * t_s}$$

Figure 3-18 illustrates this.

From the formula, it is clear that the fewer samples in the sampling function, the broader the $\sin\omega/\omega$ peak. If we create two sampling functions at 10 KHz, one with 128 impulses and another with 1024, this formula tells us that the sequence with the fewer samples is much broader in frequency than the other.

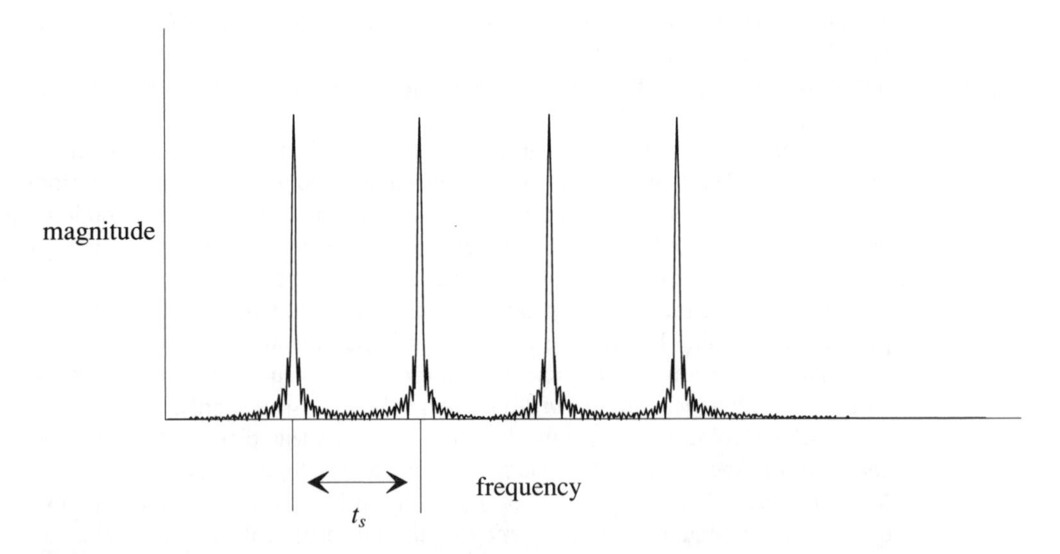


**Figure 3-18.** The spectrum of the sampling function is repeated along the frequency axis with a period equal to the sampling frequency.

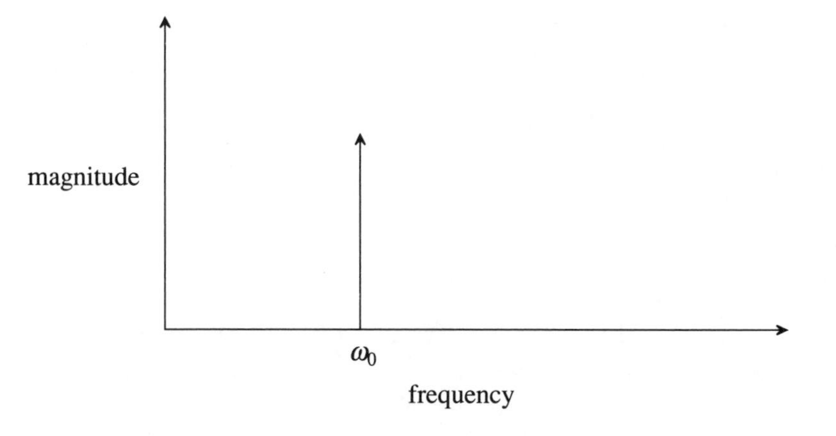

**Figure 3-19.** The ideal spectrum of a sinusoid

The ideal spectrum of a sinusoid comprises a single peak at a particular frequency. An illustration of this is given in Figure 3-19. If we convolve these two spectra, we have what is shown in Figure 3-20.

You will notice the broadening and flattening of the response. As noted before, the fewer samples in the sampling function, the broader the $\sin\omega/\omega$ peak and, therefore, the broader the resulting spectrum of the *sampled* sinusoid. As you can see, the exact peak can be difficult to find, but what happens if we have more than one frequency in the spectrum? What happens if they are at all close? See Figure 3-21 for a comparison of sampled spectra and ideal spectra for the sum of two sinusoids, one at 20 Hertz and the other at 100 Hertz, sampled at 1024 Hertz.

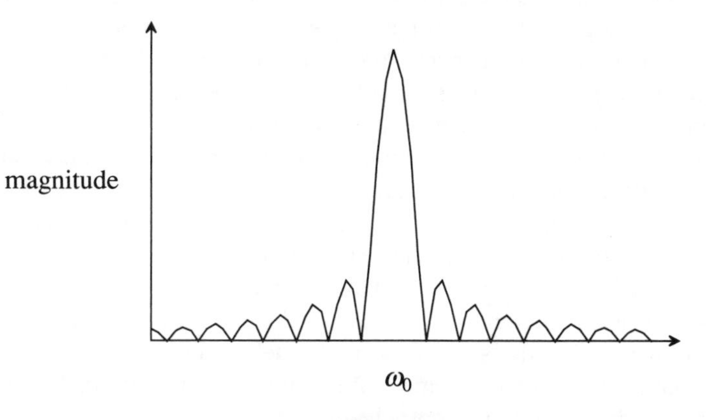


**Figure 3-20.** The result of the convolution of a rectangular sampling function with the ideal spectrum of a sinusoid

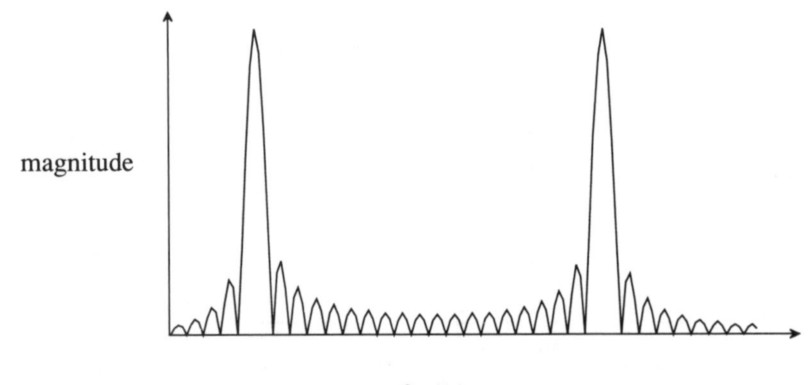


**Figure 3-21.** The spectra of two sinusoids, one ideal, the other sampled at 1024 Hertz with a rectangular window

What we are seeing here is called *spectral leakage*, discussed in the first chapter of this book as the Gibbs phenomenon. For the Fourier transform, this is caused by the fact that the signal is seen as aperiodic and the edges are interpreted as sudden divergences to zero. The discrete Fourier transform extends the truncated signal as a discontinuous periodic that exists throughout the time domain. Neither is accurate.

You can avoid this leakage by choosing a window length that contains an integer number of periods of the frequency you are sampling, In that case, you will have only one peak in the spectrum for that frequency, because the rest of the frequencies fall on zeros of the $\sin x/x$ function. If the window does not contain an integral number of periods, the sampled frequency values will miss the zeros and fall on the sidelobes instead. Since adjusting the window to fit an integral multiple of the frequency is usually not possible, this problem is usually solved or lessened with longer records and window weighting.

A number of weighting schemes have been developed for windows that attempt to minimize the ringing and overshoot that result from the discontinuous edges of the rectangular sample length. The idea is to taper the impulse response instead of simply truncating it.

The window types that we will discuss attempt to narrow the central lobe of the $\sin\omega/\omega$ envelope and attenuate sidelobe ringing. This is accomplished by tapering the window with various functions described in the next section.

Since the sidelobes grow smaller as you move away from the discontinuity at the edges of the window, it is common to measure a window by taking the ratio of the first sidelobe to the mainlobe, noting the size of the first sidelobe in dB. A good window is one whose spectrum possesses as narrow a mainlobe as possible and the smallest possible sidelobes. Unfortunately, these points are contradictory.

In the next section, definitions of various popular windows are presented. The definitions result in sequences, $w[n]$, that are multiplied point by point by the input sequence. This process shapes the resulting sequence, which is then submitted to the transform. We will deal with windows again in Chapter 4, where they will be used in conjunction with the Fourier series to produce linear phase FIR filters. Software for developing the windowing coefficients appears at the end of this chapter.

## Window Types

In the following definitions, $M$ will represent the width of the rectangular pulse, $n$ will stand for the data increment, and coefficients particular to the window are presented in the definitions. Please refer to the Mathcad document Winfft.mcd for more on the system reponses of these windows. Here are some of the more common window functions:

1. *Rectangular*—this is the default window function. It is the result of simply taking a finite length of sampled data:

$$w(n)=\begin{cases}1,\ 0\le n\le M\\0,\ \text{everywhere else}\end{cases}$$

   This window has its first sidelobe down $-13$dB from the main lobe. The system reponse for this window is given in Figure 3-22.

2. *Cosine bell*—this window is one of the first in a series that uses cosines with the rectangular window to taper the window and attenuate the sidelobes to bring the edges closer to zero. It accomplishes this by raising the cosine to the second power:

$$w(n)=\begin{cases}0.5-0.5\cos\left(2p\dfrac{n}{M}\right),\ 0\le n\le M\\0,\ \text{elsewhere}\end{cases}$$

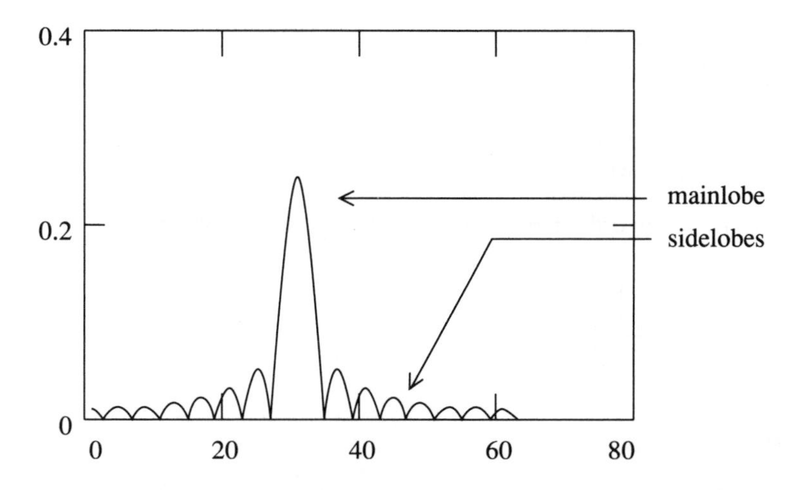


**Figure 3-22.** This is the response of a rectangular window: the mainlobe is in the center with the sidelobes rippling away from it on both sides.

The raised cosine window keeps the highest sidelobe −31dB down from the mainlobe.

3. *Hamming*—this window tries to cancel the first sidelobe of the sin$x$/$x$ function. The sidelobes can impair or obscure low-amplitude peaks in the spectrum. The definition of this window is

$$w(n)=\begin{cases}\alpha+(1-\alpha)\cos\left(2\pi\frac{n}{M}\right), & 0\le n\le M\\ 0, \text{ elsewhere}\end{cases}$$

where $N$ is the length of the vector, $n$ is the data increment, and $\alpha$=0.543478261 or $\alpha$=0.54 depending upon the precision you want or are capable of holding. The highest sidelobe in a Hamming window is −43dB below the mainlobe.

4. *Blackman*—this scheme is aimed at cancelling the effect of the third and fourth sidelobes while leaving the first. This window also brings the edges of the rectangle to zero. The definition of this window is

$$w(n)=\begin{cases}\alpha_0-\alpha_1\cos\left(2\pi\frac{n}{M}\right)+\alpha_2\cos\left(4\pi\frac{n}{M}\right), & 0\le n\le M\\ 0, \text{ elsewhere}\end{cases}$$

The Blackman window can attenuate the third and fourth sidelobes down 60dB, but broadens the central peak to about two times that of the rectangular window.

5. *Kaiser/Bessel*—this is easily the most complex of the windowing schemes offered, but one that is very popular. Its concentration is upon putting the energy into the central peak. Its definition is:

$$w(n)=\begin{cases}\dfrac{I0\left[pa\sqrt{1-\left(\dfrac{n}{\frac{M}{2}}\right)^2}\right]}{I0(pa)}\\ 0, \text{ elsewhere}\end{cases} \tag{Eq. 3-4}$$

$I_0$ is the zeroth order Bessel function of the first kind, given by

$$I_0=1+\sum_{n-1}^{\infty}\left[\frac{\left(\frac{n}{2}\right)^n}{n!}\right]^2$$

and

$$\alpha=\frac{M}{2}$$

This window is considered near optimal in concentrating the window function around $\omega=0$ in the frequency domain.The mainlobe-to-sidelobe attenuation is about the same as for the Blackman window, however.

Smaller sidelobes result in better approximations at the edges, and narrower transition bands can be achieved by increasing the width of the window, $M$. The promise of the Kaiser window is that these things are programmable. Simply by varying $M$ and $\beta$ in Equation 3-4, you change window length and shape, thereby adjusting sidelobe amplitude and mainlobe width.

Choosing the correct window for your application is not as simple as it may seem. The Kaiser/Bessel window is probably the most facile on this list and will often produce the best results. Even so, it may take several attempts to optimize a fit.

# SOFTWARE EXAMPLES

## 'Classic' FFT in C: Decimation in Time

This version of the FFT is based, loosely, on a popular method found in C, BASIC, and FORTRAN routines everywhere. In these versions, the cosine term used with the sine to produce the twiddle factors is derived using the half-angle formula instead of an explicit call

$$\cos(a) = 1 - 2\sin^2\left(\frac{a}{2}\right)$$

or

$$\sin\left(\frac{a}{2}\right) = \sqrt{\frac{1-\cos(a)}{2}}$$

Presumably, this is done to reduce the number of trigonometric functions and speed the algorithm. In the form presented here, however, this has been changed to a simple cosine, since the speed of the math coprocessor trivializes any advantage gained by the previous method. In addition, the explicit use of the cosine function increases the readability and clarity of the procedure.

The initial twiddle values are calculated at the beginning of the loop, and each succeeding iteration is accomplished with a complex multiplication at the end of each set of butterflies. As is the case with a *decimation in time FFT*, the data set is reordered as one of the first steps of the program; it splits out odd and even time samples into pairs with a bit-reversal technique that results in a correctly ordered vector at the end of the complex multiplications and divisions. To save space, the computations are done *in place*; in this way, the only space required is for the input vector itself. To simplify program management and documentation, the storage for each complex value is defined in a structure called *cnum*.

```
struct cnum
   {
   double real;
    double imag;
    };
```

```
void fft(struct cnum data[], int nn, int isign)
/*the data of length nn arrives in struct cnum data[], int isign
specifies a forward or reverse transform*/
{
    int n,mmax = 0,m,j,istep,i, mtmp;
    double wtemp,wr,wpr,wpi,wi,theta;
    double br;
    struct cnum twiddle;
    struct cnum phase;
    struct cnum temp;   /*temporary value*/

    /*bit reversal*/
    n = nn;
    mtmp = nn>>1;
    j=0;
    for (i=0;i<n;i++) {
    if (j > i) {        /*swap the complex numbers*/
          wtemp=data[j].real;
          data[j].real=data[i].real;
          data[i].real=wtemp;
          wtemp=data[j].imag;
          data[j].imag=data[i].imag;
          data[i].imag=wtemp;
    }
       m=mtmp;
       while (j >= m && m > 0) {
          j -= m;
          m >>= 1;
          }
       j += m;
    }

    /*butterflies*/
    mmax = 1;
    while (n > mmax) {
    istep=2*mmax;
    theta=6.28318530717959/(isign*istep);     /*isign determines*/
                                              /*forward*/
    phase.real = cos(theta);
    phase.imag = sin(theta);
    twiddle.real=1.0;
    twiddle.imag=0.0;
    for (m=0;m<mmax;m++) {
       for (i=m;i<n;i+=istep) {
           j=i+mmax;          /*actual butterfly*/
           temp.real=twiddle.real*data[j].real-twiddle.imag*data[j].imag;
           temp.imag=twiddle.real*data[j].imag+twiddle.imag*data[j].real;
           data[j].real=data[i].real-temp.real;
           data[j].imag=data[i].imag-temp.imag;
           data[i].real += temp.real;
```

```
                data[i].imag += temp.imag;
            }                   /*calculate new twiddle*/
            twiddle.real=(wtemp=twiddle.real)*phase.real-twiddle.imag*phase.imag;
            twiddle.imag=twiddle.imag*phase.real+wtemp*phase.imag;
        }
            mmax = istep;
        }
        if(isign < 0) return;
        br = 1.0/n;             /*this is Bracewell's scaling factor*/
        for (i = 0; i < n; i++) {      /*it may be altered for another form*/
            data[i].real *= br;
            data[i].imag *= br;
            }
    }
```

## A Faster FFT

When we are dealing with realtime data input, we need as much speed as possible. Processing speed is the limiting factor on what bandwidth we are able to handle. A standard 8086 with coprocessor requires approximately 20 milliseconds to compute a 1024 integer-only FFT, which means that it can approach 50 full-spectrum analyses per second, provided that it is doing nothing else. DSPs are currently being produced that will calculate similar length floating-point FFTs in less than 500 microseconds, permitting a theoretical maximum throughput of 2000 FFTs per second.

The next routine is used in a program found on the optional disk called Oscill.cpp, among others; the source can be found in Fftbits.asm. Oscill.exe reads data from the DACQ card, as well as reading and writing files stored on disk, for processing and analysis. This program can be used in conjuction with Mathcad to generate data and display it. See the README document on the disk for further information.

*FFT* is an assembly-language version of a decimation-in-frequency FFT; it uses either the 80387 or the built-in coprocessor on the 80486 to do the hard work of performing a floating-point FFT. Though the routine is written for use with a math coprocessor to enhance speed, it was also intended as a demonstration of the FFT in assembly language, where clarity ranked above speed. For still greater speed, this program could be rewritten using fixed point arithmetic. There is such an example written for the 80C196 microcontroller in Chapter 6.

A common technique for fast arithmetic is the use of tables. We use a table in this routine to store the twiddle factors, which are computed once at the beginning of the program. Of course, they need not be computed—they might be stored in a header file instead, or stored in ROM, depending upon the needs and proclivities of the developer.

```
#if BRACEWELL  /*results in a transform consistent with Bruce
Bracewell's definition*/
    arg=2*PI/(-ivar*N);        /*ivar is the sign of j*/
    for(i=0;i<N;i++)
        {
```

```
        W_FACTOR[i].real=cos((i*arg));
        W_FACTOR[i].imag=-sin((i*arg));
}
#else
    arg=2*PI/(ivar*N);          /*ivar is the sign of j*/
    for(i=0;i<N;i++)
        {
        W_FACTOR[i].real=cos((i*arg));
        W_FACTOR[i].imag=-sin((i*arg));
}
#endif
```

The following routine is callable from a C language program:

```
fft     proc far c uses ds es eax ebx ecx edx,
            fvector:dword, w:dword, stride:word

    local   stages:word, step:word, index:word, legdif:word, lower:word,
        upper:word, i:word, j:word, k:word, temp1r:qword, temp1i:qword,
        temp2r:qword, temp2i:qword

    pushad                          ;push registers
    fninit                          ;reset coprocessor

    mov     ax, 1024                ;define step length
    xor     dx, dx
    div     stride
    mov     step, ax

    mov     stages, 0               ;get log2 of stride
    mov     bx, 1
log2:
    inc     stages
    shl     bx, 1
    cmp     bx, stride
    jb      log2

    mov     ax, stride
    shr     ax,1
    mov     legdif, ax              ;difference between legs

    mov     i, 0

do_stages:
    mov     index, 0                ;pointer to proper w factor

    mov     j, 0

legs:
    mov     ax, j                   ;upper leg
    mov     upper, ax
```

```
        lds     si, fvector                     ;complex input array
        les     di, dword ptr w                 ;w factor

butterfly:
                                                ;butterfly
        mov     cx, index                       ;index into W_FACTOR
        shl     cx, 4                           ;correct for word length
        mov     bx, upper                       ;set upper leg
        shl     bx, 4
        mov     dx, upper
        add     dx, legdif                      ;set lower
        shl     dx, 4                           ;correct for two quad words

        fld     qword ptr [si][bx]              ;point to upper real
        fld     st                              ;duplicate for subtraction
        xchg    bx, dx                          ;point at lower        real
        fadd    qword ptr [si][bx]              ;add real parts
        fstp    templr

        fsub    qword ptr [si][bx]              ;subtract real parts
        fstp    qword ptr temp2r
        xchg    bx, dx                          ;point at upper

        fld     qword ptr [si][bx]+8
        fld     st                              ;duplicate stack
        xchg    bx, dx                          ;point at lower
        fadd    qword ptr [si][bx]+8            ;add imaginary parts
        fstp    qword ptr templi

        fsub    qword ptr [si][bx]+8            ;add
        fstp    qword ptr temp2i

        xchg    bx, cx                          ;get index into twiddles
        fld     qword ptr es:[di][bx]+8         ;W_FACTOR imaginary
        fld     qword ptr es:[di][bx]           ;W_FACTOR real
        fmul    qword ptr temp2r
        fxch
        fmul    qword ptr temp2i
        fsub
        xchg    bx, cx
        fstp    qword ptr [si][bx]              ;put in data lower real

        xchg    bx, cx
        fld     qword ptr es:[di][bx]+8         ;W_FACTOR imaginary
        fld     qword ptr es:[di][bx]           ;W_FACTOR real
        fmul    qword ptr temp2i
        fxch
        fmul    qword ptr temp2r
        fadd
        xchg    bx, cx                          ;point at lower
        fstp    qword ptr [si][bx]+8            ;put in data lower imag
```

```
    xchg    bx, dx                      ;get upper back
    fld     qword ptr templr
    fstp    qword ptr [si][bx]          ;real
    fld     qword ptr templi
    fstp    qword ptr [si][bx]+8        ;imag


    mov     ax, legdif                  ;reduce distance between arms
    shl     ax, 1
    add     ax, upper
    cmp     ax, stride
    mov     upper, ax
    jb      butterfly                   ;done with butterfly yet?

    mov     ax, index                   ;rotate index around circle
    add     ax, step
    mov     index, ax

    mov     ax, j
    inc     ax
    mov     j, ax
    cmp     ax, legdif                  ;done with legs yet?
    jb      Legs


    shr     legdif, 1
    shl     step, 1

    inc     i
    mov     ax, i
    cmp     ax, stages
    jb      do_stages

    mov     j, 0                        ;bit reversal
    mov     i, 1
decim:

    mov     ax, stride                  ;k=stride>>1
    shr     ax, 1
    mov     k, ax

reduce:
    mov     ax, k                       ;while(j>=k && k>0)
    cmp     ax, j
    jg      no_reduction
    mov     ax, j                       ;j-k
    sub     ax, k                       ;k>>1
    mov     j, ax
    shr     k, 1
    jmp     reduce

no_reduction:
```

```
        mov     ax, j                       ;j += k
        add     ax, k
        mov     j, ax

        mov     ax, i
        cmp     ax, j
        jge     chek_exit

        mov     bx, j
        mov     dx, i
        shl     bx, 4
        shl     dx, 4                       ;swap complex numbers

        fld     qword ptr [si][bx]          ;real
        fstp    qword ptr templr
        fld     qword ptr [si][bx]+8        ;imag
        fstp    qword ptr templi

        xchg    bx, dx
        fld     qword ptr [si][bx]          ;real
        xchg    bx, dx
        fstp    qword ptr [si][bx]
        xchg    bx, dx
        fld     qword ptr [si][bx]+8
        xchg    bx, dx
        fstp    qword ptr [si][bx]+8

        xchg    bx, dx
        fld     qword ptr templr
        fstp    qword ptr [si][bx]
        fld     qword ptr templi
        fstp    qword ptr [si][bx]+8

chek_exit:
        inc     i
        mov     ax, i
        mov     bx, stride
        dec     bx
        cmp     ax, bx
        jb      decim                       ;done with bit reversal?
                                            ;yes, rearrange for display
        mov     i, 1
rearrange:
        mov     ax, stride
        sub     ax, i
        mov     j, ax

        mov     bx, j
        mov     dx, i
        shl     bx, 4
        shl     dx, 4
```

```
        fld     qword ptr [si][bx]          ;real
        fstp    qword ptr templr
        fld     qword ptr [si][bx]+8        ;imag
        fstp    qword ptr templi

        xchg    bx, dx
        fld     qword ptr [si][bx]          ;real
        xchg    bx, dx
        fstp    qword ptr [si][bx]
        xchg    bx, dx
        fld     qword ptr [si][bx]+8
        xchg    bx, dx
        fstp    qword ptr [si][bx]+8

        xchg    bx, dx
        fld     qword ptr templr
        fstp    qword ptr [si][bx]
        fld     qword ptr templi
        fstp    qword ptr [si][bx]+8

        inc     i
        mov     ax, i
        mov     bx, stride
        shr     bx, 1
        cmp     ax, bx
        jb      rearrange

        popad                               ;pop registers
        ret
fft     endp
```

## Real and Complex FFTs and IFFTs

The FFT is, by nature, a complex valued operation. In a waveform, magnitude is valued by real numbers and phase is imaginary. Sometimes, however, the imaginary value is not needed, if all we are doing is finding the transform of a real input vector. It is possible to use the routines we have presented thus far to do both, by merely setting all the imaginary parts of the number to zero, thus producing a real vector—but this wastes memory. An obvious possibility is the Hartley transform, which specializes in real valued vectors. Before we get to that, though, let's look at a way to use the routines already have, including the other FFTs in this chapter, to transform two real input sequences at once.

The real and imaginary parts of the transform are separate and symmetrical. By putting the data from one real sequence into the real part of the input, and data from the other sequence into the imaginary part, we can do both at once. When we are done, we take the even and odd parts of the output as the separate outputs, and then reassemble these into the buffers the input data arrived in.

The routine that follows allows a standard FFT, like the ones that appear in this chapter, to run two real vectors simultaneously, providing twice the throughput in the same space and time. The routine here is used on the disk as part of a demonstration of performing a convolution using FFTs. It is a modification of a similar idea in Press and Vetterling, *Numerical Recipes in C* (Cambridge University Press, 1988).

```
void twinfft(struct cnum far data0[], struct cnum far data1[], struct cnum
                far ovector[], struct cnum far w[], int nn)
{
    int i, j;
    struct cnum temp0, temp1;


    for(i = 0; i<nn; i++) {
        ovector[i].real = data0[i].real;/*real part of vector one in */
                                        /*real part of input vector*/
        ovector[i].imag = data1[i].real;/*real part of vector two in*/
                                        /*imag part*/
        }

    fft(ovector, w, nn);

    data1[0].real = ovector[0].imag;    /*redistribute*/
    data0[0].real = ovector[0].real;
    data0[0].imag = data1[0].imag = 0.0;

    for(i=1;i<nn;i++) {     /*vector one is entirely contained in the*/
                            /*real part of the output*/
            /*vector two is in the imaginary part*/
        temp0.real = 0.5*(ovector[i].real+ovector[nn-i].real);
        temp1.real = 0.5*(ovector[i].real-ovector[nn-i].real);
        temp0.imag = 0.5*(ovector[i].imag+ovector[nn-i].imag);
        temp1.imag = 0.5*(ovector[i].imag-ovector[nn-i].imag);
        data0[i].real = temp0.real;
        data0[i].imag = temp1.imag;
        data0[nn-i].real = temp0.real;
        data0[nn-i].imag = -temp1.imag;
        data1[i].real = temp0.imag;
        data1[i].imag = -temp1.real;
        data1[nn-i].real = temp0.imag;
        data1[nn-i].imag = temp1.real;
        }

}
```

## The Discrete Hartley Transform

The theory behind the discrete Hartley transform was given earlier in this chapter. In this section, we present a C rendition of the algorithm for demonstration purposes only. This is not meant for actual application because, like the discrete Fourier transform, it is very time-consuming.

The discrete Hartley transform is given by

$$H(v) = N^{-1}\sum_{t=0}^{N-1} f(t)cas\left(\frac{2\pi vt}{N}\right)$$

and its inverse,

$$f(t) = \sum_{t=0}^{N-1} H(v)cas\left(\frac{2\pi vt}{N}\right)$$

In these equations, $cas\theta=\cos\theta+\sin\theta$. As you will notice, the only real difference between the discrete Hartley transform and the DFT is the lack of the multiplier $-j$. These equations are easily implemented in code, as we demonstrate in the following fragment:

```
#define CAS(z,nn) cos(PI4*z/nn)+sin(PI4*z/nn)        /*cas theta*/
#define CTWID(n,nn) cos(PI2*n/nn)         /*cosine twiddles*/
#define STWID(n,nn) sin(PI2*n/nn)         /*sine twiddles*/

void htly(double idata[], double odata[], int nn)
{
    int stages=0, j, i, step, index, leg_dif, lower, upper, k, level, z, n;
    int iter;
    double a, b;
    double temp, tempr;


    for(i = 0; i<nn/2; i++) {
        temp = fvector[(nn/2)-i] - fvector[i];
        temp *= STWID(i, nn);
        a = fvector[i] + fvector[i+(nn/2)];
        b = fvector[i] - fvector[i+(nn/2)];
        b *= CTWID(i,nn);
        b += temp;
        tbuf[i] = a;
        tbuf[i+(nn/2)] = b;
        }

    for(k = 0; k<nn/2; k++) {
        temp = 0;
        for(i = 0; i<nn/2; i++) {
            tempr = CAS(i*k,nn);
            temp += (tempr * tbuf[i]);
            }
        fvector[k*2] = temp;
        }

    for(k = 0; k<nn/2; k++) {
        temp = 0;
```

```
        for(i = 0; i<nn/2; i++) {
            tempr = CAS(i*k,nn);
            temp += (tempr * tbuf[i+nn/2]);
            }
        fvector[(k*2)+1] = temp;
        }


}
```

## The Fast Hartley Transform

As you can see, evaluation of the Hartley transform as it is written results in a lengthy operation with $N^2$ multiplications involving the cas function. Notice the similarities between the components of the FFT and the components of the FHT. Just as we did in the decimation in time FFT, the first step in the FHT is to divide the input sequence into smaller and smaller pieces, finding the DHT of the smallest sequences, then expanding and realigning them to form the DHT of the larger sequence. By doing this in an orderly manner, we can reduce the number of evaluations of the cas function to $N\log_2 N$.

To begin with, we use a method similar to that used in the FFT to reduce the number of multiplications: we break the sequence down into two-point vectors where the DHT involves only addition and subtraction:

$$DHT(a,b) = \frac{(a+b \;\; a-b)}{2}$$

If we can arrange the data so that we begin by performing this DHT on small subsequences that then become larger sequences, while we shift the data so that it always aligns, we will have accomplished our goal.

The formula that describes this decomposition is

$$H(v) = A(v) + B_e(v)\cos\left(\frac{2\pi v}{N}\right) + B_o(N-v)\sin\left(\frac{2\pi v}{N}\right)$$

Let us take an eight-point vector as an input to the FHT:

$$\{1 \;\; 2 \;\; 3 \;\; 4 \;\; 5 \;\; 6 \;\; 7 \;\; 8\}$$

The fast Hartley transform comprises two main steps: the initial permutation or bit reversal section, and the combine section, which involves shift and expansion as well as multiplication. Regardless of the length of the DHT, the first two shifts and expansions require multiplications by sines and cosines that are all either 1, $-1$, or 0. As a result, these steps reduce to additions and subtractions. That same process continues, as the granularity of the phase factors becomes finer and finer and the width of the output doubles, until we have $\log_2 N$ sets of $N$ data points. It is at this point that we make the final combination and we have the result. See Figure 3-23. (Please refer to the Mathcad document Hartley8.mcd for more illustration.)

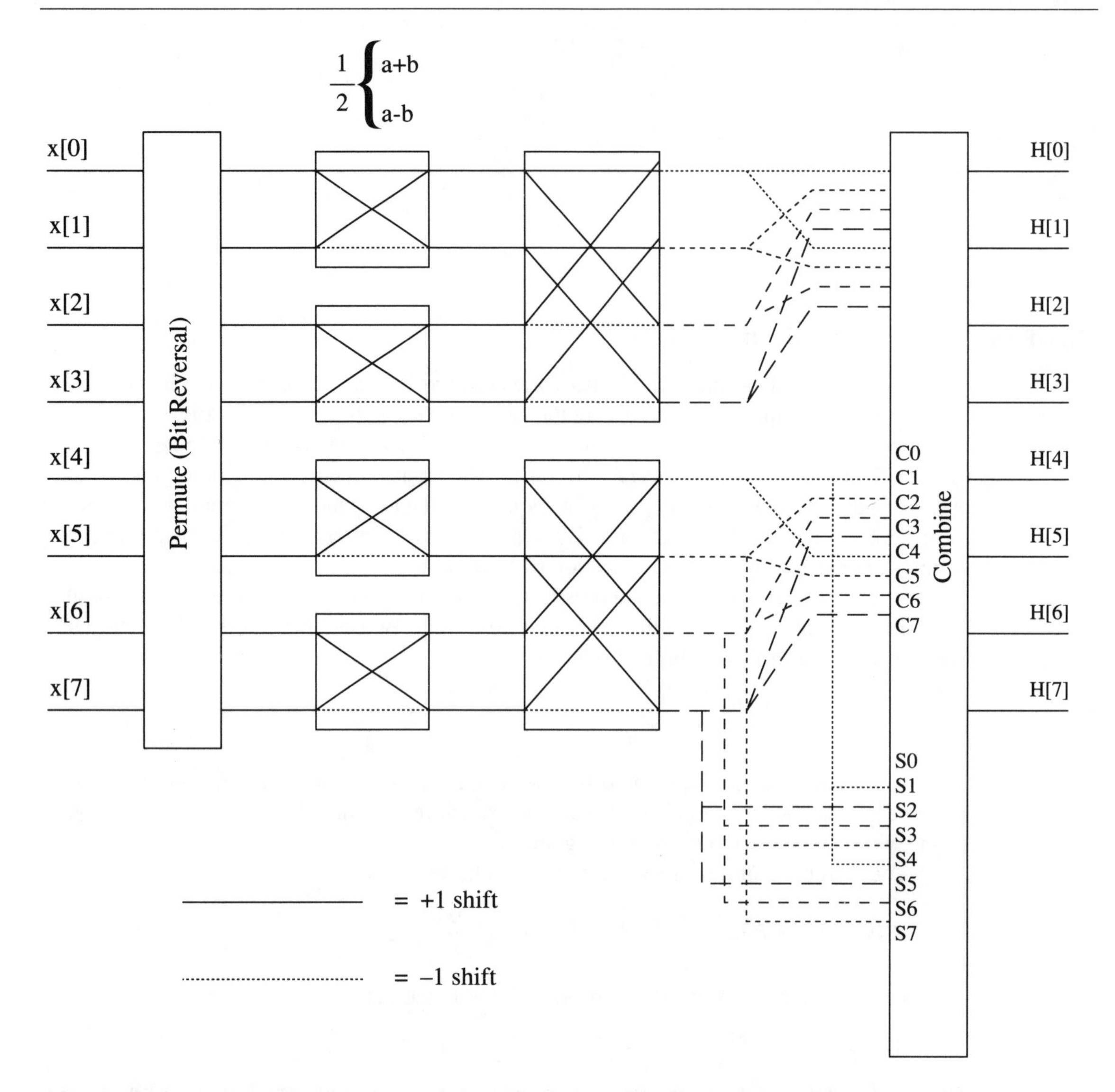


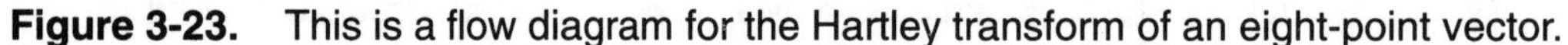

**Figure 3-23.** This is a flow diagram for the Hartley transform of an eight-point vector.

The following step-by-step discussion provides a more detailed view of the fast Hartley transform.

1. The first step is called *permute*, and is a rearrangement of the input sequence into data pairs. The idea is to separate the input data into sets of four data points and then separate them into data pairs, but this is hidden in what follows. The first

action is to take the $\log_2$ of the input sequence to determine the number of two-point vectors necessary. We then reorder that sequence based upon bit reversal:

$$\{1\ 2\ 3\ 4\ 5\ 6\ 7\ 8\} \rightarrow \{1\ 5\ 3\ 7\ 2\ 6\ 4\ 8\}$$

This operation is similar to what occurs in the opening moves of a decimation in time FFT.

2. Once the data pairs have been formed, we perform the additions and subtractions necessary to derive the simplest DHT. For example,

$$\{1\ 5\} = \frac{\{6\ -4\}}{2}$$

As you can see, the definition of the DHT of two points includes a division by two, but we will not be doing this now; it can be done more efficiently at the end. The number of divisions necessary is contained in the $\log_2$ of the record length and will be taken care of later. Table 3-7 is an illustration of the procedure just described.

We process the data pairs, and then these pairs are added via the similarity theorem and shifted to maintain the proper phase relationship to form sequences of four data points. At this point, the phase rotations called for by the shift theorem are 0, 1, and $-1$, and require only addition and subtraction.

After these two stages of taking two- and four-point DHTs, we have the results shown in Table 3-8.

3. We continue this process, except that now we are using phase factors of $\sin(2\pi/N)$ and $\cos(2\pi/N)$, where $N$ is the total number of data points involved. This operation is called *combine*, and can occur a number of times, depending upon the length of the input data sequence $(N\log_2 N)-2$. Each time it occurs, the output sequence increases in length a power of two until there are $\log_2 N$ sets of $N$ data points each.

The phase factors for an eight-point DHT are shown in Table 3-9.

| *Input data* | *Solve for DHT* | *Output data* |
|---|---|---|
| 1 | 1+5= | 6 |
| 5 | 1−5= | −4 |
| 3 | 3+7= | 10 |
| 7 | 3−7= | −4 |
| 2 | 2+6= | 8 |
| 6 | 2−6= | −4 |
| 4 | 4+8= | 12 |
| 8 | 4−8= | −4 |

**Table 3-7.**

In our example, this step occurs as two sub-sequences of four are expanded to three sequences of eight points each. The corresponding sequence numbers from each of the sets of eight points are combined to form the final output shown in Table 3-10.

| *Input from last step* | *Solve for 4 point DHT* | *Output* |
|---|---|---|
| 6 | 6 + 10 = | 16 |
| − 4 | − 4 + ( − 4) = − 8 | − 8 |
| 10 | 6 − 10 = | − 4 |
| − 4 | − 4 − ( − 4) = | 0 |
| 8 | 8 + 12 = | 20 |
| − 4 | − 4 + ( − 4) = | − 8 |
| 12 | 8 − 12 = | − 4 |
| − 4 | − 4 − ( − 4) = | 0 |

**Table 3-8.**

| *Increment* | *Cosine* | *Sine* |
|---|---|---|
| 0 | 1 | 0 |
| 1 | 0.707 | 0.707 |
| 2 | 0 | 1 |
| 3 | −0.707 | 0.707 |
| 4 | −1 | 0 |
| 5 | −0.707 | −0.707 |
| 6 | 0 | −1 |
| 7 | 0.707 | −0.707 |

**Table 3-9.**

| *Sequence number* | *Input subsequences* | *Solve for DHT∗N* | *DHT∗N* |
|---|---|---|---|
| f0 | 16 | 16 + 20C0 + 20S0 = | 36 |
| f1 | −8 | −8 + −8C1 + 0S1 = | −13.657 |
| f2 | −4 | −4 + (−4)C2 + (−4)S2 = | −8 |
| f3 | 0 | 0 + (0)C3 + (−8)S3 = | −5.657 |
| f4 | 20 | 16 + 20C4 + 20S4 = | −4 |
| f5 | −8 | −8 + (−8)C5 + (0)S5 = | −2.343 |
| f6 | −4 | −4 + (−4)C6 + (−4)S6 = | 0 |
| f7 | 0 | 0+(0)C7+(−8)S7 = | 5.657 |

**Table 3-10.**

4. Finally, we divide the 1/2 out of the results that we put off from step 2. Now, however, this is

$$\frac{1}{2^{\log_2 N}}$$

or in the case of our example, 1/8. Therefore,

$$DHT\{1\ 2\ 3\ 4\ 5\ 6\ 7\ 8\} = \{4.5\ -1.707\ -1\ -.707\ -0.5\ -0.293\ 0\ 0.707\}$$

As we indicated earlier, the DFT can be developed from the DHT by merely taking the even part of the DHT to be the real part and the negative odd part to be the imaginary part:

$$DFT\{\mathrm{Re}(v)+\mathrm{Im}(v)\} = H(v)_e + H(-v)_o$$

The power spectrum is equal to

$$Z^2 = \frac{[H(v)]^2 + [H(-v)]^2}{2}$$

## A Fast Hartley Transform Algorithm

This algorithm is a simple and straightforward implementation of the foregoing discussion. For speed, the swizzle (sine and cosine phase) factors are calculated ahead of time and stored in an array called *table*.

```
cosine = &sine[data_points];  /*sine and cosine share the same */
                              /*space, with cosine */
                              /*starting after sine*/

step = 2*PI/data_points;    /*phase angle is equal to the unit circle*/
                            /*divided by the number of data points*/
twid = 0;
for(i = 0; i<data_points; i++) {
    sine[i] = sin(twid);
    cosine[i] = cos(twid);
    twid += step;
    }
```

The data arrives at the routine in *lastbuf*, one of two buffers passed. Using two buffers was a convenient solution to the requirement that the data be processed without overwriting old data.

```
#define XCHG(a,b) tempr=(a);(a)=(b);(b)=tempr

double *fht(double lastbuf[], double nextbuf[], int nn, double table[])
{
    int stages=0, index, leg_dif, lower, upper;
    int e, r, j, i, k, n4, n2, n7, n, u, s, l, s2, s0, q, d, t;
```

```
    double temp1;
    double temp2;
    double tempr, temps, tempj, tempy, tempz;
    double *swapptr;
    double *cosine;
    double *sine;

    sine = table;      /*import table commensurate with length of input*/
    cosine = &sine[nn];


    i=1;
    do{
        stages+=1;
        i <<= 1;
        }while (i!=nn);     /*lg2 nn*/

/*hartley two-step*/


    /*the next a rearrangement by taking the remainder of the number after*/
    /*a division by two, an odd number will have a remainder of one,*/
    /*an even number will have a remainder of zero. The remainder is*/
    /*shifted into holding register (j) and used as a pointer when*/
    /*we have shifted through all the powers of two in our data length*/

    for(i = 0; i<nn; i++) {        /*nn is length of data sequence*/
        j = 0;
        r = i;
        for(k=1; k<=stages;k++) {
            s = r>>1;
            j = (j<<1) + r - (s<<1);
            r = s;
            }
        XCHG(lastbuf[i], nextbuf[j]);     /*must use two buffers to keep*/
                                          /*original data undisturbed*/
        }           /*otherwise it will be overwritten by this process*/

/*add and subtract*/

    for(i = 0; i<nn-1; i+=2) {     /* a pair of consecutive data points*/
                                   /* at a time*/
        temp1 = nextbuf[i]+nextbuf[i+1];  /*add*/
        nextbuf[i+1] = nextbuf[i]-nextbuf[i+1];       /*subtract*/
        nextbuf[i] = temp1;
        }

    for(i = 0; i<nn; i+=4) {   /*construct a sub-sequence of 4*/
                               /*consecutive data points*/
        temp1 = nextbuf[i];
        lastbuf[i]= temp1+nextbuf[i+2];
        lastbuf[i+2] = temp1-nextbuf[i+2];
```

```
            temp1 = nextbuf[i+1];
            lastbuf[i+1]= temp1+nextbuf[i+3];
            lastbuf[i+3] = temp1-nextbuf[i+3];
            }



    /*final stages*/
        n = nn;             /*length of input*/
        n2 = nn>>1;         /*half of input*/
        n4 = n2>>1;         /*one-quarter of input*/
        u = stages-1;       /*one less than the log2 of the input length*/
        s = 4;

        for(l=2; l<stages; l++) {
            s2 = s<<1;
            u--;
            s0 = 1<<(u-1);              /*raise one to 2^log2(N)-3*/
            for(q = 0; q<n;q+=s2){      /*divide input into sets equal to*/
                                        /*integer multiples of 8*/
                i = q;
                d = i+s;    /*set current pointers into sequence*/
                nextbuf[i] = lastbuf[i] + lastbuf[d];
                nextbuf[d] = lastbuf[i] - lastbuf[d];     /*solve for DHT*/
                k = --d;
                for(j = s0; j<=n4; j+=s0){
                    i++;
                    d = i+s;
                    e = k+s;
                    tempy = lastbuf[d]*cosine[j]+lastbuf[e]*sine[j];
                    tempz = lastbuf[d]*sine[j]-lastbuf[e]*cosine[j];
                    nextbuf[i] = lastbuf[i]+tempy;        /*shift*/
                    nextbuf[d] = lastbuf[i]-tempy;
                    nextbuf[k] = lastbuf[k]+tempz;
                    nextbuf[e] = lastbuf[k]-tempz;
                    k--;
                    }
                e = k+s;
                }
            swapptr = lastbuf;
            lastbuf = nextbuf;
            nextbuf = swapptr;
            s=s2;
            }

        free(sine);

        if(nextbuf == nextbuf) {  /*this algorithm switches back and forth*/
                                  /*between buffers*/
            free(nextbuf);        /*want to close the correct one and*/
                                  /*return the other*/
            return lastbuf;
```

```
        }
    else {
        free(lastbuf);
        return nextbuf;
        }

}
```

The final division by $\frac{1}{2^{\log_2 N}}$ is not done in the routine. It is required only for the FHT, not for the inverse FHT, and was kept by itself. This is purely elective; it could be placed in the routine.

```
    /*normalization*/
    if(ivar == 1) {
    for(i=0;i<data_points;i++) {
        fvector[i] *= (1.0/(double)data_points);
            }
        }
```

## A Windowing Function

The next three fragments of code, taken from WCOEF.EXE on the optional disk, demonstrate the ease with which windows coefficients are derived. The window length is determined by the length of the DFT.

A case statement is used to select the proper windowing function. The rectangular window requires no coefficients, since it is created by definition whenever finite-length data is sampled from an input signal. The number of coefficients and type are set and the main windowing function is called. The Kaiser window requires slightly different handling and uses a different function.

```
double *makewind(double data[], int nn, double cptr[], int wcoef);
        /*generalized window function*/
double *makekaiser(double data[], int nn, double cptr[], int
wcoef);/*kaiser window*/
double MBES(double value);/*modified bessel function*/
double *data;       /*input data*/
double *cptr;
double *ovector;    /*output vector*/
double *window;     /*window buffer*/

double blackman[] = {.426590, .496560, .076848};     /*window
coefficients*/
double hann[] = {.5, .5};
double hamming[] = {.54, .46};
double *coefptr;        /*pointer to windows coefficients*/
int wcoef;      /*number of coefficients for windowing function*/
int coef;       /*number of desired coefficients*/

    /*select window type and multiply window by filter coefficients*/
```

```
    switch(wtype){
        case 0:                 /*rectangular*/
            break;
        case 1:                 /*blackman*/
            wcoef = 3;
            coefptr = blackman;
            window = makewind(ovector, coef, coefptr, wcoef);
            break;
        case 2:                 /*cosine bell, von hann*/
            wcoef = 2;
            coefptr = hann;
            window = makewind(ovector, coef, coefptr, wcoef);
            break;
        case 3:                 /*hamming*/
            wcoef = 2;
            coefptr = hamming;
            window = makewind(ovector, coef, coefptr, wcoef);
            break;
        case 4:                 /*kaiser/bessel*/
            window = makekaiser(ovector, coef, coefptr, wcoef);
            break;
        default:
            perror("\nsomething went wrong...");
            quit(3);        /*exit through error handler*/
    }
```

The *MAKEWIND function requires an input vector, double data[]; the number, int wcoef; the list of coefficients, double coefptr[;] and finally, the desired number of window coefficients, int coef. Local memory is allocated for the window, which is created according to the formulae presented in the text, using the symmetry of the window for efficiency. The window coefficients are then multiplied by the input vector, double data[], and that array is returned. If you would like to get the window coefficients, the routine can easily be modified, or a vector of all ones can be passed.

```
double *makewind(double data[], int coef, double coefptr[], int wcoef,
int norm)
{

    int coef2, odd, i, j;
    double fcoef2, sum = 0.0, wind, kernel, ftwid;
    double *windc;
    int test = 0;

    if( (windc = (double *)calloc( (size_t)1,
            (size_t)sizeof(double)*coef )) == NULL )
        {
            quit(4);
        }

    coef2 = coef>>1;                /*determine odd or even*/
    odd =  coef - (coef2<<1);       /*by finding remainder from*/
```

```
        fcoef2 = (double)coef2;                 /*divide by two*/
        kernel = PI2/(double)(coef-1);          /*to make twiddles with*/
        if(!odd) kernel -= .5;

        for(i = 1; i<=coef2; i++) {
            ftwid = (double)i*kernel;       /*make index dependent twiddle*/
            wind = coefptr[0];              /*for the additional coefficients,*/
                                            /*see table*/
            for(j = 1; j<wcoef; j++)
                wind += coefptr[j]*cos(ftwid*j);
                                            /*allows for additional terms*/
            if(!odd)windc[coef2+i-1] = windc[coef2-i] = wind;
                                            /*use symmetry of window*/
            else windc[coef2+i] = windc[coef2-i] = wind;
                                            /*keep a sum for normalizing*/
            sum += wind;
                }

        sum *= 2;

        if(odd) {
            windc[coef2] = 1.0;             /*central coefficient*/
            sum += 1.0;
            }

        /*normalize*/
        if(norm) {
            for(i=0; i<coef; i++)           /*greatest value 1*/
                windc[i] /= sum;
            }


        for(i = 0; i<coef; i++)             /*apply windowing*/
            ovector[i] = windc[i]*ovector[i];

        return windc;

    }
```

The *MAKEKAISER function is passed the same data, but is a more complex function requiring more arithmetic than the others.

```
    double *makekaiser(double data[], int coef, double coefptr[], int wcoef)
    {

        int coef2, odd, i, j;
        double fcoef2, wind;
        double *windc;
        double beta = 100; /*stop band attenuation, height of first sidelobe*/
        double alpha;
        double ftmp;
        double root;
```

```
double f_ndx;
double f_coef2;
double denom;
double numer;

if( (windc = (double *)calloc( (size_t)1,
        (size_t)sizeof(double)*coef )) == NULL )
    {
        quit(4);
    }

coef2 = (coef)>>1;
f_coef2 = (double)(coef-1)/2.0;
odd =  (coef) - (coef2<<1);
fcoef2 = (double)coef2;

if(beta > 50.0) alpha = 0.1102*(beta-8.7);
else if(beta>=21.0) {
    ftmp = beta - 21.0;
    alpha = ftmp*0.07886;
    alpha += 0.5842*pow(ftmp, .4);
    }
else alpha = 0.0;

denom = MBES(alpha);


for(i = 0; i<coef; i++) {      /*only goes to argument but not equals*/

    f_ndx = (double)i;
    if(!odd) f_ndx -= .5;
```

/*

$$w(n)=\begin{cases}\dfrac{I0\left[pa\sqrt{1-\left(\dfrac{n}{\frac{M}{2}}\right)^2}\,\right]}{I0(pa)}\\ 0,\ \text{elsewhere}\end{cases}$$

*/

```
    root = alpha * sqrt(1 - (((f_ndx-f_coef2)/f_coef2)
        *((f_ndx-f_coef2)/f_coef2)));
    numer = MBES(root);
    windc[i] = numer/denom;
        }

if(odd) windc[coef2] = 1;

for(i = 0; i< coef2; i++) {         /*normalization*/
    windc[i] = windc[i]/windc[coef2];
    windc[coef2+i] = windc[i];
    }
```

```
    for(i = 0; i<coef; I++)    /*apply window*/
        ovector[i] = windc[i]*ovector[i];

    return windc;

}
```

Finally, the *modified* Bessel function of the zeroth order calculated using Horner's Rule.

```
double MBES(double value)
{
    double x, y, z, bes;
    bes = 1.0;
    y = value/2.0;
    x = y;
    z = 1.0;
```

$$/* \ I_0 = 1 + \sum_{n-1}^{\infty} \left[ \frac{\left(\frac{n}{2}\right)^n}{n!} \right]^2 \ */$$

```
    do {
        bes = bes+x*x;
        z = z+1.0;
        x = x*y/z;
        }while (x>1.0e-7);

    return bes;
}
```

# 4

# Discrete Time Filters

## INTRODUCTION

A filter, as we have already defined it, is a system that passes certain frequencies and rejects others. It has also come to mean a system that modifies selected frequencies relative to others. With discrete time filters, we are able to attain results superior to those of analog filters in almost all areas, allowing us to create nearly flat passbands, almost vertical transition bands, and close to ripple-free stopbands. In recent years, the cost of such filters has fallen greatly, so that now we can have all of these things in much shorter time and at much less expense than with a comparable analog filter. In addition, the results are consistent and do not vary or drift with components and environmental changes, as will occur with their analog counterparts.

It isn't that an entirely new technology has been discovered—actually, the basis of discrete time filtering is continous-time filtering with changes made to accommodate the difference in tools. The same rules apply to discrete time filters as to continuous-time filters in what is required to produce the results, it is just that high-order filters take up much less real estate when done with code than with hardware. Once the technology is understood, they also tend to be less complex. A low-order filter can be very much the same as a high-order filter in code, and is not sensitive to layout and parts selection.

Still, moving from continous time to discrete time involves some new problems that we did not have to deal with before. Clearly, since discrete time filters deal with sampled data, the sampling rate and the order of the filter dictate the group delay and phase delay of the system. Regardless of the technique (FIR or IIR), the filter will have a group delay dependent on the length of the response and the clock. Therefore, the rise time of the system is seriously affected by its filtering needs. Of course, resolution is also a factor; the system's response and fidelity revolve around the LSB. If it is coarse or poorly quantized, then accuracy, resolution, and fidelity suffer. Other design- and implementation-dependent questions also arise, among them the form of arithmetic chosen to do the filtering and the selection of coefficients. Finite word arithmetic can produce its own error—and sometimes output—when the hardware and the rest of the system are quiet. All of these things must be taken into consideration for selecting and designing a filter. No particular analog/digital filter (or a purely digital filter) will solve every problem.

There are two basic techniques for creating filters in discrete time. IIR (*infinite impulse response*) filters are most often generated from the same information as analog filters. Several techniques exist for directly transforming continuous-time filters to discrete time. We will discuss the *frequency sampling method* and *Fourier series method* for FIR (*finite impulse response*) development, and then concentrate on the *bilinear transform*, which maps the poles and zeros—and ultimately the transfer function—to a discrete time transfer function for IIR filters.

Most discrete time requirements involve linear phase and constant group delay, since anything else indicates distortion and makes analysis and synthesis questionable. Implementing discrete time filters with recursive techniques can result in less code, but has a deleterious effect on phase delay and group delay. If this is important in an application, and steep transition bands are also necessary, you will probably need an FIR filter of some length—and, therefore, more code. Nevertheless, you still benefit from smaller size, lower cost, and simpler maintenance than those of its analog counterpart.

FIR filters are discrete in nature and designed with discrete time techniques. This technology has become very popular in recent years, finding its way into ICs for many different applications, especially consumer applications—though there is a good deal of application-specific silicon for industrial use in realtime control, as well. In audio and video electronics, FIR filters and dedicated chips implementing transforms have become very popular.

IIR filters are compact and efficient, but suffer from some of the same problems as their analog counterparts. They can tighten code because they include poles as well as zeros in the transfer function, but they cannot have perfectly linear phase, and they can be unstable. Nevertheless, for extremely sharp transition bands and smaller code, they cannot be beat.

The purpose of this chapter is to introduce digital filtering techniques and some software for creating them. We will start by making the transition from continuous-time filtering and the Laplace transform to discrete time filtering and the Z transform. The basics of difference equations and convolution are covered, along with the design of FIR and IIR filters—these are important parts of the technology and should not be skipped. Following that, we will explore some of the most popular filtering techniques for digital filtering and discuss methods for developing coefficents and code for both of them.

The remainder of the chapter deals with code written for FIR filters and the bilinear transform, with some examples created with different techniques. If your interest is not programming, you may safely skip these. The optional disk provides complete programs for generating these coefficients, as well as Mathcad documents for a live illustration of these processes.

# BASICS OF A DISCRETE TIME SYSTEM

## Relationship between Laplace and Z Transforms

A basic tool of continous-time and discrete time processing is the Laplace transform:

$$X(s) = \int_0^\infty x(t)e^{-st}dt$$

If we set $z = e^{st} = e^{\sigma t}e^{j\omega t} = e^{\sigma t}(\cos\omega t + j\sin\omega t)$, with $s = \sigma + j\omega$ and both $\sigma$ and $\omega$ real, we have a new relationship for sampled data:

$$X(z) = \sum_{n=0}^{\infty} x[nT]z^{-n}$$

or, if we assume the sampling period, $T$, as is usually the case:

$$X(z) = \sum_{n=0}^{\infty} x[n]z^{-n}$$

This is called the *unilateral Z transform* of a discrete time sequence. The *bilateral Z transform* is

$$X(z) = \sum_{n=-\infty}^{\infty} x[n]z^{-n}$$

In this book, we will be dealing with the bilateral Z transform, which for causal systems is the same as the unilateral Z transform, $x[n] = 0$, $n < 0$.

The Z transform allows us to express a signal sampled at regular intervals as a sum of those samples, multiplied by a delay representing that point in time at which that sample was taken. For example, if we derive an equally spaced series of values from a signal $x(t)$, at intervals we will, for the moment, take to be unity, we would have

$$x(0),\ x(1),\ x(2),\ \cdots,\ x(n)$$

for the first, second, and third samples, respectively. This is equivalent to the set of samples we would get using the sampling function:

$$x[t] = x(0)\delta(t) + x(1)\delta(t-1) + x(2)\delta(t-2) + \cdots + x(n)\delta(t-n) \qquad \text{(Eq. 4-1)}$$

Taking the Z transform of Equation 4-1, we have

$$X(z) = x(0) + x(1)z^{-1} + x(2)z^{-2} + \cdots + x(n)z^{-n}$$

the function is real and $x[n] = 0$, $n < 0$, $z$ is complex. Each $z^{-1}$ in these equations corresponds to a unit time delay. This is a direct consequence of the shift theorem:

$$x(n + n_0) \rightarrow z^{n_0}X(z)$$

Therefore, a multiplication by $z^{-1}$ is equivalent to a unit delay of the function $x(t)$.

The Z transform, as you will see, is much more than just a convenience of notation, but even as such, it allows us the facility to perform simple arithmetic on otherwise unwieldy objects. For example, a sequence of samples from a signal may have the form just presented above:

$$X(z)=x(0)+x(1)z^{-1}+x(2)z^{-2}+\cdots+x(n)z^{-n}$$

while a sequence of coefficients representing filter weights might look like

$$H(z)=h(0)+h(1)z^{-1}+h(2)z^{-2}+\cdots+h(n)z^{-n}$$

and their product,

$$Y(z)=X(z)H(z)$$

which is the convolution of the two sequences.

In addition, the Z transform is directly related to both the Laplace transform and Fourier transform. Table 4-1 illustrates many of the correspondences between the Z transform and Laplace transform.

## The Z Transform and the Fourier Transform

If we set $s=j\omega$ in the Laplace transform, the transform involves the $j\omega$, or imaginary, axis *only* of the $s$-plane and becomes the Fourier transform. Accordingly, we set $z=e^{st}$ above, and making $s=j\omega$, we again have the Fourier transform (see Eq. 4-1). The Fourier transform is, in fact, the same as the Z transform evaluated on the unit circle in the $z$-plane.

Laplace transform:

$$X(s)=\int_{-\infty}^{\infty}x(t)e^{-st}dt,\quad s=j\omega$$

Discrete time Fourier transform:

$$X\left(e^{j\omega}\right)=\sum_{n=-\infty}^{\infty}x[n]e^{-j\omega n}$$

Z transform:

$$X(z)=\sum_{n=-\infty}^{\infty}x[n]z^{-n}$$

If we allow $z=e^{j\omega}$, we automatically restrict its magnitude to unity. When the absolute value of $z$ is one, $|z|=1$, we have the Fourier transform. Because $z$ is a complex variable, we deal with it on the complex plane. Figure 4-1 illustrates the relationship of the unit circle on the $z$-plane and the $j\omega$ axis on the $s$-plane.

To restate in terms of Figure 4-1, the Z transform evaluated on the unit circle is the Fourier transform: $\omega$ denotes the angle between the real axis and the vector to a point on the

| *Time domain function F(t)* | *Laplace transform F(s)* | *Z transform (continuous) F(z)* | *Z transform (discrete) F[z]* | *Sequence f[n]* |
|---|---|---|---|---|
| $u(t)$ | $\frac{1}{s}$ | $\frac{1}{1-z^{-1}}$ | $\frac{1}{1-z^{-1}}$ | $u[n]$ |
| $\delta(t)$ | 1 | 1 | 1 | $\delta[n]$ |
| $t$ | $\frac{1}{s^2}$ | $\frac{z^{-1}}{\left(1-z^{-1}\right)^2}$ | $\frac{z^{-1}}{\left(1-z^{-1}\right)^2}$ | $n$ |
| $e^{at}$ | $\frac{1}{s-a}$ | $\frac{1}{1-e^{-a}z^{-1}}$ | $\frac{1}{1-az^{-1}}$ | $a^n u[n]$ |
| $e^{-at}$ | $\frac{1}{s+a}$ | $\frac{1}{1-e^{-a}z^{-1}}$ | $\frac{1}{1-az^{-1}}$ | $-a^n u[n-1]$ |
| $te^{at}$ | $\frac{1}{(s-a)^2}$ | $\frac{e^{-a}z^{-1}}{\left(1-e^{-a}z^{-1}\right)^2}$ | $\frac{az^{-1}}{\left(1-az^{-1}\right)^2}$ | $na^n u[n]$ |
| $\sin \omega t$ | $\frac{\omega}{s^2+\omega^2}$ | $\frac{(\sin\omega)z^{-1}}{1-(2\cos\omega)z^{-1}+z^{-2}}$ | $\frac{[\sin\omega]z^{-1}}{1-[2\cos\omega]z^{-1}+z^{-2}}$ | $[\sin\omega n]u[n]$ |
| $\cos \omega t$ | $\frac{s}{s^2+\omega^2}$ | $\frac{1-[\cos\omega]z^{-1}}{1-[2\cos\omega]z^{-1}+z^{-2}}$ | $\frac{1-[\cos\omega]z^{-1}}{1-[2\cos\omega]z^{-1}+z^{-2}}$ | $[\cos\omega n]u[n]$ |
| $e^{-at}\sin \omega t$ | $\frac{\omega}{(s+a)^2+\omega^2}$ | $1\frac{\left(e^{-a}\sin\omega\right)z^{-1}}{1-\left(2e^{-a}\cos\omega\right)z^{-1}+e^{-2a}z^{-2}}$ | $1\frac{(r\sin\omega)z^{-1}}{1-(2r\cos\omega)z^{-1}+r^2z^{-2}}$ | $\left[r^2\sin\omega n\right]u[n]$ |
| $e^{-at}\cos \omega t$ | $\frac{s+a}{(s+a)^2+\omega^2}$ | $1\frac{1-\left(e^{-a}\cos\omega\right)z^{-1}}{1-\left(2e^{-a}\cos\omega\right)z^{-1}+e^{-2a}z^{-2}}$ | $1\frac{1-(r\cos\omega)z^{-1}}{1-(2r\cos\omega)z^{-1}+r^2z^{-2}}$ | $\left[r^2\cos\omega n\right]u[n]$ |

**Table 4-1.**

unit circle, beginning at $\omega=0$, $z=1$ and going through $\omega=\pi$, $z=-1$ to $\omega=2\pi$, which is the same as $\omega=0$. The actual value of $z$ goes from 1 on the real axis to $j$ on the imaginary axis, to $-1$ on the real axis again, to $-j$ on the imaginary axis, to return to 1 on the real axis where it began. Therefore, you can see that traversing the circle once, which is nothing more than rotating the radius through $2\pi$ radians, represents the entire Fourier transform. This is equivalent to $1/t_s$, where $t_s$ is the period associated with the sampling rate.

The circumference of the unit circle is equivalent to the $j\omega$ axis on the $s$-plane.

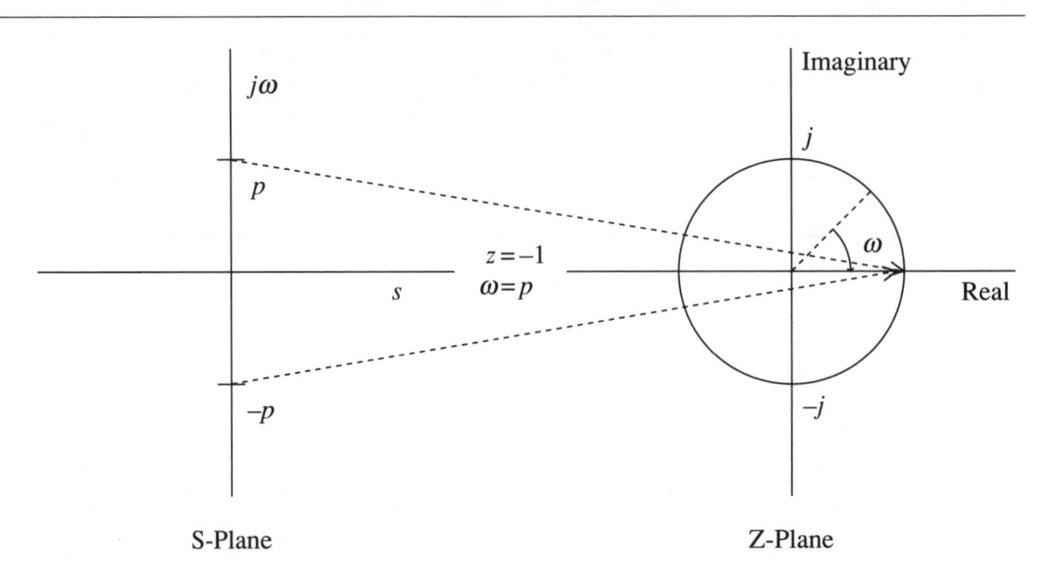


**Figure 4-1.** With $z$ restricted to a unit, the transform describes a circle of unit radius, a unit circle. Evaluated on the unit circle, the $z$ transform is the Fourier transform.

## Region of Convergence of the Z Transform

The Z transform converges on the complex plane in a system of loci called the *region of convergence*. To understand this, let us look at the Z transform in another way.

We begin by expressing the Z transform in terms of polar units. This can be very helpful when determining the convergence of the transform, since its convergence is solely dependent upon magnitude.

$$X\left(re^{j\omega}\right) = \sum_{n=-\infty}^{\infty} x[n]\left(re^{j\omega}\right)^{-n}$$

Realizing that this is an infinite sum, we can say that as long as $r \leq 1$, the transform will converge.

We can rewrite this to emphasize its connection with the Fourier transform:

$$X\left(re^{j\omega}\right) = \sum_{n=-\infty}^{\infty}\left(x[n]r^{-n}\right)e^{-j\omega n}$$

For the transform to be uniformly convergent, it must be absolutely summable, which means that the sum of the absolute value of the product of the sequence $x[n]$ and the exponential sequence $r^{-n}$ must be absolutely summable:

$$\sum_{n=-\infty}^{\infty}\left|x[n]r^{-n}\right| < \infty$$

This actually means that the Z transform might have a better chance of converging than the Fourier transform, since all that is necessary is that the sequence to be absolutely summable and that is dependent upon the absolute value of Z. Therefore, we can say that

$$X(z)=\sum_{n-\infty}^{\infty} x[n]z^{-n}$$

will converge, as long as

$$\sum_{n-\infty}^{\infty} |x[n]||z^{-n}| < \infty$$

Since it is the magnitude of $z$ that determines convergence, if a transform converges for $z$, it converges for all values of $|z|$. Using $|z|$ as the radius of a circle with its origin at the center, we can describe a circle for each value of $z$ for which we are convergent, and a ring, or annulus, for the entire region of convergence.

The following points describe the region of convergence for the Z transform:

1. The region of convergence is continuous—that is, it has no breaks.
2. Convergence of $x[n]$ depends only upon the absolute magnitude of $z$.
3. The Z transform of $x[n]$ reduces to the Fourier transform of $x[n]$, when $|z|=1$, leading to the conclusion that the Fourier transform of $x[n]$ converges absolutely only when the region of convergence of the Z transform includes the unit circle.
4. The region of convergence includes no poles, since $X(z)$ is infinite at a pole.
5. Since the absolute summability of the Z transform depends upon $x[n]r^{-n}$, a *finite duration* sequence—that is, one for which $-\infty < N_1 \leq n \leq N_2 < \infty$—will converge absolutely if each of the values of the sequence is finite. The region of convergence will include the entire $z$-plane, with the possible exception of $z=0$ or $z=\infty$.
6. If $x[n]$ is a *right-handed* sequence, one for which $n < N_1 < \infty = 0$, its region of convergence is annular, extending outward from the outermost pole of the transform.
7. Conversely, a *left-handed* sequence, $n > N_2 > -\infty = 0$, is also annular, but extends inward from the innermost pole of the transform.
8. A *two-sided* sequence is also annular, bounded by poles on either edge.

As you can see from these points, the region of convergence will appear as a ring on the $z$-plane bounded on both sides by poles. It may extend to infinity from a point on the $z$-plane, or it may extend inward to the origin for a point on the $z$-plane. Figure 4-2 illustrates the various forms of convergence of the Z transform on the complex plane.

An important form of the Z transform is the expression of a system function

$$X(z)=\frac{P(z)}{Q(z)}$$

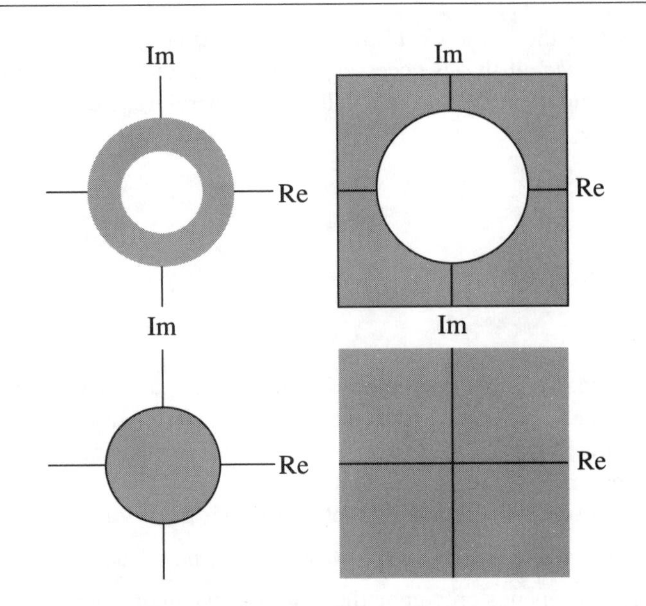


**Figure 4-2.** The region of convergence on the $z$-plane is always an annulus, as it is dependent upon the magnitude of $|r|$ only. The region may be only an annulus, it may be continuous from the origin to an outer ring, it may be bounded at a ring and extend outwardly into the entire $z$-plane, or it may cover the entire $z$-plane.

Analogous to the Laplace transform, $P(z)$ is a polynomial expression in $z$ whose roots are the zeros of the function, and $Q(z)$ is also a polynomial whose roots are the poles of the function. This is the rational expression of a system function in $z$.

## Inverse Z Transform

We will often need the inverse of a Z transform for a system function in the construction of a filter. In the most rigorous sense, this is accomplished with the contour integral and residue theorem; however, it is seldom done that way. The inverse Z transform is most often performed either by inspection or by the partial fraction technique.

As with the Laplace and Fourier transforms, the availability of tables of transform pairs makes inversion by inspection a popular and relatively easy method. Table 4-2 lists some Z transforms and their sequence pairs, as well as their functions to give a sense of what they mean.

Sometimes the expression is too complex to be listed in a table. In such case, it is often useful to expand the transform as a partial fraction whose elements *may* be listed on a table. Essentially, the process of expanding a transform as a partial fraction requires that the function be rewritten in factored form:

| *Function* | *Sequence* | *Z transform* |
|---|---|---|
| $a_n$ | $a_0\, a_1\, a_2\, a_3 \ldots$ | $a_0 + a_1 z^{-1} + a_2 z^{-2} + a_3 z^{-3} + \ldots$ |
| 1 | 1 1 1 1 | $\dfrac{1}{1-z^{-1}}$ |
| $n$ | 0 1 2 3 4 5... | $\dfrac{z^{-1}}{\left(1-z^{-1}\right)^2}$ |
| $n^2$ | 0 1 4 9 16 25... | $\dfrac{z^{-1}\left(1+z^{-1}\right)}{\left(1-z^{-1}\right)^3}$ |
| $e^{-an}$ | $1\ e^{-a}\ e^{-2a}\ e^{-3a}\ e^{-4a} \ldots$ | $\dfrac{1}{1-e^{-a}z^{-1}}$ |
| $1-e^{-an}$ | $0\ 1-e^{-a}\ 1-e^{-2a}\ 1-e^{-3a}\ 1-e^{-4a} \ldots$ | $\dfrac{\left(1-e^{-a}\right)z^{-1}}{\left(1-z^{-1}\right)\left(1-e^{-a}z^{-1}\right)}$ |
| $ne^{-an}$ | $0\ e^{-a}\ 2e^{-2a}\ 3e^{-3a}\ 4e^{-4a} \ldots$ | $\dfrac{e^{-a}z^{-1}}{\left(1-e^{-a}z^{-1}\right)^2}$ |

**Table 4-2.**

$$X(z) = \frac{\sum_{k=0}^{M} b_k z^{-k}}{\sum_{k=0}^{N} a_k z^{-k}} = \frac{b_0}{a_0} \cdot \frac{\prod_{k=1}^{M}\left(1-c_k z^{-1}\right)}{\prod_{k=1}^{N}\left(1-d_k z^{-1}\right)}$$

In this equation, $M$s represent the zeros and $N$s represent the poles. There are always the same number of poles and zeros in the finite $z$-plane; any difference, $M$–$N$, represents poles or zeros at the origin.

In the simple case, when all the poles are of the first order and there are more poles than zeros, the factored expression can be reduced to

$$X(z) = \sum_{k=1}^{N} \frac{A_k}{1-d_k z^{-1}}$$

and the coefficient $A_k$ given by

$$A_k = \left(1-d_k z^{-1}\right)X(z)\Big|_{z=d_k}$$

If there are as many or more zeros than poles, the factored expression must first be divided through to reduce the expression to one described above, plus an additive with an order of $M-N$:

$$X(z)=\sum_{r=0}^{M-N} b_r z^{-r}+\sum_{k=1}^{N}\frac{A_k}{1-d_k z^{-1}}$$

in which the values for $b_r$ are derived through long division of the numerator of the system function by the denominator:

$$X(z)=\frac{\sum_{k=0}^{M} b_k z^{-k}}{\sum_{k=0}^{N} a_k z^{-k}}$$

The process is complete when the remainder is of lesser degree than the denominator.

$$X(z)=\sum_{r=0}^{M-N} b_r z^{-r}+\sum_{k=1}^{N}\frac{A_k}{1-d_k z^{-1}}$$

As an example of this process, let us take the following transfer function:

$$H(z)=\frac{1+2z^{-1}+z^{-2}}{1-z^{-1}-2z^{-2}}=\frac{1+2z^{-1}+z^{-2}}{\left(1+z^{-1}\right)\left(1-2z^{-1}\right)}$$

We may then express this system function as

$$H(z)=b_0+\frac{A_1}{1+z^{-1}}+\frac{A_2}{1-2z^{-1}}$$

Then, to produce the additive $b_0$, we divide

$$H(z)=\frac{1+2z^{-1}+z^{-2}}{1-z^{-1}-2z^{-2}}=-.5+\frac{1.5+1.5z^{-1}}{\left(1+z^{-1}\right)\left(1-2z^{-1}\right)}$$

Solving for the coefficients,

$$A_1=\left.\frac{1+2z^{-1}+z^{-2}}{1+z^{-1}}\right|_{z^{-1}=.5}=\frac{1+1+.25}{1.5}=1.5$$

$$A_1=\left.\frac{1+2z^{-1}+z^{-2}}{1-2z^{-1}}\right|_{z^{-1}=-1}=\frac{1-2+1}{3}=0$$

resulting in the fraction

$$H(z)=-.5+\frac{1.5}{1-2z^{-1}}$$

Partial fraction expansions are very useful in expressing decomposed or second-order transfer functions in the construction of parallel IIR filters, as is described at the end of the chapter in the discussion on bilinear transformation.

Still another method of finding the inverse Z transform uses a power series expansion. To compute the inverse in this manner, the series is expanded as a sum, where all values of the sequence are in the form of coefficients of the various powers of $z-^1$.

The Z transform expanded as a power series has the form

$$X(z)=\sum_{-\infty}^{\infty}x[n]z^{-n}=\cdots+x[-2]z^2+x[-1]z+x[0]+x[1]z^{-1}+x[2]z^{-2}+\cdots$$

The most obvious examples of a power series expansion involve rational fractions in which all the poles are at $z=0$, such as

$$X(z)=z^2\left(1-\frac{3}{2}z^{-1}\right)\left(1+2z^{-1}\right)\left(1-z^{-1}\right)=z^2-\frac{1}{2}z-\frac{5}{2}+\frac{3}{2}z^{-1}$$

According to the power series, the coefficient for $z^2$ is $x[-2]$ and, in this case, is equal to 1. Following through for the rest of the coefficients, we find: $x[-1]=-1/2$, $x[0]=-5/2$, and $x[1]=3/2$.

Another very interesting and prominent transform in digital signal processing is the series

$$x(z)=\frac{1}{1-az^{-1}} \quad |z|>|a|$$

This series is expanded by synthetic division to produce

$$1+az^{-1}+a^2z^{-2}+a^3z^{-3}+\cdots$$

or, perhaps, more familiarly from the table of theorems of the Z transform

$$a^n x[n]$$

## Properties of the Z Transform

An additional aid in simplifying Z transforms lies in the theorems of the Z transform. Because the Z transform has close ties to the Fourier transform and the Laplace transform, it is probably already very clear that the properties of the Z transform play an important part in the analysis of signals, specifically in finding the inverse transforms of the more complicated expressions. A table of properties is given in Table 4-3.

## Computing in Discrete Time: Difference Equations

Analog filters are described by a differential equation; digital filters are described by a difference equation. When we were dealing with harmonic oscillators in Chapter 1, we used

| *Property* | *Function of time* | *Z transform* |
|---|---|---|
| linearity | $ax[n]+by[n]$ | $aX(z)+bY(z)$ |
| multiplication by exponential sequence | $a^n x[n]$ | $X\left(\frac{z}{a}\right)$ |
| time shifting | $x[n-m]$ | $z^m X(z)$ |
| differentiation | $nx[n]$ | $-z\frac{dX(z)}{dz}$ |
| conjugation of a complex sequence | $x*[n]$ | $X*(z*)$ |
| time reversal | $x[-n]$ | $X\left(z^{-1}\right)$ |
| convolution of sequences | $x[n]*y[n]$ | $X(z)Y(z)$ |

**Table 4-3.**

linear constant coefficient differential equations. We now introduce the *linear constant coefficient difference* equation of the form

$$\sum_{k=0}^{N} a_k y[n-k] = \sum_{k=0}^{M} b_k x[n-k] \tag{Eq. 4-2}$$

where $M$s represent the zeros of the equation and $N$s are the poles.

A discrete, linear, time-invariant system with an input $x[n]$ and an output $y[n]$ can be described by such an equation. This equation states that the computation of $y[n]$ involves the previous outputs $y[n-1]$, $y[n-2]$, and so on. Such difference equations are known as *recursive difference equations* and are rewritten from Eq. 4-2 as:

$$y[n] = \sum_{k=1}^{N} a_k y[n-k] - \sum_{k=0}^{M} b_k x[n-k]$$

$$y[n] = b_0 x[n] + b_1 x[n-1] + b_2 x[n-2] + b_3 x[n-3] + \cdots + b_k x[n-k] - a_1 y[n-1] - a_2 y[n-2] - a_3 y[n-3] - \cdots - a_k y[n-k]$$

An example of such a filter is the first-order filter given by

$$y[n] = ax[n] + (1-a)y[n-1]$$

(A live example of this filter is given in the Mathcad document diffrnce.mcd.)

If we start with $a \leq 1$ and then provide the system with a unit impulse, $y[n]$ will gradually approach 1. The time it takes to do this is relative to the value of $a$—the smaller the $a$, the longer the time constant. This kind of filter is known as an *infinite impulse response* filter, because the effect of the input never dies out completely. An example of such a filter in analog electronics might be a simple RC network. In this case, the filter is of the first order. Other orders are more frequently used.

If it happens that the coefficients $a_k$ are all zero, the equation does not involve the use of any of the previous outputs, and the equation is called a *nonrecursive difference equation.* A simple example of a a nonrecursive difference equation may be found in the *moving average filter* commonly used for elementary lowpass filtering. This filter involves choosing a width for a window on the input data, wide enough—involving enough samples—to trivialize high-frequency noise additives to any input data, then averaging the sum of the inputs in this window. For this example, we choose a moving window of five data points. Each time a new input $x[n]$ arrives, the oldest is discarded. The new sequence of five is summed and divided by five to produce a new output $y[n]$. We may write this equation as follows:

$$y[n] = \frac{x[n] + x[n-1] + x[n-2] + x[n-3] + x[n-4]}{5}$$

There are several ways this might be implemented in practice. One such way is by multiplying each new sample by 0.2, summing it with the current $y[n]$, then removing the oldest data point (now $0.2x[n-5]$) to produce a new $y[n]$. This form of filter is called a *finite impulse response* filter.

## Symbols for Block Diagrams

At this point, it is appropriate to introduce a symbolic expression for the equations we are studying. This system is used to represent express functions in Z generally, and in FIR and IIR filters in particular. Actually, any system function may be implemented in a number of ways, while still retaining the identical mathematical representation. So that different algorithmic forms can be described, we use a system of block diagrams to illustrate the manner in which the products, sums, and delays are handled.

The symbols in Figure 4-3 are used commonly in the literature for algorithmic and flow illustrations, and will be used in this book for the same purpose. These are the basic building blocks of discrete time descriptions—they correspond to the resistors, capacitors, and active components of analog systems.

Referring to the diagram, the *node* or *pickoff node* is simply a branch. No operation is performed on the input; it may be diverted to another point or separated into two branches. The *unit delay* has an output that is identical to its input delayed one sample period. The *multiplier* generates a product of a fixed constant and the input. Finally, the *summing node* or *summer* adds two or more inputs to produce a discrete time output.

Get acquainted with these symbols; using them can help simplify your designs.

# LINEAR TIME INVARIANT FILTERS IN DISCRETE TIME

It is probably apparent that a great deal of the technology for discrete time filter systems has been borrowed, or at least derived, from analog techniques. There is good reason for this:

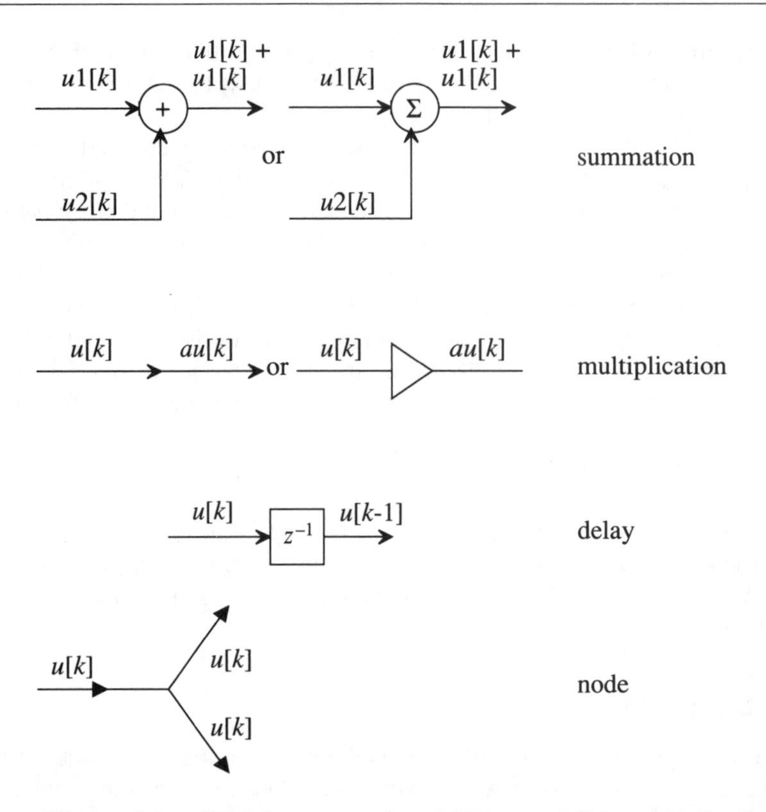


**Figure 4-3.** Part A represents addition, part B is multiplication, and part C is the unit delay.

these techniques are very mature and have a firm basis in mathematics. If you recall the general equation for a system function in continuous-time filtering, then the general equation for the discrete time system function will not surprise you:

$$H(s) = \frac{P(s)}{Q(s)} = \frac{b_0 s^n + b_1 s^{n-1} + \cdots + b_n}{a_0 s^m + a_1 s^{m-1} + \cdots + a_m} = \frac{\sum_{i=0}^{M} b_i s^i}{1 + \sum_{i=1}^{N} a_i s^i} \quad \text{(Eq. 4-3)}$$

which corresponds to the differential equation:

$$y_a(t) + \sum_{i=1}^{N} d_i \frac{d^i y_a(t)}{dt^i} = \sum_{i=1}^{M} c_i \frac{d^i x_a(t)}{dt^i} \quad \text{(Eq. 4-4)}$$

and the general equation for a linear time invariant system in the digital domain, also a rational fraction:

$$H(z)=\frac{\sum_{k=0}^{M} b_k z^{-k}}{1-\sum_{k=1}^{N} a_k z^{-k}} \qquad \text{(Eq. 4-5)}$$

By integrating both sides of Equation 4-5 and using a trapezoidal approximation for the integrals, we have

$$H(z)=\frac{\sum_{i=0}^{M} c_i \left(\frac{1}{T}\frac{1-z^{-1}}{1+z^{-1}}\right)^i}{1+\sum_{i=1}^{N} d_i \left(\frac{1}{T}\frac{1-z^{-1}}{1+z^{-1}}\right)^i}$$

This may be related to Equation 4-3 to give us the $s$ to $z$-plane mapping:

$$z=\frac{1+\frac{s}{a}}{1-\frac{s}{a}}$$

Briefly, this mapping is such that the left half of the $s$-plane is inside the unit circle on the $z$-plane, with the right half of the $s$-plane outside it; the $j\omega$ axis is the perimeter of this circle. This will be covered in more detail in the description of the bilinear transformation.

A sampled linear system can be described by the following relationship:

$$\sum_{k=0}^{N} -a_{kj} y[n-k]=\sum_{k=0}^{M} b_k x[n-k] \qquad \text{(Eq. 4-6)}$$

As you can see, this devolves directly from the system function and serves as the source for the two most common filter types in use: IIR and FIR filters.

Recalling the relationship between the Fourier transform and the Z transform, you will readily see that Equation 4-5 is a ratio of Fourier transforms. The variables $M$ and $N$ represent the zeros and poles of the function, respectively, and the coefficients $a_k$ and $b_k$ are fixed and define the filter response.

Rewriting Equation 4-6, we have

$$y[n]=\sum_{k=0}^{M} b_k x[n-k]+\sum_{k=1}^{N} a_k y[n-k]$$

This restatement of the expression describes a recursive relationship in which each new $y[n]$ is the result of current and previous inputs *and* its own previous outputs. This is called an *infinite impulse response* filter, because even though the inputs data may go to zero, the output will not do the same. This type of filter can require less code than a similar FIR filter, but cannot guarantee linear phase or constant group delay. Because the current output depends upon past inputs as well as past outputs, it has problems of instability in much the same way as the continuous-time filter.

IIR filters are not and will not be stable unless all the poles are within the unit circle. Additionally, linear phase is not truly possible unless all the poles lie on the unit circle. Finally, FIR filters are not simply concerned with magnitude response as are IIR filters; they are designed with both magnitude and phase response in mind.

If we assume that all $a_k$ coefficients in the preceding equation are zero, this places a pole at the origin, $z = -1$, and creates a new expression in which each new output depends only on present and past inputs. This is the usual form of the FIR filter:

$$y[n] = \sum_{k=0}^{M} b_k x[n-k] \qquad \text{(Eq. 4-7)}$$

with the system function

$$H(z) = \sum_{k=0}^{M} b_k z^{-k}$$

This form of the system response can be used to generate Fourier coefficients, which we convolve with the input data to produce each new output. Though the *finite impulse response* filter can be implemented with recursive or nonrecursive techniques, it is most often done in a nonrecursive manner. Figure 4-4 is an illustration of such an FIR filter.

FIR and IIR filters have each their own advantages and disadvantages—neither is right for every job. Both filters are relatively easy to design, but the FIR filter is most amenable to linear phase or constant group delay applications. Implemented as shown in Equation 4-7, it is less susceptible to the growing roundoff and truncation errors that result in noise associated with the numerical methods used. Additionally, the FIR in its *nonrecursive* form is free from limit cycling and is inherently stable. Unfortunately, the FIR filter can require much more code to meet the same response requirements as an IIR filter.

In the next sections, both forms of filter design will be covered, with software and examples to illustrate how they can be done.

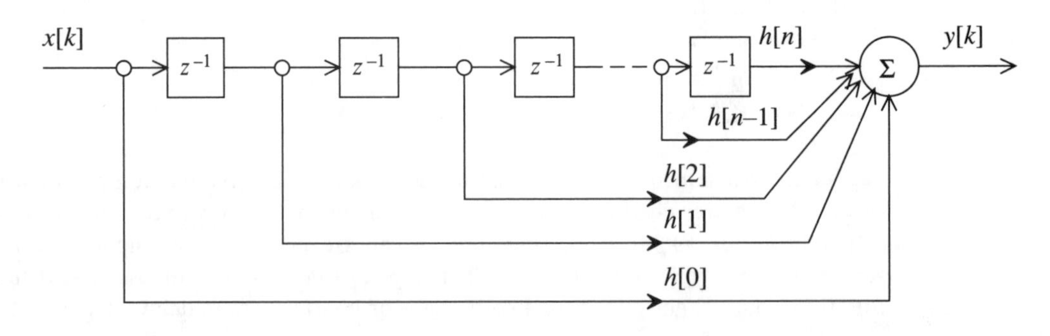


**Figure 4-4.** Illustration of an FIR filter.

## Filtering with the Fourier Transform

To get a feel for the interplay between the frequency domain and the time domain in filtering, let us review the ways in which we can use the Fourier transform as a filter, not only to modify existing signals, but also to create new ones. Please refer to the Mathcad document fcreate.mcd on the optional disk for a live demonstration of the following discussion.

In part A of Figure 4-5, we see a cosine wave and its spectrum. The cosine wave is assumed to be a continuous-time analog wave, sampled at 10 KHz, with a frame length of 64 samples. Part B is a table of the results of a DFT on the cosine, and part C is a plot of the spectrum of the cosine wave.

The frequency for each rank or incremental frequency of the spectrum is

$$f_k = \frac{k}{n} f_s$$

where $k$ is the index $n$ represents the total number of samples passed to the transform, and $f_s$ is the sampling frequency.

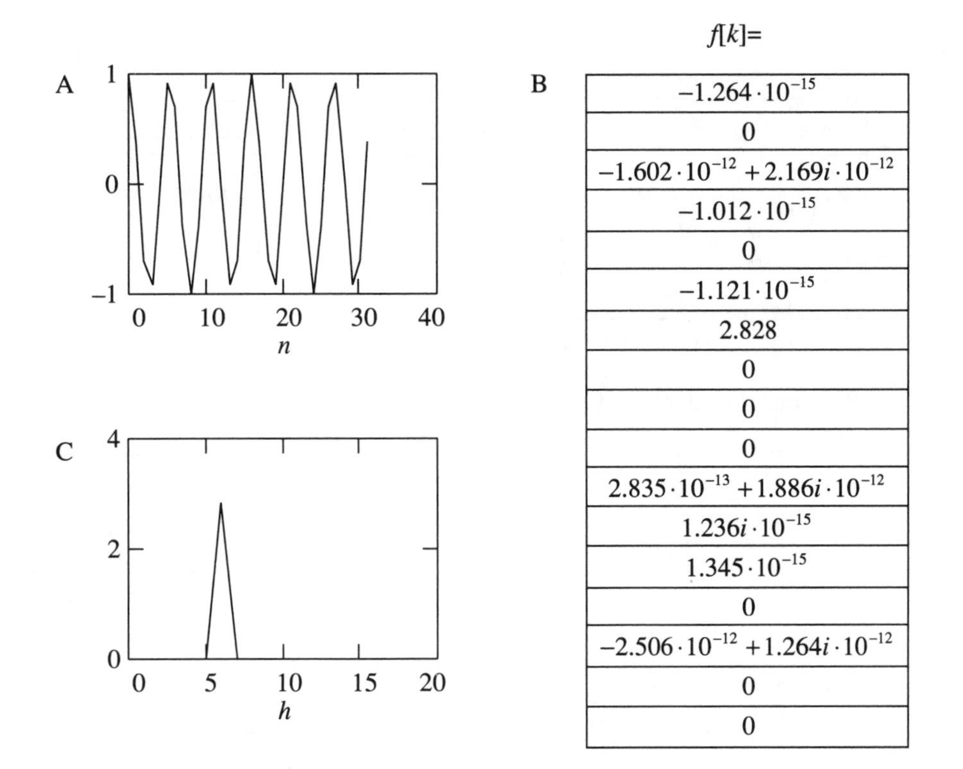


| $f[k]=$ |
| --- |
| $-1.264 \cdot 10^{-15}$ |
| 0 |
| $-1.602 \cdot 10^{-12} + 2.169i \cdot 10^{-12}$ |
| $-1.012 \cdot 10^{-15}$ |
| 0 |
| $-1.121 \cdot 10^{-15}$ |
| 2.828 |
| 0 |
| 0 |
| 0 |
| $2.835 \cdot 10^{-13} + 1.886i \cdot 10^{-12}$ |
| $1.236i \cdot 10^{-15}$ |
| $1.345 \cdot 10^{-15}$ |
| 0 |
| $-2.506 \cdot 10^{-12} + 1.264i \cdot 10^{-12}$ |
| 0 |
| 0 |

**Figure 4-5.** Part A is a representation of a cosine wave; we are sampling at 10KHz and we have 64 samples. Part B is the result of a DFT on the waveform, and part C is the spectrum of this wave.

From the spectrum we can see that we have an impulse at $k=6$ representing the contributions of the cosine at 937.5 Hertz. We could add another frequency to the original by placing another impulse in the spectrum, say at $k=4$. This is simple enough to do on the frequency domain—we simply modify the appropriate spectral coefficient. Figure 4-6 illustrates what happens if we do this.

It is just as easy to remove a frequency component as it is to add one. We can remove any evidence of the first sinusoid by placing a zero in the vector occupied by the coefficient that represents that frequency. This leads us to a simple way to remove or modify unwanted frequencies in a spectrum: either replace them with zero or multiply them with a coefficient corresponding to the result we wish to have on that frequency.

Since the spectrum is a representation of the contributions of the component frequencies of the signal we are analyzing, it seems that it would be possible to mark off the frequencies we want and don't want. We could multipy those we want by 1 and those we don't by 0 or a very small number. You will note in Figure 4-7 that the lowest frequencies are to the left and the highest frequencies are to the right. If we multiply the figure, point by point, by a vector representing a rectangle, placing 1s in the places corresponding to frequencies

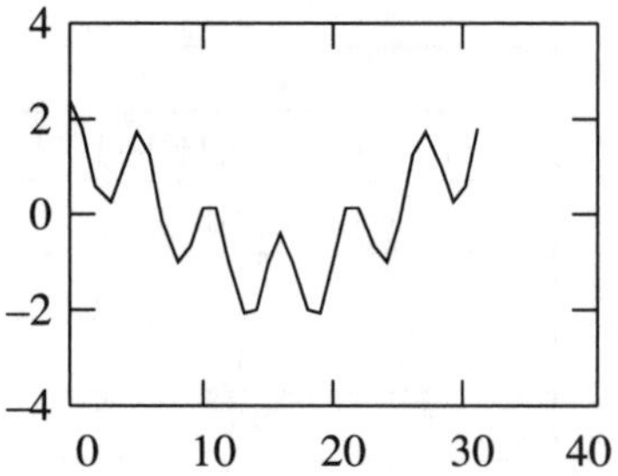

Resulting time domain effect

replace 0 with a 4

| $x_h$ |
|---|
| $-1.264 \cdot 10^{-15}$ |
| 4 |
| $-1.602 \cdot 10^{-12} + 2.169i \cdot 10^{-12}$ |
| $-1.012 \cdot 10^{-15}$ |
| 0 |
| $-1.121 \cdot 10^{-15}$ |
| 2.828 |
| 0 |
| 0 |
| 0 |
| $2.835 \cdot 10^{-13} + 1.886i \cdot 10^{-12}$ |
| $1.236i \cdot 10^{-15}$ |
| $1.345 \cdot 10^{-15}$ |
| 0 |
| $-2.506 \cdot 10^{-12} + 1.264i \cdot 10^{-12}$ |
| 0 |
| 0 |

**Figure 4-6.** Here, we have arbitrarily added an impulse to the spectrum. The result in the time domain is the sum of the two frequencies.

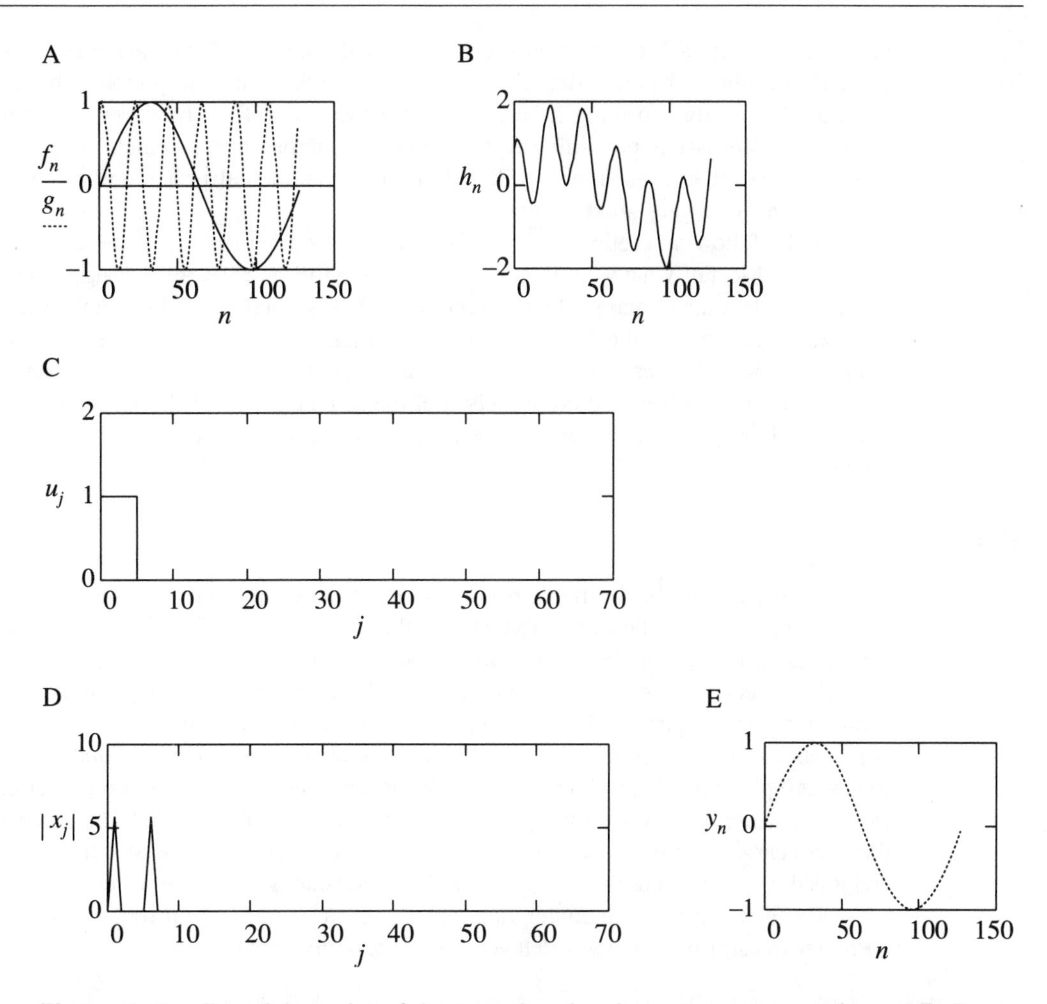


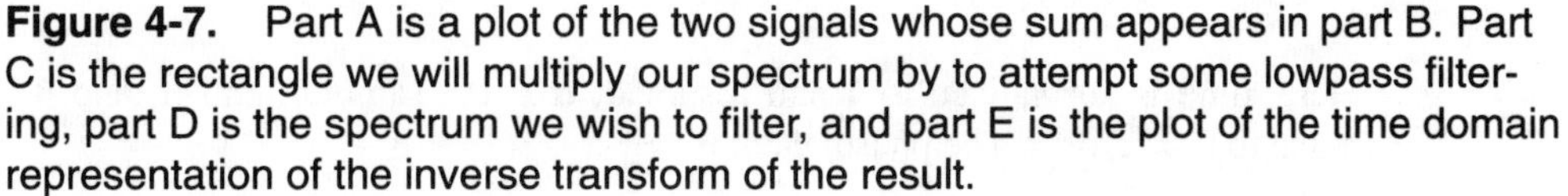

**Figure 4-7.** Part A is a plot of the two signals whose sum appears in part B. Part C is the rectangle we will multiply our spectrum by to attempt some lowpass filtering, part D is the spectrum we wish to filter, and part E is the plot of the time domain representation of the inverse transform of the result.

we wish to preserve and 0s in those places corresponding to frequencies we wish to erase, we would be simulating an *ideal brickwall* filter (see Chapter 2 for more on this). It has a passband with zero ripple passing all frequencies within it equally, a zero-length transition band, and a stopband with zero ripple attenuating all frequencies within it absolutely.

You can see from part E of Figure 4-7 that this works.

Basically, this is what we are doing when we construct a digital filter. We may construct an impulse response whose Fourier transform is a rectangle—or some other shape we choose—and convolve it with input data to qualify it for the frequencies we are after. Or, we may perform the task entirely on the frequency domain by deriving Fourier coefficients from analog filters or purely digital means, and using them as mutlipliers to shape the input spectrum before transforming it back to the time domain. This, however, is not the whole story. There are issues pertaining to the Gibb's phenomenon: placement of the cutoff frequency, linear phase, and more. Still, this is the fundamental procedure, and with refinements can provide very good results.

In the following sections, we will be discussing refinements that can make a filter perform better. Most of what we will cover will involve FIR and IIR filter design, both of which are done in the time domain. There is nothing that says it has to be done in the time domain, however—the DFT is the filter of choice for many applications. Because an FIR filter requires $N * M$ multiplications to accomplish, a long sequence can take longer than a DFT.

Regardless of the method you choose, the salient points of the following discussions concern filtering (convolutions) done on the frequency domain as well as on the time domain.

## FIR

It is most often desirable in filter design or signal processing of any sort that the phase of the signal be affected as little as possible, unless the purpose of the filter is specifically to do so. Zero phase is not attainable—there will always be *some* distortion. Linear phase, however, is possible. Filters designed with constant magnitude and linear phase can maintain phase shift proportional to frequency, thereby limiting the distortion to the signal. As we have noted before, linear phase is important for DSP work, such as speech analysis and video processing, where phase is critical. Inconstant delay can cause envelope distortion in modulated carrier signals, and it can distort the pulse shape in digital signals.The FIR filter, also known as the *transversal filter, nonrecursive filter,* or *tapped delay filter*, is an easy filter to implement, and it is the easiest to construct as a linear phase, constant group delay filter.

The FIR filter can guarantee linear phase, because the system response of the filter is a Fourier transform, in which a delayed sample is equivalent to

$$e^{-j\omega\alpha + j\beta} A\left(e^{j\omega}\right) \qquad \text{(Eq. 4-8)}$$

Here, time is normalized to 1 and assumed, $\alpha$ is the sample number, and $\beta$ is a constant equal to 0, $\pi/2$, or $\pi$. The phase, then, is

$$\theta = -\alpha\omega$$

which is clearly linear relative to $\omega$. Recalling at this point the equation for phase delay

$$\tau_p(\omega) = \frac{-\theta(\omega)}{\omega} = \frac{-(-\alpha\omega + j\beta)}{\omega} = \alpha + \frac{j\beta}{\omega}$$

and the equation for group delay

$$\tau_g(\omega) = \frac{-d}{d\omega}\theta(\omega) = \frac{-d}{d\omega}(-\alpha\omega + j\beta) = \alpha$$

we find that when $\beta$ is equal to 0 or $\pi$, the phase delay and group delay are both equal to $\alpha$. When $\beta = \pi/2$, the filter will have a linear phase response that is offset from the origin by $\pi/2$ and constant group delay.

There are four types of filters described by the *generalized linear phase system* with a frequency response:

$$H\left(e^{j\omega}\right) = A\left(e^{j\omega}\right)e^{-j\alpha\omega + j\beta}$$

where $\alpha$ and $\beta$ are constants, and $A(e^{j\omega})$ is a real function of $\omega$ called the *zero phase frequency response*, or the *amplitude response*.

A filter with a linear phase response of

$$\theta(\omega) = -\alpha\omega\,, \quad -\pi \le \omega \le \pi \qquad \text{(Eq. 4-9)}$$

will also demonstrate a constant group delay and phase delay. For a nonzero $\alpha$ to satisfy this relationship (see Eq. 4-9), it must be equal to half the length of the filter

$$\alpha = \frac{N-1}{2}$$

With group delay and phase delay equal to $\alpha$, we can say that the phase and group delay is half the length of the filter, where $N$ represents the length of the filter's impulse response vector. Filters that fit this criteria are usually divided into four groups, two even and two odd, with the axis of symmetry for $h[n]$ centered either symmetrically or antisymmetrically about the midpoint of the sequence. If it is symmetric, the sine component of exponential in Equation 4-8 vanishes; if it is antisymmetric, the cosine component disappears. Either way, the phase shift remains linear, except for an additional offset from origin of $\pi/2$ for the antisymmetric filter.

Filter type 1, as shown in Figure 4-8, satisfies this definition with a symmetric impulse response: $h[n] = h[N-n]$, for all $n$, $N$ is odd, and an axis of symmetry on a delay (tap) at $(N-1)/2$. Even symmetry gives us

$$H\left(e^{j\omega}\right) = e^{-j\omega\frac{(N-1)}{2}}\left(\sum_{k=0}^{\frac{N-1}{2}} a[k]\cos\omega k\right)$$

$$a[0] = h\left[\frac{N-1}{2}\right]$$

$$a[k] = 2h\left[\frac{(N-1)}{2} - k\right],\ k = 1,2,3,\cdots,\frac{N-1}{2}$$

Filter type 2, shown in Figure 4-9, also fits this equation with a symmetric impulse response: $h[n] = h[N-n]$, for all $n$, but with an even length, with the axis of symmetry between delays $(N-2)/2$ and $N/2$. In this case, the frequency response is

$$H\left(e^{j\omega}\right) = e^{-j\omega\frac{(N-1)}{2}}\left(\sum_{k=1}^{\frac{N}{2}} b[k]\cos\left[\omega\left(k - \frac{1}{2}\right)\right]\right)$$

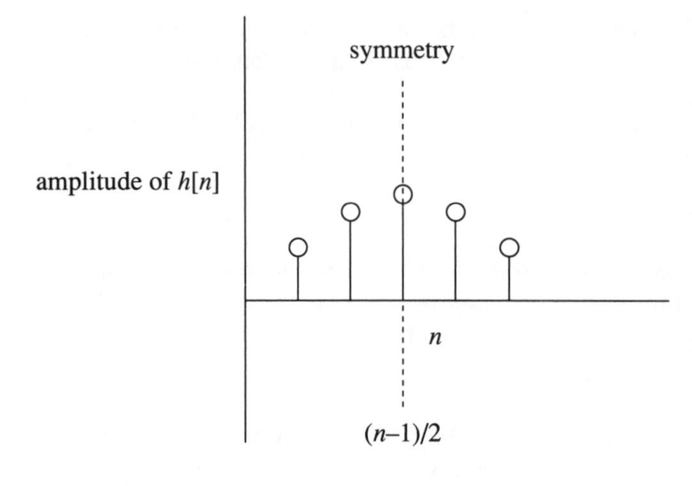


$$h[n]=h[(n-1)-n]$$

**Figure 4-8.** Type 1 linear phase.

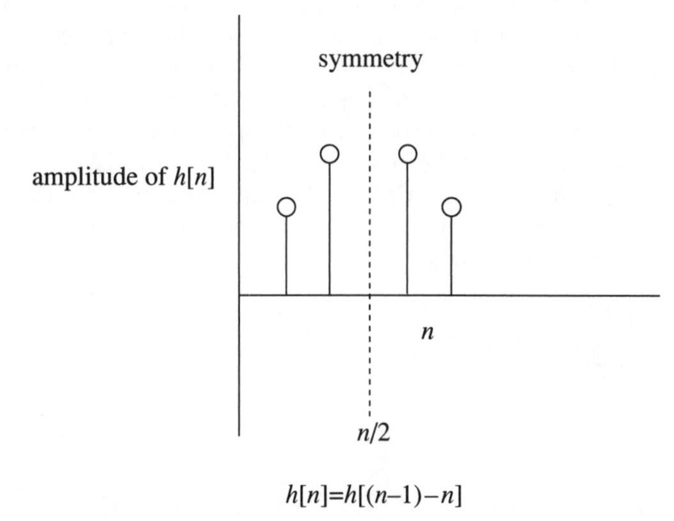

**Figure 4-9.** Type 2 linear phase.

$$b[k]=2h\left[\frac{(N)}{2}-k\right],\ k=1,2,3,\cdots,\frac{N}{2}$$

The next group of filter types may be said to have constant group delay but not constant phase delay. If we ignore any discontinuities resulting from the additions of constant phase, $\beta$, in all or part of $|\omega| < \pi$. This group will have linear phase plus a constant 90 degree component:

$$\theta(\omega) = j\beta - \alpha\omega\,,\ \ \beta = \frac{\pi}{2},\ \text{and } 0 < \omega < \pi \qquad \text{(Eq. 4-10)}$$

The two filters in this category are both antisymmetric: $h[n] = -h[N-n]$, for all $n$.

Filter type 3, shown in Figure 4-10, has $N$ odd and centered on a delay. Its frequency response is

$$H\left(e^{j\omega}\right) = je^{-j\omega\frac{(N-1)}{2}}\left(\sum_{k=1}^{\frac{N-1}{2}} c[k]\sin\omega k\right)$$

$$c[k] = 2h\left[\frac{(N-1)}{2} - k\right],\ k = 1,2,3,\cdots,\ \frac{N-1}{2}$$

Finally, filter type 4 (Figure 4-11) has an even length; it is centered between delays and has a frequency response

$$H\left(e^{j\omega}\right) = je^{-j\omega\frac{(N-1)}{2}}\left(\sum_{k=1}^{\frac{N}{2}} d[k]\sin\left[\omega\left(k - \frac{1}{2}\right)\right]\right)$$

$$d[k] = 2h\left[\frac{(N)}{2} - k\right],\ k = 1,2,3,\cdots,\ \frac{N}{2}$$

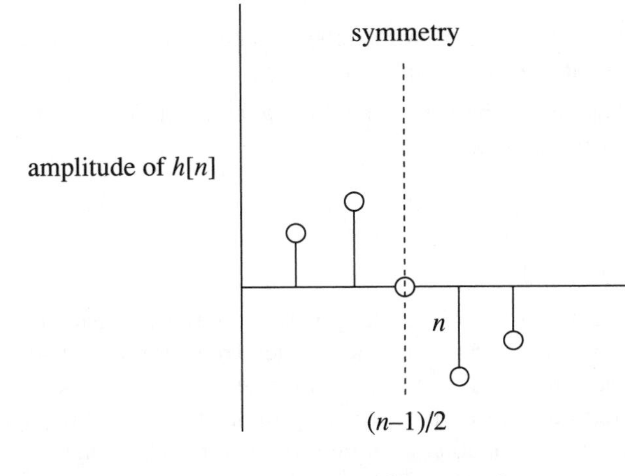


**Figure 4-10.** Type 3 linear phase.

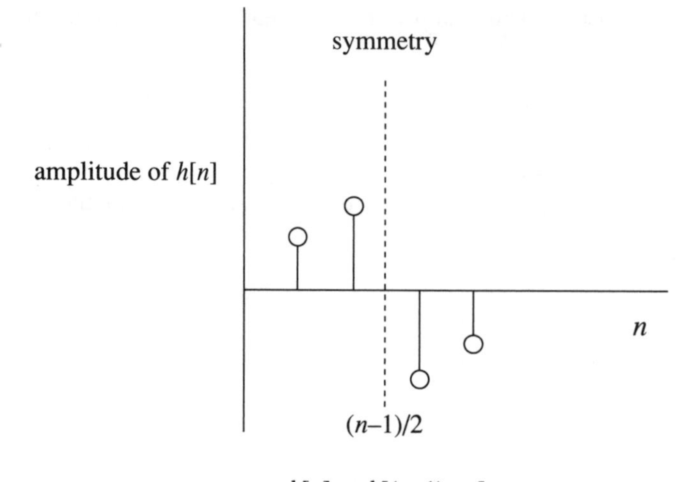


$h[n] = -h[(n-1)-n]$

**Figure 4-11.** Type 4 linear phase.

The fact that these filter types exploit the benefits of even or odd symmetry and are real allows them to be computed more easily. In the rest of this chapter we will be dealing primarily with type 1 FIR filters.

Usually, the symmetry of the filter is chosen such that the cutoff frequencies fall between sample points. Symmetric coefficients are effective for lowpass filters, and anti-symmetric filters can generally be used for highpass and derivative purposes. Here are a two caveats, however:

1. Types 3 and 4 should not be used for lowpass and bandstop filtering because of their offset around the origin ($\beta = \pi/2$ in Eq. 4-10).
2. Types 2 and 3 have odd symmetry around $\omega = \pi$, and therefore should not be used for highpass or bandstop filtering.

## FIR Filter Design Techniques

### Frequency Sampling Method

There are many popular methods of FIR filter design. The intuitive approach to filter design described in the section entitled Filtering with the Fourier Transform is one such approach. It is known as the *frequency sampling method* of FIR filter design. The design procedure begins by describing a step function on the frequency domain that fits the parameters we want. We then transform that step function back into its corresponding impulse response on the time domain, where it fits neatly into an FIR convolution.

The main problem with using the step function in the frequency domain are the discontinous edges, which can cause ringing, as described by the Gibb's phenomenon and in the earlier sections on windowing. The ringing looks like ripple in both the passband and stopband, and can be smoothed by including one or more samples of the transition band in the frequency sequence to be transformed back to the time domain or by windowing. These samples give some slope to the edges and ease the ringing.

Actually, any DFT can be used to derive the impulse response coefficients from the frequency response for the FIR filter. There are, however, a few direct interpolating functions for FIR design using the frequency sampling technique. One such function is given in

$$X(z) = \frac{1-z^{-n}}{N} \sum_{k=0}^{N-1} \frac{X\left(e^{jk2\frac{\pi}{N}}\right)}{1-e^{jk2\frac{\pi}{N}} z^{-1}}$$

Another set tailored to provide constant group delay for type 1 through 4 linear phase FIR filters is given in Table 4-4.

Another very popular method is known as the *Remez exchange method* of FIR design. The goal here is to minimize the differences between the ideal filter and its FIR approximations. The algorithm is based upon the fact that the frequency response of a linear phase filter can be expressed as a sum of cosines. This is changed to a Chebyshev approximation; the best approximation to the magnitude response has equiripple behavior. No method produces the perfectly rectangular response represented in the ideal filter—there is always some ripple in the passband and stopband, and there is always a nonzero transition band. The details of this algorithm are available in many books on digital filtering (please see the bibliography). There are several public domain implementations of this procedure.

| *Linear phase filter type* | *h(n) =* |
|---|---|
| 1. symmetric, odd | $\frac{1}{N}\left(A[0]+\sum_{k=1}^{M} 2A[k]\cos\left[\frac{2\pi(n-M)k}{N}\right]\right)$ |
| 2. symmetric, even | $\frac{1}{N}\left(A[0]+\sum_{k=1}^{\frac{N}{2}-1} 2A[k]\cos\left[\frac{2\pi(n-M)k}{N}\right]\right)$ |
| 3. antisymmetric, odd | $\frac{1}{N}\left(\sum_{k=1}^{M} 2A[k]\sin\left[\frac{2\pi(M-n)k}{N}\right]\right)$ |
| 4. antisymmetric, even | $\frac{1}{N}\left(A\left[\frac{N}{2}\right]\sin[\pi(M-n)]+\sum_{k=1}^{\frac{N}{2}-1} 2A[k]\sin\left[\frac{2\pi(M-n)k}{N}\right]\right)$ |

**Table 4-4.**

## The Fourier Series Method of FIR Filter Design

The Fourier series method in association with windowing is one of the most popular methods for determining FIR filter coefficients. It is easy to use and implement and it works well—with two caveats. First, it does not give precise control over passband and stopband edge frequencies; second, passband and stopband ripple are not distributed equally, but concentrated near the band edges.

In this section, we will first develop the procedure for deriving the coefficients for the filter of our choice. Using the Fourier series, we expand the frequency response of an ideal filter to the desired order. Since the Fourier series is infinite and we are truncating it somewhat short of that, we are in effect multiplying our frequency response by a rectangle—something that can add unwanted high-frequency ringing and distortion to our results. As a result, the response is usually windowed to eliminate as many of these problems as possible. We will first discuss the development of the Fourier coefficients and then review windowing as it applies to the coefficients of an FIR filter.

The frequency response of the basic form of a frequency selective filter is periodic, as one might expect, and we can therefore represent it as a Fourier series. To create such a filter, it is only necessary to know what kind of response we wish—that is, lowpass bandpass, highpass, or notch—and how many taps are necessary.

The transfer function can be any form of the Fourier series. We will choose the exponential form:

$$Hd(\omega) = \sum_{n=-\infty}^{\infty} C_n e^{jn\omega T}, \quad |n| < \infty$$

We rewrite the formula slightly to introduce an element $v$:

$$Hd(\omega) = \sum_{n=-\infty}^{\infty} C_n e^{jn\pi v}$$

where $v = f/F_N$ and $F_N$ is the Nyquist frequency given by $F_N = F_s/2$. Since $v$ is a ratio of the cutoff frequency to the Nyquist frequency, its value will be less than 1.

To simplify our efforts again, we will use the odd/even property of the series to write the formula for the coefficients as an even function:

$$C_n = \int_0^1 H(\omega) \cos n\pi v \, dv, \; n \geq 0$$

We can then define formulas for the desired frequency response by setting the range of integration for the bandwidth of the particular form we are after and then integrating. Table 4-5 lists the formulas for common filter formats:

As an example of how this might work, let us develop a type 1 lowpass filter with 64 coefficients, a cutoff frequency of 5 KHz, and a sampling frequency of 40 KHz. (Please refer to the Mathcad document fsfilt.mcd for a live example.)

1. Determine the kind of filter desired.

| *Type* | *Coefficients given by* |
|---|---|
| lowpass | $C_n = \frac{\sin n\pi v}{n\pi}, \; \lvert n\rvert > 0$ <br> $C_0 = v$ |
| highpass | $C_n = -\frac{\sin n\pi v}{n\pi}, \; \lvert n\rvert > 0$ <br> $C_0 = 1 - v$ |
| bandpass | $C_n = \frac{\sin n\pi v_2 - \sin n\pi v_1}{n\pi}, \; \lvert n\rvert > 0$ <br> $C_0 = 1 - v_2 - v_1$ <br> $v_1$ = lower cutoff $v_2$ = upper cutoff |
| bandstop | $C_n = \frac{\sin n\pi v_1 - \sin n\pi v_2}{n\pi}, \; \lvert n\rvert > 0$ <br> $C_0 = 1 - v_2 - v_1$ <br> $v_1$ = lower cutoff $v_2$ = upper cutoff |

**Table 4-5.**

In this case, it will be a lowpass filter. According to Table 4-5, its general transfer function is

$$C_n = \frac{\sin n\pi v}{n\pi}$$

$C_0$ must be defined separately, because a division by zero is indeterminate. Using l'Hospital's rule, we arrive at

$$C_0 = v$$

2 Define $v$.

The ratio of the cutoff frequency to the Nyquist frequency is represented by $v$. Since $F_N = F_s/2$—that is, the Nyquist frequency is equal to the sampling frequency divided by 2—we make $F_N = 20000$ Hertz.

$$v = \frac{f_c}{F_N} = \frac{5000}{20000} = .25$$

3. Solve.

The coefficients are symmetrical about the center point. A type 1 filter has an even length. Therefore,

$$N = 63, \quad n = 0 \cdots \frac{N-1}{2}$$

The symmetry point is then $n = 31$.

$$H_{lp} = \frac{\sin(n\pi v)}{n\pi},\ C_0 = v$$

Figure 4-12 shows that the coefficients we derive from this formula represent half of a mirror-image response. Notice the ringing manifested in the sidelobes—this is the result of the rectangular window used to develop the coefficients. We are truncating the response at 64 coefficients. The coefficients will need to be rearranged with $C_0 = h_{31}$ for our use. This will have the effect of constructing the response as

$$h[n]\left(n=0\cdots\frac{N-1}{2}\right) = \frac{N-1}{2} - n \quad \text{and} \quad h[n]\left(n=\frac{N-1}{2}\cdots N-1\right) = \frac{N-1}{2} + n$$

Then, taking the DFT of $h[n]$ for the frequency response, we have the lowpass characteristic illustrated in Figure 4-13.

The same ringing contained in the sidelobes of the system response is now ringing on the frequency response, as you can see in Figure 4-13. The ringing is the well-known Gibbs phenomenon, caused by our truncation of the response at 64 coefficients. This is our first, or default, window: the rectangle. It is defined as

$$w[n] = 1,\ n = 0\cdots N$$
$$w[n] = 0,\ n < 0 \text{ or } n > N$$

the length is equal to the number of coefficients we wish. This kind of truncation deprives the approximation of all the frequency components necessary to truly complete the wave-

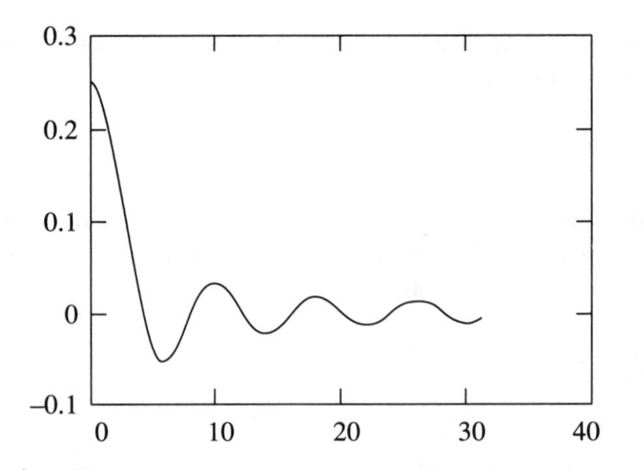


**Figure 4-12.** This is a plot of half the system response for half the coefficients, The plot shows that 0 to 31 is half the system response we need, starting at the peak of the sin*x*/*x* curve, or the mainlobe, falling to 0, and oscillating there in sidelobes as it gradually settles to zero.

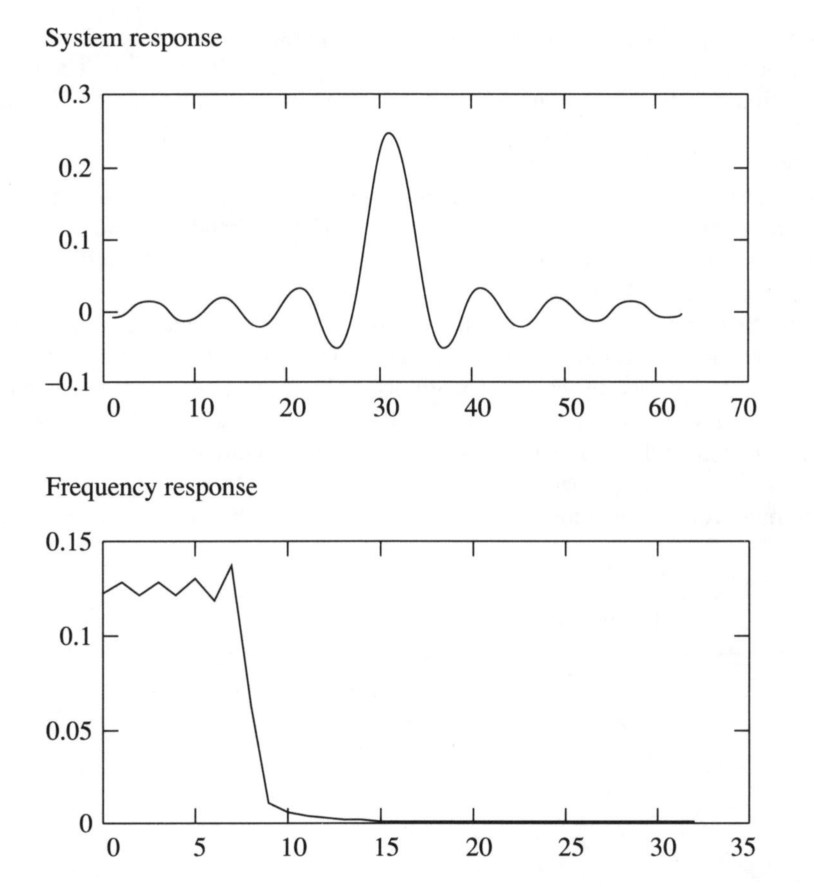


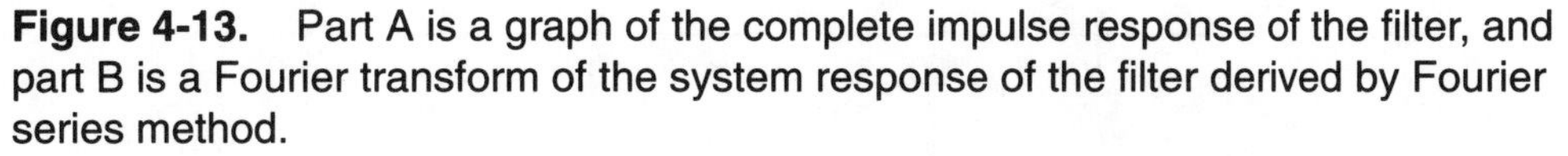
**Figure 4-13.** Part A is a graph of the complete impulse response of the filter, and part B is a Fourier transform of the system response of the filter derived by Fourier series method.

form. As a result, we are left with what the coefficients we have will produce. Having more coefficients will help, but it will never completely remedy the situation.

As a remedy, we can use the the same windowing functions discussed in Chapter 3 to attenuate the sidelobes somewhat and improve the performance of the filter. As before, the window coefficients are applied to the filter with a point-to-point multiplication:

$$C_w = C_n w[n]$$

In the paragraphs and illustrations that follow, we will be discussing the various windows mentioned in Chapter 3. There are two important aspects of these windows that need attention: the transition width and the attenuation of the sidelobes. The transition width of the window is important in determining the effect of the window upon the frequency response of the signal it is applied to. Recall that windowing a signal convolves the desired frequency response of the signal with the Fourier transform of the window. Therefore, a narrow transition width can produce sharp transitions or spectral peaks in the frequency domain, but if it is not combined with a good deal of attenuation in the side lobes, the result will still be smeared and oscillatory.

The transition band width for the Hanning or raised cosine, window (see Figure 4-14) is $4\pi/N$, where $N$ is the number of coefficients used. The stopband attenuation is 50dB.

The Hamming window (see Figure 4-15) has a transition band width of $8\pi/N$ and a stopband attenuation of 45dB.

In the Blackman window (see Figure 4-16), the transition band width is $12\pi/N$ and the stopband attenuation is 80dB. Though the Blackman window has the most attenuation on the sidelobes, it also has the broadest mainlobe, which indicates reduced selectivity. The selectivity of this window can be enhanced by increasing the number of coefficients.

Finally, there is the Kaiser/Bessel filter depicted in Figure 4-17, a window whose performance is comparable to that of the Blackman window. As mentioned in chapter three, this is the most versatile window in this group. It is popular because it can be configured to

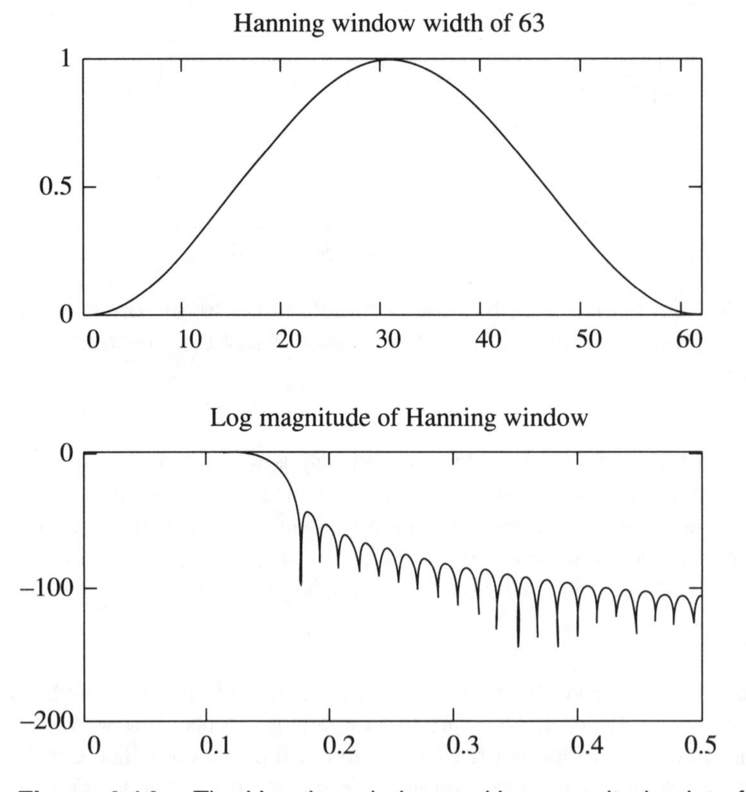


**Figure 4-14.** The Hanning window and log magnitude plot of frequency response.

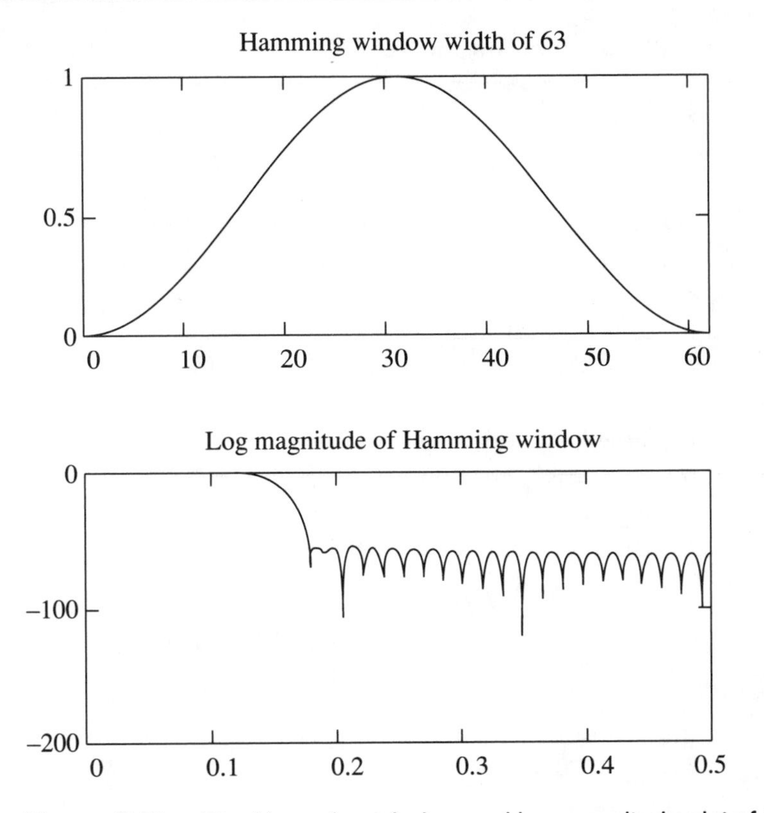


**Figure 4-15.** The Hamming window and log magnitude plot of frequency response.

give the same response of any of the other windows here and better. This window method does require some experimentation to use most effectively in any particular application, but it is also the most facile.

## Convolution

### Convolution in the Frequency Domain

Convolution and multiplication form a Fourier transform pair. Performing either function on one domain is equivalent to performing the corresponding function in the other domain. This means that one could perform a complex convolution in the time domain with a simple multiplication in the frequency domain:

$$x[t] * y[t] = DFT^{-1}[X[\omega]Y[\omega]]$$

The FFT requires a good deal of computation time, but you may be surprised to note that even though an FFT requires so many multiplications and divisions to produce a result,

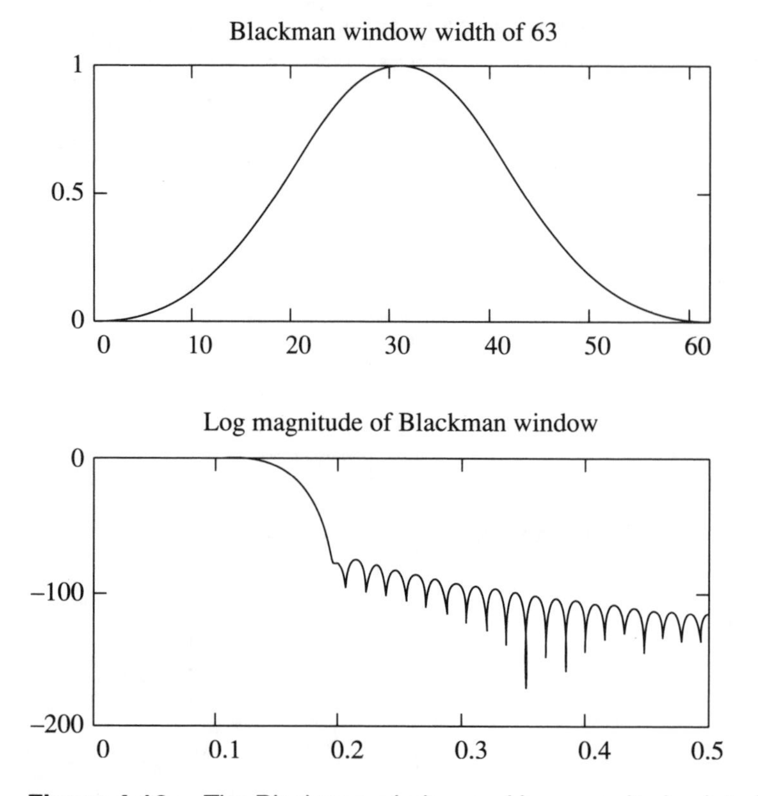


**Figure 4-16.** The Blackman window and log magnitude plot of frequency response.

it may still be faster than the same convolution done in the time domain. It generally requires—depending upon whether it is a real or complex convolution—two FFTs, a point-by-point multiplication, and an IFFT for a convolution. As with the convolution in the time domain, the length of the result is $L = (N + M) - 1$ components. That is, the length of the result is one less than the sum of the components of each of the input vectors.

In the frequency domain, $N$ and $M$ are equal to the *nonzero* portion of the two input vectors. To perform the convolution without aliasing, each vector must be extended to $L$ elements by padding with zeros. This is because the periodicity within the DFT effectively wraps both input vectors around a cylinder. To avoid spillover and resultant aliasing, the length must be equal to $L$ elements.

Figure 4-18 illustrates the difference between a properly sized convolution and one that overlaps. The convolution in the right is too long by one count and displays a nonzero offset. This is the direct result of aliasing as the rectangles extend around the cylinder and over one another.

At greater than 20 taps, the FFT begins to be more efficient than the time domain convolution. The FFT can be of greater benefit when we are convolving real-only functions,

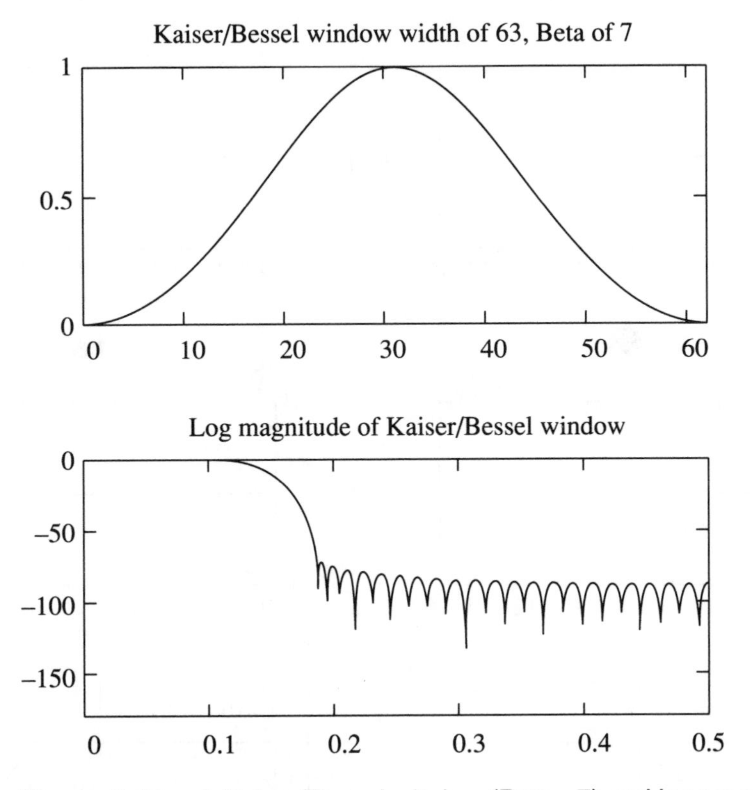


**Figure 4-17.** A Kaiser/Bessel window (Beta = 7) and log magnitude plot of frequency response.

since two FFTs can be performed at once. There is even more savings in filtering with fixed filter coefficients, because the FFT of the filter vector need only be done once and saved—the only FFTs are then the ones performed on the data vector. In fact, if the filter is known in advance, the entire algorithm can be tailored to the necessary length of the FFT—that is, the window can be adjusted to make $N+M$ a power of two. To compute the difference, consider that a time domain convolution always requires $N*M$ multiplications, while the FFT needs $2N\log_2(N/2)$ multiplications, where $N>>M$.

Suppose we take the Fourier coefficients for the filter described in the previous section, multiply it by a Blackman window, and take its Fourier transform. The results are illustrated in Figure 4-16.

As you can see, the filter defines a frequency response that includes all $\omega \le \omega_c$ and excludes all $\omega > \omega_c$. Now, let us take a signal

$$x[n] = \cos(2\pi f_0 t) + \cos(2\pi f_1 t)$$

with two peaks in its DFT, each representing one of the frequency components.

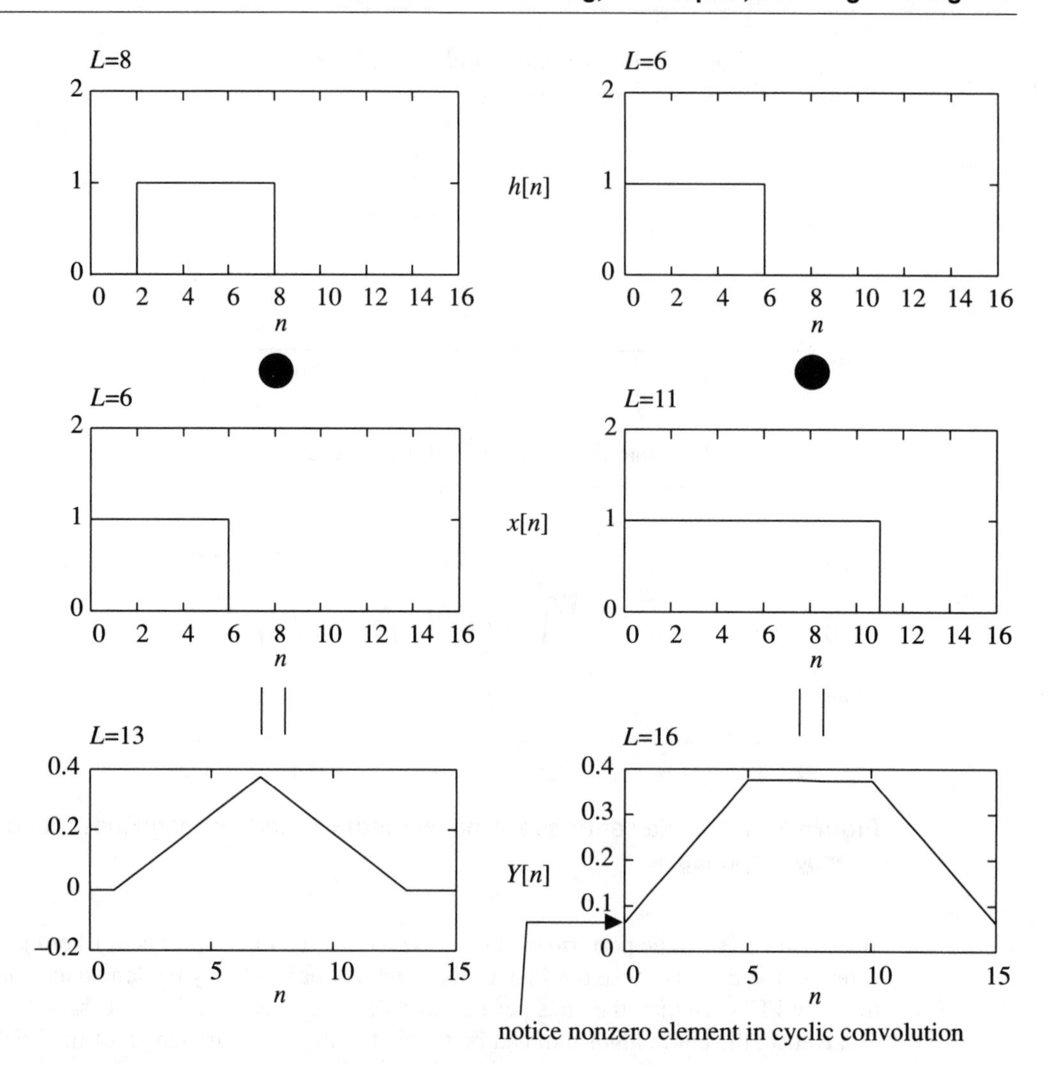


**Figure 4-18.** Part A is a convolution as it might be performed in the frequency domain; part B illustrates the effects of circular convolution when there is overlap.

A convolution involving this filter and the given input signal would require the following steps:

1. Take the DFT of the impulse response of the filter function and the input vector.
2. Determine the length of the FFTs involved by summing the nonzero elements of the input vector and the filter vector and finding the smallest power of 2 greater than that sum.

3. Pad each vector with zeros to attain that length.
4. Multiply the two vectors point by point.
5. Perform an inverse transform of the result—this is the filtered function.

Time domain convolution will require $N*M$ multiplications. When the signal window or number of filter taps is small enough, this is the preferred method. Convolution in the frequency domain is the preferred method for non-realtime processing, and for realtime processing when the record lengths support it. It is purely a matter of economy and speed. The time domain convolution *is* the FIR filter. The next section on time domain convolution illustrates how it is done.

## Convolution in the Time Domain

The process of convolving two data sequences in the time domain is precisely the same as the multiplication of polynomials. If we were to multiply the following two polynomials,

$$a_0 + a_1x + a_2x^2 + a_3x^3 + \cdots$$

and

$$b_0 + b_1x + b_2x^2 + b_3x^3 + \cdots$$

we would have the product

$$a_0b_0 + (a_0b_1 + a_1b_0)x + (a_0b_2 + a_1b_1 + a_2b_0)x^2 + (a_0b_3 + a_1b_2 + a_2b_1 + a_3b_0)x^3 \cdots$$

For convenience, we can call these sums of products by a new name to simplify the representation:

$$c_0 = a_0b_0$$
$$c_1 = (a_0b_1 + a_1b_0)$$
$$c_2 = (a_0b_2 + a_1b_1 + a_2b_0)$$
$$c_3 = (a_0b_3 + a_1b_2 + a_2b_1 + a_3b_0)$$

Probably the most well-known way of describing the serial product is by demonstration; we will use the pair of polynomials just presented. Find a strip of paper and write the coefficients to the first polynomial on its lower edge, so that they appear on the paper like this:

$$a_0 \quad a_1 \quad a_2 \quad a_3 \quad \cdots \quad a_n$$

Take another strip of paper and write the coefficients for the other polynomial on it so that the two pieces of paper can be placed next to one another with the coefficients lined up:

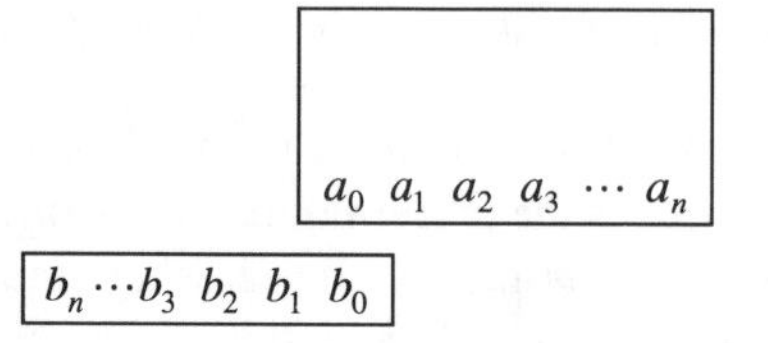


On another piece of paper, write down the first value output by the convolution:

$$c_0 = a_0 b_0$$

Move the piece of paper to the right one place:

$a_0\ a_1\ a_2\ a_3\ \cdots\ a_n$

$b_n \cdots b_3\ b_2\ b_1\ b_0$

Write down the second value output:

$$c_1 = a_0 b_1 + a_1 b_0$$

Again, move the paper one place to the right:

$a_0\ a_1\ a_2\ a_3\ \cdots\ a_n$

$b_n \cdots b_3\ b_2\ b_1\ b_0$

and write down the next output value:

$$c_2 = a_0 b_2 + a_1 b_1 + a_2 b_0$$

continue until the bottom piece of paper has moved past all the coefficients in the upper row. The length of the resulting sequence will equal the sum number of components in the individual sequences minus one.

This, then, is the same process indicated by the sum of products in Equation 4-7. It is probably also clear that an algorithm implementing this procedure would not prove too complex. Examples of such algorithms appear at the end of this chapter and on the optional disk.

### Deconvolution in the Frequency Domain

*Deconvolution* is very useful for removing known noise or deterioration from a signal to arrive at the original system response. It is also used in the examination of unknown systems: a known signal can be divided out of the resulting signal to find the impulse response of the unknown. A good example of its use is in image enhancement, when the particular distortion, blurring, or corruption is known or can be estimated.

It is performed as the inverse to convolution in the frequency domain. Take the DFT of the corrupted image and divide through by the DFT of the estimated or known corruption:

$$Y(\omega)=\frac{Y'(\omega)}{X(\omega)}$$

where $Y'(\omega)$ is the corrupted image and $X(\omega)$ is the DFT of the corruption. Improvement found in this manner can be heuristic in providing an improved estimate and in correcting the problem. Refer to Figure 4-19 for an illustration of this process.

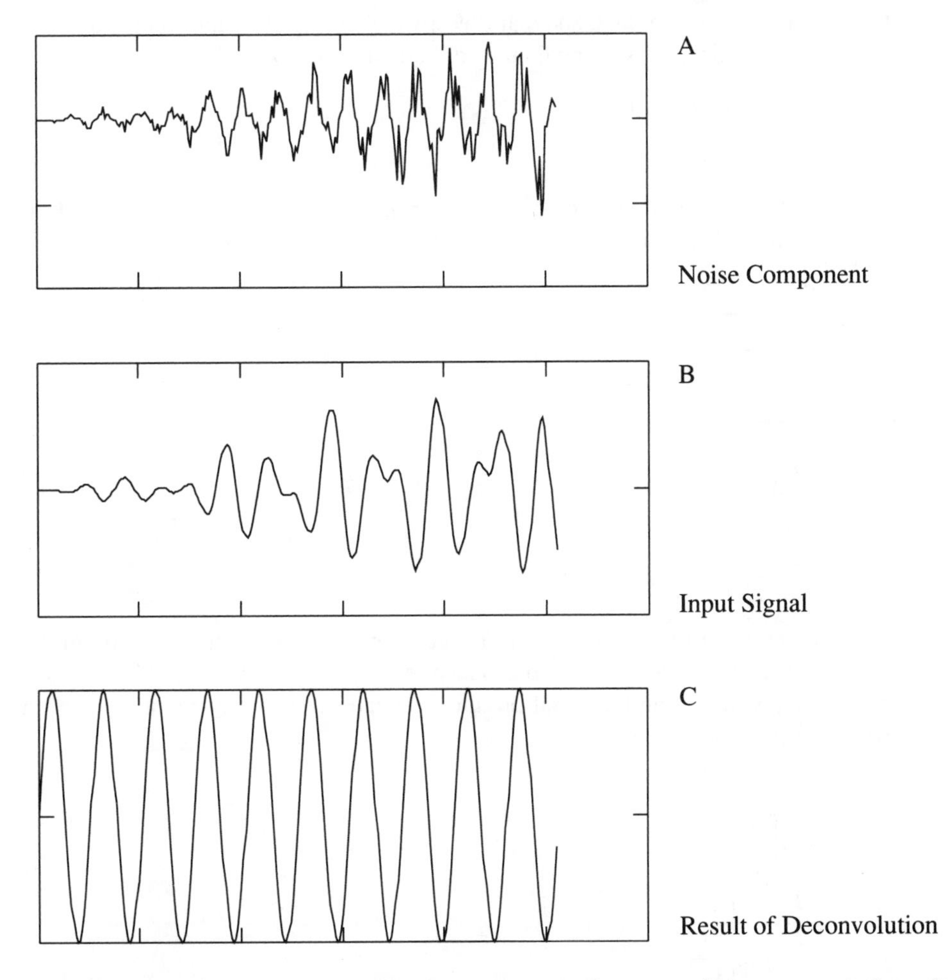


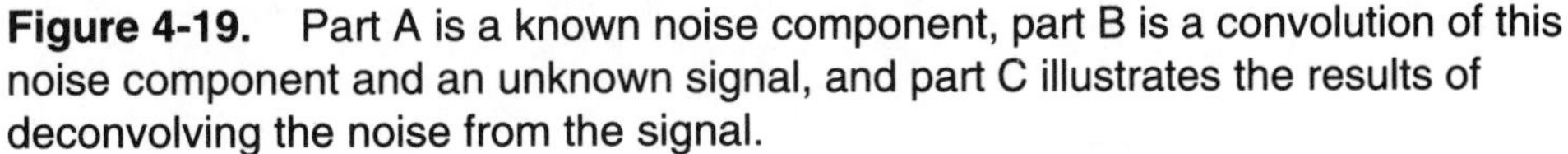

**Figure 4-19.** Part A is a known noise component, part B is a convolution of this noise component and an unknown signal, and part C illustrates the results of deconvolving the noise from the signal.

The caveats for using this method lie in numerical methods. Division is noisy—it involves subtraction and can be responsible for the introduction of high-frequency elements. Also, one must be careful that the divisor has no zeros, because the function can diverge dramatically. This is often compensated for with an artifical floor that prevents it from crossing zero.

### Deconvolution in the Time Domain: Serial Quotient

The *serial quotient* is the time domain analog to the deconvolution involving DFTs that was previously discussed. When dividing polynomials, synthetic division is normally used, but its implementation on a computer is cumbersome—and in this case, unnecessary. There is a method of deriving the $a_n$ coefficients from the other two that is similar to the method we used to perform the serial product, and that is easily and efficiently implemented on a computer.

Let us begin by assuming that we already have the $a_n$ and $c_k$ coefficients:

$$c_k = 3\ 5\ 10\ 11\ 14\ 11\ 9$$

$$a_n = 1\ 1\ 2\ 2\ 3$$

Write these on a piece of paper, lining up each coefficient rank by rank with a space between large enough to slide another small strip of paper through:

| $a_n$ | $b_m$ | $c_k$ |
|---|---|---|
| 1 | | 3 |
| 1 | | 5 |
| 2 | | 10 |
| 2 | | 11 |
| 3 | | 14 |
| | | 11 |
| | | 9 |
| | | |

In serial multiplication, we introduce data samples incrementally, multiply all the coefficients, and sum the results to produce the next value, which in this case is $c_k$. Since we are really just reversing the serial product, we can say that the first unknown must then be equal to $c_0/a_0$, which is 3.

| $a_n$ | $b_m$ | $c_k$ |
|---|---|---|
| 1 | 3 | 3 |
| 1 | 2 | 5 |
| 2 | | 10 |
| 2 | | 11 |
| 3 | | 14 |
| | | 11 |
| | | 9 |

The next is found by multiplying out the products of $a_n$ and $b_m$ that have thus far been found, which in this case is only the first value: 3. We sum the products and subtract that sum from the next value of $c_n$. Then divide by $a_n$ for $b_m$.

We already know that the length of the convolution sum is equal to the sum of the number of coefficients in $a_n$ and $b_m$ minus 1. Therefore, to find the length of the unknown $b_m$, subtract the number of coefficients in $a_n$ from $c_k$ and add 1. In the example, the number of coefficients in $b_m$ is 3.

## Discrete Time: Infinite Impulse Response Filters

The general form for the output of an IIR filter obeys the general equation for a discrete time system function given in Equation 4-5. Unlike the FIR filter, the IIR filter possesses both poles and zeros corresponding to the basic difference equation of the discrete time system:

$$y[n] = \sum_{n=1}^{N} a_n[k-n] + \sum_{m=0}^{M} b_m x[k-m] \quad \text{(Eq. 4-11)}$$

This expression says that the output $y[n]$ is equal to a combination of the current and previous $N$ inputs, plus $M$ of the previous outputs. In order for us to have an IIR filter, at least one of the $a_n$ must remain after all possible cancellations, and it must be nonzero. Of course, for the filter to be stable, all the poles must be within or on the unit circle.

### Impulse Invariant Method

Based upon what we have seen of FIR design, it is natural to look for some way to use what we know of analog filter techniques that would help us design IIR filters composed of poles and zeros. We know that the impulse response of a system or network completely characterizes the network—so why not sample the impulse response of a known or obtainable continuous-time filter and use that data?

We can. What we need is an analog prototype transfer function, $H_a(s)$.

1. Factor the transfer function for the prototype analog filter, as with the general equation

$$H_a(s) = \frac{\sum_{i=0}^{M} b_i s^i}{1 + \sum_{i=1}^{N} a_i s^i}$$

Express the result as a partial fraction expansion, combining complex conjugate pairs:

$$H_a(s) = \sum_{i=1}^{N} \frac{A_i}{s - s_i} \quad \text{(Eq. 4-12)}$$

As before, $s_i$ are the poles of $H_a(s)$, and

$$A_i = \left[(s - s_i)H_a(s)\right]\Big|_{s=s_i}$$

The impulse response of the prototype filter (the inverse DFT) is formed from Equation 4-12 as

$$h_a(t) = \sum_{i=1}^{N} A_i e^{s_i t} u(t)$$

from which we can write the unit-sample response:

$$h[n] = \sum_{i=1}^{N} A_i T e^{s_i T n} u[n]$$

Of course, $T$ represents the sample period, meaning that $h[n]$ is a scaled version of $h_a(t)$ sampled at regular and equally spaced intervals, $nT$.

2. Map this expression to the $z$-plane to obtain a discrete time transfer function:

$$H(z) = \sum_{i=1}^{N} \frac{TA_i}{1 - e^{s_i t} z^{-1}}$$

The poles map according to the relationship

$$z_i = e^{s_i t}$$

Since the frequency response of the discrete time filter is related to the continuous-time filter by

$$H\left(e^{j\omega}\right) = \sum_{k=-\infty}^{\infty} H_z\left(j\frac{\omega + 2\pi k}{T}\right)$$

and

$$j\frac{\omega + 2\pi k}{T}$$

maps from the imaginary axis of the $s$-plane to the circumference of the unit circle on the $z$-plane, this procedure can produce a discrete frequency response that is an aliased version of the analog frequency response. It can do this because rarely can you guarantee that

$$H_a = (j\omega) \text{ for } |\omega| \geq \frac{\pi}{T}$$

Still, this method has uses in well-limited responses, and it is instructive in the transformation from analog filters having both poles and zeros to discrete time filters.

### The Bilinear Transform

Probably the most popular technique for developing IIR filters is the *bilinear transform*, the BLT. (For more detail and a live illustration, please refer to the Mathcad document BLT.mcd.)

The BLT transforms an analog filter transfer function into the system function for a digital filter with the equality

$$s = \frac{2\left(1 - z^{-1}\right)}{T\left(1 + z^{-1}\right)}$$

where $T$ is the sampling interval.

Actually, the factor $2/T$ is nothing more than a scaling factor, meant to produce a ratio of the cutoff frequency to Nyquist frequency. Multiplied by $2\pi f$, it reduces to a fraction of $\pi$. Depending upon your purposes, you may choose to omit or replace it.

The process of producing a system function in $z$ is a simple one:

1. Find or derive an analog transfer function in $s$ for the filter you desire.
2. Replace all $s$ with

$$\frac{2\left(1 - z^{-1}\right)}{T\left(1 + z^{-1}\right)}$$

3. Simplify the transfer function to a ratio of polynomials in $z$.
4. Finally, divide through by the highest power of $z$.

As an example, we will take a simple Butterworth lowpass filter with the transfer function

$$H_a(s) = \frac{\omega_c^2}{s^2 + \sqrt{2}\omega_c s + \omega_c^2}$$

The sampling frequency, $f_s$, will be 40 KHz, and the desired cutoff frequency, $\omega_c$, will be 5 KHz. Replacing $s$ with

$$\frac{2\left(1 - z^{-1}\right)}{T\left(1 + z^{-1}\right)}$$

results in

$$H(z) = \frac{\omega_c^2}{\left[\left(\frac{2}{T}\right)\left(\frac{1 - z^{-1}}{1 + z^{-1}}\right)\right]^2 + \frac{2\sqrt{2}\omega_c}{T}\left(\frac{1 - z^{-1}}{1 + z^{-1}}\right) + \omega_c^2}$$

Simplify:

$$H(z)=\frac{\omega_c^2T^2\left(1+z^{-1}\right)^2}{4\left(1-z^{-1}\right)^2+2\sqrt{2}\omega_cT\left(1-z^{-1}\right)\left(1+z^{-1}\right)+\omega_c^2T^2\left(1+z^{-1}\right)^2}$$

$$H(z)=\frac{.6169\left(1+z^{-1}\right)^2}{4\left(1-z^{-1}\right)^2+2.2214\left(1-z^{-1}\right)\left(1+z^{-1}\right)+.6169\left(1+z^{-1}\right)^2}$$

$$H(z)=\frac{0.6169+1.2336z^{-1}+0.6169z^{z-2}}{2.39552z^{-2}-6.7662z^{-1}+6.8383}$$

Simplifying still further:

$$H(z)=\frac{0.09021+0.18042z^{-1}+0.09021z^{-2}}{0.3503z^{-2}-0.98946z^{-1}+1}$$

Replacing the coefficients in Equation 4-11 with the values arrived at in the transfer function, we have the difference equation:

$$y[n]=-0.3503y[n-2]+0.98946y[n-1]+0.09022x[n-2]+0.1804x[n-1]+0.09021 \quad \text{(Eq. 4-13)}$$

### BLT for Second-Order Analog Sections

Second-order analog sections were used in Chapter 2 to create larger filters. The same is often done in the digital domain. If you have a second-order section you are translating to discrete time, there are some simple relationships that can help.

A second-order analog section is represented by the following function:

$$H_a(s)=\frac{\delta_0+\delta_1 s+\delta_2 s^2}{1+\alpha_1 s+\alpha_2 s^2}$$

and a digital second-order section has the form

$$H(z)=\frac{b_0+b_1z^{-1}+b_2z^{-2}}{1+a_1z^{-1}+a_2z^{-2}}$$

The coefficients for the digital second-order section can be derived from the analog section using the following relationships:

$$b_0=\left(\frac{4}{c}\right)\left(\frac{\delta_0}{\left(\frac{2}{T}\right)^2}+\frac{\delta_1}{\left(\frac{2}{T}\right)}+\delta_2\right)$$

$$b_1 = \left(\frac{8}{c}\right)\left(\frac{\delta_0}{\left(\frac{2}{T}\right)^2} - \delta_2\right)$$

$$b_2 = \left(\frac{4}{c}\right)\left(\frac{\delta_0}{\left(\frac{2}{T}\right)^2} - \frac{\delta_1}{\left(\frac{2}{T}\right)} + \delta_2\right)$$

$$a_1 = \left(\frac{8}{c}\right)\left(\frac{1}{\left(\frac{2}{T}\right)^2} - \alpha_2\right)$$

$$a_2 = \left(\frac{4}{c}\right)\left(\frac{1}{\left(\frac{2}{T}\right)^2} + \frac{\alpha_1}{\left(\frac{2}{T}\right)} + \alpha_2\right)$$

$$c = \left(\frac{4}{\left(\frac{2}{T}\right)^2} + \frac{4\alpha_1}{\left(\frac{2}{T}\right)} + 4\alpha_2\right)$$

As a special case, the numerators of Butterworth and Chebyshev filters reduce to

$$\frac{4\alpha_0\left(1+z^{-1}\right)^2}{a^2 c}$$

since $\delta_1$ and $\delta_2$ are both equal to zero. In addition, the $k$th second-order section has the following reductions for $\alpha$:

$$\alpha_1 = 2\sin\left(\frac{(2k+1)\pi}{2N}\right),\ \alpha_2 = 1$$

### Factored System Functions

Throughout this book, we have been dealing with the poles and zeros of system function in factored form. It is just as easy to develop the necessary coefficients in this manner as in the complete rational form.

We know that $(s-s_p)=0$ and $(s-s_z)=0$ will provide us with the poles and zeros in a factored transfer function. Knowing this, we can readily find the corresponding poles and zeros on the $z$-plane by making the same substitution we made earlier with the transfer function:

$$s=\frac{2\left(1-z^{-1}\right)}{T\left(1+z^{-1}\right)}$$

Thus,

$$\left(\frac{2}{T}\frac{1-z^{-1}}{1+z^{-1}}\right)-s_p=0$$

will yield poles on the $z$-plane, and

$$\left(\frac{2}{T}\frac{1-z^{-1}}{1+z^{-1}}\right)-s_z=0$$

will yield the zeros. The first step is to simplify the equations slightly to give us

$$z_p=\frac{2+s_pT}{2-s_pT} \qquad \text{(Eq. 4-14)}$$

and

$$z_z=\frac{2+s_zT}{2-s_zT}$$

Placing the results of these equalities into the familiar equation for factored system functions yields

$$H(z)=H_0\frac{(z+1)^{N-M}\prod_{m=1}^{M}(z-z_m)}{\prod_{n=1}^{N}(z-z_n)}$$

where $z_m$ are the zeros and $z_n$ are the poles. When there are more poles than zeros in the transfer function, the factor $(z+1)^{N-M}$ produces zeros at $-1$ on the $z$-plane for the zeros at infinity on the $s$-plane.

This form is especially nice for providing poles and zeros that can be paired for cascade filter structures. We will discuss these in upcoming sections and in Chapter 5.

## Frequency Warping

Once we have transformed an analog filter to a discrete time filter using the BLT, we quickly find that the filter we have created in the digital domain has different characteristics than the original prototype we developed in the time domain. Even though

$$\left(\frac{2}{T}\frac{1-z^{-1}}{1+z^{-1}}\right)=s$$

is a one-to-one mapping of the complex $s$-plane to the $z$-plane, it is not linear everywhere. This can lead to obvious problems in filters where cutoff frequencies are critical, such as notch filters.

Actually, the mapping corresponds to that of taking the arctangent of a circle

$$\omega_a = \frac{2}{T}\tan\frac{\omega_d}{2}$$

and

$$\omega_d = 2\tan^{-1}\frac{\omega_a T}{2}$$

Here $\omega_d$ is a value normalized to the region between 0 and $\pi$ in the $z$-plane, while $\omega_a$ is equivalent to $2\pi f$ in the $s$-plane.

In Figure 4-20, we see how the analog frequencies are compressed to fit in an area between 0 and $\pi$ in the $z$-plane. It is also interesting and important to note that the relative difference between the analog and digital frequencies decreases as the sampling rate goes up. Linear phase is not possible with this transformation, because linear phase forms a sawtooth that the arctangent output above can not track. Where IIR filters are required and phase is important, it is possible to use an allpass digital filter to linearize the problematic sections of the passband.

If we wish to create a filter with the same cutoff frequencies as an analog prototype, we need to *prewarp* the analog frequencies so that they will fall where we need them in the $z$-plane. Therefore, in our example we needed a cutoff of 5KHz with a sampling frequency of 40KHz. This corresponds to a $z$-plane frequency of $5000\pi/20000 = \pi/4$. Substituting this into our formula,

$$\omega_a = \frac{2}{\frac{1}{40000}}\tan\frac{\frac{\pi}{4}}{2} = 8000\tan\frac{\pi}{8} = 33137 \text{ Radians, or } 5274 \text{ Hertz}$$

This, then, is the frequency $\omega_c$ in the analog prototype above.

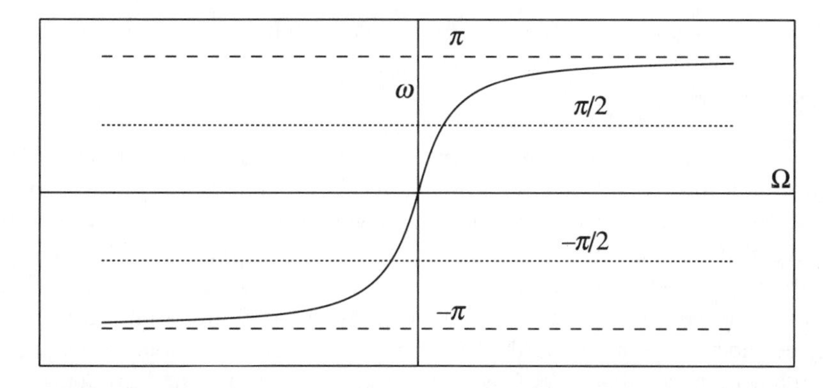


**Figure 4-20.** The mapping of the $s$-plane to the $z$-plane is one-to-one but highly nonlinear after $f_s/4$.

### Magnitude and Phase

Computing the magnitude and phase of the digital poles derived from the BLT can be performed by substituting the equality $s = \sigma + j\omega$ back into the equation for the pole (see Eq. 4-12), resulting in

$$z_p = \frac{2 + \sigma T + j\omega T}{2 - \sigma T - j\omega T}$$

Magnitude then becomes

$$|z_p| = \sqrt{\frac{(2+\sigma T)^2 + (\omega T)^2}{(2-\sigma T)^2 + (\omega T)^2}}$$

and phase is

$$\theta(z_p) = \tan^{-1}\left(\frac{\omega T}{2+\sigma T}\right) - \tan^{-1}\left(\frac{-\omega T}{2-\sigma T}\right)$$

## Frequency Transformation

Just as with analog filters, it is sometimes desirable to transform the transfer function from one type of filter (lowpass, highpass, or whatever) to another, even if it is as simple as transforming a lowpass to another lowpass with a different cutoff. There are methods for performing the transformations, but unfortunately, they are not always easy.

In order to perform the required mapping, we must come up with a rational function of $z^{-1}$ that maps the inside of the unit circle of the new system function to the inside of the unit circle of the existing function. The following relationship describes such a function:

$$G(z^{-1}) = \pm \prod_{k=1}^{N} \frac{z^{-1}\alpha_k}{1 - \alpha_k z^{-1}}$$

Table 4-6 lists the appropriate transformation formulas.

# FILTER STRUCTURES

Up to now, all the arithmetic we have done on analog functions and discrete functions has assumed infinite precision. This is not the case with discrete time systems. We deal here with finite-length arithmetic, which is arithmetic with a fixed precision or resolution. Many DSPs have a 16-bit bus and 16-bit word length—this is the limit of their resolution. Floating-point processors only offer some relief; their values must still fit in a finite-length space, just as fixed point values. The main benefit of floating-point arithmetic is often speedier development and greater dynamic range. We will not go into the details of finite-length arithmetic here—that is for Chapter 5. We mention it only to draw some attention to the importance of the filter structures we are about to discuss.

| *Type* | *Transformation cutoff frequencies are between 0 and π* | *Parameters* $\omega_c^p$ = cutoff of prototype; $\omega_c^p$ = cutoff of target; $\omega_1$ = desired lower cutoff; $\omega_u$ = desired upper cutoff |
|---|---|---|
| Lowpass | $z^{-1} \to \frac{z^{-1}-\alpha}{1-\alpha z^{-1}}$ | $\alpha = \frac{\sin\left(\frac{\omega_c^p - \omega_c^t}{2}\right)}{\sin\left(\frac{\omega_c^p + \omega_c^t}{2}\right)}$ |
| Highpass | $z^{-1} \to -\frac{z^{-1}+\alpha}{1+\alpha z^{-1}}$ | $\alpha = \frac{\cos\left(\frac{\omega_c^p + \omega_c^t}{2}\right)}{\cos\left(\frac{\omega_c^p - \omega_c^t}{2}\right)}$ |
| Bandpass | $z^{-1} \to -\frac{z^{-2} - \frac{2\alpha k}{k+1} z^{-1} + \frac{k-1}{k+1}}{\frac{k-1}{k+1} z^{-2} - \frac{2\alpha k}{k+1} z^{-1} + 1}$ | $\alpha = \frac{\cos\left(\frac{\omega_u + \omega_l}{2}\right)}{\cos\left(\frac{\omega_u - \omega_l}{2}\right)}$ <br> $k = \cot\left(\frac{\omega_u - \omega_l}{2}\right)\tan\left(\frac{\omega_c^p}{2}\right)$ |
| Bandstop (notch) | $z^{-1} \to \frac{z^{-2} - \frac{2\alpha k}{k+1} z^{-1} + \frac{1-k}{k+1}}{\frac{1-k}{k+1} z^{-2} - \frac{2\alpha k}{k+1} z^{-1} + 1}$ | $\alpha = \frac{\cos\left(\frac{\omega_u + \omega_l}{2}\right)}{\cos\left(\frac{\omega_u - \omega_l}{2}\right)}$ <br> $k = \tan\left(\frac{\omega_u - \omega_l}{2}\right)\tan\left(\frac{\omega_c^p}{2}\right)$ |

**Table 4-6.**

One system function can be implemented in a number of ways. There are direct forms that implement the function from the transfer function by inspection. An example of such a function is the difference equation developed directly from the transfer function for a Butterworth filter we used in the section on the bilinear transform:

$$H(z) = \frac{0.09021 + 0.18042z^{-1} + 0.09021z^{-2}}{0.3503z^{-2} - 0.98946z^{-1} + 1}$$

becomes

$$y[n] = -0.3503y[n-2] + 0.98946y[n-1] + 0.09022x[n-2] + 0.1804x[n-1] + 0.09021$$

There are problems with this, however, that usually make it an unacceptable solution. First, there is wasted time in implementing a filter in this form, which results in negative phase delay and longer group delay. This is almost always to be avoided in realtime control.

Another problem exists that in many systems makes this form unworkable: the noise from the roundoff and truncation necessary in finite-word length arithmetic. This will be covered much more deeply in Chapter 6, but for now please note that this is a prevalent problem in both floating-point processors and fixed-point, though the problem is often less serious in floating-point.

The next few sections will deal with structures for FIR and IIR filters, as well as what might be done to enhance their performance and overcome some of these difficulties.

## FIR CASCADE STRUCTURES

There are no complete solutions to problems of noise resulting from finite-length arithmetic, but there are some things that can be done to ameliorate the situation. Of the two filter forms, FIR filters have fewer problems arithmetically than IIR filters, but they take longer to compute. There are several basic forms of the FIR filter, among them the direct form derived from the transfer function, the cascade form using second-order sections to build a filter, and the lattice network, a popular form for its robust behavior despite coefficient variations.

The cascade form of the FIR filter uses the factored form of the transfer function to create the filter, in much the same way as in analog cascade filters. This allows separate sections to be tuned individually. Different scaling factors can be used to minimize the chances of saturation or overflow, and there are fewer roundoff/truncation problems because the coefficients are in factored form. Figure 4-21 is a block diagram of an FIR filter function.

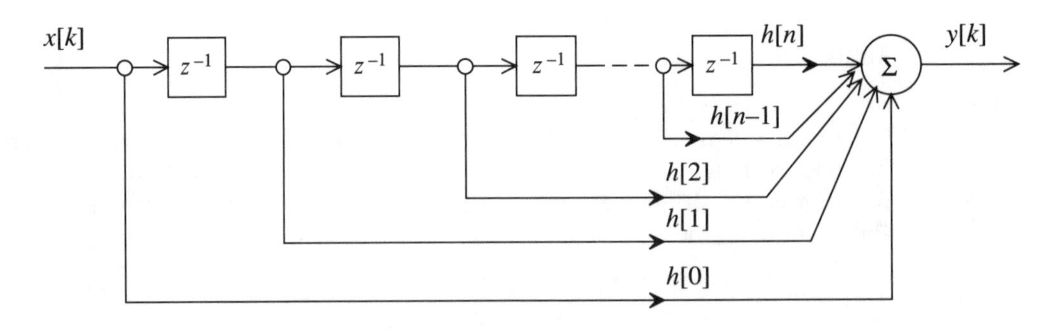


**Figure 4-21.** Block diagram for FIR filter.

## IIR Structures

### Cascade

One transfer function can result in many structures for IIR filters. Initially, a difference equation can be generated from the system function simply by inspection. This method of deriving a difference equation is known as *direct form 1*.

An example of the reasonable use of this form might be found in the generation of a sinusoid from the system function:

$$H(z) = \frac{b_0 z^{-1}}{1 - a_0 z^{-1} - a_1 z^{-2}}$$

which is easily rearranged to form the difference equation:

$$y[n] = a_0 y[n-1] + a_1 y[n-2] + b_0 x[n-1]$$

where

$$a_0 = 2\cos\omega t$$

$$a_1 = -1$$

$$b_0 = \sin\omega t$$

This is not necessarily so for more complex functions, however. Figure 4-22 illustrates the standard direct form 1 expression of another, more complex difference equation. This structure is two systems in cascade. It will take $2N$ delays to perform the computation,

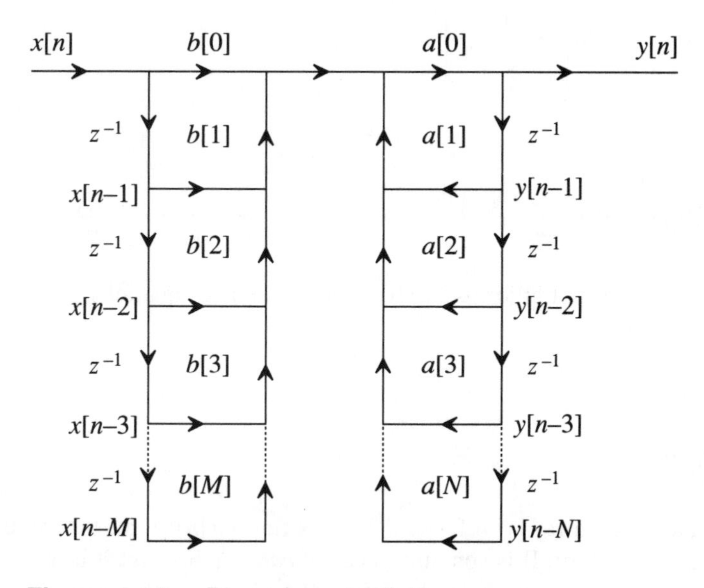


**Figure 4-22.** Direct form 1 IIR filter structure.

because each element of the equation is represented just as the equation is written—the duplication is obvious.

If we reorder the computations and merge these two systems by pairing parallel delays, and then arranging that each pair of delays at the same level takes the same input, we have *direct form II*. This is a common variant that requires only half the delays.

To explain, let us view the structure in Figure 4-23 as two systems in cascade. Then we can rewrite the system function for the IIR filter as the product of two Z transforms:

$$H(z) = \frac{\sum_{k=0}^{M} b_k z^{-k}}{1 - \sum_{k=1}^{N} a_k^{-k}} = H_2(z)H_1(z) = \left( \frac{1}{1 - \sum_{k=1}^{N} a_k^{-k}} \right) \left( \sum_{k=0}^{M} b_k z^{-k} \right)$$

which is a straightforward decomposition of the system function. Now, rearranging the order of muliplication, the structure in Figure 4-23 may also be represented by the product of two system functions, as:

$$H(z) = H_1(z)H_2(z) = \left( \sum_{k=0}^{M} b_k z^{-k} \right) \left( \frac{1}{1 - \sum_{k=1}^{N} a_k^{-k}} \right)$$

Separating each of the functions out, we could write

$$W(z) = X(z)H_2(z) = X(z) \left( \frac{1}{1 - \sum_{k=1}^{N} a_k^{-k}} \right)$$

and

$$Y(z) = W(z)H_1(z) = W(z) \left( \sum_{k=0}^{M} b_k z^{-k} \right)$$

Now, taking the inverse tranforms, we have the difference equations

$$w[n] = \sum_{k=1}^{N} a_k w[n-k] + x[n]$$

$$y[n] = \sum_{k=0}^{M} b_k w[n-k]$$

This allows us to get away with fewer delays, because $w[n]$ stores the same signal for two sets of delays. Direct form II is sometimes called *canonic* because it is an implementa-

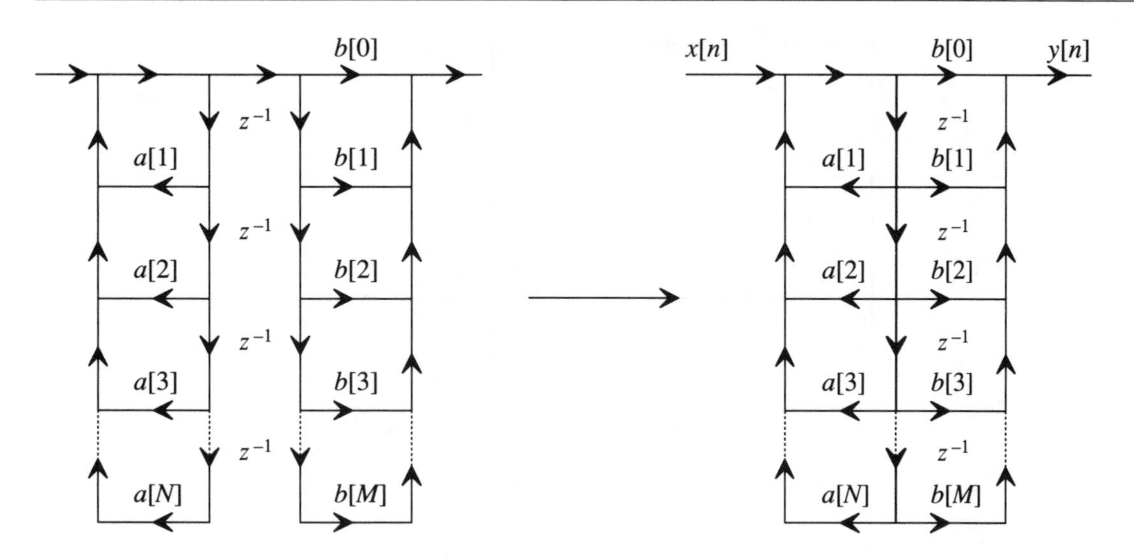

**Figure 4-23.** Direct form II IIR filter.

tion with the fewest delays. An example of the coding necessary for such a procedure may be found at the end of the chapter in the section entitled "Canonic IIR."

Factoring or decomposing transfer functions into second-order systems and then cascading them is very effective in reducing truncation and roundoff noise in IIR filters, or in any iterative finite-length arithmetic sequence. Just as in analog systems, any transfer function of an order greater than 2 can be written as a product of second-order functions and possibly a first-order function. For the IIR filter, this can provide lattitude for individually tuning each section to avoid quantization noise from the coefficients and overflow by matching poles that are close to each other.

### Parallel Sections

Parallel structures are the least susceptible to roundoff errors and overflow problems. Paralleling sections produces a sum; Figure 4-24 is an example of such a parallel structure. It is possible to take the system function of a high-order system and rewrite the equation as a sum of partial fractions—that is, first- and second-order sections. In this way, it is perfectly possible to express a high-order system function or filter as a sum of lower-order filters.

Such a structure could have a transfer function in the following form:

$$H(z) = \sum_{i=0}^{N} \frac{b_{i0} + b_{i1} z^{-1}}{1 + a_{i1} z^{-1}}$$

A method of solving for partial fraction expansions was presented in the section on "The Inverse Z Transform" earlier in this chapter.

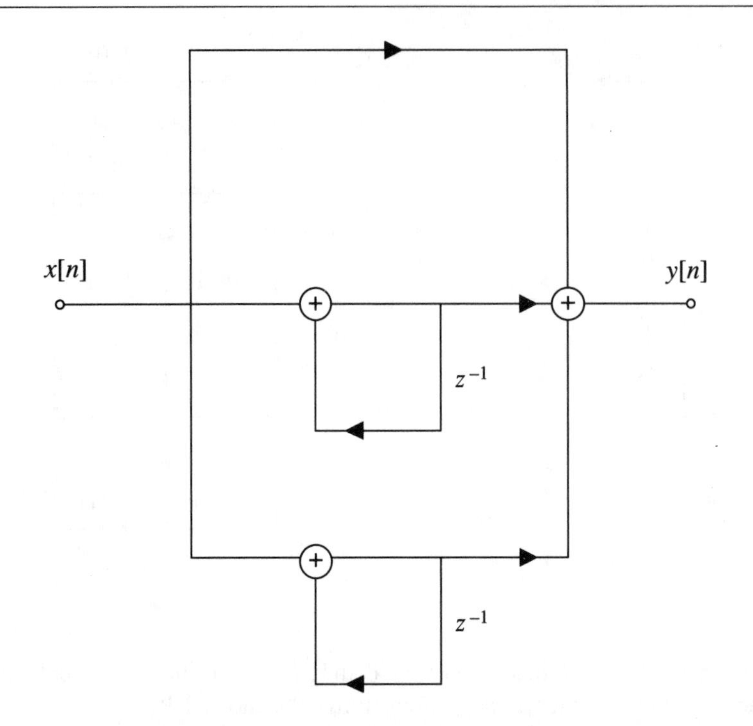


**Figure 4-24.** Parallel structure of first-order sections.

### Lattice Networks

The lattice structures for both the FIR and IIR filters were developed from analog filter technology. They are used in such applications as signal modeling, spectrum estimation, adaptive filtering, and speech processing. They have the advantage that the frequency response is not as sensitive to small variations in the coefficients as the standard form filters, though they are not as computationally efficient as the cascade or direct form implementations.

### The FIR Lattice Network

The $k$-parameters can be found with the following equation:

$$a_{(r-1)i} = \frac{a_{ri} - k_r a_{r(r-i)}}{1-k_r^2}, \quad i=0 \ldots r-1,\ r=N,\ N-1 \ldots 1,\ |k_r| \neq 1,\ k_r = a_{rr} \tag{Eq. 4-15}$$

We will let $r$ be the number of sections and $N$ the order of the filter; $a$ are the coefficients that produce the following relationships (please refer to Figure 4-25):

$$Y_n(z) = \sum_{i=0}^{N} a_i z^{-i}$$

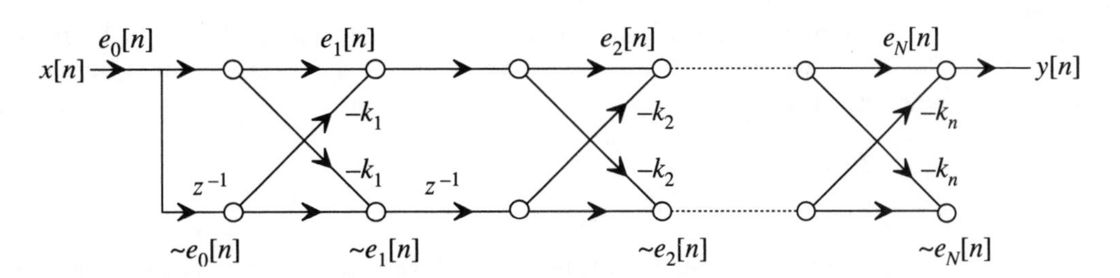


**Figure 4-25.** The general form of an FIR lattice network, the $k_n$ parameters are referred to as reflection coefficients of $k$-parameters.

$$E_n(z) = \sum_{i=0}^{N} a_{N-i} z^{-i}$$

Together, these two sets of equations relate the impulse response of the network to the $k$-parameters.

As an example of the derivation of a simple lattice network, assume a system function

$$Y(z) = 1 + .4z^{-1} - .6z^{-2}$$

in a two-section lattice network. This means that $r = 2$, and that the transfer function may be rewritten as

$$Y(z) = a_{20} + a_{21}z^{-1} + a_{22}z^{-2}, \quad a_{20} = 1,\ a_{21} = .4,\ a_{22} = .6$$

According to Equation 4-7, $k_r = a_{rr}$, so $k_2 = -.5$ and

$$a_{10} = \frac{a_{20} - k_2 a_{22}}{1 - k_2^2} = 1$$

and

$$a_{11} = \frac{a_{21} - k_2 a_{21}}{1 - k_2^2} = .25$$

The coefficients at the delays can be calculated as

$$y_i(n) = y_{i-1}(n) - k_i e_{i-1}(n-1)$$

$$e_i(n) = k_i y_{i-1}(n) + e_{i-1}(n-1)$$

## The IIR Lattice Network

The performance of the IIR lattice filter is somewhere between the FIR and IIR direct form designs. It is more stable than the IIR alone, and usually has fewer coefficients than a comparable FIR filter.

An initial implementation of an IIR lattice network might simply be an all-pole version of the previous FIR all-zero network. This is the inverse of the FIR network with $N$ poles instead of zeros. The $k$-parameters yield the appropriate delay coefficients with

$$y_{i-1}(n) = y_i(n) - k_i e_{i-1}(n-1)$$

$$e_i(n) = k_i y_{i-1}(n) + e_{i-1}(n-1)$$

Figure 4-26 illustrates a lattice network with both poles and zeros. Notice that it combines a ladder network with the all-poles structure we have already discussed.

$$c_i = a_i - \sum_{r=i+1}^{N} c_r b_{r(e-i)},\ i=0\ldots N$$

uses numerator ($a_i$) and denominator ($b_i$) coefficients to derive a new set of coefficients for the ladder network.

## NUMERICAL METHODS

With discrete time filters, the way you handle numbers becomes important. The greater the word length or resolution of the processing system, the smaller the errors. Still, quantization errors, both at the hardware level and in the selection of filter coefficients can produce more noise than the system is designed to remove. It is important that consistent and well thought-out methods be used to minimize this effect. Here is a short list of points to be aware of in designing digital filters:

1. It is simply true that the accuracy and resolution of any calculation can never be greater than the most accurate value or number with the least resolution. Regardless of the resolution employed by your system, whether you use single or double floating-point types, if you are using an eight-bit A/D converter, your numbers are only good to eight bits, at best. In general, the greater the resolution of your input and your coefficients, the closer your system function will operate to the ideal you have worked out on paper.
2. Roundoff noise also enters the system at the A/D converter, introducing a level of doubt into each conversion equal to one-half LSB. These errors are amplified by

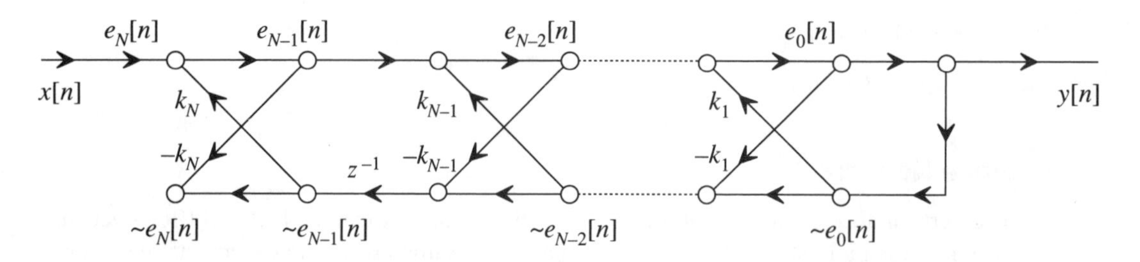


**Figure 4-26.** A generalized IIR lattice network.

the arithmetic of the system (especially in recursive systems involving subtraction) and deteriorate the quality of the output.

3. Quantizing the coefficients, poles, and zeros of an ideal system function effectively changes their values and moves them (poles and zeros) to new locations. Again, this is greatly influenced by the resolution of the system you are using. An eight-bit system will only have $-128$ to $+127$ places for poles to exist, which is very different from a linear continuous system.

4. Two different kinds of limit cycling can occur in direct form architectures. The first results from overflows in intermediate calculations. A large value may result from a sum of products in an IIR filter implementation, for instance, and what should have been a large positive number can become a large negative number. This number can be fed back to cause still another number sign reversal.

   The second form of limit cycling results from quantization. A signal smaller than a quantization step can be rounded up to a higher level, causing low-level amplitude fluctuations or oscillations. A higher level of input usually trivializes these effects or makes them go away completely. This is sometimes the cause of audio noise in the absence of input.

   Limit cycling in cascade form is usually filtered out by each succeeding section. If the frequency of a limit cycle falls near the resonant frequency of a succeeding section, it can mean very large fluctuations in output. This problem is eliminated by arranging the filter sections so that this cannot happen or is minimized.

For more detail on the numerical methods involved in computer processing, please refer to Chapter 5.

# SOFTWARE FOR DISCRETE TIME FILTERS

## Convolution in Frequency Domain

As explained in the section entitled "Convolution and the Fourier Transform," convolution in the frequency domain is performed as a product of two DFTs of sufficient length to avoid aliasing. The following program fragment from FCON.EXE on the optional disk illustrates one possible implementation of this process.

An important part of this code is the determination of the length of the target DFT. A convolution of $N$ samples and $M$ coefficients will have a length of $(N+M)-1$ data points. The program reads in two files from the disk, one containing the input data and the other containing the vector we wish to convolve with it. Each data point is counted and summed, less one:

```
max_data = data_points0+data_points1-1;
```

Then, we determine the smallest power of two greater than *max_data* by counting the powers of two in *max_data* and producing one more than that:

```
N=1;
do{
    N<<=1;
    }while(max_data>>=1);

max_data=N;
```

That number then becomes *max_data*. The input vectors are zero-padded to attain the necessary length, the twiddle factors are produced in the usual manner, and the two input vectors are passed to TWINFFT, the routine introduced in the last chapter for performing two *real* FFTs at once.

```
w = W_FACTOR;

twinfft(ivector0, ivector1, ovector, w, N);
```

Here, *ivector0* and *ivector1* are the two input vectors, *ovector* is a scratchpad, *w* are the twiddle factors and *N* is the size of the DFT required for the convolution. The two resulting spectra are multiplied, point by point, and then the inverse DFT is taken of the product.

```
for(i = 0;i<=max_data; i++) {      /* point by point multiplication*/
    ovector[i] = cmult(ivector0[i], ivector1[i]);
    }

isign=-1;  /*invert the direction of the DFT by inverting all the*/
           /*twiddle factors*/
for(i=0;i<1024;i++)
    {
    W_FACTOR[i].real*=isign;
    W_FACTOR[i].imag*=isign;
    }

fft(ovector, w, N)     /*inverse DFT*/
```

The result is a time domain sequence representing the convolution of the two input sequences.

This will serve as a demonstration and is a possible scenario for an adaptive filtering program, where the DFT can be used to track a time-varying frequency response. Most of the time, the filter's impulse response is known ahead of time and fixed. In this case, we really only need to perform the FFT on the input sequence, multiply it by the known filter spectrum, and then invert the transform for the output. This may be necessary and faster if the filter and the input are complex data.

Either way, the process is the same. The most important part of the algorithm is the determination of the necessary length of the required DFT for convolution without aliasing.

## Generating FIR Filter Coefficients with Fourier Series

There is a program on the optional disk called FCOEF.EXE that generates the Fourier coefficients for lowpass, highpass, bandpass, and notch filters. They can be of any length and are all linear phase.

```
char *filter_type[] = {
    "lowpass",
    "bandpass",
    "highpass",
    "notch"
};
```

The user must specifiy the following variables:

```
    int coef;       /*the number of desired coefficients*/
    float lcutoff, hcutoff, sampling;     /*low cutoff frequency, high*/
                               /*cutoff frequency, and the sampling rate*/

    /*normalize to a value between 0 and 1*/
    lcutoff = (lcutoff)/(sampling/2);
    hcutoff = (hcutoff)/(sampling/2);
    sampling = 2*PI*sampling;

    /*determine whether there is an even or odd number of coefficients*/
    coef2 = coef>>1;
    odd =  coef - (coef2<<1);

    /*select high, low bandpass or notch and create filter
coefficients*/
    switch(lambda) {
        case 0:         /*lowpass*/
            ovector[(coef-1)/2] = hcutoff;    /*set the symmetry point*/
                                              /*if odd length*/
            for(i = 1; i<coef2; i++) {        /*uses symmetry of filter*/
                                              /*coefficients*/
                ovector[coef2+i] = ovector[coef2-i] =
                    sin(ovector[coef-1/2]*i*PI)/(i*PI);
                }
        case 1:         /*bandpass*/
            ovector[coef-1/2] = hcutoff-lcutoff;/*set the symmetry*/
                                               /*point if odd length*/
            for(i = 1; i<coef2; i++) {        /*uses symmetry of filter*/
                                              /*coefficients*/
                ovector[coef2+i] = ovector[coef2-i] =
                    (sin(hcutoff*i*PI)-sin(lcutoff*i*PI))/(i*PI);
                }
            break;
        case 2:         /*highpass*/
            ovector[coef-1/2] = 1-(lcutoff); /* uses symmetry of*/
                                             /*filter coefficients */
            for(i = 1; i<coef2; i++) {     /*uses symmetry of filter*/
                                           /*coefficients*/
                ovector[coef2+i] = ovector[coef2-i] =
                    -1*fabs(sin(ovector[coef2]*i*PI)/(i*PI));
                }
            break;
```

```
case 3:          /*notch*/
    ovector[coef-1/2] = 1-(hcutoff-lcutoff);
    for(i = 1; i<coef2; i++) {
        ovector[coef2+i] = ovector[coef2-i] =
            (sin(lcutoff*i*PI)-sin(hcutoff*i*PI))/(i*PI);
        }
    break;
default:
    printf("\nsomething went wrong...");
    exit(-1);
}
```

## Sum of Products

A familiar element in both FIR and IIR filters is the *sum of products*. This is the dot product of the filter coefficients with the input vector that occurs in both kinds of filters and many other places. The basic form that you will see in routines that follow is built from the equation

$$y[n] = \sum_{k=0}^{M} b_k x[n-k]$$

The following code fragment is an example in C:

```
for(k=0; k<length; k++) {
    sum=0.0;
    for(i=0; i<ncoef; i++) sum=sum+coef[i]*data[beginning+i];
    result_vector[k]=sum;      /* ncoef multiplies */
    beginning++;
}
```

There is an enhancement to this basic routine that can be made for filters with symmetric coefficients. The input data at the points of coincidence is summed (antisymmetric filters will negate the second data point) and then multiplied. This not only improves the speed, but saves one noisy multiplication:

```
for(k=0; k<length; k++) {
    sum=0.0;
    for (i=0; i<center; i++) {
        reflect=data[end-i];  /*reflected data*/
        if(type < 0)  reflect=-reflect;        /*antisymmetric*/
                                               /*filters*/
        sum=sum+coef[i]*(data[beginning+i]+reflect);/* ncoef/2
```

## An FIR Implementation: Convolution in the Time Domain

The next example demonstrates the sum of products as it might be used in an FIR filter; it is from VCON.C on the optional disk. The format of this filter is really suitable for static processing or postprocessing, as might be done with data files. The following routine is appropriately called CONVOLVE:

```
The prototype:
void convolve(double data0[], int nn, double data1[], int ncoef,
    double result_vector[], int dcm, int type, int smtry);
```

The routine is passed an array of filter coefficients, *double coef[]*; the number of coefficients, *int ncoef*; a data array, *double data[]*; the length of the data array, *int nn*; an output array, *double result_vector[]*;filter type, *int type*; and symmetry, *int symmetry*.

```
void convolve(double coef[], int ncoef, double data[], int nn, double
result_vector[],
        int type, int symmetry)
{

    int center, odd, k, i, end, length, beginning;
    double sum, reflect;

    length = nn+1;
    beginning = 0;
    end=ncoef-1;

        /************************************/
        /*symmetrical filter coefficients*/
    if(symmetry) {      /* exploit the symmetry in the filter response*/

        center=ncoef>>1;
        odd=1;
        if((center<<1) == ncoef) odd=0;    /*determine odd or even*/

        for(k=0; k<length; k++) {
            sum=0.0;
            for (i=0; i<center; i++) {
                reflect=data[end-i];   /*reflected data*/
                if(type < 0)  reflect=-reflect;        /*antisymmetric*/
                                                      /*filters*/
                sum=sum+coef[i]*(data[beginning+i]+reflect);/* ncoef/2*/
                                                          /*multiplies */
                }
            result_vector[k]=sum;
            if(type >= 0 && odd > 0) {
                result_vector[k]=sum+coef[center]*data[beginning+center];
                }
            beginning++;
            end++;
            }
        return;
    }
```

```
        /*************************************/
        /*asymmetrical filter coefficients*/

    /*typical filter is N*M multiplications with an ultimate length*/
    /*of N+M-1 data points; this handling is merely point by point*/
    /*since we can take advantage of none of the simplicities of */
    /*symmetry or antisymmetry*/

    for(k=0; k<length; k++) {
        sum=0.0;
        for(i=0; i<ncoef; i++) sum=sum+coef[i]*data[beginning+i];
        result_vector[k]=sum;      /* ncoef multiplies */
        beginning++;
    }
}
```

## A Realtime FIR Filter

Though the previous example demonstrates FIR techniques, it probably isn't as useful for realtime processing as the next example. This code fragment illustrates a sampled data FIR and is available on the optional disk in FIR.C. This filter operates on a sample-by-sample basis, as it might in a sampled data system.

This code and the next example involving the IIR filter incorporate a *synthesized* circular buffer. The buffer is created in C. It has no hardware, like the DSP chips, but allows you to write routines without manually moving data around or having to come up with individual routines for each filter you write. This circular buffer is based upon the same technique that the DSP chips use: modular arithmetic to compute and index into a data space.

Using a circular buffer instead of a fixed buffer means that when data comes to the end of the buffer, it wraps to the beginning again. It is ideal for processing of this sort, because data is always being flowed into the buffer and only requires increments of the data pointer to find the next location or push data down.

This implemetation requires an index *ndx* that is incremented during each pass. It is added to the current loop counter value, *I*, and divided by the size of the buffer. The remainder is the current pointer into the buffer. You will note that the operations of the routine are identical regardless of where it is in the buffer, and that the buffer is extensible.

The fragment we present here is contained within a while loop that reads data from an input source until that source no longer has any data. There are two initializations that must occur outside the loop:

```
ncoef = cdata_points-1;     /*ncoef starts at 0, cdata_points*/
                                                /*starts at 1*/
ndx = ndata_points = 0;     /*index for input and index to output*/
                            /*array*/
```

The two variables cdata_points and ncoef both represent the number of FIR coefficients read from a file, these are used to maintain the circular buffer. There are two index variables, ndx helps maintain the buffer and ndata_points handles the output array.

Within the loop, the code looks like this:

```
do{
    sum = 0.0;
    wn[ndx] = data_point;

    for(i=0; i<=ncoef; i++) {
            /*calculate local data location in circular buffer*/
        if((remainder = ndx+i-cdata_points) < 0) remainder = ndx+i;
            /*compute the next sum*/
        sum=sum+coefficients[i]*wn[remainder];
        }

    if(++ndx > ncoef) ndx = 0;      /*roll over to stay in buffer*/
                                    /*and not repeat*/

    result_vector[ndata_points++] = sum; /*this is the output*/
    }while(...);
```

## The Direct Form IIR Filter

This is the filter obtained from the transfer function by inspection, which is the reason it is known as the *Direct form*. We will use for our example a second-order function, also known as a *biquad* because it has a quadratic in both the numerator and the denominator:

$$H(z) = \frac{Y(z)}{X(z)} = \frac{\sum_{k=0}^{2} b_k z^{-k}}{1 - \sum_{k=1}^{2} a_k z^{-k}} = \frac{b_0 + b_1 z^{-1} + b_2 z^{-2}}{1 - a_1 z^{-1} + a_2 z^{-2}}$$

We can then write the difference equation,

$$y[n] = \sum_{k=0}^{2} b_k\, x(n-k) + \sum_{k=1}^{2} a_k y[n-k] = b_0 x[n] + b_1 x[n-1] + a_1 y[n-1] + a_2 y[n-2]$$

Code for this form of the difference equation can be written directly from the equation, however, it requires more delays and two buffers to implement. This is a direct form I IIR filter.

A direct form II IIR filter requires only one buffer, and is actually written from the following interpretation of the basic equation:

$$w[n] = \sum_{k=1}^{N} a_k\, w[n-k] + x[n]$$

$$y[n] = \sum_{k=0}^{M} b_k\, w[n-k]$$

The sampled data arrives as *data_point* and is placed in the variable *sum*. From there, the sum of products for the poles is performed using *sum* and the circular buffer holding the history of *wn*. The result of that operation is made the most recent *wn*, and the process moves to perform the sum of products for the zeros. This code uses the same circular buffer mechanisms as the last example of FIR processing.

The following variables must be initialized before the loop is begun. direct form II is canonic, that is it requires the least delays to implement, therefore it possesses only one buffer whose size is equal to the greatest number of coefficients.

```
cdata_points = max(pcoef, zcoef); /*size buffer to largest number*/
                                  /*of coefficients*/
ncoef = cdata_points-1;
sum = ndx = ndata_points = 0;     /*initialize indices*/

do{
    /*cdata_points are coefficients, xn is the input data point*/
    /*direct form II uses only one buffer*/
    /*first the poles*/

    sum = data_point;  /*input from xn*/

    for(i=0, k=(cdata_points-pcoef); i<pcoef; k++, i++) {
        if((remainder = k+ndx-cdata_points) < 0) remainder = k+ndx;
        sum=sum+coefficients[i]*wn[remainder];  /*oldest data first*/
        }
    I--;    /*make it equal to ncoef for next loop*/
            /*place result of this SOP in buffer*/
    if((remainder = k+ndx-cdata_points) < 0) remainder = k+ndx;
    wn[remainder] = sum;              /*new sum*/

    /*then the zeros*/
    sum = 0.0;
    for(k=zcoef; k<=cdata_points; i++, k++) {
        if((remainder = k+ndx-cdata_points) < 0) remainder = k+ndx;
        sum=sum+coefficients[i]*wn[remainder];
        }                                  /*new index to output*/
    if(++ndx > ncoef) ndx = 0;

    result_vector[ndata_points++] = sum;

}while(...);
```

This code can be found on the optional disk in IIR.C. More examples of FIR and IIR filters follow in the next chapters.

# 5

# Using Digital Signal Processing

## INTRODUCTION

We live in an infinite, continuous world, but our tools are finite and lacking in resolution. This limitation of our tools make its imprint on our work by adding noise to a filter that is supposed to remove noise; lack of resolution in the quantization of filter coefficients can move poles and zeros, and turn a benign, narrowband filter into an oscillator.

The digital computer is capable of only finite precision—the more finite, the more coarse the approximation of any filter we might make. But even if the DSP we designed were capable of infinite precision and our I/O was not, we might still have noise—no amount of precision and accuracy is going to make up for a coarse input.

In this chapter, we will be dealing with some of the more practical aspects of DSP technology: how to move what we have developed from the ideal to the real world. We will discuss both floating-point and fixed-point, but will spend most of the time with fixed-point arithmetic both in calculations and in I/O processing, since that is still the most economical and popular approach to signal processing for size, speed, and price.

We will present some rules of thumb and caveats, but once this is done, most of the work is up the reader to design for *finite-length* arithmetic and simulate the results well. For those interested only in a theoretical viewpoint to signal processing, this chapter may not be necessary. But for anyone interested in actually producing a system based on this technology, this chapter is a must.

### Precision and Accuracy

In order to prepare for a discussion of numerical methods, there are two important things to know about precision and accuracy. First, there is a widespread confusion about precision or resolution and accuracy. Accuracy is really only a measure of the *rightness* of a solution—and there are an infinite number of degrees of rightness. These *degrees* of rightness might be likened to the number of bits a solution is expressed in. For example, the value of $\pi$ may be expressed accurately as 3, or if we increase the precision, 3.1. Increasing it still more, we might write 3.14; a popular calculator uses the value 3.14159265359. None of these is inaccurate, it is just that each is expressed with a different precision.

Second, no calculation is more accurate than its least accurate argument. As an example of what this means, let us take the value of $\pi$ used by that calculator, invert it, and multiply it by $\pi$. At infinite precision, the answer would have to be 1, but without rounding, we will not get that result on the calculator. Because the numbers are truncated short of their full value, we only have an approximation of their values, and the answer reflects that approximation. This is exactly what happens to our arithmetic when we move from the ideal, continuous-time world to the finite world of the computer.

## FINITE-LENGTH ARITHMETIC

Finite-length arithmetic is *any* arithmetic with a fixed or limited precision. Because the mantissas of the floating-point formats discussed above and the fixed-point formats to be introduced here are of a given length, both *floating-point* and *fixed-point* are considered finite-length arithmetic. Floating-point software and hardware take care of so much of the maintenance for you, delivering delightfully exact appearing answers that it is easy to believe that they are not affected by problems of fixed length, but they are. In the next few sections, we will explore fixed length arithmetic and much of what we say will pertain to floating-point math; most will concern fixed-point only. This is important information, because these topics deal with maintaining and enhancing the precision of the work you do.

### Fixed-Point or Floating-Point

One of the first decisions to be made in an application is whether to use a fixed-point or floating-point processor. It is probably no surprise to anyone that floating-point processors are more costly than fixed-point processors—a major argument in favor of fixed-point arithmetic. Another argument often used is that 16-bit data, which is a common DSP buswidth, represents about 90dB of data, and this is plenty for almost any real-world application. It is true that given the proper tools, 15 bits of precision, using the Q15 format (to be discussed below) will adequately represent five decimal digits, and the integer format is the ideal format for A/Ds and D/As.

Floating-point will, however, allow us to move it over a large range. Some applications require a dynamic range that exceeds that represented by this narrow buswidth by several times before falling to a range reasonable for 16 bits. Signals can cover several orders of magnitude in a matter of seconds. Examples of signals of this sort are records of seismic activity, which decay as much as 100dB in seconds. Power calculations and matrix manipulations can also require a greater dynamic range. Compensating for gain coefficients or signals with a large dynamic range with fixed-point arithmetic can be cumbersome. These things make floating-point very attractive, since they obviate the necessity for any of this handling. Also, applications written in floating-point can require less development time, and usually need far less unique code.

There are points to both arguments, and there are places for each form of arithmetic. We will be spending much of this chapter studying fixed-point arithmetic, since it is the most popular and economical choice available today, and since it is the language of the I/O with which the system must deal regardless of the choice for processor arithmetic.

### Floating-Point

Floating-point arithmetic is no magic answer. It is true that it can be beneficial in performing arithmetic because it takes care of so much of the housekeeping for you, but in the end, it has many of the same limitations as fixed-point arithmetic. There are only so many bits. In the mid 1970s and with revisions in the '80s, the IEEE issued a standard known as IEEE 754, so that a program that produced certain results on one machine using a certain software package would produce similar results on the same or other machine with another software package. It offered some basic definitions for the sizes of floating-point words, how much precision should be contained in each, and what the ranges are for these same words.

Basically, IEEE 754 defined two formats, as shown in Table 5-1. First, a single-precision floating-point word uses a total of 32 bits—1 bit for sign, 8 bits for an exponent, and a 23-bit fractional part that, together with the sign bit, constitutes the mantissa. Second, a double-precision word uses 64 bits—1 bit for sign, 11 bits for the exponent, and 52 bits for the fractional part. Actually, the fractional part is one more than what is stated above, owing to the *hidden bit* assumed to occupy the place just left of the binary point. The format is *signed magnitude*, which means that the fractional part represents the magnitude of the number, whether positive or negative, and requires the sign bit to give it meaning. This form has less range than 2's complement, but allows us the possibility of the hidden bit, which would otherwise be absorbed by a sign bit. It also presents us with the paradox of having two possible values of zero: +0 and −0.

The floating-point number can be represented mathematically by

$$h = sign * 2^a \sum_{n=0}^{N} b_n 2^{-n}$$

| *Single-precision floating-point* | | |
|---|---|---|
| S | (MSB)EEEEEEEE(LSB) | (MSB)FFFFFFF...FFFFF(LSB) |
| SIGN = 1 bit | EXPONENT = 8 bits | FRACTION = 23 bits |

| *Double-precision floating-point* | | |
|---|---|---|
| S | (MSB)EEE...EE(LSB) | (MSB)FFFFFFF...FFFFF(LSB) |
| SIGN = 1 bit | EXPONENT = 11 bits | FRACTION = 52 bits |

**Table 5-1.**

The standard also proposes certain minimums concerning the precision of the *intermediate* values—the values used within the library or function itself when performing calculations. Even though you may be using only single-precision arguments, these number are expanded internally to 32 bits and 64 bits for double precision, while the exponent is also expanded to 11 bits for single precision and 15 for double precision. Basically, this provides for decimal range of $10^{38}$ to $10^{-38}$ for single precision and $10^{308}$ to $10^{-308}$ for double precision arithmetic.

Alternate formats have been proposed, and there is validity to arguments that the IEEE standard is inefficient, but arguing is of little value when so much silicon already conforms (at least to some degree) to that format, probably including the math coprocessor chip in your computer. This is not necessarily true for all DSPs, however—Texas Instruments has adopted a modified form of this fomat for its systems, with an instruction that automatically converts a value from one system to the other embedded in silicon for the effected DSPs.

Despite the range of the floating-point word, it possesses only so much precision. It does, however, attempt to maintain as high a precision as possible by adjusting the exponent so that as many of the available bits as possible are occupied by data. Any problems that occur are not particular to floating-point arithmetic or the standard—they are common to arithmetic of finite length, and they are mainly given to precision.

## Fixed Point

To begin with, let us introduce a type of fixed-point format: Q15, or 1.15. In this format, the most significant bit (MSB) is a sign bit, followed by an imaginary binary or heximal point, followed by 15 bits of mantissa normalized to 1 and expressed as a fraction. There is no attempt to equal 1 exactly except in the negative—see Table 5-2.

When two Q15 numbers are multiplied, the result is a Q30 number in which the binary point follows the top two most significant bits, as you might expect. Of course, this form of doing arithmetic is extendable to other word lengths as well.

The Q15 mantissa can be expanded as

$$h = \sum_{n=0}^{N} b_n 2^{-n} \qquad \text{(Eq. 5-1)}$$

or

$$h = b_0 2^0 + b_1 2^{-1} + b_2 2^{-2} + \cdots + b_n 2^{-n}$$

where $b$ is a binary value, either 1 or 0. It is sometimes easier to see if a table of fractional values are given—see Table 5-3.

| *Q15 or 1.15 fixed-point format* | | |
|---|---|---|
| S | Δ | (MSB)FFFFFFF...FFFFF(LSB) |
| SIGN = 1 bit | heximal or binary point | FRACTION = 15 bits |

**Table 5-2.**

| | |
|---|---|
| $2^{-1}$ | .5 |
| $2^{-2}$ | .25 |
| $2^{-3}$ | .125 |
| $2^{-4}$ | .0625 |
| $2^{-5}$ | .03125 |
| $2^{-6}$ | .015625 |
| $2^{-7}$ | .0078125 |
| $2^{-8}$ | .00390625 |
| $2^{-9}$ | .001953125 |
| $2^{-10}$ | .0009765625 |
| $2^{-11}$ | .00048828125 |
| $2^{-12}$ | .000244140625 |
| $2^{-13}$ | .0001220703125 |
| $2^{-14}$ | .00006103515625 |
| $2^{-15}$ | .00003051758125 |

**Table 5-3.**

Expressing a number in fractional binary form is a simple matter of forming a vector, $b_n$, of 1s and 0s such that when expanded in Equation 5-1, their sum is equal to the number we are trying to approximate. As an example, take the very common decimal value, 0.1, as shown in Table 5-4.

As a number in 1.15 format, this would look like

0.000110011001100

(the binary point shown is imaginary).

Besides illustrating the format for 1.15, there is another important point here. Rational decimal numbers are not necessarily rational binary numbers and vice versa. The coefficients we created for our filters will probably *not* make the transition to binary exactly—what we will have is an approximation, accurate to the precision we choose.

In order to preserve as much of the integrity of the filter as possible, we need to preserve as much precision as possible. One of the most common precisions chosen for DSP work is 16 bits. In 1.15, or Q15, format, this gives us a range of −1 to +.999969482422 in .000030517578125 increments (our LSB). This gives us a dynamic range of 90dB.

The bottom line is that when we quantize our coefficients, we do not have the same filter we designed. Examine Table 5-5, which involves the quantized coefficients for a simple

| $2^{-1}$ | $2^{-2}$ | $2^{-3}$ | $2^{-4}$ | $2^{-5}$ | $2^{-6}$ | $2^{-7}$ | $2^{-8}$ | $2^{-9}$ | $2^{-10}$ | $2^{-11}$ | $2^{-12}$ | $2^{-13}$ | $2^{-14}$ | $2^{-15}$ |
|---|---|---|---|---|---|---|---|---|---|---|---|---|---|---|
| 0 | 0 | 0 | 1 | 1 | 0 | 0 | 1 | 1 | 0 | 0 | 1 | 1 | 0 | 0 |

**Table 5-4.**

*Quantization of filter coefficients (truncated)*

| *n* | *Ideal coefficients* | *16-bit quantization* | *14-bit quantization* | *10-bit quantization* |
|---|---|---|---|---|
| 0,20 | 0.0000000007307677 | 0 | 0 | 0 |
| 1,19 | 0.0000004813459621 | 0 | 0 | 0 |
| 2,18 | 0 | 0 | 0 | 0 |
| 3,17 | −0.000125481405920 | −0.0001220703125 | −0.01220703125 | 0 |
| 4,16 | −0.001088169510676 | −0.001083374023438 | −0.0103759765625 | −0.0009765625 |
| 5,15 | −0.003321665867754 | −0.003311157226563 | −0.0032958984375 | −0.0029296875 |
| 6,14 | 0 | 0 | 0 | 0 |
| 7,13 | 0.030585310471953 | 0.03057861328125 | 0.03057861328125 | 0.0302734375 |
| 8,12 | 0.159154943091895 | 0.159149169922 | 0.159118652344 | 0.158203125 |
| 9,11 | 0.107354626950032 | 0.107345581054688 | 0.1072998046875 | 0.1064453125 |
| 10 | 0.25 | 0.25 | 0.25 | 0.25 |

**Table 5-5.**

21-tap symmetrical lowpass filter. This filter is designed using the Fourier series method described in Chapter 4 and windowed with a Kaiser window.

As you can see, the poorer the precision, the coarser the filter coefficients. Not only is it important the we quantize our filter coefficients into great enough precision, but it is also important that the intermediate values be kept as long as possible. If you are not working in floating-point, where all of this is taken care of for you, here are some simple facts to remember about space requirements:

1. Addition of $b$-bits and $b$-bits can always generate a result $b$-bits$+1$ long. For $N$ additions, allow $\log_2 N$ extra bits to accumulate a result to guarantee no overflow.
2. Subtraction of $b$-bits from $b$-bits will require $b$-bits for the result.
3. The product of a multiplication of $b$ and $b$ will require $2b$ bits for the result. In a recursive filter, this number of bits will be required after every iteration.
4. The result of a division will require as many bits as the quotient.

## Fixed-Point Arithmetic: Creating and Interpreting 1.15

Often, especially in embedded systems, the filter coefficients and input values will already be in fractional form. A/D converters naturally communicate in binary form, and that data will need no translation. For the purpose of this discussion, let us assume that the filter coefficients or input values are not in fractional form, that you have computed them using infinite precision and wish now to move to a fixed-point fractional form.

Let us assume also that we have found a value by which we can divide each coefficient that will prevent overflows in any sum-of-products or integral routine we have in mind. Here is an outline of steps that will produce a Q15 value:

1. In the beginning, we ignore the sign of the value and deal only with the magnitude. The first step in converting an infinite precision number to Q15 is to divide by this *scaling factor*. For example, we have a filter coefficient, 23.9959677, and a scaling factor, 49.2224345—division yields 0.487500627382.
2. Multiply this value by $2^{15}$: $32768 * 0.487500627382 = 15974.4205581$.
3. Round to an integer by whatever means you choose, avoiding simple truncation. In this case, the fractional part is less than .5, so we will round down to 15974.
4. Convert this number, 15974, to hex: 3e66h. Now if we needed a negative number, instead of a postive one, we would invert, or NOT, the magnitude to give us: 0c199h, then add 1 for 0c19ah- this is a 2's complement negative number. To produce the 2's complement negative of any number, it is only necessary that you invert all the bits and add 1.

If the need arises for a Q15 number to be converted back, it is a simple matter of dividing the decimal equivalent of the hex number by $2^{15}$, $15974/32768 = 0.48748779$. If the number is negative, convert it to positive by inverting all the bits and subtracting 1.

Fixed-point arithmetic differs from floating-point, in that it is up to the programmer to maintain the position of the binary point, and to choose the proper handling for the results. Here is a short list of things to remember:

1. Always provide enough space for the result of an operation, plus extended bits. Multiplication requires for a result the sum of the bits in both operands, division can require the number in the largest operand, addition will require the number of bits in the largest operand +1, and subtraction will require the number of bits in the largest operand.

   Using a format such as Q15, the length of the arithmetic word is standardized; it is best to see that the results are standard too. Local operations should be done with the greatest precision possible; intermediate registers for Q15 can be kept at 32 bits, which allows for any single operation you might employ. Iterative operations can overflow if scaling is not correct. Without a check for overflow and the ability to saturate, this could result in wild oscillations at the output of the routine.
2. In Q15 format, it is possible to use the instructions for signed arithmetic that are usually a part of a microprocessor's instruction set. The result register of a multiplication will be 32 bits in Q30 format—the two MSBs will be sign bits, a single left shift and the number can be stored, without any loss of precision as a Q15 value. You gain little by not using the signed instruction and taking care of sign yourself, unless you are not using a fixed format like Q15.
3. Arithmetic right shifts differ from logical right shifts, in that the sign bit is always repeated in the MSB of the operand—for example, 10000000b>>1 = 11000000b. An arithmetic right shift performs a division by some power of two on the target number and produces a result that is always rounded toward zero. On a negative number, it rounds up; the same shift on a positive number rounds down.

## SOURCES OF ERROR IN FINITE-LENGTH ARITHMETIC: RANGE

In many processes where recursion is used, such as IIR filters, there can be many iterations of multiplication and accumulation (addition) and always the danger of overflow. To minimize problems of range, it is usually wise to constrain the arguments for an operation to fractional parts whose maximum possible sum is less than 1, so that there is no need to worry about the proper location of the binary point or overflow. This is often done by requiring the result of an operation to be $|y[n]| < 1$. The worst case is determined by

$$y[n] = s = \sum_{n=0}^{N} h[n]$$

Then either the gain, $A$, of the operation (filter) or all arguments are scaled:

$$\frac{A}{s} \text{ or } \frac{h[n]}{s}$$

More specifically,

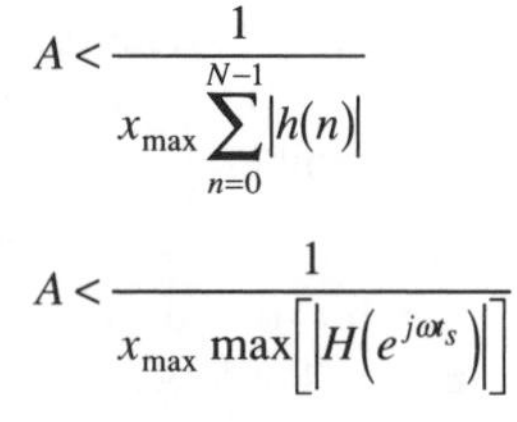

broadband $$A < \frac{1}{x_{\max} \sum_{n=0}^{N-1} |h(n)|}$$

narrowband signal $$A < \frac{1}{x_{\max} \max\left[\left|H\left(e^{j\omega x_s}\right)\right|\right]}$$

This guarantees $|y[n]| < 1$ and that it will fit in the proper format.

It is important to note that scaling the filter gain also scales the filter coefficients by implication—this can degrade frequency response because of higher quantization errors. It is often more efficient to scale the input to the filter.

For broadband signals, a general rule of thumb for this scaling is that it be one-fourth the *rms signal level*, $\sigma_x$. If we make $A\sigma_x = 1/4$, there is little chance of clipping (saturating). This makes the *noise to signal ratio* ($-SNR$) in dB:

$$NSR(\text{dB}) = 10\log\left(\frac{\sigma_\varepsilon}{A\sigma_x}\right)^2 = 10\log\left(\frac{4}{3}2^{-2b}\right) = -6b + 1.24\text{dB}$$

or, for sine wave calculations:

$$NSR(\text{dB}) = -6b + 1.76\text{dB}$$

where $b$ represents the precision of the argument in bits.

## QUANTIZATION OF FILTER COEFFICIENTS AND INTERMEDIATE VALUES

Keeping filter coefficients as precise as possible is important for filters, especially for IIR filters, since quantization errors could move a pole from within the unit circle to the unit circle or outside it, resulting in small scale oscillations (on it) or large scale oscillations (outside it). It is not difficult to see how a coefficient that is already near unity could be rounded to unity. This could result in a *limit cycle*, small output oscillation about the LSB. This, actually, is a technique for creating high-fidelity oscillators.

Cascade and parallel forms of the IIR filter are most common, because the coefficient quantization for each pair of complex conjugate poles is independent from any other, making them less likely to degrade. This is not usually the case with direct form filters. The noise estimation for an IIR filter is

$$\sigma_o^2 = \frac{2^{-2b}}{12} \int_{-\pi}^{\pi} \left| H\left(e^{j\omega}\right) \right|^2 d\omega$$

FIR filters are inherently immune to instability, since they are of finite length and always absolutely summable. The noise in an FIR filter is proportional to the number of taps involved and is independent of the coefficient values:

$$\sigma^2 = (N+1)\left(\frac{2^{-2b}}{12}\right)$$

It is also possible to round the results of a calculation before going on to contain the word size, but this can have deleterious effects of its own. Here are a few points to keep in mind:

1. To maintain the greatest precision and ultimate accuracy, all the intermediate arithmetic in a calculation should be kept as long (use as many bits) as possible. Rounding can then be performed upon the result of a calculation.
2. The precision is maintained by using *extended* arithmetic on all intermediate values and rounding only at the end. Extended arithmetic consists of using the basic word size we are working with and *adding* auxiliary bits—a minimum of one—to maintain the accuracy of the process. These extended bits increase the precision of the arguments for operations such as subtraction that can steal significance from your value, especially when the arguments are very close.
3. The *width* of any rounding error is dependent upon the LSB of the word size we are rounding *into*. It is usually $2^{-b}$ (equal to the least significant bit). Ideally, this rounding should have no bias whatsoever, but this depends upon the rounding scheme used.

   *Truncation*, sometimes called *chopping*, is simply the process of deleting the least significant bits that are not part of the value we mean to pass on. This has an error width of $2^{-b}$, but the error is biased entirely below the truncation line touching it when there is no error. This method is not recommended because of the noise it creates. The total error can be proportional to to the number of filter taps. Figure 5-1 illustrates how truncation biases the error below the truncation line.

   The rounding scheme in many math coprocessors and math libraries defaults to *round to nearest*. In this format, the extended bits or guard bits are checked before

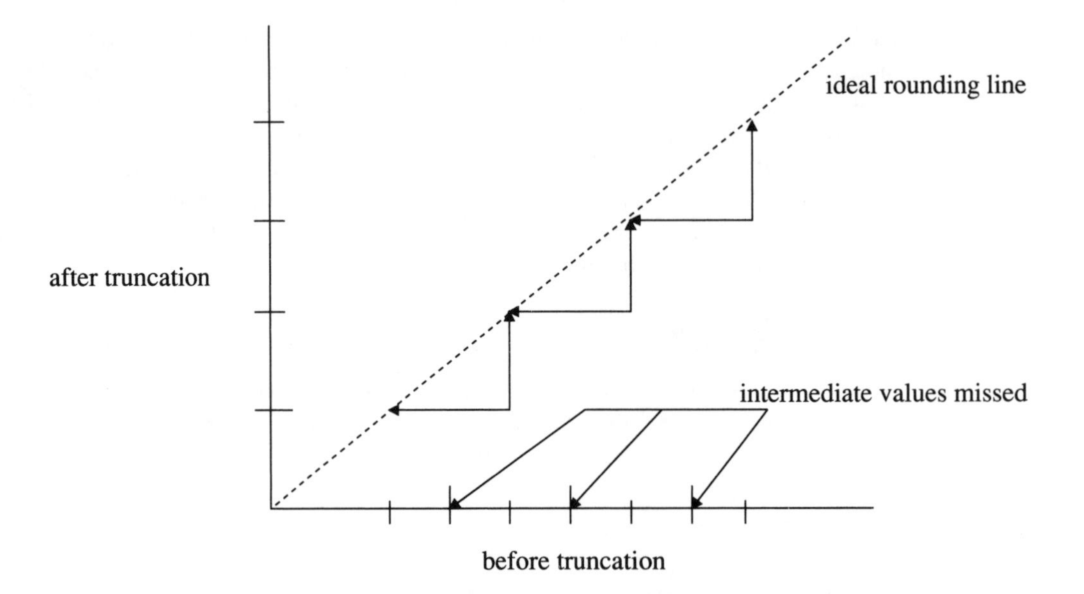


**Figure 5-1.** In truncation, the width of the error is $2^{-b}$, but its bias pushes the mean down.

truncating. If they are greater than half, they are rounded up; if they are less than half, they are rounded down; if they are exactly one-half, they are rounded toward even.

*Unbiased rounding* is the rounding you learned in school, in which the *guard* or extended bits are examined. If they are greater than one-half the result is rounded up, and if they are less than one-half, they are rounded down; if the extended bits are exactly one-half, they are rounded up half the time and down half the time. This results in an error width of $2^{-b}$, but a mean centered about the ideal rounding line. As opposed to truncation, the total error in unbiased rounding is proportional to $\sqrt{N}$, where $N$ represents the number of filter taps. Figure 5-2 illustrates unbiased rounding.

4. Another problem in calculations that cannot always be forseen is overflow. This can be avoided with *saturation* arithmetic. Counters will overflow or underflow, sometimes causing horrendous results. Suppose we have an eight-bit counter and it arrives at its maximum count 0FFH, another clock will cause it to roll over and output 0H—this can result in oscillations. The same can happen in the opposite direction. If the same counter is counting down—it arrives at 0H, and the next clock sends it to 0FFH. Again, there are oscillations.

   Saturation arithmetic refuses to allow a register or value to exceed its maximum or minimum. When the number increases to its maximum, 0FFH, regardless of what positive value is added to it, it never exceeds 0FFH. The same is true in the other direc-

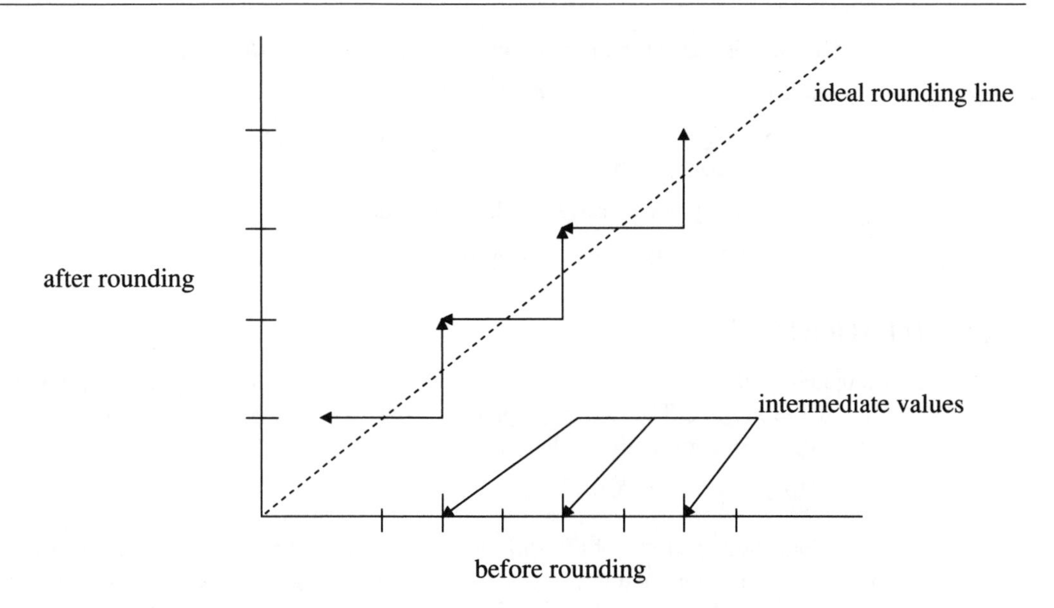


**Figure 5-2.** With unbiased rounding, the error is centered about the ideal rounding line.

tion. The results of a calculation involving saturation may not be correct, but at least they cannot exceed a certain value and will not cause wild oscillations in your output.

Avoiding roundoff error and overflow is obviously the best way to go, but the possibilities for a real-world situation that would allow that are very low. How much precision is really needed to quantize real world values? At least 16 bits, as you can see from Table 5-2 on quantizing coefficients. More is better; always allow the appropriate number of extended bits for the results of the operations of the filter and for rounding.

## I/O QUANTIZATION

Given a monotonic A/D converter with reasonable linearity, input quantization noise is uncorrelated *white* noise of zero mean (no bias) and variance ( $\sigma_\varepsilon^2$ ):

$$\sigma_\varepsilon^2 = \int_{-\frac{\Delta}{2}}^{\frac{\Delta}{2}} \frac{1}{\Delta}\varepsilon^2\, d\varepsilon = \frac{\Delta^2}{12} = \frac{2^{-2b}}{12} \qquad \text{(Eq. 5-2)}$$

The $\varepsilon$ is for error and $\Delta = 2^{-b}$. This is the same relationship that exists for rounding. It is an important formula, because it provides a baseline for the DSP system—whenever the output noise of the system exceeds Equation 5-2, the DSP is *adding* noise to the system and information is being lost.

If you wish to get the maximum usable data from an I/O device:

1. Maximize the input signal for the highest SNR.
2. Use the greatest A/D bandwidth possible, adjusting amplifier gains to use the entire bandwidth.
3. Filter any noise above the desired input frequencies.
4. Minimize any harmonic distortion.

# COMPUTATIONAL DELAY

Computational delay shows up as phase delay in a system, thereby affecting the phase margins of the system. This delay or phase shift is the product of the actual delay time, the bandwidth frequency, and 360 degrees:

$$phase_{delay} = T_d * f * 360$$

Contributing to this phase shift are all the conversion delays in the system, in addition to any software; this includes any phase delay resulting from anti-aliasing filters. For real-time control, and minimum phase shift, anti-aliasing filter orders should be kept at as low an order as possible. Lower-order filters may be tolerated in the presence of noise if the transducers on the input to the system have a known and limited bandwidth themselves and the sampling rate is high enough (six to ten times the Nyquist frequency for realtime control). The greater the delay from the input to the system response, the less realtime is the control, and the greater the chance of instability.

# COMPUTATIONAL AIDS

## Division

One of the longest arithmetic operations on a microcomputer is division. Few DSPs offer a complete division instruction for either floating-point or integer operations—more often, they supply instructions that produce a multistep division.

Here is a list of common division algorithms:

1. On a very simple processor, integer division is often implemented with simple subtraction. This process is often called *nonrestoring division*. For a positive division, the divisor is continually subtracted from the dividend, a counter is incremented, and the dividend replaced by the result, until the result is negative. This signals the end of the process. The counter is decremented to remove the subtraction that caused the overflow, and the divisor is added back to the dividend to make the value positive again. We have our result. The counter is the quotient, and the final value in the dividend is the remainder.
2. If the processor has a division instruction, one can use that to implement a division that can be used on longer dividends (not divisors!) by starting the division with the most significant bits and proceeding to the least, using the remainder

from each division as the MSB of the next dividend. Using a 16-bit divide, as we have on the many processors, the algorithm is implemented as follows:

a. Clear the 16-bit dividend register.

b. Move the 8 MSBs of the number you wish to divide into the lower 8 bits of the register used to hold the dividend.

c. Divide this dividend by the 8-bit divisor.

d. Store the result in the lower 8 bits of the result register.

e. If there are no remaining dividend bits, you are done—exit the routine.

f. Move or shift the remainder into the 8 MSBs of the register used for the dividend.

g. Move or shift the 8 MSBs of the remaining dividend into the lower 8 bits of the register used as the dividend.

h. Shift the result register to the left 8 bits.

i. Continue with step b.

3. Sometimes, the divisor is longer than hardware will support. In this case, we will offer here two algorithms that have proved beneficial in the past for deriving quotients for Q15 and Q30 (or larger) operands.

   The first is Newton's method. With this method, we perform the division by inverting the divisor and multiplying—a much faster process and one much easier to implement on a multiword basis. The inversion is found by writing the inversion process as a linear function with a root on the X axis. Given an initial estimate of the inverse, you arrive at a better estimate by evaluating the function, and its first derivative for the slope of the curve, at that point. The intersection of the tangent line with the X axis produces the better estimate. The process is iterated until the estimate is within a prescribed error. This results in the iterative equation

$$x[i] = x[i-1] * (2 - divisor * x[i-1])$$

   However, this can be improved for more speed by eliminating the divisor from the equation and using three equations:

$$x[n] = x[n-1] * (2 - y[n])$$
$$y[n-1] = y[n]$$
$$y[n] = y[n-1] * (2 - y[n-1])$$

   The arguments to this algorithm are expected to be in 2's complement format. We need an initial estimate in the inverse, which can come from a table in ROM—it need only be close. Another way might be to produce an estimate based upon the number of significant bits in the divisor: $2^{-significant\ bits}$. Whichever method you choose, multiply this initial estimate by the divisor to get $y[n]$. Follow this with iterations of the set of equations above to the necessary precision of the result, expecting the accuracy to double with each iteration.

4. Another very useful and fast technique for division is *interpolation*. This can produce a very small and fast routine for very long arguments:
   a. Break the operands into word sizes (this is machine-dependent).
   b. Perform an initial division on the entire dividend using only the most signficant word (MSW) of the divisor.
   c. Save this result.
   d. Take again the MSW of the divisor and increment it.
   e. Perform the same division again.
   f. Subtract the second quotient from the first.
   g. Multiply the result of this subtraction by the remaining bits of the divisor above. This is a fractional multiplication and produces a fractional result.
   h. Subtract the result of this multiplication from the division performed in b. Iterate for greater precision.

## Pseudo-random Number Generator

A useful pseudo-random number generator:

$$x[n] = (ax[n-1] + c) \bmod m$$

where $a = 1664525$ and $c = 32767$. As you can see, this is a simple operation based upon a modular division; the modulus is usually related to the word size of the platform it is running on. As simple as it appears, it can easily go wrong, forming a linear dependence. If this happens, it will start producing patterns and no longer be a pseudo-random number generator.

The values suggested for the equation are based upon values proposed by Donald Knuth in *SemiNumerical Algorithms*, Vol. 2. This is an excellent work with information on many topics, including numerical methods. For more information on pseudo-random number generators, please refer to this book.

## Polynomial Evaluation: Horner's Rule

A simple but valuable computing trick is known as *Horner's rule*, or *Horner's method*. It involves the simplification of the solutions of polynomials, such as

$$f(x) = x^4 + 3x^3 - 6x^2 + 2x + 1$$

Normally, this might be evaluated by computing each power of $x$, multiplying by each coefficient, and performing each addition and subtraction in the order given in the equation. Some might compute $x$, saving each power to use it again to compute the next power. There is a simpler way.

This method alternates the mulitplication and addition operations in such a way that a $N$-degree equation requires only $N-1$ multiplicatons and $N$ additions (including subtractions) to evaluate. For the equation above,

$$f(x) = x(x(x(x+3)-6)+2)+1$$

In the case of a polynomial with only one term, such as an exponential series:

$$x, x^2, x^4, x^8, x^{16}, x^{32}, \ldots$$

Using Horner's rule, this is computed with five multiplications by simply squaring the previous term.

Horner's method is based upon nested grouping of the elements of expression, as demonstrated above, and can be systematized as follows:

$$f(x) = \sum_{k=0}^{n} C_, x^k = C_n x^n + C_{n-1} x^{n-1} + \cdots + C_1 x^1 + C_0$$

$$u_k = x u_{k+1} + C_k, \quad k = n, n-1, \ldots, 0$$

$$u_{n+1} = 0$$

Using these relationships, $u_0 = f(x)$ can be determined recursively:

$$u_n = C_n, \quad u_{n-1} = xC_n + C_{n-1}, \quad u_{n-2} = x(xC_n + C_{n-1}) + C_{n-2}$$

and

$$u_0 = \sum_{k=0}^{n} C_, x^k$$

## Useful Approximations

Often it will be necessary to generate a transcendental for the purposes of calculation. Usually, this is based upon a power series, such as:

$$\sin(x) = x - \frac{x^3}{3!} + \frac{x^5}{5!} - \frac{x^7}{7!} + \frac{x^9}{9!} \cdots + (-1)^{n+1} \frac{x^{2n-1}}{2n-1}$$

$$\cos(x) = 1 - \frac{x^2}{2!} + \frac{x^4}{4!} - \frac{x^6}{6!} + \frac{x^8}{8!} \cdots + (-1)^{n+1} \frac{x^{2n-2}}{2n-2}$$

$$\tan(x) = x + \frac{x^3}{3} + \frac{2x^5}{15} + \frac{17x^7}{315} + \cdots$$

$$e^x = 1 + x + \frac{x^2}{2!} + \cdots + \frac{x^n}{n!} + \cdots$$

$$\ln(1+x) = x - \frac{x^2}{2} + \frac{x^3}{3} - \cdots + (-1)^{n+1} \frac{x^n}{n} + \cdots$$

Usually, the expression is truncated at a point beyond the necessary precision required for the computation or system, and the divisors are precomputed and stored in RAM or ROM as coefficients. The computations can then be performed using Horner's method.

Similar results can also be obtained using polynomial expansions. A few examples follow:

$$\sin(x) = 3.140625x + 0.02026367x^2 - 5.325196x^3 + 0.5446778x^4 + 1.800293x^5$$

$$a\tan(x) = 0.318253x + 0.003314x^2 - 0.130908x^3 + 0.068542x^4 - 0.009159x^5$$

$$sqrt(x) = 1.454895x - 1.34491x^2 + 1.106812x^3 - 0.536499x^4 + 0.1121216x^5 + 0.2075806$$

$$2\log_{10}(x) = 0.8678284(x-1) - 0.4255677(x-1)^2 + 0.2481384(x-1)^3 - 0.1155701(x-1)^4 + 0.0272522(x-1)^5$$

$$\ln_e(x) = 0.9991150(x-1) - 0.4899597(x-1)^2 + 0.2856751(x-1)^3 - 0.1330566(x-1)^4 + 0.03137207(x-1)^5$$

Many more approximations of several types are contained in two very useful books: *Handbook of Mathematical Functions*, by Abramowitz and Stegun, and *Computer Approximations*, by Hart et al.

## Tables in ROM

Another very fast method of function approximation is known as *table lookup*. Tables of the functions are precomputed and stored in ROM for use by the program. These tables, in conjunction with interpolation, can produce very high resolution results in short order. The major problem is the space the table requires in ROM.

In the next few sections, we will discuss table methods, including interpolation for higher resolution. Table lookup is easily understood and easily implemented. The FFT routine at the end of this chapter illustrates most of the techniques that will be described.

In addition to table lookup methods for active processing, ROM and RAM may be used for preprocessing data to and from I/O. This can involve anything from automatic scaling to interpolation and nonlinear functions. We will present a program that can be used to implement these functions and write a standard Intel hex file for easy programming.

## Tables: Interpolation

Along with tables goes interpolation. It is seldom possible or economical to enter a table in ROM or RAM with the resolution necessary for the purposes of a DSP application. The remaining resolution may be attained with interpolation.

In the section on division, a method was discussed involving multiplication by the inverse of the divisor. To find the inverse, we needed an initial estimate—this is a reasonable application for table lookup because the inverse is a linear function and easily interpolated. Let us say that we are working with 8-bit numbers and we are using a 16-entry table that begins in low memory with 0 and increases progressively through the next 16 locations. If we use the 4 MSBs of our operand as a pointer, we can access a value in the table that is an approximation of the inverse of the 8-bit number.

1. Use the four most significant bits of your operand to point into the table for an initial estimate.

2. Increment your pointer and retrieve the next estimate.
3. Subtract the initial estimate from the second.
4. Multiply the least significant bits by the difference derived above to get the rest of the approximation.

This will work for any function with a reasonable slope.

## Tables: Sine and Cosine

A table is a list or array of solutions separated by a fixed resolution. Using the table is a simple matter of creating a pointer based upon an input argument into the table and retrieving a specific solution. Differences among methods usually evolve because of the varying needs of forming the initial pointer into the table. Sometimes it is a simple linear function, like a multiplication table, where no special handling is required at all. Other times, such as with this example, the input argument must be scaled properly to point to the correct location. The FFT routine at the end of the chapter demonstrates a typical table lookup method, and this should be studied for specific examples of tables involving powers, logs, sines, and cosines. Below, we examine an enhanced table algorithm that can be used for certain functions involving periodicity and symmetry, as the trigonometric functions.

Sine and cosine tables typically occupy separate tables in RAM or ROM, as they do in the FFT example code for the 80C196 at the end of the chapter. However, they may be collapsed into a single table that is traversed in two directions. This follows from the trigonometric property, $\cos = \pi/2 - \sin$. This algorithm is based upon $\pi/2$, though it is perfectly reasonable to assume that the table could be divided into smaller pieces with only a moderate increase in complexity.

The algorithm follows this pattern:

1. Determine whether we are solving for a sine or cosine—save this as a flag.
2. Reduce the input argument to an angle less than 360 degrees. If the angle is negative, add 360 degrees to find the positive complement. Save any remainder generated by this division as a fractional part for a later linear interpolation step.
3. Determine which quadrant of the circle the angle is by dividing a copy of the argument by 90 degrees—save this result as a pointer.
4. To solve for a cosine: if the pointer from step 3 is

   0 Go to step 5.

   1 Set a bit to signify a negative result, form the 2's complement of the angle, add 180 degrees to make the argument positive again, and go to step 6.

   2 Set a bit to signify a negative result, subtract 180 degrees from the argument to reduce it to less than 90 degrees, and go to step 5.

   3 Form the 2's complement of the angle, add 360 to point into the first 90 degrees, and go to step 6.

To solve for a sine: if the pointer from step 3 is

0 Form the 2's complement of the argument and add 90 degrees; go to step 6.

1 Subtract 90 degrees to bring it back into the scope of the table; go to step 5.

2 Form the 2's complement of the argument, add 270 degrees, set a bit to indicate a negative result, and go to step 6.

3 Set a bit to indicate a negative result, subtract 270 degrees from the argument, and go to step 5.

5. Use the argument to point into the table and retrieve the value, this is the base. Check to see if there was a remainder from the division in step 2. If not, go to step 7.
   - If there was a remainder in step 2, *increment* the pointer into the table and retrieve the next value.
   - Subtract the base value from this one and multiply the result by the fractional part saved from step 2.
   - Add this product to the base.
6. Use the argument to point into the table and retrieve the value—this is the base. Check to see if there was a remainder from the division in step 2. If not, go to step 7.
   - If there was a remainder in step 2, *decrement* the pointer into the table and retrieve the next value.
   - Subtract the base value from this one and multiply the result by the fractional part saved from step 2.
   - Add this product to the base.
7. If a bit was set earlier to indicate a negative result, form the 2's complement of the table value and leave.

The table used in this algorithm has 90 entries:

```
;sines (degrees)
    0ffffh, 0fff6h, 0ffd8h, 0ffa6h, 0ff60h, 0ff06h,
    0fe98h, 0fe17h, 0fd82h, 0fcd9h, 0fc1ch, 0fb4bh,
    0fa67h, 0f970h, 0f865h, 0f746h, 0f615h, 0f4d0h,
    0f378h, 0f20dh, 0f08fh, 0eeffh, 0ed5bh, 0eba6h,
    0e9deh, 0e803h, 0e617h, 0e419h, 0e208h, 0dfe7h,
    0ddb3h, 0db6fh, 0d919h, 0d6b3h, 0d43bh, 0d1b3h,
    0cf1bh, 0cc73h, 0c9bbh, 0c6f3h, 0c41bh, 0c134h,
    0be3eh, 0bb39h, 0b826h, 0b504h, 0b1d5h, 0ae73h,
    0ab4ch, 0a7f3h, 0a48dh, 0a11bh, 09d9bh, 09a10h,
    09679h, 092d5h, 08f27h, 08b6dh, 087a8h, 083d9h,
    08000h, 07c1ch, 0782fh, 07438h, 07039h, 06c30h,
    0681fh, 06406h, 05fe6h, 05bbeh, 0578eh, 05358h,
    04f1bh, 04ad8h, 04690h, 04241h, 03deeh, 03996h,
    03539h, 030d8h, 02c74h, 0280ch, 023a0h, 01f32h,
    01ac2h, 0164fh, 011dbh, 00d65h, 008efh, 00477h,
    0h
```

This algorithm may be applied to any function with similar symmetries and periodicities—and, of course, the table can be rewritten for radians as well as degrees. The optional disk contains a file, TABLE.TBL, with this and some other useful tables. Figure 5-3 may be useful in helping you understand the mechanics of this algorithm.

## Circular Buffers: Software

Circular buffers in software are created, as they are in hardware, with modulo arithmetic. If you have a choice over the size of the buffer, a circular buffer becomes a trival matter when you size it to a power of two. In this case, an index can be made to be combined with an array pointer that will allow you to step through your buffer, wrapping neatly when you reach the end to begin again at the start. This is accomplished by incrementing or decrementing the index, as you choose, then anding it with a word $2^b - 1$ before combining it with the array pointer.

For example, suppose we desire a rotary or circular buffer comprising 32 words. We set a base address in RAM with the five least significant bits set to zero. We clear an index variable that we can add to the base address to increment through the buffer. Each time we address the buffer, we add the index to the base address. We must be careful to add only the five least significant bits to the base address, and we take care of this by anding the index

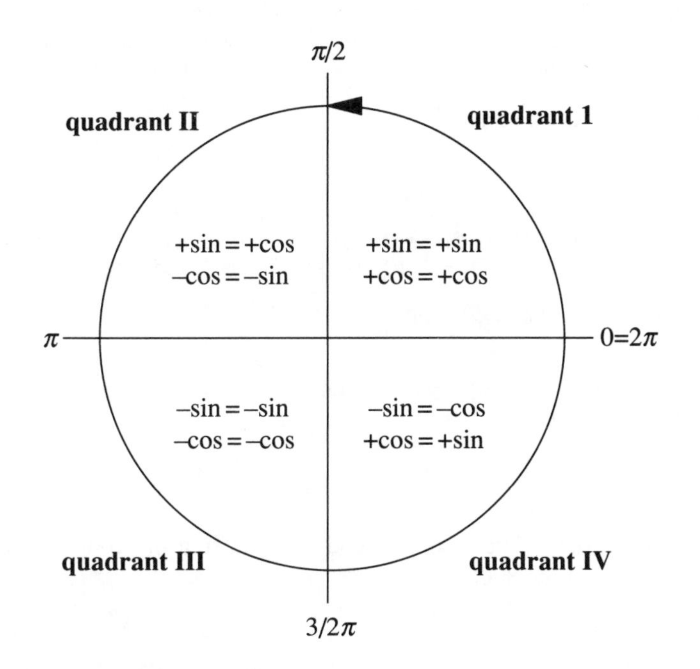


**Figure 5-3.** The symmetry of these functions may be used to simplify their generation.

with 1fh after each increment. This way, when the index rolls over, it clears to zero and we start again at the beginning of the buffer:

```
int *circbuf=0x4000;
int circndx=0;
for(;;) {
    *(circbuf || cirndx)=0;
    cirndx++;
    cirndx &= 0x1f;
}
```

It is usually not the case, however, that the number of taps in the filter in the FIR filter you need or the number of delays in the IIR filter you are designing comes out to a perfect power of two. A circular buffer can still be implemented using modular techniques—an example of such an implementation was presented in Chapter 4 in the C coding of both the FIR and IIR filters. The current loop counter for the sum of products is added to an index representing the current base of the circular buffer, the length of the buffer is then subtracted from that sum. If the result is less than zero, the sum is our new index, if it is greater than zero, the result of the subtraction is our new index.

This method will work for any length of buffer and is certainly faster and more economical than moving data around each time new data arrives. The code fragment below illustrates the procedure:

```
ncoef = buffer_size-1;      /*this is a subscript 0 system*/
ndx = 0;     /*base index for buffer, increments after each new*/
             /*output is computed, wraps when it sees the end */
```

Within the loop, the code looks like this:

```
do{
    for(i=0; i<=ncoef; i++) {

        /*calculate local data location in circular buffer*/
        if((buff_ndx = ndx+i-buffer_size) < 0) buff_ndx = ndx+i;

        /*compute the next SOP*/

        }
    /*roll over to stay in buffer and not repeat*/
    if(++ndx > ncoef) ndx = 0;

    /* output*/
    }while(...);
```

This method is also used in the assembly code for the sum of products routine that appears later in this chapter for the 80C196.

## Bit Reversal: Software

This is accomplished, again, via a table in ROM or RAM. A table of bit-reversed addresses based on linear addresses the size of the FFT is stored in ROM. Values of $n$ in $x[n]$ are used as indices into that table. There will be $2^n$ entries in the table, where $n$ is equal to $\log_2$*Vector Length* for either the data or twiddle factors, depending upon the kind of FFT you are doing. See the FFT routine for an example.

## Hardware Support: Generating ROM Tables

Tables, such as the sine table on page 280, are usually generated in some convenient manner and placed in an array or matrix, which is then added to the program in a header or include file.

Sometimes, however, you may wish to handle the operation more directly, as when you are constructing a table in ROM that will actually be used as an I/O device. Suppose, for instance, you are using quadrature signals, sine and cosine, to encode position, either linear or circular. Analog signals fed to A/D converters can generate address bits for ROM lookup—a processor can then use the ROM as a parallel input source with angular position precomputed. This can provide a very fast method of preprocessing data.

Of course, it can also be used for postprocessing position data from a processor, turning linear position into angular data such as phase. ROMs used for active data transformation can enhance throughput on slower processors.

The next program automates the process of generating ROM data from functions. By placing the function in the code, this program will write an Intel hex file in standard format for use in programming ROMs.

The example function is the arctangent function, which produces a signed magnitude $-\pi \leq x \leq \pi$. It will write a hex file to disk that can be programmed into any ROM large enough.

```
#include <process.h>
#include <io.h>
#include <conio.h>
#include <stdio.h>
#include <math.h>
#include <float.h>
#include <stdio.h>
#include <stdlib.h>
#include <fcntl.h>
#include <sys\types.h>
#include <sys\stat.h>
#include <malloc.h>
#include <errno.h>
#include <time.h>
#include <string.h>

#define TRUE 1
#define FALSE 0
#define PI 3.14159265359

char *buf;
char __huge *codebuf;
char __huge *codeptr;
char __huge *code;
```

```
unsigned int count = 0x1000;
int ad_buf, wrt = 0, speak = 0;

char hexletters[] =
{'0','1','2','3','4','5','6','7','8','9','A','B','C','D','E','F',
        '10', '11', '12', '13', '14', '15', '16', '17', '18', '19',
        '1A', '1B', '1C', '1D', '1E', '1F', '20', '21', '22', '23',
        '24', '25', '26', '27', '28', '29', '2A', '2B', '2C', '2D',
        '2E', '2F', '30'};


int disk_write(char *, unsigned int);



main(int argc, char **argv)
{
    int x, y, tempx, tempy;
    int r = 0;
    double delta_y = PI/32.0;
    char xdeg[256];
    char ydeg[256];
    char *row = xdeg;
    char *col = ydeg;
    int i, j, dcnt;
    char chksm;
    int record, checksum;
    char data[0x20];
    unsigned prom = 0xffff;

    union{
        char nib[2];
        unsigned int angle;
        }temp;

    union{
        signed int bytes[2];
        unsigned long tmp;
        char nibbles[4];
        signed int full;
        }address;

    union{
        int address;
        char nibbles[2];
        }eprom;


    if ( argc >= 2 ) {                          /*set internal flags for*/
        for(i=1; i < argc; i++) {       /*read and write*/
```

```
            if (!strcmp(argv[i], "-w")) {
                wrt = TRUE;
                continue;
                }
            if (!strcmp(argv[i], "-s")) {
                speak = TRUE;
                continue;
                }
            printf("\nunimplemented flag");
            }
        }

    if(speak) {        /*write intermediate file*/
        ad_buf = open( "tstdata", _O_BINARY | O_WRONLY | O_CREAT |
                    O_TRUNC | _O_APPEND, S_IREAD | S_IWRITE );
        if( ad_buf == -1 ){
            perror("\nopen failed");
            exit(-1);
            }
        }

    /* allocate a file buffer.*/
    if( (buf = (char *)malloc( (size_t)count )) == NULL )   {
        perror("\nnot enuf memory");
        exit(-1);
        }


    /* allocate a buffer for the hex code.*/
    if( (code = codeptr = codebuf = (char *)halloc( 65536,
            sizeof(char))) == NULL )   {
        perror("\nnot enuf memory");
        exit(-1);
        }

/*
function:
this is where the function that generates the data for the prom goes
in this case it will generate an arctangent table
*/


    for(x = 0; x < 128; x++) {
        for(y = 0; y < 256; y++) {
            if((y < 0x80)) {
                tempy = 0x7f - y;  /*these are to remedy wierd*
                tempx = 0x7f - x;  /*crossover effects*/
                }
            else {
                tempy = 0xff - y;
                tempx = x;
                }
```

```
            if(tempx == 0)
                *codebuf = (unsigned char)floor((atan2((double)tempy,
                    (double).1)*(64/PI)));
            else *codebuf = (unsigned char)floor((atan2((double)tempy,
                (double)tempx)*(64/PI)));

            *codebuf &= 0x1f;   /*has to change*/
            if(y & 0x80) *codebuf |= 0x40;
            *codebuf++;
            }
        }

    for(x = 128; x < 256; x++) {
        for(y = 0; y < 256; y++) {
            if(y < 0x80) {
                tempx = 0xff - x;
                tempy = y;
                }
            else {
                tempy = y - 0x7f;
                tempx = x - 0x80;
                }

            if(tempx == 0)
                *codebuf = (unsigned char)floor((atan2((double)tempy,
                    (double).1)*(64/PI)));
            else *codebuf = (unsigned char)floor((atan2((double)tempy,
                (double)tempx)*(64/PI)));

            *codebuf &= 0x1f;   /*has to change*/
            *codebuf |= 0x80;
            if(y & 0x80) *codebuf |= 0x40;
            *codebuf++;
            }
        }
/*
end of function
*/

    if(speak) {        /*this is the raw data from the function*/
        if(!disk_write(codeptr, 32766)) perror("\n can't write array");
        }
    close( ad_buf );
    free( buf );


/*create file to INTEL HEX file*/
    if(wrt) {
        ad_buf = open( "atan.hex", O_TEXT | O_WRONLY | O_CREAT | O_TRUNC,|
                                S_IREAD | S_IWRITE );
        if( ad_buf == -1 ){
            perror("\nopen failed");
            exit(-1);
            }
```

```
        }

    /* allocate a file buffer.*/
    if( (buf = (char *)malloc( (size_t)count )) == NULL )   {
        perror("\nnot enuf memory");
        exit(-1);
        }
    if(wrt) {
        sprintf(buf,"\n");
        if(!disk_write(buf, strlen(buf))) perror("\ncan't write newline");
        }

    printf("\n.");

/*
to satisfy the requirements of the standard
we must emit the following code
*/
    if(wrt) {
        sprintf(buf,"\n:020000020000FC",record,address);
        if(!disk_write(buf, strlen(buf))) perror("\n can't write end of
record");
        }
/*
this is where we convert the numerical data to ascii code for the
hex file.
*/

    for(address.bytes[1] = 0; address.bytes[1] < 256;
                address.bytes[1]++){
        for(address.bytes[0] = 0; address.bytes[0] < 256;
                address.bytes[0]+=16) {
            record = 0x10;
            chksm = 0x0;
            for(dcnt = 0; dcnt <= 31;) {  /*each line will be 32 chars*/
                chksm += *code;    /*checksum is a sum of binary*/
                temp.nib[1] = (*code >> 4) & 0xf;
                temp.nib[0] = *code++ & 0xf;/*separate nibbles*/
                data[dcnt++] = hexletters[temp.nib[1]];
                data[dcnt++] = hexletters[temp.nib[0]];
                }           /*get ascii representation*/
            checksum = 0x10; /*record*/
            checksum += address.nibbles[0];   /*add address*/
            checksum += address.nibbles[2];
            checksum += chksm;
            checksum = 0x100 - checksum;
            checksum = checksum & 0xff;        /*truncate checksum*/
            data[0x20] = 0x0;
            if(wrt) {                       /*put it in the file*/
                eprom.nibbles[0] = address.nibbles[0];
                eprom.nibbles[1] = address.nibbles[2];
                sprintf(buf,"\n:%.2X%.4X00", record, eprom.address);
```

```
                    if(!disk_write(buf, strlen(buf))) perror("\ncan't write
                                                     code");
                    disk_write(data, 32);
                    sprintf(buf,"%.2X",checksum);
                    if(!disk_write(buf, strlen(buf))) perror("\ncan't write
                                                     code");
                    printf("\nxcnt = %d ycnt = %d", address.bytes[1],
                            address.bytes[0]);
                    }
                }
        }
/*
finally, we must sign off with this
*/

    if(wrt) {
        sprintf(buf,"\n:00000001ff\r\n",record,address);
        if(!disk_write(buf, strlen(buf))) perror("\n can't write end of
                                                     record");
        }

    close( ad_buf );
    free( buf );

    return 0;
}



int disk_write(char *buf, unsigned int count)
{

    if(count = write( ad_buf, buf, count ) == - 1) {
        printf("\ncouldn't write");
        return FALSE;
        }
    return TRUE;
}
```

## DIGITAL SIGNAL PROCESSES ON A NON-DSP

There are also those occasions when you may need or wish to use some of the techniques we have discussed here for general-purpose processors, such as the one in the machine you are using now, or perhaps an embedded microcontroller.

DSPs are fast making it possible to create realtime filters and controls that rival and surpass equivalent analog units. But not all signal processing requirements must be fast. Many can be handled by more general-purpose processors, such as the 80C196. Applications that fit this group are some video and audio processing and a great deal of post-processing and passive processing needs. In the next chapter, we will examine two popular DSP chips, their architecture, and some code written for them. Next, we will look at DSP code written to run on a popular microcontroller, the 80C196.

By whatever name, the methods involved in digital signal processing are common to many levels of electronic hardware and software. When to filter and how to filter are important in many applications, as is the proper use of an A/D to benefit from its entire bandwidth. Moving average filtering and median filtering have long been part and parcel of the embedded systems toolbox.

We begin with the familiar sum of products embedded in an FIR filter. This routine presents an example of the sum of products and software circular buffer implemented on the 80C196.

```
COEF    equ     $
    org     4000h
XN      equu    $

    oseg    24h
AX:     dsw     1       ;general purpose register
BX:     dsw     1       ;general purpose register
XNPTR:  dsw     1       ;pointer to input data
COEFPTR:    dsw     1       ;base pointer into coefficient table
CINDEX0:    dsw     1   ;index into coefficient table
CINDEX1:    dsl     1
    dividend    equ     1
    NDX         equ     1
TEMP:   dsw     1   ;temporary register for intermediate results
MRR:    dsl     1       ;multiply register
ACCUM:  dsl     1       ;accumulate register
ICNTR:  dsw     1       ;number of taps of FIR filter

length  equ     32

;we will assume an infinite loop
;values preinitialized before the start of the SOP

fir:
    ld  COEFPTR, #COEF      ;beginning of coefficient table
    ld  XNPTR, #XN      ;point to oldest input value
    ld  CINDEX0, #length;index for circular buffer

get_new_data:
;base pointer for the circular buffer must be adjusted after every SOP
;we will assume here that this has already happened once
    dec     CINDEX0             ;move base of circular buffer

;actual SOP
new_val:
    ld      ICNTR, #length      ;length of sum
    ld      XNPTR[CINDEX0], port1  ;input new data point
```

```
sop:
    add     CINDEX1, CINDEX0, ICNTR    ;add loop counter to buffer index
    sub     TEMP, CINDEX1, #length     ;get difference
    jnc     new_index          ;if it is negative, we have it in CINDEX1
    ld      CINDEX1, TEMP          ;otherwise put it there

new_index:
    ld      AX, COEFPTR[NDX];circular buffer
    mul     MRR, AX, XNPTR[NDX] ;multiply coefficient by z-n
    add     ACCUM, MRR      ;accumulate
    jv      SAT             ;if there is an overflow
    djnz    ICNTR, sop ;if we are done, go home
    cmp     CINDEX0, R0         ;are we at zero yet?
    jnz     continue_loop
    ld      CINDEX0, #length;reinitialize base index
continue_loop:
    sjmp    get_new_data    ;go get new data for loop
    ret                     ;return with sum in ACCUM

sat:
    jge     psat            ;positive overflow
    clr     ACCUM
    ret                     ;return with saturation
psat:
    ld      ACCUM, #7fffh
    ret
```

The following routine is an FFT written for the 80C196. You will see in it examples of software bit reversal, the use of tables, and interpolation of sine, cosine, log, and square root. This routine is a good example of how a digital signal processing routine can be ported to a general-purpose micro with minimal loss.



```
RSEG
    EXTRN   port1, zero, error


OSEG    at  24h
    TMPR:   dsl     1   ;temp reg, Real
    TMPI:   dsl     1   ;temp reg, Imaginary
    TMPR1:  dsl     1   ;temp reg1, Real
    TMPI1:  dsl     1   ;temp reg1, Imaginary
    XRTMP:  dsl     1   ;temp data reg, Real
    XITMP:  dsl     1   ;temp data reg, Imaginary
    XRRK:   dsl     1
    XRRNK:  dsl     1
    XIRK:   dsl     1
    XIRNK:  dsl     1
```

```
    diff    equ     xrrk    :long   ;table difference for square root
    sqrt    equ     xrrnk   :long   ;square root
    log     equ     xrrnk   :long   ;10 log magnitude^2
    nxtloc  equ     xirk    :long   ;next location in the table

    WRP     equ     xrrk    :word   ;multiplication factor, Real
    WIP     equ     xrrk+2  :word   ;multiplication factor, Imaginary
    PWR     equ     xrrnk   :word
    IN_CNT  equ     xrrnk+2     :word
    NDIV2   equ     xirk    :word   ;n divided by 2 (0<n<N)*2

    KPTR:   dsw     1   ;K for counter *2 to index words
    KN2:    dsw     1   ;KPTR+NDIV2
    N_SUB_k:dsw     1   ;N-K *2 to index words
    RK:     dsw     1   ;bit reversed pointer of KPTR
    RNK:    dsw     1   ;bit reversed pointer of N_SUB_K
    SFT_CNT:dsw     1
    LOOP_CNT:dsb    1
    ptr     equ     kn2     :word

DSEG

EXTRN   FFT_MODE        ;FFT_MODE: mode for FFT input and graphing
EXTRN   XREAL, XIMAG        ;XREAL, XIMAG: home addresses for 32 16-bit
signed
            ;entries for real and imaginary numbers
EXTRN   FFT_OUT             ;FFT_OUT: starting address for 32 word array
            ;of magnitude information.

    OUTR:   dsw     32  ;real component of fft
    OUTI:   dsw     32  ;imaginary component of waveform

PUBLIC  OUTR, OUTI


CSEG    at  2280h
PUBLIC  fft_calc        ;starting point for fft algorithm
EXTRN   scale_factor        ;shift factor used to prevent overflow when
averaging
            ;fft outputs

;           ;START FOURIER CALCULATIONS
            ;400 INITIALIZATION OF LOOP
FFT_CALC:
    clrb    error       ;indication only
    ldb     port1, #00000001b

    clrvt
    ldb     loop_cnt, #1
    ldb     shft_cnt, #4
    ld      ndiv2, #32
            ;410 K=0
```

```
OUT_LOOP:
    xorb    port1, #00000100b
    clr     kptr        ;indication only
                ;420 IF LOOP>EXP THEN 700
    cmpb    loop_cnt, #5    ;32=2^5
    bgt     UNWEAVE

MID_LOOP:
    clr     in_cnt      ;430 INCNT=0
                ;440 CALCULATIONS BEGIN HERE

in_loop:
    add     in_cnt, #2 ;450 INCNT-INCNT+1
                ;460 P=BR(INT(K/(2^SHIFT)))
    ld      pwr, kptr
    shr     pwr, shft_cnt  ;calculate multiplication factors
    andb    pwr, #11111110b
    ld      pwr, brev[pwr]
                ;470 WRP=WR(P) : WIP=WI(P) : KN2=K+N2
gw:
    ld      wrp, wr[pwr]
    ld      wip, wi[pwr]
    add     kn2, kptr, ndiv2

                ;complex multiplication follows
                ;480 TMPR=(WRP*XR(KN2)-WIP*KI(KN2))/2
gm:
    mul     tmpr, wrp, xreal[kn2]
    mul     tmpi, wip, ximag[kn2]
    sub     tmpr+2, tmpi+2
                ;490 TMPI=(WRP*XI(KN2)+WIP*XR(KN2))/2
    mul     tmpr1, wrp, ximag[kn2]
    mul     tmpi, wip, xreal[kn2]
    add     tmpi+2, tmpr1+2
                ;using the high byte only of a signed multiply
                ;provides an effective divde by two
    BVT     ERR1        ;branch on error in complex multiplications

    ld      tmpr1, xreal[kptr];500 TMPR1=XR(K)/2
    shra    tmpr1, #1   ;TMPI1=XI(K)/2
    ld      tmpi1, ximag[kptr]
    shra    tmpi1, #1

gr2:                ;510 XR(KN2)=TMPR1-TMPR
    sub     xrtmp, tmpr1, tmpr+2
    st      xrtmp, xreal[kn2]

gx2:                ;520 XI(KN2)=TMPI1-TMPI
    sub     xitmp, tmpi1, tmpi+2
    st      xitmp, ximag[kn2]
                ;530 XR(K)=TMPR1+TMPR
    add     xrtmp, tmpr1, tmpr+2
    st      xrtmp, xreal[kptr]
                ;540 XI(K)=TMPI1+TMPI
```

```
gx:
    add     xitmp, tmpil, tmpi+2
    st      xitmp, ximag[kptr]

    BVT     ERR2
                ;560 K=K+1
ik:
    add     kptr, #2
                ;570 IF INCNT<K2 THEN GOTO 450
    cmp     in_cnt, ndiv2
    blt     IN_LOOP
                ;580 K=K+N2
    add     kptr, ndiv2
                ;590 IF K<N1 THEN GOTO 430
    cmp     kprt, #62
    blt     MID_LOOP
                ;600 LOOP-LOOP+1 : N2=N2/2
    incb    loop_cnt    ;605 SHIFT=SHIFT+1
    shra    ndiv2, #1
    decb    shft_cnt
                ;610 GOTO 400
    br      OUT_LOOP

ERR1:
    ldb     error, #01 ;overflow error, 1st set of calculations
    ret
ERR2:
    ldb     error, #02 ;overflow error, 2nd set of calculations
    ret



                ;700 POST-PROCESSING AND REORDERING STARTS HERE
UNWEAVE:
    ldb     port1, #00000010b
                ;720 FOR K=0 TO 31
    clr     kptr
    ld      n_sub_k, #64
                ;740 XIBRK=XI(BR(K)) : KRBRK=KR(BR(K))
UN_LOOP:
    ld      rk, brev[kptr]
    ld      xrrk, xreal[rk]
    ext     xrrk
    ld      xirk, ximag[rk]
    ext     xirk
                ;750 KIBRNK=XI(BR(N-K)) : KRBRNK=XR(BR(N-K))
    ld      rnk, brev[n_sub_k]
    ld      xrrnk, xreal[rnk]
    ext     xrrnk
    ld      xirnk, ximag[rnk]
    ext     xirnk
                ;760 TI=(XIBRK+XIBRNK)/2
```

```
ar:
    add     tmpi, xirk, xirnk
    ld      tmpi+2, xirnk+2
    addc    tmpi+2, xirk+2
    shral   tmp, #1         ;16 bit result in tmpi
                ;770 TR=(XRBRK-XRBRNK)/2
    sub     tmpr, xrrk, xrrnk
    ld      tmpr+2, xrrk+2
    subc    tmpr+2, xrrnk+2
    shral   tmpr, #1    ;16 bit result in tmpr
                ;780 XRT-(XRBRK_XRBRNK)/4
    add     xrtmp, xrrk, xrrnk
    ld      xrtmp+2, xrrk+2
    addc    xrtmp+2, xrrnk+2
    shll    xrtmp, #14  ;32 bit result in xrtmp
                ;790 XIT=(XIBRK-XIBRNK)/4
    sub     xitmp, xirk, xirnk
    ld      xitmp+2, xirk+2
    subc    xitmp+2, xirnk+2
    shll    xitmp, #14  ;32 bit result in xitmp
                ;multiply with provide effective divide by two
                ;800 OUTR=(XRT+TI*COSFN(K)/2-TR*SINFN(K)/2)
mr:
    mul     tmprl, tmpr, sinfn[kptr]
    mul     tmpil, tmpi, cosfn[kptr]
    add     xrtmp, tmpil
    addc    xrtmp+2, tmpil+2
    sub     xrtmp, tmprl
    subc    xrtmp+2, tmprl+2
    st      xitmp+2, outi[kptr]
                ;OUTI = imaginary output values
                ;830 MAG=SQR(OUTR*OUTR+OUTI*OUTI)
                ;Get magnitude of vector
GET_MAG     :
    ld      tmpr, xrtmp+2
    ld      tmpi, xitmp+2

    mul     tmpr, tmpr ;tmp=tmpi**2 +tmpr**2
    mul     tmpi, tmpi
    add     tmpr, tmpi
    addc    tmpr+2, tmpi+2

    bbc     FFT_MODE, 2, CALC_SQRT
                ;CALCULATE 10 LOG MAGNITUDE^2
;Output=512*10*LOG(x) x=1,2,3 ... 64k

CALC_LOG:
    clr     shft_cnt
    norml   tmpr, shft_cnt      ;normalize and get normalization factor
    cmpb    shft_cnt, #15
    jle     LOG_IN_RANGE;jump if shft_cnt <= 15
```

```
        clr     log
        br      LOG_STORE

LOG_IN_RANGE:
        add     shft_cnt, shft_cnt, shft_cnt  ;make shft_cnt a pointer
        ldbze3  ptr, ptr, ptr
        add     ptr, #LOG_TABLE-256
                    ;ptr=table_offset (offset-tmpr+3)
                    ;use -256 since tmpr+3 is always >= 128
        ld      log, [ptr]+
        ld      nxtloc, [ptr]  ;linear interpolation

        sub     nxtloc, log     ;nxtloc = next log-log

        ldbze   diff, tmpr+2    ;diff+1=nxtloc*tmpr+2/256
        mulu    diff, nxtloc

        shrl    diff, #8    ;log=log+diff/256
        add     log, diff
        shr     log, #5         ;8192/32*20log(x)=256*20log(x)

        addc    log, log_offset[shft_cnt]
                    ;add log of normalization factor
                    ;log(M*N)=log M+log N

LOG_STORE:
        shr     log, #SCALE_FACTOR
        addc    log, zero  ;divide to prevent overflow during
        add     log, FFT_OUT[kptr]
                    ;averaging of outputs
        st      log, FFT_OUT[kptr]

        BR      ENDL

CALC_SQRT:
        clr     shft_cnt
        norml   tjpr, shft_cnt      ;normalize and get normaliztion factor

        jne     SQRT_IN_RANGE  ;jump if tmpr >0
        st      zero, sqrt_2
        br      SQRT_STORE

SQRT_IN_RANGE:
        ldbze   ptr, tmpr_3     ;most significant byte is table pointer
        add     ptr, ptr, ptr
        add     ptr, #SQ_TABLE-256
                    ;ptr=table+offset (offset=tmpr+3)
                    ;use -256 since tmpr+3 is always >=128
        ld      sqrt, [ptr]+
        ld      nxtloc, [ptr]  ;linear interpolation

        sub     nxtloc, sqrt   ;nxtloc=sqrt-next sqrt
```

```
    ldbze   diff, tmpr+2    ;diff+1=nxtloc*tmpr+2/256
    mulu    diff, nxtloc

    ldbze   diff, diff+1    ;sqrt=sqrt+delta (diff < OFFH)
    add     sqrt, diff

    add     shft_cnt, shft_cnt, shft_cnt

    mulu    sqrt, tab_sqr[shft_cnt]
                ;divide by normalization factor
                ;mulu acts as divide since if tab2=OFFFFH
                ;sqrt would remain essentially unchanged

SQRT_STORE:
    shr     sqrt_2, #SCALE_FACTOR
    addc    sqrt_2, zero    ;divide to prevent overflow during
    add     sqrt_2, FFT_OUT[kptr]
                ;averagin of outputs
    st      sqrt_2, FFT_OUT[kptr]

                ;END OF LOOP

ENDL:
    add     kptr, #2
    sub     n_sub_k, #2
    bne     UN_LOOP
    ret



    CSEG    AT 3800H
                ;2*bit reversal value
BREV:
DCW     2*0, 2*16, 2*8, 2*24, 2*4, 2*20, 2*12, 2*28
DCW     2*2, 2*18, 2*10, 2*26, 2*6, 2*22, 2*14, 2*20
DCW     2*1, 2*17, 2*9, 2*25, 2*5, 2*21, 2*13, 2*29
DCW     2*3, 2*19, 2*11, 2*27, 2*7, 2*23, 2*15, 2*31

SINFN:
DCW     0, 3212, 6393, 9612, 12539, 15446, 18204, 20787
DCW     23170, 25329, 27245, 28898, 30273, 31356, 32137, 32609
DCW     32767, 32609, 32137, 31356, 30273, 28898, 27245, 25329
DCW     23170, 20787, 18204, 15446, 12539, 9512, 6393, 3212
DCW     0, -3212, -6393, -9512, -12539, -15446, -18204, -20787
DCW     -23170, -25329, -27245, -28898, -30273, -31356, -32137, -32609
DCW     -32767, -32609, -32137, -31356, -30273, -28898, -27245, -25329
DCW     -23170, -20787, -18204, -15446, -12539, -9512, -6393, -3212
DCW     0
```

```
COSFN:
DCW     32767, 32609, 32137, 31356, 30273, 28898, 27245, 25329
DCW     23170, 20787, 18204, 15446, 12539, 9612, 6393, 3212
DCW     0, -3212, -6393, -9512, -12539, -15446, -18204, -20787
DCW     -23170, -25329, -27245, -28898, -30273, -31356, -32137, -32609
DCW     -32767, -32609, -32137, -31356, -30273, -28898, -27245, -25329
DCW     -23170, -20787, -18204, -15446, -12539, -9512, -6393, -3212
DCW     0, 3212,  6393, 9612, 12539, 15446, 18204, 20787
DCW     23170, 25329, 27245, 28898, 30273, 31356, 32137, 32609
DCW     32767

    ;WR=COS(K*2PI/N)
WR:
DCW     32767, 32137, 30273, 27245, 23170, 18204, 12539, 6393
DCW     0, -6393, -12539, -18204, -23170, -27245, -30273, -32137
DCW     -32767, -32137, -30273, -27245, -23170, -18204, -12539, -6393
DCW     0, 6393, 12539, 18204, 23170, 27245, 30273, 32137
DCW     32767

    ;WI=-SIN(K*2PI/N)
WI:
DCW     0, -6393, -12539, -18204, -23170, -27245, -30273, -32137
DCW     -32767, -32137, -30273, -27245, -23170, -18204, -12539, -6393
DCW     0, 6393, 12539, 18204, 23170, 27245, 30273, 32137
DCW     32767, 32137, 30273, 27245, 23170, 18204, 12539, 6393
DCW     0

    ;655435/(square root of 2**SHFT_CNT)  ; 0<=SHFT_CNT<32
TAB_SQR:
;  1   2   4   8   16  32  64  128
DCW     65535, 46340, 32768, 23170, 16384, 11585, 8192, 5793

;   256 512 1024 2048 4096 8192 16384 32768
DCW     4096, 2896, 2048, 1448, 1024, 724, 512, 362

;   65536   131072  262144  524288...
DCW     256,     181,     128,     91,      64,      45,      32,      23
DCW     16,      11,      8,  6,  4,  3,  2,  1

    ;   square root of n*2**24 N=128, 129, 130,... 255
SQ_TABLE:
DCW     46341,  46522,  46702,  46881,  47059,  47237,  47415,  47591
DCW     47767,  47942,  48117,  48291,  48465,  48637,  48809,  48981
DCW     49152,  49322,  49492,  49661,  49830,  49998,  50166,  50332
DCW     50499,  50665,  50830,  50995,  51159,  51323,  51486,  51649
DCW     51811,  51972,  52134,  52294,  52454,  52614,  52773,  52932
DCW     53090,  53248,  53405,  53562,  53719,  53874,  54030,  54185
DCW     54340,  54494,  54647,  54801,  54954,  55106,  55258,  55410
DCW     55561,  55712,  55862,  56012,  56162,  56311,  56459,  56608
DCW     56756,  56903,  57061,  57198,  57344,  57490,  57636,  57781
DCW     57926,  58071,  58215,  58359,  58503,  59646,  58789,  58931
DCW     59073,  59215,  59357,  59458,  59639,  59779,  59919,  60059
```

```
DCW     60199, 60338, 60477, 60615, 60754, 60891, 61029, 61166
DCW     61303, 61440, 61576, 61712, 61848, 61984, 62119, 62254
DCW     62388, 62532, 62657, 62790, 62924, 63047, 63190, 63323
DCW     63455, 63587, 63719, 63850, 63982, 64113, 64243, 64374
DCW     64504, 64634, 64763, 64893, 65022, 65151, 65280, 65408

    ;16384*10*LOG(N/128)    N=128, 129, ... 256
LOG_TABLE:
DCW     0, 554,    1103,   1648,   2727,   3260,      3789
DCW     4314,  4835,  5353,  5866,  6376,  6883,  7386,  7885
DCW     8381,  8873,  9362,  9848,  10330, 10810, 11286, 11758
DCW     12228, 12695, 13158, 13619, 14076, 14531, 14963, 15432
DCW     15878, 16321, 16762, 17200, 17635, 18067, 18497, 18925
DCW     19349, 19772, 20191, 20609, 21024, 21436, 21846, 22254
DCW     22660, 23063, 23464, 23862, 24259, 24653, 25045, 25435
DCW     25822, 26208, 26592, 26973, 27353, 27730, 28106, 28479
DCW     28851, 29220, 29588, 29954, 30318, 30680, 31040, 31399
DCW     31755, 32110, 32463, 32815, 33165, 33512, 33859, 34203
DCW     34546, 34887, 35227, 35565, 35902, 36236, 36570, 36901
DCW     37232, 37560, 36887, 38213, 38537, 38860, 39181, 39501
DCW     39819, 40136, 40452, 40766, 41079, 41390, 41700, 42009
DCW     42316, 42622, 42927, 43230, 43533, 43833, 44133, 44431
DCW     44729, 45024, 45319, 45612, 45905, 46196, 46486, 46774
DCW     47062, 47348, 47633, 47917, 48200, 48482, 48763, 49042
DCW     49321

    ;512*10*LOG(2**(15-N))    N=0,1,2,3,... 15
    ;512*10*LOG(0.5)   N=16, 17, 18, ... 31
DCW     23119, 21578, 20037, 18495, 16954, 15413, 13871, 12330
DCW     10789, 9248,  7706,  6165,  4624,  3083,  1541,  0
END
```

# 6

# DSP Hardware and Software

## INTRODUCTION

Application-specific hardware has become increasingly popular as the prices for developing and manufacturing the devices have dropped. These devices find their places in every area of the electronics marketplace, from high-fidelity audio and video products to devices for controlling an automobile engine. Their magic comes from the fact that they do not try to do everything; their hardware is geared to a specific task, at which they excel.

An application-specific device is usually defined as a hardware specifically designed for a particular need. It can be a full custom design or a standard part with custom hardware or a gate array. The digital signal processor is not strictly an application-specific device, though it can be if custom hardware is added to it. The DSP is, however, at least genre-specific. It is built to tackle the tasks associated with signal processing with special address generators to implement rotary buffers and bit reversal, accumulators with extended precision for the iterative loops of FIR and IIR filters, and a very fast arithmetic core tuned to the needs of the mathematics involved in signal processing. The DSP is specific to *signal* processing; its elements and architecture are tuned to that environment and that purpose. If this same processor is turned to a general-purpose task, such as becoming the core of a personal computer, it loses its advantage both in code size and speed. It is not designed for the file handling or database functions so common on microcomputers and personal computers.

DSP chips have a common purpose, and thus have many of the same functions. The real comparisons that can be done fruitfully are those that relate directly to the application the chips are being considered for. As die sizes shrink, chips get faster and more capable. It would be useless to press a point about a specific processor's speed or singular capability, when that advantage is likely to vanish within months of broadcasting it. The purpose for this chapter is to examine the trends in DSP and the capabilities of the group. We will present specific examples of code to see how the processes are being performed by the chips and note the fact that there are common themes and trends throughout the group.

The real differences among the chips usually concern a specific area within a specific area, such as low cost, special hardware, or built-in A/Ds with filters tuned for audio applications or speech processing. Regardless which you choose, you will still find a certain set of capabilities provided for the applications the devices are basically designed for: FIR and IIR filters and FFT processing.

In this chapter, we will look at two processors: the TMS320C30 from Texas Instruments and the ADS2101 from Analog Devices. Both are good examples of DSPs, and both provide the basic services necessary to the genre. As the technology changes so quickly, there seems little purpose in an in-depth analysis of each chip. Our goal will be to provide an overview of what these chips can do, and hopefully enough information to aid in selecting a chip for your application, whether it is one of these or another altogether. If you wish a more complete understanding of both the architecture and instruction sets of these chips, refer to the manuals available from the manufacturers themselves. The code provided here is workable, and can be used with very little modification.

The facilities of the DSP make it suitable and advantageous for a wide variety of embedded applications requiring speed and arithmetic power. Most chips are based upon the *Harvard*[1] architecture which separates data and program buses into two areas, allowing data and program to be retrieved simultaneously. Both high-end and low-end machines are based upon this idea. Both the DSPs presented in this chapter are Harvard, and so is the lowly 8051 used so commonly in embedded applications.

In many machines, there are a multiplicity of buses and memory areas. Data can be read from two areas at once for an operation, and can be stored again in as little as two instruction cycles—and this at 60 to 100 MHz. This makes once unthinkable jobs perfectly reasonable, and puts an enormous amount of power into the hands of the engineer.

## DESIRABLE FEATURES IN A DIGITAL SIGNAL PROCESSOR

Speed and precision are both important attributes of a DSP, but architecture is also an important consideration. Basic DSP applications, such as FIR and IIR filters, as well as the FFT, use certain operations again and again. With any realtime processing, it is essential to minimize any overhead in repeated operations. The aim of the DSP chip is to do this; the choice of architecture determines how it is done.

From what we know about signal processing, we have made a list of features we believe are desirable in a DSP. Not every DSP will have all these features: some are very narrowly defined for certain applications; some are reduced implementations for economic reasons. Of

[1]Harvard is an alternative to the Von Neumann architecture, in which program and data share the same space and sequential accesses are required to get the same information. The 80486 is based upon the Von Neumann architecture.

course, the needs of the application should always be your guide, the sole basis upon which this list is formed is optimum performance. We present this list for information and discussion:

1. Since the bulk of the effort in a signal processing product is arithmetic, one of the first features we might specify in a DSP is a *bus width* great enough to accommodate *double precision numbers*. This would include the external bus as well as the internal bus, since throughput can suffer if data bottle necks at the chip I/O.
2. We also need an *ALU* wide enough (double precision) to minimize roundoff error. It should perform all standard arithmetic instructions, such as addition and subtraction with some division capability, and should have provisions for nonlinear functions such as arithmetic rounding and saturation. All data moves and arithmetic should be single-cycle. It should support preshifting and postshifting for scaling to prevent overflow. The logic capability should include AND, OR, EXCLUSIVE OR, and NEGATION.

   In addition, there should be a complete set of register files for fast storage and retrieval of intermediate values.
3. There should be a fast multiply-accumulator that multiplies, accumulates (adds or subtracts), and stores in one cycle. This unit should support extended precision of signed (2's complement), unsigned, integer, and fractional data. It should also possess some sort of rounding, preferably unbiased rounding, saturation capability, and lookahead overflow detection. In addition, it needs multiple registers on the output for holding temporary values and pipelining results.
4. There should be a *barrel shifter* for single cycle shifts. This allows for rounding and scaling, but exponent detection, normalization, denormalization[2], and *block floating point*[3] are also a definite bonus.

---

[2] Normalization and denormalization are floating-point support instructions, as well as aids in the use of table lookup methods. For fixed to floating-point conversions (IEEE), a number is considered normalized when it has been shifted left or right until its leftmost 1 is just to the left of the imaginary binary point. The shifts are counted and, together with a bias, are computed to produce an exponent for the purpose of creating a floating-point number. Denormalization is the process of shifting a floating-point mantissa right or left the number of counts in the exponent with the bias removed. This produces a fixed-point number. In a DSP, the normalization instruction may cause a register to be left-shifted until its MSB is a 1, or it may leave the register untouched and simply calculate the number of counts required in the shift. As for fixed- to floating-point conversion, the TMS320C30 actually has an instruction for that, though the floating-point it produces is not strictly IEEE-compatible.

[3] Block floating-point derives its exponent from the largest member of an array of numbers. This exponent becomes the exponent of the entire array, controlling its scale just as the exponent in a standard floating-point number. In this scheme, the exponent of each member of the data array is examined as it passes through exponent detection logic. The least-negative exponent is latched and named to the entire array. This form of arithmetic takes no bits away from the mantissa for scaling purposes, allowing maximum dynamic range to the data block. Since the dynamic range for each block is the same, arithmetic can be perfomed on this array of numbers like fixed-point arithmetic.

5. The process should be highly *pipelined*—that is, the operations of the CPU must be split, as much as possible, into smaller sequential tasks that can be executed in parallel by different units within the processor. With this technique, the slowest task sets the throughput rate, and the delay through the system is equal to the sum of the all the delays associated with the task. This *parallelism* allows for multiplication and accumulation, as well as addressing and retrieving the next instruction at the same time. Other features in this category include *no-overhead* loops and jumps.

   For speed, the processor should be tuned for simultaneity and single cycle execution.
6. There should be a *specialized address generator* capable of handling rotary buffers of various sizes and bit reversals of various sizes. This would remove some of the overhead from these often iterative procedures.
7. There should be a bus structure that allows simultaneous access to data and program memory with separate address generation units, making the CPU capable of retrieving both instructions and data in the same clock cycle.

## ARCHITECTURES OF SPECIFIC CHIPS

### 320C30

Overview:

- Very fast instruction cycle
- Two on-chip 1K×32 bit dual-access RAM
- 64×32 bit instruction cache
- Bit instructions, 24-bit address
- Floating-point and integer multiplier
- Extended precision accumulators
- On-chip DMA controller for concurrent I/O
- Two and three operand instructions
- Two serial ports
- Two 32-bit timers
- Bit barrel shifter
- Two auxiliary register arithmetic units to generate two operand addresses in a single cycle
- 40-bit extended precision registers for accumulation

The C30 comprises a CPU, RAM, ROM and cache, several bus systems, peripherals such as timers and serial ports, and a DMA unit.

Comprising a multiplier, an ALU, a 32-bit barrel shifter, auxiliary register arithmetic units, and a CPU register file, the central processing unit is capable of floating point, integer, and logical operations. The multiplier can perform a 24-bit integer and 32-bit floating-point multiplication with no difference in speed, with a 50ns instruction cycle that is currently one of the fastest available. Instructions are pipelined, allowing multiple operations to occur at the same time, including multiplication, data and instruction fetch, and ALU actions.

The ALU will handle 32 bits of integer and logical data, with extended precision for floating-point data of 40 bits. Parallel operation is facilitated by the multiple bus structure that allows for two operands from memory and two from the internal register file to be accessed simultaneously, which means that up to four multiplication and accumulation operations can occur in single cycle.

The address generators, or auxiliary register arithmetic units can generate two addresses in a single cycle and perform such nonlinear address operations as displacements, indexing, and circular and bit-reverse addressing.

The chip possesses seven buses:

1. Two program buses: PADDR and PDATA.
2. Three data buses: DADDR1, DADDR2, and DDATA. DADDR1 and DADDR2 are 24 bits each; each can access memory at the same time. DDATA is 32 bits and carries data to the CPU over the two buses, providing the means for up to two data memory accesses every machine cycle.
3. DMA buses: DMAADDR and DMADATA. These buses perform memory accesses in parallel; memory accesses occur on the program and data buses.

In addition, there is a CPU register file. These may be used as general-purpose registers, but have special functions as well. There are some extended precision registers ideally suited for holding extended precision arithmetic; some support special addressing modes, and others are useful for memory and CPU management.

The C30 can address up to 16 million 32-bit words. The CPU has on-board RAM arranged in two blocks of 1K by 32-bit words. There is a ROM block on the C30 of 4K by 32-bit words. There is also a 64 by 32-bit cache. Fast on-board RAM, multiple buses, parallelism of architecture and the ability to pipeline allow this processor to perform many operations simultaneously.

Of course, there should be a fast instruction cycle and the ability to perform floating-point arithmetic in a single cycle—this can produce smaller, faster code blocks. Most of the standard floating-point utilities are built into hardware, such as addition, subtraction, multiplication, comparisons, load exponent, and load mantissa. Division is not supported on the chip in either integer or floating-point, presumably because of the time this process normally takes, but this can easily be implemented in integer format with subtraction and in floating point with inversion (Newton's method) followed by multiplication—see Chapter 5 for more information. Floating-point arithmetic comes at a cost, however—this processor is much more expensive than integer-only units.

Among the parallel instructions available are addition, subtraction, and multiplication with store or load of additional additions and subtractions.

The peripherals available on the C30 are two 32-bit timers and two bidirectional serial ports that can also be configured as timers, or whose pins can be used for general I/O.

The DMA (direct memory access) is a self-contained unit permitting reads and writes to external memory independently of the CPU. The unit contains its own address generators and registers, and can perfom block or single word transfers to or from memory. This can be very useful for getting data from sampling A/D's or to D/A's without interfering with the ongoing operations of the CPU in processing the data it already has.

There are two serial ports that may be used for interprocessor communications, as well as direct serial control and communication with serial A/D's and D/A's.

### TMS320C30 Code

The instruction set supports a rich set of conditionals, including jumps and branches with no-overhead looping. Memory locations may be pre- and post-incremented within the same instruction, and almost all the instructions execute in one cycle.

The syntax for TMS320C30 assembly language is much more traditional than that of the ADSP2101, though it definitely reflects enhancements derived from high-level languages:

```
*++SP   ;increment SP and use incremented SP as address
*SP--   ;use SP as address and decrement SP
x<<y    ;shift x to the left y bits
x>>y    ;shift x to the right y bits
```

Unlike many forms of assembly language, the TMS320C30 does not use algebraic notation. The format is

```
        Opcode      Source, Destination
```

The TMS320C30 is a 32-bit floating-point/integer processor with, among other things, a 32-bit ALU, 8 40-bit extended precision registers (R0-R7) for accumulation, a 32-bit block size register (BK) for circular addressing, and a 32-bit repeat counter (RC) for looping operation. It supports addressing for any register, three-operand addressing, parallel addressing, immediate addressing, and conditional branch addressing.

This processor has a highly parallel architecture. Some of the instructions can occur in pairs and will be executed in parallel. Instructions of this kind include parallel loading of registers, parallel arithmetic operations, and arithmetic/logical instructions in parallel with a store. Each instruction is entered as a separate source statement, and the second statement is preceded by two vertical bars: ||.

Circular addressing required for filters and transforms is controlled by the BK register and an auxiliary register that must point into the circular buffer. It is performed using modular arithmetic, adding or subtracting a step to or from the auxiliary register:

```
*AR0++(5)% ;AR0 = 0
*AR0++(2)% ;AR0 = 5
*AR0--(3)% ;AR0 = 1
*AR0++(6)% ;AR0 = 4
*AR0--%    ;AR0 = 4 -step back the same amount
*AR0       ;AR0 = 3
```

Note the use of the percent sign to represent a circular buffer operation.

An example of the pervasive sum-of-products operation in TMS320C30 assembly language might take the form:

```
MPYF3    *AR0++(1),*AR1++(1)%,R0 ; h(N-1-i) * x(n-(N-1-i)) -> R0
              ||  ADDF3   R0,R2,R2.
```

MPYF3 is a three-operand version of the floating-point muliplication. AR0 and AR1 are used as pointers to filter coefficients and input values; each is post-incremented with AR1 used in circular mode. The result is stored in R0. In this example, the multiplication is performed simultaneously with a three-operand addition of R0 and R2, and the result is deposited in R2.

When a parallel instruction is executed, there must be some predetermined order. In the sum-of-products above, we have a multiplication and addition—note that they both use the R0 register. Obviously, if we load the R0 register with the result of the multiplication before the addition, we will have a different result than if we performed the addition first. As with the ADSP2101, all registers are read at the beginning of the cycle and loaded at the end—this means that the addition above is using the contents of the R0 register before it is loaded from the multiplication.

Filtering, matrix, and vector arithmetic are often implemented with instructions such as the foregoing, but repeated many times. The TMS320C30 offers two repeat modes, block and single instruction. These are dependent upon three control registers: RS (repeat start address), RE (repeat end address), and RC (repeat counter). The first two control registers are set and controlled by the code, while the last is set by the user. A block repeat might be set up as follows:

```
LDI     @ADDR, AR0 ;load AR0 from ADDR
LDI     AR0, AR1   ;load AR0 into AR1
ADDI    63, AR1    ;add immediate 63 with AR1
LDI     31, RC     ;put 31 into the repeat counter
RPTB    EXCH       ;repeat the block of code between this
    .       ;location and the label EXCH.
    .
    .
EXCH    ...
```

The single instruction repeat is similar to the block repeat, except that it is not interruptible, and it has the advantage that it is retrieved only once from memory. The following example of an FIR filter demonstrates many of the features mentioned above, including the single instruction repeat mode.

```
;                   SUBROUTINE  F I R
;
; EQUATION: y(n) = h(0) * x(n) + h(1) * x(n-1) + ... + h(N-1) * x(n-(N-1))
;
; TYPICAL CALLING SEQUENCE:
;
;loadAR0
;loadAR1
;loadRC
;loadBK
;CALLFIR
;
;
; ARGUMENT ASSIGNMENTS:
;argument | function
;---------+-----------------------
;AR0   | address of h(N-1)
;AR1   | address of x(N-1)
;RC| length of filter - 2 (N-2)
;BK| length of filter (N)
;
; REGISTERS USED AS INPUT: AR0, AR1, RC, BK
; REGISTERS MODIFIED: R0, R2, AR0, AR1, RC
; REGISTER CONTAINING RESULT: R0
;
;
; PROGRAM SIZE: 6 words
;
; EXECUTION CYCLES: 11 + (N-1)
;
;==================================================================
;
       .global FIR
;                  ; initialize R0:
FIR        MPYF3   *AR0++(1),*AR1++(1)%,R0    ; h(N-1) * x(n-(N-1)) -> R0
       LDF 0.0,R2      ; initialize R2.
;
; filter ( 1 <= i < N)
;
       RPTS        RC  ; setup the repeat single.
       MPYF3       *AR0++(1),*AR1++(1)%,R0 ; h(N-1-i) * x(n-(N-1-i)) -> R0
||     ADDF3   R0,R2,R2
                   ; multiply and add operation
                   ;
       ADDFR0,R2,R0        ; add last product
                   ;
                   ; return sequence
                   ;
       RETS            ; return
;
; end
;
.end
```

The next example is of an IIR filter implemented as second-order cascaded systems or *biquads*.

```
; IIR2 == IIR FILTER (N > 1 BIQUADS)
;
;
; EQUATIONS: y(0,n) = x(n)
;         for (i = 0; i < N; i++)
;         {
;         d(i,n) = a2(i) * d(i,n-2) + a1(i) * d(i,n-1) + y(i-1,n)
;         y(i,n) = b2(i) * d(i,n-2) + b1(i) * d(i,n-1) + b0(i) * d(i,n)
;         }
;         y(n) = y(N-1,n)
;
; TYPICAL CALLING SEQUENCE:
;
;loadR2
;loadAR0
;loadAR1
;loadIR0
;loadIR1
;loadBK
;loadRC
;CALLIIR2
;
;
; ARGUMENT ASSIGNMENTS:
;argument | function
;---------+----------------------
;R2| input sample x(n)
;AR0   | address of filter coefficients (a2(0))
;AR1   | address of delay node values (d(0,n-2))
;BK| BK = 3
;IR0   | IR0 = 4
;IR1   | IR1 = 4*N-4
;RC| Number of biquads (N) - 2
;
; REGISTERS USED AS INPUT: R2, AR0, AR1, IR0, IR1, BK, RC
; REGISTERS MODIFIED: R0, R1, R2, AR0, AR1, RC
; REGISTER CONTAINING RESULT: R0
;
; PROGRAM SIZE: 17 words
;
; EXECUTION CYCLES: 23 + 6N
;
;=========================================================================
;
    .global    IIR2
;
IIR2    MPYF3      *AR0, *AR1, R0          ; a2(0) * d(0,n-2) -> R0
        MPYF3      *++AR0(1), *AR1--(1)%, R1; b2(0) * d(0,n-2) -> R1
;
```

```
        MPYF3      *++AR0(1), *AR1, R0    ; a1(0) * d(0,n-1) -> R0
||      ADDF3R0, R2, R2                   ; first sum term of d(0,n).
;
        MPYF3   *++AR0(1), *AR1--(1)%, R0 ; b1(0) * d(0,n-1) -> R0
||      ADDF3R0, R2, R2             ; second sum term of d(0,n).
;
        MPYF3*++AR0(1), R2, R2          ; b0(0) * d(0,n) -> R2
||      STF R2, *AR1--(1)%              ; store d(0,n); point to d(0,n-2).
;
;
        RPTB       LOOP                   ; loop for 1 <= i < N
;
        MPYF3*++AR0(1), *++AR1(IR0), R0   ; a2(i) * d(i,n-2) -> R0
||      ADDF3R0,R2,R2                     ; first sum term of y(i-1,n)
;
        MPYF3      *++AR0(1), *AR1--(1)%, R1; b2(i) * d(i,n-2) -> R1
||   ADDF3R1,R2,R2                  ; second sum term of y(i-1,n)
;
        MPYF3*++AR0(1), *AR1, R0        ; a1(i) * d(i,n-1) -> R0
||   ADDF3R0, R2, R2                    ; first sum term of d(i,n).
;
        MPYF3*++AR0(1), *AR1--(1)%, R0; b1(i) * d(i,n-1) -> R0
||   ADDF3R0, R2, R2                    ; second sum term of d(i,n).
;
        STF R2, *AR1--(1)%              ; store d(i,n); point to d(i,n-2).
LOOPMPYF3*++AR0(1), R2, R2              ; b0(i) * d(i,n) -> R2
;
; final summation
;
        ADDFR0,R2                       ; first sum term of y(N-1,n)
        ADDF3R1,R2,R0                   ; second sum term of y(N-1,n)
;
        NOP     *AR1--(IR1)             ; return to first biquad
        NOP     *AR1--(1)%              ; point to d(0,n-1)
;
; return sequence
;
        RETS                            ; return
;
; end
;
.end
```

Reprinted by Permission of Texas Instruments.

The final example here incorporates another important feature of the TMS320C30, bit reversal. The bit reversing operation is indicated with a B:

```
*AR1++(IR0)B
```

```
;
;GENERIC PROGRAM TO DO A RADIX-2 REAL FFT COMPUTATION IN 320C30.
;
; THE PROGRAM IS TAKEN FROM THE PAPER BY SORENSEN ET AL., JUNE 1987 ISSUE OF
; THE TRANSACTIONS ON ASSP.
; THE (REAL) DATA RESIDE IN INTERNAL MEMORY.  THE COMPUTATION
; IS DONE IN-PLACE.  THE BIT-REVERSAL IS DONE AT THE BEGINNING OF THE
; PROGRAM.
; THE TWIDDLE FACTORS ARE SUPPLIED IN A TABLE PUT IN A .DATA SECTION.  THIS
;DATA IS INCLUDED IN A SEPARATE FILE TO PRESERVE THE GENERIC NATURE OF
;THE PROGRAM.  FOR THE SAME PURPOSE, THE SIZE OF THE FFT N AND LOG2(N)
;ARE DEFINED IN A .GLOBL DIRECTIVE AND SPECIFIED DURING LINKING.
;THE LENGTH OF THE TABLE IS N/4 + N/4 = N/2.
;
;
    .GLOBL  FFT         ; ENTRY POINT FOR EXECUTION
        .GLOBL  N         ; FFT SIZE
        .GLOBL  M         ; LOG2(N)
        .GLOBL  SINE          ; ADDRESS OF SINE TABLE

    .BSS    INP,1024      ; MEMORY WITH INPUT DATA

        .TEXT

;INITIALIZE

      .WORD FFT     ; STARTING LOCATION OF THE PROGRAM

      .SPACE100     ; RESERVE 100 WORDS FOR VECTORS, ETC.

FFTSIZ    .WORD N
LOGFFT    .WORD M
SINTAB    .WORD SINE
INPUT     .WORD INP

FFT:      LDP   FFTSIZ  ; COMMAND TO LOAD DATA PAGE POINTER

; DO THE BIT-REVERSING AT THE BEGINNING

    LDI     @FFTSIZ,RC      ; RC=N
    SUBI    1,RC            ; RC SHOULD BE ONE LESS THAN DESIRED #
    LDI     @FFTSIZ,IR0
    LSH     -1,IR0          ; IR0=HALF THE SIZE OF FFT=N/2
    LDI     @INPUT,AR0
    LDI     @INPUT,AR1

    RPTB    BITRV
    CMPI    AR1,AR0         ; XCHANGE LOCATIONS ONLY
    BGE     CONT            ;IF AR0<AR1
    LDF     *AR0,R0
||  LDF     *AR1,R1
    STF     R0,*AR1
```

```
||  STF      R1,*AR0
CONT    NOP      *AR0++
BITRV   NOP      *AR1++(IR0)B

;LENGTH-TWO BUTTERFLIES

    LDI     @INPUT,AR0      ; AR0 POINTS TO X(I)
    LDI     IR0,RC          ; REPEAT N/2 TIMES
    SUBI    1,RC            ; RC SHOULD BE ONE LESS THAN DESIRED #

        RPTBBLK1
    ADDF    *+AR0,*AR0++,R0     ; R0=X(I)+X(I+1)
    SUBF    *AR0,*-AR0,R1       ; R1=X(I)-X(I+1)
BLK1    STF      R0,*-AR0       ; X(I)=X(I)+X(I+1)
||  STF     R1,*AR0++       ; X(I+1)=X(I)-X(I+1)

; FIRST PASS OF THE DO-20 LOOP (STAGE K=2 IN DO-10 LOOP)

    LDI     @INPUT,AR0  ; AR0 POINTS TO X(I)
    LDI     2,IR0       ; IR0=2=N2
    LDI     @FFTSIZ,RC
    LSH     -2,RC       ; REPEAT N/4 TIMES
    SUBI    1,RC        ; RC SHOULD BE ONE LESS THAN DESIRED #

    RPTB    BLK2
    ADDF    *+AR0(IR0),*AR0++(IR0),R0; R0=X(I)+X(I+2)
    SUBF    *AR0,*-AR0(IR0),R1       ; R1=X(I)-X(I+2)
    NEGF    *+AR0,R0          ; R0=-X(I+3)
||  STF     R0,*-AR0(IR0)        ; X(I)=X(I)+X(I+2)
BLK2    STF      R1,*AR0++(IR0)          ; X(I+2)=X(I)-X(I+2)
||  STF     R0,*+AR0          ; X(I+3)=-X(I+3)

; MAIN LOOP (FFT STAGES)

    LDI     @FFTSIZ,IR0
    LSH     -2,IR0    ; IR0=INDEX FOR E
    LDI     3,R5      ; R5 HOLDS THE CURRENT STAGE NUMBER
    LDI     1,R4      ; R4=N4
    LDI     2,R3      ; R3=N2
LOOP    LSH     -1,IR0    ; E=E/2
    LSH     1,R4      ; N4=2*N4
    LSH     1,R3      ; N2=2*N2

;INNER LOOP (DO-20 LOOP IN THE PROGRAM)

    LDI     @INPUT,AR5; AR5 POINTS TO X(I)
INLOP   LDI     IR0,AR0
    ADDI    @SINTAB,AR0; AR0 POINTS TO SIN/COS TABLE
    LDI     R4,IR1    ; IR1=N4

    LDI     AR5,AR1
    ADDI    1,AR1     ; AR1 POINTS TO X(I1)=X(I+J)
    LDI     AR1,AR3
    ADDI    R3,AR3    ; AR3 POINTS TO X(I3)=X(I+J+N2)
```

```
     LDI     AR3,AR2
     SUBI    2,AR2      ; AR2 POINTS TO X(I2)=X(I-J+N2)
     ADDI    R3,AR2,AR4; AR4 POINTS TO X(I4)=X(I-J+N1)

     LDF     *AR5++(IR1),R0      ; R0=X(I)
     ADDF    *+AR5(IR1),R0,R1; R1=X(I)+X(I+N2)
     SUBF    R0,*++AR5(IR1),R0  ; R0=-X(I)+X(I+N2)
||   STF     R1,*-AR5(IR1)  ; X(I)=X(I)+X(I+N2)
     NEGF    R0       ; R0=X(I)-X(I+N2)
     NEGF    *++AR5(IR1),R1      ; R1=-X(I+N4+N2)
||   STF     R0,*AR5      ; X(I+N2)=X(I)-X(I+N2)
     STF     R1,*AR5      ; X(I+N4+N2)=-X(I+N4+N2)


; INNERMOST LOOP

     LDI     @FFTSIZ,IR1
     LSH     -2,IR1           ; IR1=SEPARATION BETWEEN SIN/COS TBLS
     LDI     R4,RC
     SUBI    2,RC             ; REPEAT N4-1 TIMES

     RPTB    BLK3
     MPYF    *AR3,*+AR0(IR1),R0      ; R0=X(I3)*COS
     MPYF    *AR4,*AR0,R1        ; R1=X(I4)*SIN
     MPYF    *AR4,*+AR0(IR1),R1      ; R1=X(I4)*COS
||   ADDF    R0,R1,R2         ; R2=X(I3)*COS+X(I4)*SIN
     MPYF    *AR3,*AR0++(IR0),R0    ; R0=X(I3)*SIN
     SUBF    R0,R1,R0         ; R0=-X(I3)*SIN+X(I4)*COS !!!
     SUBF    *AR2,R0,R1       ; R1=-X(I2)+R0 !!!
     ADDF    *AR2,R0,R1       ; R1=X(I2)+R0 !!!
||   STF     R1,*AR3++        ; X(I3)=-X(I2)+R0 !!!
     ADDF    *AR1,R2,R1       ; R1=X(I1)+R2
||   STF     R1,*AR4--        ; X(I4)=X(I2)+R0 !!!
     SUBF    R2,*AR1,R1       ; R1=X(I1)-R2
||   STF     R1,*AR1++        ; X(I1)=X(I1)+R2
BLK3    STF     R1,*AR2--        ; X(I2)=X(I1)-R2

     SUBI    @INPUT,AR5
     ADDI    R3,AR5           ; AR5=I+N1
     CMPI    @FFTSIZ,AR5
     BLED    INLOP            ; LOOP BACK TO THE INNER LOOP
     ADDI    @INPUT,AR5
     NOP
     NOP

     ADDI    1,R5
     CMPI    @LOGFFT,R5
     BLE     LOOP

END        BR    END           ;BRANCH TO ITSELF AT THE END
        .END
```

Reprinted by Permission of Texas Instruments.

## Analog Devices ADSP-2101

Overview:

- Very fast instruction cycle
- Single cycle access to both data and program memory
- Parallel architecture
- Timer
- Two serial ports
- Zero overhead looping and single cycle branching
- Two independent data buses
- Two data address generators providing support for circular buffers and bit reversal
- 16×32 bit barrel shifter with support for floating-point implementations
- Multiplier-accumulator
- Arithmetic-logic unit (ALU)

The ADSP-2101 is a 16-bit fixed point unit. It contains an ALU capable of a full set of arithmetic and logical operations. There is a multiplication/accumulation unit that performs single-cycle multiplication alone or with addition and subtraction, and does so in extended precision arithmetic. Also included is a barrel shifter that performs a wide array of arithmetic and logical operations. The CPU has two address generators and five buses: the program memory address bus, the program memory data bus, the data memory address bus, the data memory data bus, and the result bus.

The ALU performs a standard set of arithmetic and logical instructions including addition, subtraction, negations increment, decrement, and absolute value. In addition, there are two division primitives that allow multiple cycle division in which each quotient bit is computed separately. The logic functions supported by the AD2101 are AND, OR, XOR, and NOT. The ALU is 16 bits wide.

The multiplier-accumulator (MAC) is capable of high-speed multiplication, multiplication with summation, multiplication with cumulative subtraction, saturation, and clear to zero. The inputs to the MAC are each 16 bits wide. The 32-bit product goes to a 40-bit adder/subtracter, which can add to, subtract from, or pass directly to the multiplier result register.

Logical and arithmetic shifts, normalization, denormalization, exponent derivation and block floating-point operations are performed by the barrel shifter. The ouput of the barrel shifter is 32 bits.

Two data address generators can access memory simultaneously; each data memory address generator can maintain up to four address pointers. One of the address generators can access both data and program memory. Both address generators support indirect addressing capabilities and modulo arithmetic for circular buffers. One of the generators can do bit reversal.

The program sequencer is capable of overhead free looping and single cycle branching. A single level of pipelining provides that instructions can be fetched and loaded during one processor cycle, then executed in the following cycle. The sequencer also allows single cycle conditional jumps, subroutine calls, and returns.

Of the five internal buses, four leave the chip. The program and data memory address buses are each 14 bits wide, allowing for 16k of code and data, the program data bus is 24 bits wide and the data memory data bus is 16 bits wide.

The ADSP-2101 also contains one 16-bit timer for interrupt generation that is self loading and equipped with a pre-scaler. There are also two complete serial ports on-board for interfacing directly with other DSPs, A/Ds, D/As, and most codecs.

### ADSP-2101 Code

ADSP2101 assembly language is easy to read. Whole words and expressions are used instead of the cryptic abbreviations so common in the field. Actually, it appears very much like a high-level language using algebraic syntax for arithmetic operations and data moves. The sources and destinations are spelled out explicitly, so the meaning of each line is much clearer than one might expect. Of course, there is no penalty for this—it is simply the coding mechanism of the assembly language itself. Each program statement assembles into a 24-bit instruction and executes in one cycle.

There is a very high degree of parallelism in the structure of the processor. All of the following combinations can be executed in one cycle:

- Any ALU, MAC, or barrel shifter operation
- Any register-to-register move
- Any data memory read or write
- Computations with any data register-to-data register move
- Computations with any memory read or write
- Computations with a read from two memories

Most all these instructions can be executed conditionally, and may be combined with data transfers in a single cycle.

Probably the most basic tool of signal processing is the *sum of products*, which involves a multiply-and-accumulate pair. On the ADSP2101, this can be implemented as follows:

```
MR = MR+MX0*MY0(SS)
```

MR is the result register of the multiplier/accumulator, which in this case receives the sum of its current contents and the product of two input registers. The parenthetical expression is a modifier dictating how the operands are to be treated. RND indicates that the signed result is to be rounded, SS means that both are signed, and UU means that both are unsigned. Then there are combinations of the the last two indicating that one or the other is signed or unsigned: SU or US.

To make this more efficient, it is possible to combine this sum of products instruction with *two* memory fetches:

```
MR=MR+MX0*MY0(SS), MX0=DM(I0,M1), MY0=PM(I4,M5);
```

The ADSP2101 reads registers at the beginning of the cycle and writes at the end, so the statement above means that MR, MX0, and MY0 would be read at the beginning of the cycle, and the results calculated and written at the end. The new operands fetched at the end of the cycle then overwrite the previous values in each of these registers.

Most of the registers are 16 bits. Some of these registers are concatenated, such as MR0, MR1, and MR2, as well as SR0 and SR1. The MR register, the multiplier result register, is an extended register for arithmetic. It comprises MR0, which is 16 bits, MR1 (16 bits), and the most significant bits MR2 (8 bits). SR0 and SR1 are each 16 bits. After a sum-of-products routine, a value can easily be truncated by simply loading the ouput register with MR1, the middle 16 bits of the result register.

It is appropriate to introduce another important feature of the ADSP2101 at this point: the *circular buffer*. With proper declarations, such as

```
.VAR/DM/CIRC coefficients[128];
```

The linker is instructed to place a circular buffer at the proper address boundaries. In addition, the L register must be initialized with the length of the buffer, the address of the buffer must be passed to an index register, such as I0, and a value set for the stride of the increment, M0 = 1. Now a statement:

```
MX0 = DM(I0,M0);
```

will load MX0 from the address pointed to by I0 (the circular buffer), and then I0 will be incremented by the value in M0, 1. In a loop, this instruction will cause the pointer to step through the buffer and wrap automatically until the loop ends. You will see examples of these mechanisms in the following code fragments.

The first example is an FIR filter. Initially, MR is cleared and MX0 and MY0 are loaded with the first data and coefficient values.

The sequence

```
MR=MR+MX0*MY0(SS), MX0=DM(I0,M1), MY0=PM(I4,M5);
```

is executed N−1 times, until CE, which means *count expired*. This sum of products is done in extended precision, 40 bits, using simultaneous program and data memory accesses to retrieve the filter coefficents [MY0=PM(I4,M5)] and x[n] [MX0=DM(I0,M1)]. This leaves the final multiplication and accumulation, which is performed with rounding enabled. Depending upon the status of MV, the overflow flag, MR is saturated at a positive or negative limit. I0 and I4 are implemented as circular buffers.

In the following example, you will see from the header that the computation time is N−1+5+2 cycles. With an instruction cycle of 125 nanoseconds, a 64-tap FIR filter would execute in 8.75 microseconds.

```
.MODULE fir_sub;
{
        FIR Transversal Filter Subroutine

        Calling Parameters
        I0 --> Oldest input data value in delay line
        L0 = Filter length (N)
        I4 --> Beginning of filter coefficient table
        L4 = Filter length (N)
        M1,M5 = 1
        CNTR = Filter length - 1 (N-1)

        Return Values
        MR1 = Sum of products (rounded and saturated)
        I0 --> Oldest input data value in delay line
        I4 --> Beginning of filter coefficient table

        Altered Registers
        MX0,MY0,MR

        Computation Time
        N - 1 + 5 + 2 cycles

        All coefficients and data values are assumed to be in 1.15 format.
}

.ENTRY  fir;

fir:MR=0, MX0=DM(I0,M1), MY0=PM(I4,M5);
        DO sop UNTIL CE;
sop:            MR=MR+MX0*MY0(SS), MX0=DM(I0,M1), MY0=PM(I4,M5);
        MR=MR+MX0*MY0(RND);
        IF MV SAT MR;
        RTS;
.ENDMOD;
```



The next example is a direct form II IIR filter, implementing the difference equation expressed as

$$y[n] = \sum_{k=0}^{M} b_k x[n-k] + \sum_{k=1}^{N} a_k y[n-k]$$

The zeros are represented by the $b_k$ coefficients and the poles by the $a_k$ coefficients. First, the poles are computed using the same technique as in the previous FIR example—that is, simultaneous fetches from program and data memory and a multiplication and accumulation. In a sense, all we have done is to concatenate two FIR filter subroutines, the one that computes the poles and the one that computes the zeros.

A new instruction appears in this routine: the MODIFY instruction. This instruction causes the one register to be added to another, then handled with the modulus arithmetic associated with the circular buffers. This causes the buffer address to wrap if exceeds L.

```
.MODULE diriir_sub;

{
        Direct Form IIR Filter Subroutine

        Calling Parameters
        MR1 = Input sample ( x[n] )
        MR0 = 0
        I0 --> Delay line buffer current location ( x[n-1] )
        L0 = Filter length
        I5 --> Feedback coefficients (a's)
        L5 = Filter length - 1
        I6 --> Feedforward coefficients (b's)
        L6 = Filter length
        M0 = 0
        M1,M4 = 1
        M2 = 2
        CNTR = Filter length - 2
        AX0 = Filter length - 1

        Return Values
        MR1 = output sample ( y[n] )
        I0 --> delay line current location ( x[n-1] )
        I5 --> feedback coefficients
        I6 --> feedforward coefficients

        Altered Registers
        MX0,MY0,MR

        Computation Time
        (N - 2) + (N - 1) + 10 + 4 cycles  (N = M = Filter order)

        All coefficients and data values are assumed to be in 1.15
format.
}
.ENTRY  diriir;

diriir: MX0=DM(I0,M1), MY0=PM(I5,M4);
        DO poleloop UNTIL CE;
poleloop:        MR=MR+MX0*MY0(SS), MX0=DM(I0,M1), MY0=PM(I5,M4);
        MR=MR+MX0*MY0(RND);
        CNTR=AX0;
        DM(I0,M0)=MR1;
        MR=0, MX0=DM(I0,M1), MY0=PM(I6,M4);
        DO zeroloop UNTIL CE;
zeroloop:        MR=MR+MX0*MY0(SS), MX0=DM(I0,M1), MY0=PM(I6,M4);
        MR=MR+MX0*MY0(RND);
    MODIFY (I0,M2);
        RTS;
.ENDMOD;
```

The last example is a radix 4 decimation in frequency FFT routine. As with most fast routines for calculation of an FFT on a DSP, the twiddle factors are tabularized. The ADSP2101 defaults to bit reversal involving 14 bits, though it can operate on a smaller number if you desire. In this example, we are bit-reversing 10 bits, requiring registers I0, I1, I2, and I3 to be initialized to H#0000, H#2000, H#0008, and H#2008 and setting the M0 register to H#0010.

```
.MODULE/RAM/boot=0/ABS=0       main;
.CONST  N=1024;          {number of samples in FFT}
.CONST  Nx2=2048;
.VAR/DM     inplace[Nx2];       {inplace array contains original input}
                   {and also hold intermediate results}
.INIT       inplace:<input.dat>;  {load inplace array  with data}
.GLOBAL     inplace, output;
.EXTERNAL   fft;              {FFT routine}



        SI=0;       {these 3 lines reset the system }
        DM(H#3fff)=SI;          {control register and the data }
        DM(H#3FFE)=SI;          {memory control register to allow}
                        {zero state memory access}
        CALL fft;

TRAPPER:    JUMP TRAPPER;
.ENDMOD;

{_____________________________________________________________________
        FFT4_1024.DSP

        Optimized complex 1024-Point Radix-4 DIF FFT
        This routine uses a modified radix-4 algorithm to unscramble
results as they are computed. The results are thus in sequential order

        Complex data is stored as x0 (real, imag),y0 (real, imag), x1
(real, imag), y1 (real,imag),...

        Butterfly trems:
        xa = 1st real input leg        x'a = 1st real output leg
        xb = 2nd real input leg        x'b = 2nd real output leg
        xc = 3rd real input leg        x'c = 3rd real output leg
        xd = 4th real input leg        x'd = 4th real output leg
        ya = 1st imag input leg        y'a = 1st imag output leg
        yb = 2nd imag input leg        y'b = 2nd imag output leg
        yc = 3rd imag input leg        y'c = 3rd imag output leg
        yd = 4th imag input leg        y'd = 4th imag output leg
```

```
        pointers
        I0 --> xa,xc
        I1 --> xb,xd
        I2 --> ya,yc
        I3 --> yb,yd
        w0 (= Ca = Sa = 0)
        I5 --> w1 (1st Cb, - pi/4 for Sb)
        I6 --> w2 (2nd Cc, - pi/4 for Sc)
        I7 --> w3 (3rd Cd, - pi/4 for Sd)

        Input:
        Inplace[2*N] normal order, interleaved real,imag.

        Output:
        Inplace[2*N] digit reversed, interleaved real,imag.

        Computation Time (cycles)
        setup  = 9
        stage 1= 7700 = 20+256(30)
        stage 2= 7758 = 18+4(15+64(30))
        stage 3= 7938 = 18+16(15+16(30))
        stage 4= 8658 = 18+64(15+4(30))
        stage 5= 5140 = 20+256(20)

        Total 37203 cycles * 80ns/cycle = 2.97624ms

_____________________________________________________________________}

.CONST      N=1024;
.CONST      Nx2=2048;
.CONST      Nov2=512;
.CONST      Nov4=256;
.CONST      Nov8=384;
.VAR/DM/CIRC        cos_table[1024];
.VAR/DM         m3_space;   {memory space used to store M3 values}
                    {when M3 is loaded with its alternate value}
.VAR/DM         bfy_count;
.INIT           cos_table:<cos1024.dat>;      {N cosine values}
.EXTERNAL       inplace;
.ENTRY      fft;

fft: M4=-Nov4;          {-N/4 = -90 degrees for sine}
 L0=0;
 L1=0;
 L2=0;
 L3=0;
 L5=%cos_table;
 L6=%cos_table;
 L7=%cos_table;
 SE=0;
```

```
{________________________________Stage 1________________________________}


stage1:        I0=^inplace;        {in->Xa,Xc}
        I1=^inplace+Nov2;       {in+N/2->Xb,Xd}
        I2=^inplace+1;          {in+1 ->Ya,Yc}
        I3=^inplace+Nov2+1;          {in+N/2+1 ->Yb,Yd}
        I5=^cos_table;
        I6=^cos_table;
        I7=^cos_table;

        M0=N;         {N,skip forward to dual node}
        M1=-N;       {-N,skip back to primary node}
        M2=-N+2;          {-N+2, skip to next butterfly}
        M3=-2;      {because we have modified the middle branches}
                {of bfly, pointers for I0 require more}
                {complex manipulation, using M3}
        M5=Nov4+1;      {N/4 + groups/stage*1, Cb Sb offset}
        M6=Nov4+2;      {N/4 + groups/stage*2, Cc Sc offset}
        M7=Nov4+3;      {N/4 + groups/stage*3, Cd Sc offset}
        AX0=DM(I0,M0);                 {get first Xa}
        AY0=DM(I0,M1);                 {get first Xc}
        AR=AX0-AY0, AX1=DM(I2,M0);          {Xa-Xc,get first Ya}
        SR=LSHIFT AR(LO), AY1=DM(I2,M1);  {SR1=Xa-Xc,get first Yc}
        CNTR=Nov4;                  {Bfly/group, stage one}
{Middle 2 branches of butterfly are reversed.}
{This alteration, done in every stage, results in bit-reversed}
{outputs instead of digit-reversed outputs.}
        DO stg1bfy  UNTIL CE;
         AR=AX0+AY0, AX0=DM(I1,M0);
             {AR=xa+xc, AX0=xb }
         MR0=AR, AR=AX1+AY1;
             {MR0=xa+xc, AR=ya+yc}
         MR1=AR, AR=AX1-AY1;
             {MR1=ya+yc, AR=ya-yc}
         SR=SR OR LSHIFT AR (HI), AY0=DM(I1,M1);
                     {SR1=ya-yc, AY0=xd}
         AF=AX0+AY0, AX1=DM(I3,M0);
             {AF=xb+xd, AX1=yb}
         AR=MR0+AF, AY1=DM(I3,M1);
                 {AR=xa+xb+xc+xd, AY1=yd}
         DM(I0,M0)=AR, AR=MR0-AF;
             {output x'a=(xa+xb+xc+xd), AR=xa+xc-xb-xd}
         AF=AX1+AY1, MX0=AR;
             {AF=yb+yd, MX0=xa+xc-xb-xd}
         AR=MR1+AF, MY0=DM(I6,M4);
             {AR=ya+yc+yb+yd, MY0=(Cc)}
         DM(I2,M0)=AR, AR=MR1-AF;
             {output y'a, AR=ya+yc-yb-yd}
         MR=MX0*MY0(SS), MY1=DM(I6,M6);
                 {MR=(xa+xc-xb-xd)(Cc), MY1=(Sc)}
```

```
          MR=MR+AR*MY1(RND);
                {MR=(xa-xb+xc-xd)(Cc)+(ya-yb+yc-yd)(Sc)}
          DM(I0,M2)=MR1, MR=AR*MY0(SS);
                {output x'c=xa-xb+xc-xd)(Cc)+(ya-yb+yc-yd)(Sc)}
                {MR=(ya+yc-yb-yd)(Cc)}
          MR=MR-MX0*MY1(RND), MY0=DM(I5,M4);
                {MR=(ya+yc-yb-yd)(Cc)-(xa+xc-xb-xd)(Sc), MY0=(Cb)}
          DM(I2,M2)=MR1, AR=AX0-AY0;
                {output y'c=(ya+yc-yb-yd)(Cc)-(xa+xc-xb-xd)(Sc)}
                {AR=xb-xd}
          AY0=AR, AF=AX1-AY1;
                {AY0=xb-xd, AF=yb-yd}
          AR=SR0-AF, MY1=DM(I5,M5);
                {AR=xa-xc-(yb-yd), MY1=(Sb)}
          MX0=AR, AR=SR0+AF;
                {MX0=xa-xc-yb+yd, AR=xa-xc+yb-yd}
          SR0=AR, AR=SR1+AY0;
                {SR0=xa-xc+yb-yd, AR=ya-yc+xb-xd}
          MX1=AR, AR=SR1-AY0;
                {MX1=ya-yc+xb-xd, AR=ya-yc-(xb-xd)}
          MR=SR0*MY0(SS), AX0=DM(I0,M0);
                {MR=(xa-xc+yb-yd)(Cb), AX0= xa of next bfly}
          MR=MR+AR*MY1(RND), AY0=DM(I0,M1);
                {MR=(xa-xc+yb-yd)(Cb)+ (ya-yc-xb+xd)(Sb)}
                {AY0= xc of next bfly}
          DM(I1,M0)=MR1, MR=AR*MY0(SS);
                {ouput x'b=(xa-xc+yb-yd)(Cb)+ (ya-yc-xb+xd)(Sb)}
                {MR=ya-yc-xb+xd)(Cb)}
          MR=MR-SR0*MY1(RND), MY0=DM(I7,M4);
                {MR=(ya-yc-xb+xd)(Cb)-(xa-xc+yb-yd)(Sb)}
                {MY0=(Cd)}
          DM(I3,M0)=MR1, MR=MX0*MY0(SS);
                {ouput y'b= (ya-yc-xb+xd)(Cb)-(xa-xc+yb-yd)(Sb)}
                {MR=(xa-yb-xc+yd)(Cd)}
          MY1=DM(I7,M7), AR=AX0-AY0;
                 {MY1=(Sd), AR=xa-xc }
          MR=MR+MX1*MY1(RND), AX1=DM(I2,M0);
                {MR=(xa-yb-xc+yd)(Cd)+(ya+xb-yc-xd)(sd)}
                {AX1=ya of next bfly}
          DM(I1,M2)=MR1, MR=MX1*MY0(SS);
                {output x'd=(xa-yb-xc+yd)(Cd)+(ya+xb-yc-xd)(sd)}
                {MR=(ya+yb-yc-yd)(Cd)}
          MR=MR-MX0*MY1(RND), AY1=DM(I2,M1);
                {MR=(ya+yb-yc-yd)(Cd)-(xa-xc-yb+yd)Sd}
                {yc of next bfly}
stg1bfy:        DM(I3,M2)=MR1, SR=LSHIFT AR(LO);
                {output y'd=(ya+xb-yc-xd)Cd-(xa-xc-yb+yd)Sd}
                {SR0=ya-yc of next bfly}
```

```
{______________________________Stage 2______________________________}

stage2:     I0=^inplace;        {in-> Xa,Xc}
        I1=^inplace+128;    {in+N/8-> Xb,Xd}
        I2=^inplace+1;          {in+1 -> Ya,Yc}
        I3=^inplace+129;    {in+N/8+1 -> Yb,Yd}
        M0=Nov4;            {N/4,skip forward to dual node}
        M1=-Nov4;          {-N/4,skip back to primary node}
        M2=-Nov4+2;             {-N/4+2, skip to next butterfly}
        M3=384;            {N*3/8,  skip to next group}
        M5=Nov4+4;          {N/4 +groups/stage*1, Cb Sb offset}
        M6=Nov4+8;          {N/4 +groups/stage*2, Cc Sc offset}
        M7=Nov4+12;             {N/4 +groups/stage*3, Cd Sd offset}
        SI=64;      {Bfy/group, save counter for inner loop}
        DM(bfy_count)=SI;
        CNTR=4;            {groups/stage}
        CALL mid_stg;       {do stage 2}

{______________________________Stage 3______________________________}

stage3: I0=     ^inplace;       {in          -> Xa,Xc}
          I1=^inplace+32;  {in+N/32-> Xb,Xd}
          I2=^inplace+1;        {in+1-> Ya,Yc}
          I3=^inplace+33;  {in+N/32+1   -> Yb,Yd}
          M0=64;          {N/16,skip forward to dual node}
          M1=-64;         {-N/16,skip back to primary node}
          M2=-62;         {-N/16+2, skip to next butterfly}
          M3=96;           {N*3/32,  skip to next group}
        DM(m3_space)=M3;    {M3_space is temporary storage}
                    {space needed because M3 is used}
                    {in 2 contexts and will alternate}
                    {in value}
          M5=Nov4+16;         {N/4 +groups/stage*1, Cb Sb offset}
          M6=Nov4+32;         {N/4 +groups/stage*2, Cc Sc offset}
          M7=Nov4+48;         {N/4 +groups/stage*3, Cd Sd offset}
          SI=16;             {Bfy/group, save counter for inner loop}
          DM(bfy_count)=SI;
          CNTR=16;         {groups/stage}
          CALL mid_stg;         {do stage 3}

{______________________________Stage 4______________________________}

stage4:         I0=^inplace;{in        -> Xa,Xc}
          I1=^inplace+8; {in+N/128-> Xb,Xd}
          I2=^inplace+1; {in+1         -> Ya,Yc}
          I3=^inplace+9; {in+N/128+1   -> Yb,Yd}
          M0=16;         {N/64,skip forward to dual node}
          M1=-16;          {-N/64,skip back to primary node}
          M2=-14;          {-N/64+2, skip to next butterfly}
          M3=24;         {N*3/128, skip to next group}
        DM(m3_space)=M3;    {M3_space is temprary storage}
                    {space needed because M3 is used}
```

```
                {in 2 contexts and will alternate}
                {in value}
            M5=Nov4+64;      {N/4 +groups/stage*1, Cb Sb offset}
            M6=Nov4+128;{N/4 +groups/stage*2, Cc Sc offset}
            M7=Nov4+192;{N/4 +groups/stage*3, Cd Sd offset}
            SI=4;          {Bfly/group, save counter inner loop}
            DM(bfy_count)=SI;
            CNTR=64;       {groups/stage}
            CALL mid_stg;  {do stage 4}


{_________________Last Stage__No Multiplies___________________________}


laststage: I4=^inplace;         {in ->Xa, Xc}
        I5=^inplace+2;          {in+N/512 ->Xb, Xd}
        I6=^inplace+1;          {in+1 ->Ya, Yc}
        I7=^inplace+3;          {in+N/512+1 ->Yb, Yd}
        M4=4;          {N/256, skip forward to dual node}
        M0=H#0010;     {This imodify value is used to perform bit-}
                {reverse as the final results are written}
                {out. The derivation of this value}
                {is explained in the text.}
        I0=H#0000;     {These base address values are derived}
        I2=H#20000;        {for output at address 0000}
        I1=H#0008;
        I3=H#2008;

        L4=0;          {This last stage has no twiddle factor}
        L5=0;          {multiplication}
        L6=0;          {Because the output addresses are bit-}
        L7=0;          {reversed, the I's M's & L's are reassigned}
                {and reinitialized}

          AX0=DM(I4,M4);   {first Xa}
          AY0=DM(I4,M4);   {first Xc}
          CNTR=Nov4;           {groups/stage}
        ENA BIT_REV;       {all data accesses using I0..I3 are}
                {bit_reversed}

{Middle 2 branches of butterfly are reversed.}
{This alteration, done in every stage, results in bit-reversed}
{outputs instead of digit-reversed outputs.}

        DO laststgbfy UNTIL CE;
         AR=AX0-AY0, AX1=DM(I6,M4);        {AR=xa-xc, AX1=ya}
         SR=LSHIFT AR(LO), AY1=DM(I6,M4);     {SR0=xa-xc, AY1=yc}
         AR=AX0+AY0, AX0=DM(I5,M4);    {AR=xa+xc, AX0=xb}
         MR0=AR, AR=AX1+AY1;        {MR0=xa+xc, AR=ya+yc}
         MR1=AR, AR=AX1-AY1;        {MR1=ya+yc, AR=ya-yc}
```

```
        SR=SR OR LSHIFT AR (HI), AY0=DM(I5,M4);
                         {SR1=ya-yc, AY0 xd}
        AF=AX0+AY0, AX1=DM(I7,M4);{AF=xb+xd, AX1=yb}
        AR=MR0+AF, AY1=DM(I7,M4);         {AR=xa+xc+xb+xd, AY1=yd}
        DM(I0,M0)=AR, AR=MR0-AF;       {output x'a=xa+xc+xb+xd}
                         {AR=xa+xc-(xb+xd)}
        DM(I1,M0)=AR, AF=AX1+AY1;         {output x'c=xa+xc-(xb+xd)}
                         {AF=yb+yd}
        AR=MR1+AF;         {AR=ya+yb+yc+yd}
        DM(I2,M0)=AR, AR=MR1-AF;       {output y'a=ya+yc+yb+yd}
                         {AR=ya+yc-(yb+yd)}
        DM(I3,M0)=AR, AR=AX0-AY0;         {output y'c=ya+yc-(yb+yd)}
                         {AR=xb-xd}
        AX0=DM(I4,M4);          {AX0=xa of next group}
        AF=AX1-AY1, AY1=AR;          {AF=yb-yd, AY1=xb-xd}
        AR=SR0+AF, AY0=DM(I4,M4);         {AR=xa-xc+yb-yd}
                         {AY0=xc of next group}
        DM(I1,M0)=AR, AR=SR0-AF;       {output x'b=xa-xc+yb-yd}
                         {AR=xa-xc-(yb-yd)}
        DM(I1,M0)=AR, AR=SR1-AY1;         {output x'd=xa-xc-(yb-yd)}
                         {AR=ya-yc+(xb-xd)}
        DM(I2,M0)=AR, AR=SR1+AY1;         {output y'b=ya-yc+(xb-xd)}
                         {AR=ya-yc-(xb-xd)}
laststgbfy:  DM(I3,M0)=AR;          {output y'd=ya-yc-(xb-xd)}
       DIS BIT_REV;        {shut-off bit reverse mode}
        RTS;       {end and exit from FFT subroutine}


{_______________________Subroutine for middle stages____________________}


mid_stg:   DO midgrp UNTIL CE;
        I5=^cos_table;
        I6=^cos_table;
        I7=^cos_table;
        AX0=DM(I0,M0);                 {get first Xa}
        AY0=DM(I0,M1);                 {get first Xc}
        AR=AX0-AY0, AX1=DM(I2,M0);          {Xa-Xc,get first Ya}
        SR=LSHIFT AR (LO), AY1=DM(I2,M1); {SR1=Xa-Xc,get first Yc}
        CNTR=DM(bfy_count);        {butterflies/group}
       M3=-2;                  {M3 is loaded with the value}
                          {required for pointer manipulation}

{Middle 2 branches of butterfly are reversed.}
{This alteration, done in every stage, results in bit-reversed}
{outputs instead of digit-reversed outputs.}

        DO midbfy UNTIL CE;
        AR=AX0+AY0, AX0=DM(I1,M0);
              {AR=xa+xc, AX0=xb }
        MR0=AR, AR=AX1+AY1;
```

```
        {MR0=xa+xc, AR=ya+yc}
MR1=AR, AR=AX1-AY1;
        {MR1=ya+yc, AR=ya-yc}
SR=SR OR LSHIFT AR (HI), AY0=DM(I1,M1);
        {SR1=ya-yc, AY0=xd}
AF=AX0+AY0, AX1=DM(I3,M0);
        {AF=xb+xd, AX1=yb}
AR=MR0+AF, AY1=DM(I3,M1);
        {AR=xa+xb+xc+xd, AY1=yd}
DM(I0,M0)=AR, AR=MR0-AF;
        {output x'a=(xa+xb+xc+xd), AR=xa+xc-xb-xd}
AF=AX1+AY1, MX0=AR;
        {AF=yb+yd, MX0=xa+xc-xb-xd}
AR=MR1+AF, MY0=DM(I6,M4);
        {AR=ya+yc+yb+yd, MY0=(Cc)}
DM(I2,M0)=AR, AR=MR1-AF;
        {output y'a, AR=ya+yc-yb-yd}
MR=MX0*MY0(SS), MY1=DM(I6,M6);
        {MR=(xa+xc-xb-xd)(Cc), MY1=(Sc)}
MR=MR+AR*MY1(RND), SI=DM(I0,M2);
        {MR=(xa-xb+xc-xd)(Cc)+(ya-yb+yc-yd)(Sc)}
      {SI is a dummy to cause a modify(I0,M2)}
DM(I0,M2)=MR1, MR=AR*MY0(SS);
        {output x'c=xa-xb+xc-xd)(Cc)+(ya-yb+yc-yd)(Sc)}
        {MR=(ya+yc-yb-yd)(Cc)}
MR=MR-MX0*MY1(RND), MY0=DM(I5,M4);
        {MR=(ya+yc-yb-yd)(Cc)-(xa+xc-xb-xd)(Sc), MY0=(Cb)}
DM(I2,M2)=MR1, AR=AX0-AY0;
        {output y'c=(ya+yc-yb-yd)(Cc)-(xa+xc-xb-xd)(Sc)}
        {AR=xb-xd}
AY0=AR, AF=AX1-AY1;
        {AY0=xb-xd, AF=yb-yd}
AR=SR0-AF, MY1=DM(I5,M5);
        {AR=xa-xc-(yb-yd), MY1=(Sb)}
MX0=AR, AR=SR0+AF;
        {MX0=xa-xc-yb+yd, AR=xa-xc+yb-yd}
SR0=AR, AR=SR1+AY0;
        {SR0=xa-xc+yb-yd, AR=ya-yc+xb-xd}
MX1=AR, AR=SR1-AY0;
        {MX1=ya-yc+xb-xd, AR=ya-yc-(xb-xd)}
MR=SR0*MY0(SS), AX0=DM(I0,M0);
        {MR=(xa-xc+yb-yd)(Cb), AX0= xa of next bfly}
MR=MR+AR*MY1(RND), AY0=DM(I0,M3);
        {MR=(xa-xc+yb-yd)(Cb)+ (ya-yc-xb+xd)(Sb)}
        {AY0= xc of next bfly}
DM(I0,M2)=MR1, MR=AR*MY0(SS);
        {ouput x'b=(xa-xc+yb-yd)(Cb)+ (ya-yc-xb+xd)(Sb)}
        {MR=ya-yc-xb+xd)(Cb)}
MR=MR-SR0*MY1(RND), MY0=DM(I7,M4);
        {MR=(ya-yc-xb+xd)(Cb)-(xa-xc+yb-yd)(Sb)}
        {MY0=(Cd)}
DM(I2,M2)=MR1, MR=MX0*MY0(SS);
```

```
                    {ouput y'b= (ya-yc-xb+xd)(Cb)-(xa-xc+yb-yd)(Sb)}
                    {MR=(xa-yb-xc+yd)(Cd)}
        MY1=DM(I7,M7), AR=AX0-AY0;
                  {MY1=(Sd), AR=xa-xc }
        MR=MR+MX1*MY1(RND), AX1=DM(I2,M0);
                    {MR=(xa-yb-xc+yd)(Cd)+(ya+xb-yc-xd)(sd)}
                    {AX1=ya of next bfly}
        DM(I1,M2)=MR1, MR=MX1*MY0(SS);
                    {output x'd=(xa-yb-xc+yd)(Cd)+(ya+xb-yc-xd)(sd)}
                    {MR=(ya+yb-yc-yd)(Cd)}
        MR=MR-MX0*MY1(RND), AY1=DM(I2,M1);
                    {MR=(ya+yb-yc-yd)(Cd)-(xa-xc-yb+yd)Sd}
                    {yc of next bfly}
midbfy:           DM(I3,M2)=MR1, SR=LSHIFT AR(LO);
                    {output y'd=(ya+xb-yc-xd)Cd-(xa-xc-yb+yd)Sd}
                    {SR0=ya-yc of next bfly}
       M3=DM(m3_space);
                  {modifier M3 is loaded with skip to}
                  {next group_count and is used in the}
                  {next four instructions}
        MODIFY (I0,M3);
        MODIFY (I1,M3);        { point to next group }
        MODIFY (I2,M3);        { of butterflies }
midgrp:           MODIFY (I3,M3);
       RTS;           { return to middle stage calling code}
.endmod;
```

Reprinted by Permission of Analog Devices.

# 7

# Two Final DSP Applications

## INTRODUCTION

DSP applications are virtually unlimited, so any attempt to offer a detailed coverage of this subject can only be inadequate. Much of the reason lies in the fact that DSP is not simply a new device but the increasing ability to perform analog functions in the digital domain, often with greater efficiency and reliability and at less cost. Many traditionally analog functions are now being performed digitally—a familiar example is high-fidelity stereophonic recording and recreation—but more are coming all the time. As the speed of the devices increases, scientists and engineers are looking for more and more to this technology to replace older and less reliable techniques.

In addition to the traditional functions, there are applications opening up that were never even considered for analog implementation because similar or parallel technologies were not ready, or because analog electronics offered no solution. This includes fields such as medicine, where DSP is used in imaging, radiography, and various forms of tomography. There are also uses in image processing, where, coupled with artificial intelligence techniques, it is used for pattern recognition for automated placement work, sorting, fingerprint identification, handwriting analysis, computer aided design, engineering, and manufacturing.

Here is a short list of current applications created by or using DSP: We have arithmetic functions such as high-speed polynomial approximation, waveform generation, and new musical instruments. Instrumentation, such as spectrum analyzers and oscilloscopes, is being produced with greater bandwidth and accuracy and lower cost. Robotics benefits, as do automobiles with better and greater engine control and more intelligent environmental control. We have voice mail and fine audio equipment for recording, mixing, enhancing, and reproducing sounds. Data encryption has become faster and more reliable, as have radar and sonar. High-speed mathematical computers and add-ons are becoming quite common. Graphic image generation in 2D and 3D, CAD, CAM, and CAE appear in almost every major company. Cameras, transducers, and sensors have been linearized and made "smart" with DSP. There are translating telephones in the works, as well as text-to-voice converters for the blind. There are prosthetic ICs for sight and hearing, diagnostic aids, ultrasonic imaging, and low-dose x-rays. Industrially, the speed of the DSP chips has spawned a generation of very high-speed and intelligent controllers for highly facile control loops. Telecommunications benefits with products that remove echos from telephone lines and satellite links, as well as fast DTMF processing and speech synthesis and compression.

As diverse as these applications may seem, they use the same methods descibed in this book. Of course, we could not give examples of all of these applications in this book, especially with any detail. It is difficult even to write about a class of applications, such as image processing, without becoming deeply involved in the fascinating details of the subject and overrunning the scope of the book. Therefore, we chose to offer general applications that demonstrate the basic processes of digital image processing with their comcomitant parts in analog electronics. Earlier in the book, we presented code for spectral analysis and filtering. On the optional disk, there is much more with the Mathcad documents. Finally, with the DACQ card, you have the ability to process real-world signals yourself. In this chapter, we would like to present two final applications that are both useful and intriguing. The coverage will be brief, but if either subject is interesting to you, please consult the bibliography for further references.

These two examples include the use of filters, convolution, correlation, and spectral analysis. We begin with spectral analysis.

## DEDICATED SPECTRAL ANALYSIS: THE GOERTZEL ALGORITHM

Thus far, we have spent much time writing about spectral analysis and presenting various ways of deriving and using the data from the two domains. Though the DFT is the senior method for time-to-frequency and frequency-to-time domain transformation, there are other ways to accomplish the same thing, at least in certain respects. We covered the DFT and FFT and the Hartley transform, and mentioned the sine and cosine transforms. In addition to these, there are many variations of the DFT that are useful under certain conditions, as well as new techniques such as the Wavelet transform. In this section, we will cover a particular interpretation of the DFT that is used in such as applications as DTMF decoding and others where less than a complete spectrum is required. This technique is called Goertzel's algorithm.

This algorithm is not a complete DFT, it computes just two spectral points at a time. It is most useful for limited spectral analysis, where it can prove to be quite fast, compared to computing a complete FFT. It is interesting because of its speed in computing small numbers of spectral points and the simplicity with which it does it by using a linear filter.

We know from the discussion in Chapter 3 that the DFT calculates each index point of the spectrum individually, but this requires $N^2$ complex multiplications and $N(N-1)$ complex additions, where $N$ is the length of the input sequence (see Chapter 3). Using Goertzel's algorithm, $2N$ multiplications and $4N$ additions produce two DFT values.

The derivation of the algorithm has been defined by varying means, chiefly by rewriting the DFT as a polynomial and factoring it:

$$X[k]=\sum_{n=0}^{N-1}x[n]z^n\Big|_{z=W_N{}^k} \equiv P(z)\big|_{z=W_N{}^k}=\left(z-W_N{}^k\right)\left(z-W_N{}^{-k}\right)Q(z)+R(z)$$

or by expressing the computation of the DFT as a linear filter:

$$y_k[n]=\sum_{m=0}^{N-1}x[m]W_N{}^{-k(n-m)}u[n-m] \qquad \text{(Eq. 7-1)}$$

We begin by acknowledging the following identity:

$$W_N{}^{-kN}=e^{j\left(\frac{2\pi}{N}\right)Nk}=e^{j2\pi k}=1$$

which merely states the perodicity of the basic sequence, $W_N^{-kN}$. Multiplying this factor by the DFT produces the result in Equation 7-1, a convolution. This is especially clear if we rewrite the second half of the equation in the form of an impulse response:

$$h_k[n]=W_N{}^{-k(n)}u[n]$$

yielding:

$$y_k[n]=\sum_{m=0}^{N-1}x[m]h_k[n-m]$$

This convolution produces the value of a single point DFT at $k$ or $2\pi k/N$.

Moving to the frequency domain, we can write the system function for this filter as

$$H_k[z]=\frac{1}{1-W_N{}^{-k}z^{-1}} \qquad \text{(Eq. 7-2)}$$

which has a single pole at $\omega_k=2\pi k/N$ on the unit circle. Now, we could compute this as either the convolution described previously or as difference equation based upon the system function:

$$y_k[n]=W_N{}^{-k}y_k[n-1]+x[n]$$

But both the convolution and the difference equation involve complex arithmetic, which can be avoided by multiplying the denominator and numerator by the complex conjugate of the denominator of Equation 7-2:

$$1-W_N{}^kz^{-1}$$

This results in a new system function with two poles. Only one complex operation is necessary at the end of the computation:

$$H_k[z]=\frac{1-W_N{}^kz^{-1}}{1-2\cos(2\pi\frac{k}{N})z^{-1}+z^{-2}}$$

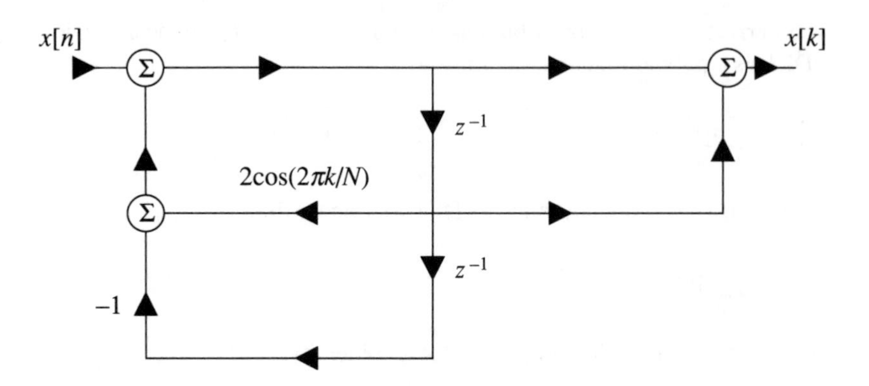


**Figure 7-1.** Flow diagram for two-pole Goertzel algorithm.

This system function yields two difference equations. First, an intermediate one to compute the poles:

$$\gamma_k[n] = 2\cos\frac{2\pi k}{N}\gamma_k[n-1] - \gamma_k[n-2] + x[n]$$

followed by a single iteration of the next equation for the zero:

$$y_k = \gamma_k[n] - W_N{}^k \gamma_k[n-1]$$

As we stated earlier, this set of equations yields two DFT points due to the symmetry of the factor $W_N{}^k$—thus we have $X[k]$ and $X[N-k]$.

The program GOERTZEL.C reads a file of complex data points and produces $X[k]$. As you can see, it implements the two difference equations directly. Please refer to the diagram in Figure 7-1 for an illustration of the Goertzel algorithm.

```
#include <stdio.h>
#include <math.h>
#define N 8
#define Pi 3.141592653589793

struct cnum {
    double real;
    double imag;
    };


void main()
{
    long T;
    int ndata_points;
    double c, s, cc, a1, a2, b1, b2, temp, norm, k=2; /*index of*/
                                                       /*spectral point*/
```

```
    struct cnum y[N+1] = {0.0};
    struct cnum x[N+1] = {0.0};
    FILE *xn;

    xn = fopen("xn.prn", "r++");
    ndata_points = 0;
    do{
        fscanf(xn, "%lf ", &x[ndata_points].real);
        fscanf(xn, "%lf ", &x[ndata_points].imag);
        x[ndata_points].imag *= -1;
        ndata_points++;
        }while(!feof(xn));      /*read in complex data points*/

    norm = N;
    norm = sqrt(norm);      /*to normalize the result*/
    c=cos(2*Pi*k/N);    /*generate the twiddle factors*/
    s=sin(2*Pi*k/N);
    cc = c+c;       /*2*cosine*/
    a2 = 0;
    a1 = x[0].real;         /*get first real value*/
    b1 = 0;
    b2 = x[0].imag;
    for (T=1; T<N; T++)     {
        temp = a1;
        a1 = cc*a1-a2+x[T].real;
        a2 = temp;
        temp = b1;
        b1 = cc*b1-b2+x[T].imag;
        b2 = temp;
            }   /* γ_k[n] = 2cos(2πk/N) γ_k[n-1] - γ_k[n-2] + x[n] */

    y[0].real = (c*a1-a2-s*b1)/norm;        /* y_k = γ_k[n] - W_N^k γ_k[n-1] */
    y[0].imag = (s*a1+c*b1-b2)/norm;
    printf ("\n%10g    %10g", y[0].real, y[0].imag);
    fclose (xn);
}
```

## ADAPTIVE FILTERING

Adaptive filtering is an attractive technique for commercial and industrial applications. It permits the control or interface to change as the environment changes. There are several popular uses for adaptive filters, involving adaptive prediction for speech reconstruction and synthesis, channel equalization for overcoming the deleterious effects of long lines to signals, noise cancellation for removing environmental noise, echo cancellation for removing the echo from long lines, and satellite transmission of signals. Digital communications receivers in which adaptive filters provide equalization for intersymbol interference, as well as channel identification, and system modelling in which an adaptive filter is used as a model to estimate the characteristics of an unknown system.

This example is excellent because it brings together two important DSP techniques: convolution (in the form of a linear filter) and correlation.

Correlation is an interesting and powerful tool in signal processing, in that it provides solutions, or at least some support, in the processing of signals that are subject to complex, broadband noise that cannot be easily described with simple mathematical expressions, tables, or rules. This noise generally thought of as a stochastic sequence, and is handled by statistical means with probability density functions. These signals are described in terms of averages, and cross-correlation and autocorrelation are often used to summarize their properties.

On the time domain, the process of correlation differs from convolution, in that neither sequence is reversed before the serial multiplication. Therefore, correlation can be expressed as follows:

$$C_{xy}[n] = \sum_{n=1}^{N} x[n]y[n]$$

One can see how closely related convolution and correlation are; since this is a discrete time function, a simple change of sign in the second factor would result in $y[N-n]$ and a convolution.

Another pertinent and useful theorem we would like to recall at this point is Parseval's theorem. Parseval's thereom states that the average squared magnitude of the signal is equal to the average squared magnitude of the transform. This is expressed as

$$\int_{-\infty}^{+\infty} |x(t)|^2 dt = \int_{-\infty}^{+\infty} |X(\omega)|^2 d\omega$$

The power, or energy, in a complex noise signal lacks focus, as we mentioned above, so whatever the total area of the signal, it is spread across the spectrum, whereas any signal components with definite frequency elements will concentrate their energies in peaks representing those frequency components. This can often allow the detection of even weak signals in noisy environments, such as mechanical systems and those suffering from the effects of finite word length arithmetic.

In autocorrelation, one correlates a signal with itself—in fact, that correlation will look very much like the integral above. This is a powerful tool for digging signal strength out of a channel buried in noise. If the noise is truly white and random, the signal alone will be left; if the noise actually has a strong frequency component, that frequency will correlate as well, and that information can be used to remove it.

Cross-correlation is the correlation of a signal with another:

$$C_{xy}(t) = \int_{-\infty}^{+\infty} x(\tau)y(t+\tau)d\tau$$

If the two signals are related in time, peaks will be evident in $C_{xy}(t)$. This is useful for finding the relationship between signal elements in a system, and is often used to quantify systems stimulated with white noise to produce an impulse response or frequency response.

With this understanding of these two devices, let us move to the creation of an adaptive filter. There are really two fundamental ways of accomplishing an adaptive filter: by means of an MSE (mean square error) adaptive algorithm, and by the more popular LMS (least mean square) algorithm. Since the LMS algorithm has its roots in the MSE, we will begin there.

An adaptive filter might be constructed from either an FIR or IIR filter. However, it is dangerous to update the poles of an IIR filter in real-time, because it is possible that they could move outside the unit circle. Therefore, we choose as our filter an asymmetrical FIR, though antisymmetric and lattice structures are equally possible. The code in the following example can be easily modified for use with any one of the three.

The FIR filter kernel is, then,

$$y[n] = \sum_{k=0}^{N-1} h[k]x[n-k]$$

Here, $h[k]$ represents the filter coefficients, which we will allow to be adjustable in the upcoming discussions.

If a digital filter is to adapt to a changing environment, the filter coefficients must change to meet those conditions. This requires a model, $d(n)$ (often called the *desired* signal), against which we compare the *output* of our filter, $y(n)$, generating an *error*, $e(n)$:

$$e(n) = d(n) - y(n) \qquad \text{(Eq. 7-3)}$$

Using mean square error (MSE) as the criterion to be minimized in updating the filter coefficients, we say:

$$\varepsilon = E\left[e^2(n)\right] \qquad \text{(Eq. 7-4)}$$

in which $E[.]$ is the Expectation operator.

Now, if we replace $y[n]$ in Equation 7-3 with the expression for the filter we will be using and then solve for $\varepsilon$,

$$\varepsilon = E\left[d^2(n)\right] - 2E\left[d(n)x(n)\right]\vec{w}^T(n) + E\left[x^2(n)\right]\vec{w}^T(n)\vec{w}(n) \qquad \text{(Eq. 7-5)}$$

In this equation, $w(n)$ is the filter weight vector. We find two correlations. In the final term of Equation 7-5, there is the autocorrelation of the input sequence representing the sample-by-sample correlation of the input signal:

$$r_{xx}[k] = E\left[x(n)x^T(n)\right] = \sum_{n=0}^{M} x(n)x(n+k), \;\; 0 \le k \le N-1 \text{ and } 0 \le M \le N-1$$

where $T$ indicates the transpose of the vector. In the middle term, we have the cross-correlation between the desired sequence and the input sequence:

$$r_{dx} = E\left[d(n)\vec{x}(n)\right] = \sum_{n=0}^{M} d(n)x(n-k), \;\; 0 \le k \le N-1 \text{ and } 0 \le M \le N-1$$

This looks like a standard convolution because of the $x(n-k)$ factor. Actually, however, this simply aligns the samples of the input with the desired signal.

Simplifying Equation 7-5, we have

$$\varepsilon = E\left[d^2(n)\right] + \vec{w}^T(n)\vec{w}(n)r_{xx} - 2\vec{w}^T(n)r_{dx}$$

We can solve for an optimum set of weights by differentiating:

$$\frac{\partial \varepsilon}{\partial \vec{w}(n)} = 0$$

The result of this operation is an equation that equates the convolution of a vector of optimum filter weights and the autocorrelation of the input signal with the cross-correlation of the input with the desired signal:

$$r_{dx}(m) = r_{xx}\vec{w}^T(n) = \sum_{k=0}^{N-1} w(k) r_{xx}(k-m), \quad m = 0,1,\ldots,N-1$$

This, of course, requires the lengthy process of autocorrelating the input signal and cross-correlating the desired signal with the input signal in order to solve for the optimum filter coeffients, all of which leads us to find a simpler (and faster) way.

Another manner of obtaining the desired result without having to perform these calculations is known as the LMS (least mean square) algorithm. This is the most popular manner of producing an adaptive filter, and is the basis of the code presented in this section. This algorithm involves what is known as the *steepest descent* method, in which each succeeding approximation to the optimum filter weight vector is produced as a sum of the current weights and a proportion of the derivative of the mean square error with respect to the current filter coefficients:

$$\vec{w}(n+1) = \vec{w}(n) - \frac{\partial\left[e^2(n)\right]}{\partial \vec{w}(n)}$$

which, by differentiating, can be restated:

$$\vec{w}(n+1) = \vec{w}(n) - 2\beta e(n)\vec{x}(n) = w(n) - \beta e(n)x(n-k)$$

Notice that the 2 in the middle of the equation is absorbed into the proportionality constant in the final equality.

As you might guess, $\beta$ controls the rate of convergence of the algorithm. The larger the value of $\beta$, the more rapid the convergence, but the greater the possibility of instability. It is, therefore, nice to know that there is a way to determine a value for $\beta$ that will be stable and yield the fastest possible convergence. A stable proportionality coefficient is given by

$$0 < \beta < \frac{1}{10NP_x}$$

where $N$ is the length of the FIR filter and $P_x$ is the average power of the input signal:

$$P_x \approx \frac{1}{M+1}\sum_{n=0}^{M} x^2(n) = \frac{r_{xx}(0)}{M+1}$$

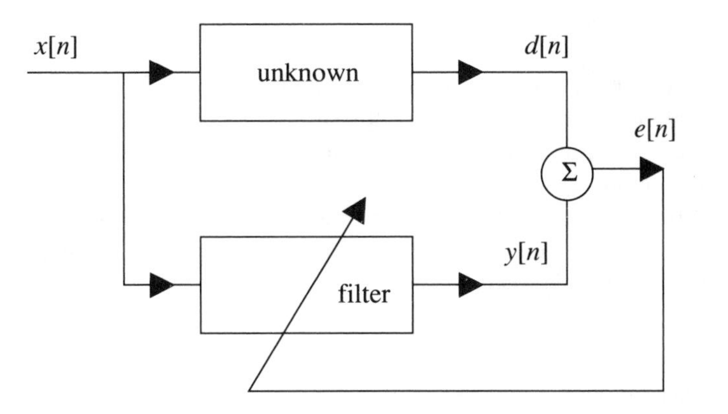


**Figure 7-2.** Adaptive filter.

Please note the simplicity of the adaptive filter function illustrated in Figure 7-2.

The following code is from a file on the optional disk called ADPT.C; it forms the kernel for an adaptive filter.

In this fragment, *dn_points* are the number of data points to be processed, *Xn*[] is an array containing the data points, *Y*[] is the output array. The filter coefficients are stored in *weights*[]. *Dn*[] is an array containing the desired output, *error* is the difference between the desired and actual output, and *beta* is variable controlling the rate of convergence of the filter.

```
for (i=0; i<dn_points; i++)    {
    x[0]=Xn[dn_points];
    Y[dn_points]=0;
    for(j=0; j<=N; j++)
        Y[dn_points]+=(weights[j]*x[j]);    /*assymetrical transversal*/
    error=Dn[dn_points]-Y[dn_points];       /*establish error*/
    for(j=N; j>=0; j--)    {
        weights[j] = weights[j]+(2*beta*error*Xn[j]);/*approximate*/
                                            /*new filter coefficients*/
        if(j!=0) x[j]=x[j-1];
    }
```

Finally, we will list and briefly discuss some of the more common uses of the adaptive filter.

## System Modelling or Identification

A very interesting application of the adaptive filter is the identification and modelling of unknown systems. To explain, let us suppose that we have an unknown system we wish to model or identify. We will use an FIR to approximate the unknown system. The same signal is fed to both the FIR filter and the unknown system. The output of the unknown system is called the desired signal, $d[n]$, and is compared with the output of our FIR filter to generate the error signal, $e[n]$, used to modify the filter coefficients.

## Adaptive Prediction

Adaptive prediction is used extensively in speech applications, such as coding the waveform for later compression, regeneration, or synthesis. This form of the adaptive filter uses the correlation between adjacent samples in the input sequence to produce a prediction error that is smaller than the input signal.

In this application, we call the input sequence the desired signal, $d[n]$. A delayed form of this signal is also the input to the adaptive filter. The ouput of the adaptive filter is summed with the input sequence to create an error signal to correct the adaptive weights in the filter.

## Adaptive Channel Equalization

The adaptive filter is used to correct for channel distortion that causes intersymbol interference over such communication channels as the telephone. Channel distortion can be especially deleterious at high data rates, where the influence of the transmitted symbol can extend beyond the time interval used to represent it. Signals are then smeared or overlayed and data is distorted. This can come from the effect of the line itself or from filters that are also used on the line to correct or separate the channels from one another.

Usually, there is a short training period of about one second before the actual data is sent. During this time, known or random data is transmitted so the filter can get a preliminary estimate of the line. Next, the actual data is transmitted and the filter coefficients adjusted to maintain as high a quality of results as possible.

In this context, the input to the filter is the received signal plus any channel noise, the desired signal $d[n]$ is the detected signal, the output of the filter $y[n]$ is the equalized signal used to detect the data, and the error signal $e[n]$ is the negative sum of the detected signal and the filter output.

## Echo Cancellation

Echo cancellation is used in telephonic communications links to remove unwanted echoes resulting from mismatches in the media involved. This is often caused by an impedance mismatch in the coupling (transformer) between the transmit and receive side in full duplex lines. Due to the mismatch, some of the transmitted signal leaks back into the receive channel and is perceived as echo. In modem communications, an echo canceller—an adaptive FIR filter—is used to reduce this interference.

In this mode, the input to the adaptive filter is the far-end signal, and the desired signal $d[n]$ is the echo of the far-end signal summed with the near-end signal. The ouput of the filter is subtracted (negative sum) with this signal, $d[n]$, to produce an error signal, $e[n]$, to correct the filter coefficients. In this way, the echo is approximated in the adaptive filter and subtracted from the return signal.

# Glossary

**analog signals** Another name for continuous-time signals. More specifically, these are signals with infinite resolution in time and magnitude.

**autocorrelation** Cross-correlation applied to one signal only. It provides information about the time variation of a signal.

**bandwidth (filter)** This is the difference between the lowest frequency and the highest frequency at which the response finally falls to 3db down from its peak value as it leaves the passband.

**basis function** A collection of elements of a linear space that spans that space and is linearly independent.

**center frequency (filter)** The is the geometric mean of the upper and lower 3db cutoff frequencies. This is expressed as

$$f_{center} = \sqrt{f_{lower} * f_{upper}}$$

**characteristic equation** An equation derived from a linear differential equation that is made equal to zero and has only one variable. The coefficient and power of this variable in each term correspond to the coefficient and power of the derivative in the original equation.

**complex frequency** $s = \sigma + j\omega$.

**corner frequency** Cutoff frequency for a filter.

**correlation** An operation performed on two functions of the same variable measuring the same property. See *autocorrelation, cross-correlation.*

**critically damped** Flattest response (balanced ration of capacitor and inductor). See *damping.*

**cross-correlation** The sum of the scalar products of two signals in which the signals are displaced in time with respect to one another. This displacement is the independent variable of the cross-correlation function. This produces a measure of the similarity between two signals—it can be used to detect time-shifted or periodic similarities.

**cutoff frequency** Point at which the response begins to fall off significantly. Frequency drops 3db (0.707 of its peak value as it leaves passband), determined by the product of the inductor and capacitor values.

**damping** Set by ratio of inductor and capacitor; an index to its tendency towards oscillation. Practical values range from 0 to 2; 1.414 to give maximum flatness without overshoot. See *underdamped, critically damped, highly damped/overdamped.*

**decibel** $20\log_{10}\left(\frac{E_2}{E_1}\right)$, $E_1$ and $E_2$ represent two signal amplitudes.

**degree** The degree of an equation is the power of the highest ordered derivative after all possible algebraic reduction.

**Dirac delta function** Also known as the *impulse* function, usually denoted $\delta(t)$ and depicted as a vertical arrow at the origin. Named for Paul Dirac (1902–1984), a quantum physicist who used impulse functions extensively in his work. This impulse is described as having zero width and infinite height, such that its area is equal to 1.

**domain of a function** The set of admissible values is called the domain of a function.

**dot product** Scalar product of the magnitudes of any two vectors and the cosine of the angle between them.

**dynamic range** Ratio of the largest allowable input signal to the smallest. The smallest signal is defined as one that is discernible above the noise floor. This ratio is a measure of noise.

**ENOB** Effective number of bits. Used in the evaluation of A/D and D/A converters to determine how many bits are truly available for data.

**filter tap** A tap is a delay element, $z^{-1}$. The signal at each filter tap is weighted by a *tap coefficient,* and the resulting products are summed to create the output.

**forcing function** A function representing an energy source.

**group delay** The time interval, in milliseconds, required for an input pulse to appear at the output of the filter.

**highly damped/overdamped** Load resistor dominates (small capacitor, large inductor). In filter equations, damping is the inverse of $Q$, and so strongly effects bandwidth. If the damping is high, the bandwidth is broad and the $Q$ is low; if the damping is light, $Q$ is high and bandwith is narrow. See *damping*.

**homogenous equation** A linear differential equation whose sum is zero. An example of this might be the equation for a simple harmonic oscillator:

$$\alpha_0 \frac{d^2 i}{di^2} + a_1 \frac{di}{dt} + a_2 i = 0$$

**immitance** Combination word meaning impedence and admittance. Coined because these two are so often examined together.

**initial conditions** The starting conditions of an oscillator. What was occuring at $t<0$.

**limit cycling** When output of a system moves between two outputs states only. This can occur when the resolution of the system is smaller than required by the operating point. For example, 5.11 cannot be achieved with two decimal digits, therefore, the filter output will cycle betwen 5.2 and 5.1 as it tries to approximate that value.

Limit cycling also results from intermediate overflows that produce large negative numbers from what should have been large positive results. This number can then be fed back to produce still more sign changes.

**linear** An equation is linear if the dependent variable and all of its derivatives are of the first degree.

**linear independence** No function in a subject series can be formed from a linear combination of other functions in the series.

**Linear space** A class of functions, all having the same domain, such that if two functions belong to a class, their sum does also. If $f$ is a member of the class, every scalar multiple of $f$ is, also.

**neper** A dimensionless number meaning attenuation. It is a measure of a voltage ratio or a current ratio under certain specified conditions. It is one of the values in the complex number $\gamma = \alpha + j\beta$. It is the real part of a complex frequency.

**normalized filter** A filter design for analysis with a cutoff frequency of one radian per second and an impedance of one ohm. 1 ohm $*$ 1 farad $=$ 1 rad/sec.

**octave** Frequency doubles with each octave.

**operator** Operators, such as $D$ or $L$, are used to replace much longer or more awkward expressions. The operator $D$ usually stands for differentiation with respect to time; $L$ may be anything.

**order** *Mathematics*: number of the highest ordered derivative. In

$$\alpha_0 \frac{di}{dt} + \alpha_1 i = 0\,,\quad \frac{di}{dt}$$

is the highest ordered derivative, and it is of the first degree.

*filters:* describes falloff: second-order lowpass falls off as square of frequency, and so on. Often determined by energy storage units such as capacitors: second-order means two capacitors.

**ordinary differential equation** An equation containing only total and no partial derivatives.

**periodic** A function or signal is periodic if a waveform can be synthesised by repeating a particular cycle over and over at regular intervals—that is, for a fixed positive number $t$, $f(x+t)=f(x)$ for all $x$.

**phase** The fraction or part of a complete cycle that has elapsed from a particular reference point.

**pole** Values for which the ratio of two polynomials becomes infinite.

***Q*** Inverse of the bandwidth of a single filter structure. Ratio of the filter or sample bandwidth to its center frequency, reactance to resistance, inductance to capacitance. A measure of bandwidth and peaking.

**radian frequency** This is $\omega$. $\omega = 2\pi f$ radians per second.

**resonance** A frequency $\omega$ at which individual the reactances in are equal and opposite, leaving the total reactance of the circuit zero. In an LC circuit, we have at resonance:

$$\omega_0 L = \frac{1}{\omega_0 C}$$

where $\omega_0$ is equal to the frequency of resonance multiplied by $2\pi$. The frequency of resonance is given by:

$$f_0 = \frac{1}{2\pi\sqrt{LC}}$$

**restoring force** A force exerted on a mass which acts in the direction of returning it to the equilibrium point.

**scaling** Impedance level is increased by multiplying all resistors and dividing all capacitors by the desired factor. Frequency is shifted inversely by multiplying all frequency-determining resistors or all frequency-determining capacitors by the desired factor.

**settling time** This is the time elapsed from the application of a perfect step input to the time the amplifier output enters and remains within a specified error band. This includes the time from the initial input through any overshoot and ringing to the final value.

**sinad** Signal to noise and distortion ratio. This is the ratio between the RMS amplitude of the fundamental frequency and the RMS amplitudes of all the other frequencies expressed in dB. This usually results from nonlinearities in the filter's amplifiers or the nonlinearities in the A/D and D/As.

**sinc** $\frac{\sin x}{x}$, the ideal sampling function.

**slew rate** This is the largest possible rate of change in an operational amplifier's output.

**SNR** Signal to noise ratio. This is the Rms signal to quantization noise ratio given by $SNR = 6.02N + 1.76\text{dB}$, where $N$ is the number of bits truly realized in the A/D as determined by test. For example, A/D may be providing only 11 bits of data, once the noise caused by poor stopband attenutation and converter particular errors are subtracted off. In that case, $SNR = 6.02(11) + 1.76\text{dB} = 67.98\text{dB}$.

**superposition** In linear systems, the principle of superposition states that the total response of a linear network is identical to that found by taking each contributing energy source alone, with all others removed, and summing the individual responses.

**Total harmonic distortion (THD)** The ratio of the RMS sum of the input harmonics to the RMS amplitude of the fundamental frquency. This is usually expressed as

$$THD = 20 \log \frac{\sqrt{\sum \left(V_1^2 \ldots V_n^2\right)}}{V_0}$$

where $V_1 \ldots V_n$ represent the amplitudes of the input harmonics through $n$, and $V_0$ is the RMS amplitude of the fundamental frequency.

**transfer function** The function relating the transform of a quantity at an input to the transform of another quantity at an output is called a transfer function:

$$H(s) = \frac{E_{out}(s)}{E_{in}(s)}$$

**underdamped** Peaking (capacitor very large compared to inductor). See *damping*.

**zero** The value for which the ratio of two polynomials approaches zero.

# Bibliography

Abramowitz, M., and Stegun, I.A., ed. *Handbook of Mathematical Functions.* New York: Dover Publications, Inc., 1964.

Acton, F.S. *Numerical Methods that Usually Work.* Washington D.C.: Mathematical Association of America, 1970.

Bracewell, R.N. *The Fourier Transform and Its Applications.* New York: McGraw-Hill, 1986.

Burrus, C.S., and Parks, T.W. *DFT/FFT and Convolution Algorithms.* New York, NY: John Wiley and Sons, 1984.

Chassaing, Rulph, *Digital Signal Processing with C and the TMS320C30.* New York, NY: John Wiley and Sons, 1992.

Davis, H.F. *Fourier Series And Orthogonal Functions.* New York: Dover, 1963.

Gonzales, Rafael C., and Wintz, Paul. *Digital Image Processing.* Reading MA: Addison-Wesley, 1977.

Hamming, R. W. *Numerical Methods for Engineers and Scientists*, New York: McGraw-Hill, 1962.

Hart et al. *Computer Approximations.* New York: John Wiley & Sons, 1968.

Higgins, Richard J. *Digital Signal Processing in Vlsi.* New Jersey: Prentice-Hall, 1990.

Hildebrand, E.B. *Introduction to Numerical Analysis.* New York: Dover, 1956.

Horden, Ira. *An FFT Algorithm for MCS-96 Products Including Supporting Routines and Examples, AP-275.* Intel, 1988.

Huelsman, L.P. *Active and Passive Analog Filter Design.* NewYork: McGraw-Hill, 1993.

Ingle, V.K., and Proakis, J.G. *Digital Signal Processing Laboratory Using the ADSP-2101 Microcomputer.* New Jersey: Prentice-Hall, 1991.

Lancaster, Don. *Active-Filter Cookbook.* Indiana:Sams, 1975.

Lin, Kun-Shan, ed. *Theory, Algorithms, and Implementations, Volume 1.* Texas Instruments: Prentice-Hall, 1986.

Lindley, Craig A., *Practical Image Processing in C.* New York: John Wiley & Sons, 1991.

Mar, Amy, ed. *Digital Signal Processing Applications Using the ADSP-2100 Family.* New Jersey: Prentice-Hall, 1992.

Mitra and Kaiser, ed. *Handbook for Digital Signal Processing.* New York: John Wiley & Sons, 1993.

Oppenheim, Alan V., and Shaffer, R.W. *Digital Signal Processing.* New Jersey: Prentice-Hall, 1985.

Oppenheim, Alan V., and Shaffer, R.W. *Discrete-Time Signal Processing.* New Jersey: Prentice Hall Inc, 1989.

Papamichalis, Panos, ed. *Theory, Algorithms, and Implementations, Volume 2.* Texas Instruments: Prentice-Hall, 1990.

Papamichalis, Panos, ed. *Theory, Algorithms, and Implementations, Volume 3.* Texas Instruments: Prentice-Hall, 1990.

Pavlidis, Theo. *Algorithms for Graphics and Image Processing.* Rockville, MD: Computer Science Press, 1982.

Press, Flannery, Teukolsky, and Vetterling. *Numerical Recipes in C: The Art of Scientific Computing.* New York: Cambridge University Press, 1988.

Roragaugh, C.B. *Digital Filter Designer's Handbook.* New York: McGraw-Hill, 1993.

*TMS320C3X.* Texas Instruments: Prentice Hall, 1991

Van Dam, A., and Foley, J.D. *Fundamentals of Interactive Computer Graphics*. Reading, MA: Addison-Wesley, 1983.

Wong and Ott. *Function Circuits: Design and Applications*. New York: McGraw-Hill, 1975.

# About the Software

## *Practical DSP Modeling, Techniques, and Programming in C*

### WHAT IS ON THIS DISK?

There are 19 subdirectories in this collection containing four sorts of files: Mathcad documents, data files, binaries and source code. The Mathcad files are stored according to chapter and are in directories CH1, CH2, CH3, CH4, and CH5. The files mentioned in those chapters will be found in those directories.

The program files and associated data files are stored according to project names in the following directories (These files are almost entirely in C with two in assembly language):

**ADAPTIVE** contains source, binaries and sample data files for an adaptive filter.

**ASMFFT** contains source, binaries and sample data files for an assembly language Fourier transform routine.

**CFFT** contains source, binaries and sample data files for a Fourier transform in C.

**CONVOLVE** contains source, binaries and sample data files for two kinds of convolution, one done in the time domain and the other in the frequency domain.

**DPY** contains the source and executable for the screen plotting program.

**FIR** contains source, binaries and sample data files for an FIR filter written in C.

**FILTER** contains source, binaries and sample data files for two programs. The first program develops filter coefficients for and FIR filter using the Fourier series. The second also produces filter coefficients with the Fourier series, but also applies various window weighting schemes to these coefficients.

**IIR** contains source, binaries and sample data files for an IIR filter written in C.

**GOERTZEL** contains source, binaries and sample data files for an interesting alternative to the Fourier transform based upon a convolution.

**HARTLEY** contains source, binaries and sample data files for both the Hartley transform and the Fast Hartley transform.

**DACQ** contains source, binaries and sample data files for the PC based oscilloscope program.

**Note:** one program in particular is called by almost all of the others, and that is DPY.EXE. It is a plotting routine used to display the results of many of the programs. In order for the individual programs to find DPY.EXE, the environmental variable DPY must be set in the autoexec.bat to point at the DPY directory:

```
set dpy=(drive):\(path)
```

For example, if your store all of your programs in the CODE directory on drive C, then the variable might be set as:

```
set dpy=c:\code\dpy
```

Each directory contains a READ.ME explaining the protocol for the program(s) therein. Some of the directories include Mathcad documents to aid in the generation of test data for the programs.

All of the programs on this disk will write to Mathcad and Excel compatible files. These programs can read those same formats permitting one to create vectors in any of the programs for use for any of the others.

## GETTING STARTED

### System Requirements

The companion disk for *Practical DSP Modeling, Techniques, and Programming in C* requires the following computer system hardware and software:

- 386 or 486 CPU *with* math coprocessor
- 584k of free memory
- at least 3 megabytes of free hard disk space
- VGA color graphics adaptor
- DOS 5.0 or later
- text editor
- ANSI C/C++ compiler—(this code was originally compiled with Microsoft C/C++ 7.0)

## Making a Backup Copy

Before you install the *Practical DSP Modeling, Techniques, and Programming in C* software, we strongly recommend that you make a backup copy of the original disk. Remember, however, the backup disk is for your personal use only. Any other use of the backup disk violates copyright law. Please take the time now to make the backup, using the following procedure:

1. Insert the *Practical DSP Modeling, Techniques, and Programming in C* disk into drive A: of your computer (assuming that your floppy disk drive is drive A:)
2. At the C:\> prompt, type DISKCOPY A: A: and press Enter

You will be prompted through the steps to complete the disk copy. When you are through, remove the new copy of the disk and label it immediately. Remove the original *Practical DSP Modeling, Techniques, and Programming in C* disk and store it in a safe place.

# INSTALLING THE SOFTWARE

The *Practical DSP Modeling, Techniques, and Programming in C* disk contains all the necessary files in a compressed format. The default installation settings will create a directory called **DSP** and 18 subdirectories. To install the files, please do the following:

1. Assuming you will be using drive **A** as the floppy drive for your diskette, at the A:> prompt type INSTALL.
2. Follow the instructions displayed by the installation program. At the end of the process you will be given the opportunity to review the README file for more information about the software,

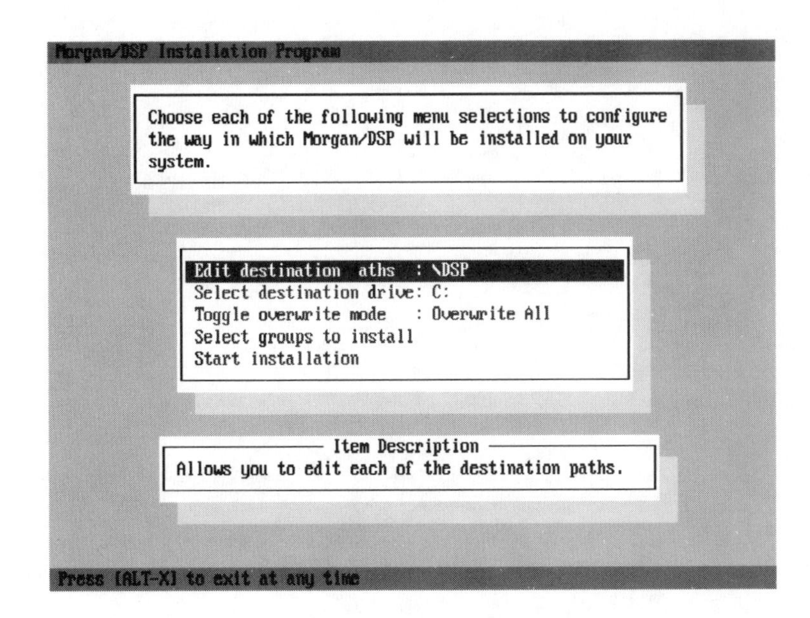

After the *Practical DSP Modeling, Techniques, and Programming in C* installation is complete, remove the disk and store it in a safe place

## In Case of Trouble - User Assistance and Information

John Wiley & Sons, Inc. is pleased to provide assistance to users of the *Practical DSP Modeling, Techniques, and Programming in C* software package. Should you have questions regarding the installation or use of this package, please call our technical support number at 212-850-6194 weekdays between 9 am and 4 pm Eastern Time.

To place orders for additional copies of this book, including the software, or to request information about other Wiley products, please call 800-879-4539.

# Appendices

**APPENDIX A**
**12-Bit + Sign Data Acquisition System with Self-Calibration** 353

**APPENDIX B**
**Dual FET-Input, Low Distortion Operation Amplifier** 391

**APPENDIX C**
**8th Order Continuous-Time Active Filter** 405

**APPENDIX D**
**Voltage-Output, 12-Bit DACs with Internal Reference** 425

National Semiconductor

March 1993

# LM12454/LM12H454/LM12458/LM12H458 12-Bit + Sign Data Acquisition System with Self-Calibration

## General Description

The LM12454, LM12H454, LM12458, and LM12H458 are highly integrated Data Acquisition Systems. Operating on just 5V, they combine a fully-differential self-calibrating (correcting linearity and zero errors) 13-bit (12-bit + sign) analog-to-digital converter (ADC) and sample-and-hold (S/H) with extensive analog functions and digital functionality. Up to 32 consecutive conversions, using two's complement format, can be stored in an internal 32-word (16-bit wide) FIFO data buffer. An internal 8-word RAM can store the conversion sequence for up to eight acquisitions through the LM12(H)458's eight-input multiplexer. The LM12(H)454 has a four-channel multiplexer, a differential multiplexer output, and a differential S/H input. The LM12(H)454 and LM12(H)458 can also operate with 8-bit + sign resolution and in a supervisory "watchdog" mode that compares an input signal against two programmable limits.

Programmable acquisition times and conversion rates are possible through the use of internal clock-driven timers. The reference voltage input can be externally generated for absolute or ratiometric operation or can be derived using the internal 2.5V bandgap reference.

All registers, RAM, and FIFO are directly addressable through the high speed microprocessor interface to either an 8-bit or 16-bit databus. The LM12(H)454 and LM12(H)458 include a direct memory access (DMA) interface for high-speed conversion data transfer.

## Key Specifications ($f_{CLK}$ = 5 MHz; 8 MHz, H)

- Resolution: 12-bit + sign or 8-bit + sign
- 13-bit conversion time: 8.8 μs, 5.5 μs (H) (max)
- 9-bit conversion time: 4.2 μs, 2.6 μs (H) (max)
- 13-bit Through-put rate: 88k samples/s (min); 140k samples/s (H) (min)
- Comparison time ("watchdog" mode): 2.2 μs (max); 1.4 μs (H) (max)
- ILE: ±1 LSB (max)
- $V_{IN}$ range: GND to $V_A{}^+$
- Power dissipation: 30 mW, 34 mW (H) (max)
- Stand-by mode: 50 μW (typ)
- Single supply: 3V to 5.5V

## Features

- Three operating modes: 12-bit + sign, 8-bit + sign, and "watchdog"
- Single-ended or differential inputs
- Built-in Sample-and-Hold and 2.5V bandgap reference
- Instruction RAM and event sequencer
- 8-channel (LM12(H)458), 4-channel (LM12(H)454) multiplexer
- 32-word conversion FIFO
- Programmable acquisition times and conversion rates
- Self-calibration and diagnostic mode
- 8- or 16-bit wide databus microprocessor or DSP interface

## Applications

- Data Logging
- Instrumentation
- Process Control
- Energy Management
- Robotics
- Inertial Guidance

## Connection Diagrams

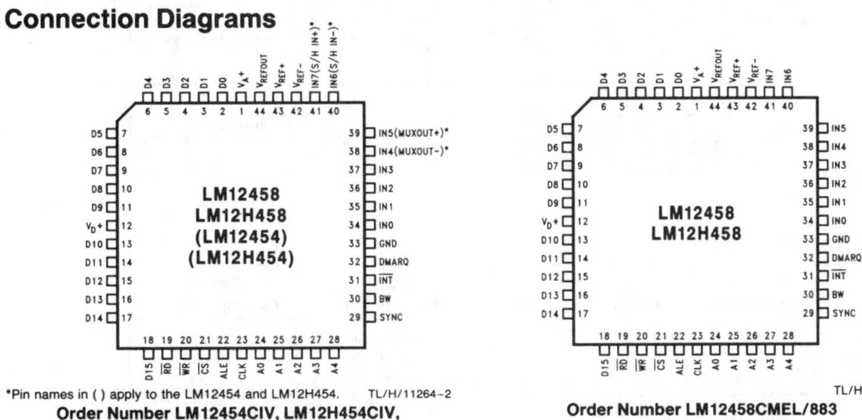


*Pin names in ( ) apply to the LM12454 and LM12H454. TL/H/11264-2

**Order Number LM12454CIV, LM12H454CIV, LM12458CIV or LM12H458CIV**
**See NS Package Number V44A**

TL/H/11264-34

**Order Number LM12458CMEL/883 or LM12H458CMEL/883**
**See NS Package Number EL44A**

## Functional Diagrams

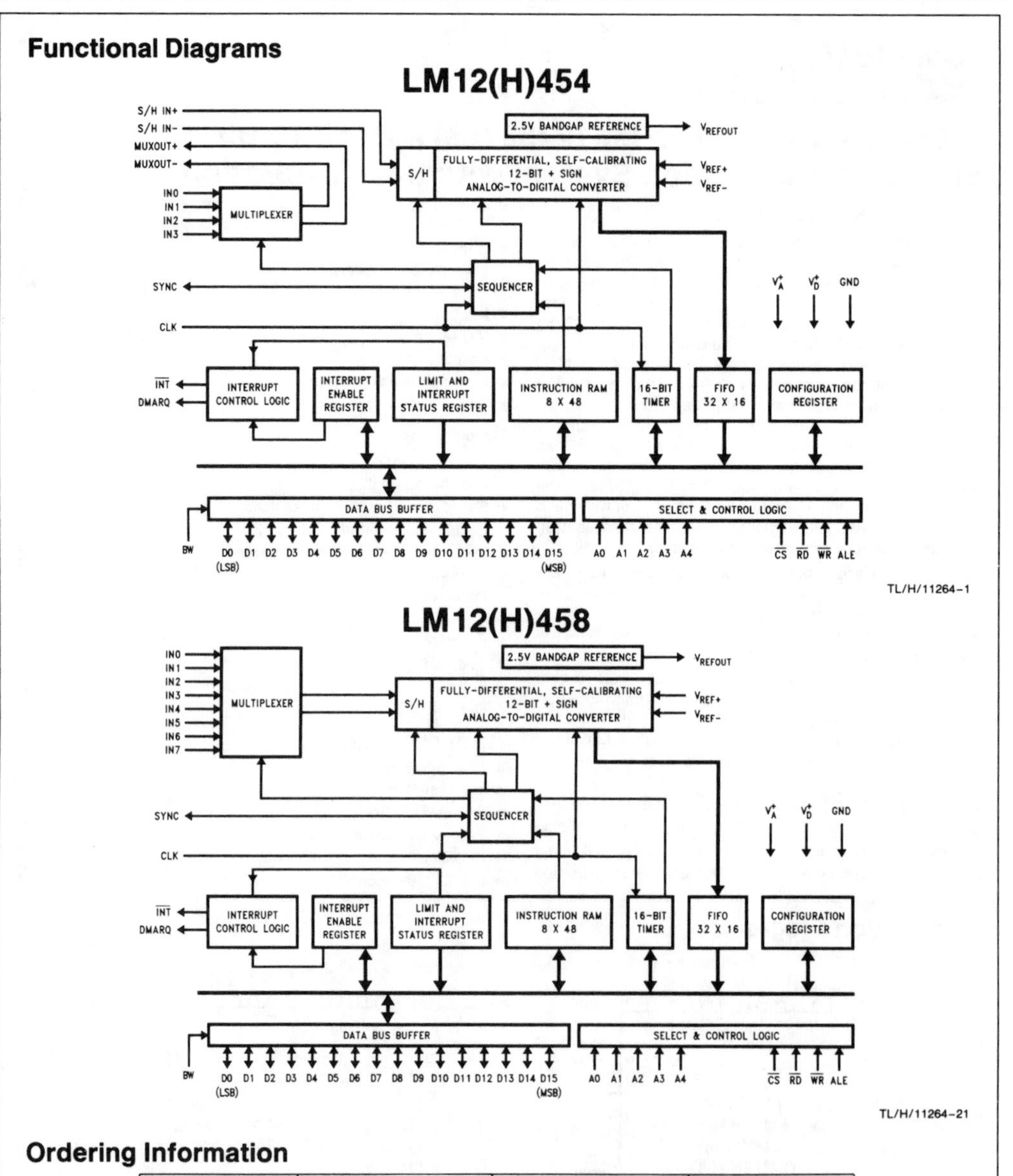


## Ordering Information

| Guaranteed Clock Freq (min) | Guaranteed Linearity Error (max) | Order Part Number | See NS Package Number |
|---|---|---|---|
| 8 MHz | ±1.0 LSB | LM12H454CIV | V44A |
| | | LM12H458CIV | V44A |
| | | LM12H458CMEL/883 | EL44A |
| 5 MHz | ±1.0 LSB | LM12454CIV | V44A |
| | | LM12458CIV | V44A |
| | | LM12458CMEL/883 | EL44A |

## Absolute Maximum Ratings (Notes 1 & 2)

**If Military/Aerospace specified devices are required, please contact the National Semiconductor Sales Office/Distributors for availability and specifications.**

| | |
|---|---|
| Supply Voltage ($V_A^+$ and $V_D^+$) | 6.0V |
| Voltage at Input and Output Pins except IN0–IN3 (LM12(H)454) and IN0–IN7 (LM12(H)458) | $-0.3V$ to $V^+ + 0.3V$ |
| Voltage at Analog Inputs IN0–IN3 (LM12(H)454) and IN0–IN7 (LM12(H)458) | GND $-$ 5V to $V^+ + 5V$ |
| $\lvert V_A^+ - V_D^+ \rvert$ | 300 mV |
| Input Current at Any Pin (Note 3) | ±5 mA |
| Package Input Current (Note 3) | ±20 mA |
| Power Dissipation ($T_A = 25°C$) V Package (Note 4) | 875 mW |
| Storage Temperature | $-65°C$ to $+150°C$ |
| Lead Temperature V Package, Infrared, 15 sec. | $+300°C$ |
| ESD Susceptibility (Note 5) | 1.5 kV |
| LM12(H)458CMEL/883 | 2.0 kV |

See AN-450 "Surface Mounting Methods and Their Effect on Product Reliability" for other methods of soldering surface mount devices.

## Operating Ratings (Notes 1 & 2)

| | |
|---|---|
| Temperature Range ($T_{min} \le T_A \le T_{max}$) | |
| LM12(H)454CIV/LM12(H)458CIV | $-40°C \le T_A \le 85°C$ |
| LM12(H)458CMEL/883 | $-55°C \le T_A \le 125°C$ |
| Supply Voltage | |
| $V_A^+$, $V_D^+$ | 3.0V to 5.5V |
| $\lvert V_A^+ - V_D^+ \rvert$ | $\le 100$ mV |
| $V_{IN+}$ Input Range | GND $\le V_{IN+} \le V_A^+$ |
| $V_{IN-}$ Input Range | GND $\le V_{IN-} \le V_A^+$ |
| $V_{REF+}$ Input Voltage | $1V \le V_{REF+} \le V_A^+$ |
| $V_{REF-}$ Input Voltage | $0V \le V_{REF-} \le V_{REF+} - 1V$ |
| $V_{REF+} - V_{REF-}$ | $1V \le V_{REF} \le V_A^+$ |
| $V_{REF}$ Common Mode Range (Note 16) | $0.1\ V_A^+ \le V_{REFCM} \le 0.6\ V_A^+$ |

## Converter Characteristics

The following specifications apply to the LM12454, LM12H454, LM12458, and LM12H458 for $V_A^+ = V_D^+ = 5V$, $V_{REF+} = 5V$, $V_{REF-} = 0V$, 12-bit + sign conversion mode, $f_{CLK} = 8.0$ MHz (LM12H454/8) or $f_{CLK} = 5.0$ MHz (LM12454/8), $R_S = 25\Omega$, source impedance for $V_{REF+}$ and $V_{REF-} \le 25\Omega$, fully-differential input with fixed 2.5V common-mode voltage, and minimum acquisition time unless otherwise specified. **Boldface limits apply for $T_A = T_J = T_{MIN}$ to $T_{MAX}$**; all other limits $T_A = T_J = 25°C$. (Notes 6, 7, 8, 9 and 19)

| Symbol | Parameter | Conditions | Typical (Note 10) | Limits (Note 11) | Unit (Limit) |
|---|---|---|---|---|---|
| ILE | Positive and Negative Integral Linearity Error | After Auto-Cal (Notes 12, 17) | ±1/2 | **±1** | LSB (max) |
| TUE | Total Unadjusted Error | After Auto-Cal (Note 12) | ±1 | | LSB |
| | Resolution with No Missing Codes | After Auto-Cal (Note 12) | | **13** | Bits (max) |
| DNL | Differential Non-Linearity | After Auto-Cal | | **±1/2** | LSB (max) |
| | Zero Error | After Auto-Cal (Notes 13, 17)<br>LM12H454<br>LM12H458 | ±1/2 | **±1**<br>**±1.5**<br>**±1.5** | LSB (max) |
| | Positive Full-Scale Error | After Auto-Cal (Notes 12, 17)<br>LM12(H)458CMEL | ±1/2 | **±2**<br>**±2.5** | LSB (max) |
| | Negative Full-Scale Error | After Auto-Cal (Notes 12, 17)<br>LM12(H)458CMEL | ±1/2 | **±2**<br>**±2.5** | LSB (max) |
| | DC Common Mode Error | (Note 14) | ±2 | **±3.5** | LSB (max) |
| ILE | 8-Bit + Sign and "Watchdog" Mode Positive and Negative Integral Linearity Error | (Note 12) | | **±1/2** | LSB (max) |
| TUE | 8-Bit + Sign and "Watchdog" Mode Total Unadjusted Error | After Auto-Zero | ±1/2 | **±3/4** | LSB (max) |
| | 8-Bit + Sign and "Watchdog" Mode Resolution with No Missing Codes | | | **9** | Bits (max) |
| DNL | 8-Bit + Sign and "Watchdog" Mode Differential Non-Linearity | | | **±1/2** | LSB (max) |
| | 8-Bit + Sign and "Watchdog" Mode Zero Error | After Auto-Zero | | **±1/2** | LSB (max) |
| | 8-Bit + Sign and "Watchdog" Positive and Negative Full-Scale Error | | | **±1/2** | LSB (max) |

## Converter Characteristics

The following specifications apply to the LM12454, LM12H454, LM12458, and LM12H458 for $V_A+ = V_D+ = 5V$, $V_{REF+} = 5V$, $V_{REF-} = 0V$, 12-bit + sign conversion mode, $f_{CLK} = 8.0$ MHz (LM12H454/8) or $f_{CLK} = 5.0$ MHz (LM12454/8), $R_S = 25\Omega$, source impedance for $V_{REF+}$ and $V_{REF-} \leq 25\Omega$, fully-differential input with fixed 2.5V common-mode voltage, and minimum acquisition time unless otherwise specified. **Boldface limits apply for $T_A = T_J = T_{MIN}$ to $T_{MAX}$**; all other limits $T_A = T_J = 25°C$. (Notes 6, 7, 8, 9 and 19) (Continued)

| Symbol | Parameter | Conditions | Typical (Note 10) | Limits (Note 11) | Unit (Limit) |
|---|---|---|---|---|---|
| | 8-Bit + Sign and "Watchdog" Mode DC Common Mode Error | | ±1/8 | | LSB |
| | Multiplexer Channel-to-Channel Matching | | ±0.05 | | LSB |
| $V_{IN+}$ | Non-Inverting Input Range | | | **GND**<br>**$V_A+$** | V (min)<br>V (max) |
| $V_{IN-}$ | Inverting Input Range | | | **GND**<br>**$V_A+$** | V (min)<br>V (max) |
| $V_{IN+} - V_{IN-}$ | Differential Input Voltage Range | | | **$-V_A+$**<br>**$V_A+$** | V (min)<br>V (max) |
| $\frac{V_{IN+} - V_{IN-}}{2}$ | Common Mode Input Voltage Range | | | **GND**<br>**$V_A+$** | V (min)<br>V (max) |
| PSS | Power Supply Sensitivity (Note 15): Zero Error<br>Full-Scale Error<br>Linearity Error | $V_A+ = V_D+ = 5V \pm 10\%$<br>$V_{REF+} = 4.5V$, $V_{REF-} = GND$ | ±0.2<br>±0.4<br>±0.2 | **±1.75**<br>**±2** | LSB (max)<br>LSB (max)<br>LSB |
| $C_{REF}$ | $V_{REF+}/V_{REF-}$ Input Capacitance | | 85 | | pF |
| $C_{IN}$ | Selected Multiplexer Channel Input Capacitance | | 75 | | pF |

## Converter AC Characteristics

The following specifications apply to the LM12454, LM12H454, LM12458, and LM12H458 for $V_A+ = V_D+ = 5V$, $V_{REF+} = 5V$, $V_{REF-} = 0V$, 12-bit + sign conversion mode, $f_{CLK} = 8.0$ MHz (LM12H454/8) or $f_{CLK} = 5.0$ MHz (LM12454/8), $R_S = 25\Omega$, source impedance for $V_{REF+}$ and $V_{REF-} \leq 25\Omega$, fully-differential input with fixed 2.5V common-mode voltage, and minimum acquisition time unless otherwise specified. **Boldface limits apply for $T_A = T_J = T_{MIN}$ to $T_{MAX}$**; all other limits $T_A = T_J = 25°C$. (Notes 6, 7, 8, 9 and 19)

| Symbol | Parameter | Conditions | Typical (Note 10) | Limits (Note 11) | Unit (Limit) |
|---|---|---|---|---|---|
| | Clock Duty Cycle | | 50 | <br>**40**<br>**60** | %<br>% (min)<br>% (max) |
| $t_C$ | Conversion Time | 13-Bit Resolution, Sequencer State S5 *(Figure 11)* | 44 ($t_{CLK}$) | **44 ($t_{CLK}$) + 50 ns** | (max) |
| | | 9-Bit Resolution, Sequencer State S5 *(Figure 11)* | 21 ($t_{CLK}$) | **21 ($t_{CLK}$) + 50 ns** | (max) |
| $t_A$ | Acquisition Time | Sequencer State S7 *(Figure 11)* Built-in minimum for 13-Bits | 9 ($t_{CLK}$) | **9 ($t_{CLK}$) + 50 ns** | (max) |
| | | Built-in minimum for 9-Bits and "Watchdog" mode | 2 ($t_{CLK}$) | **2 ($t_{CLK}$) + 50 ns** | (max) |
| $t_Z$ | Auto-Zero Time | Sequencer State S2 *(Figure 11)* | 76 ($t_{CLK}$) | **76 ($t_{CLK}$) + 50 ns** | (max) |
| $t_{CAL}$ | Full Calibration Time | Sequencer State S2 *(Figure 11)* | 4944 ($t_{CLK}$) | **4944 ($t_{CLK}$) + 50 ns** | (max) |
| | Throughput Rate (Note 18) | <br>LM12H454, LM12H458 | 89<br>142 | **88**<br>**140** | kHz<br>(min) |
| $t_{WD}$ | "Watchdog" Mode Comparison Time | Sequencer States S6, S4, and S5 *(Figure 11)* | 11 ($t_{CLK}$) | **11 ($t_{CLK}$) + 50 ns** | (max) |

## Converter AC Characteristics

The following specifications apply to the LM12454, LM12H454, LM12458, and LM12H458 for $V_A+ = V_D+ = 5V$, $V_{REF+} = 5V$, $V_{REF-} = 0V$, 12-bit + sign conversion mode, $f_{CLK} = 5.0$ MHz, $R_S = 25\Omega$, source impedance for $V_{REF+}$ and $V_{REF-} \leq 25\Omega$, fully-differential input with fixed 2.5V common-mode voltage, and minimum acquisition time unless otherwise specified. **Boldface limits apply for $T_A = T_J = T_{MIN}$ to $T_{MAX}$**; all other limits $T_A = T_J = 25°C$. (Notes 6, 7, 8, 9 and 19) (Continued)

| Symbol | Parameter | Conditions | Typical (Note 10) | Limits (Note 11) | Unit (Limit) |
|---|---|---|---|---|---|
| DSNR | Differential Signal-to-Noise Ratio | $V_{IN} = \pm 5V$ | | | |
| | | $f_{IN} = 1$ kHz | 77.5 | | dB |
| | | $f_{IN} = 20$ kHz | 75.2 | | dB |
| | | $f_{IN} = 40$ kHz | 74.7 | | dB |
| SESNR | Single-Ended Signal-to-Noise Ratio | $V_{IN} = 5\ V_{p\text{-}p}$ | | | |
| | | $f_{IN} = 1$ kHz | 69.8 | | dB |
| | | $f_{IN} = 20$ kHz | 69.2 | | dB |
| | | $f_{IN} = 40$ kHz | 66.6 | | dB |
| DSINAD | Differential Signal-to-Noise + Distortion Ratio | $V_{IN} = \pm 5V$ | | | |
| | | $f_{IN} = 1$ kHz | 76.9 | | dB |
| | | $f_{IN} = 20$ kHz | 73.9 | | dB |
| | | $f_{IN} = 40$ kHz | 70.7 | | dB |
| SESINAD | Single-Ended Signal-to-Noise + Distortion Ratio | $V_{IN} = 5\ V_{p\text{-}p}$ | | | |
| | | $f_{IN} = 1$ kHz | 69.4 | | dB |
| | | $f_{IN} = 20$ kHz | 68.3 | | dB |
| | | $f_{IN} = 40$ kHz | 65.7 | | dB |
| DTHD | Differential Total Harmonic Distortion | $V_{IN} = \pm 5V$ | | | |
| | | $f_{IN} = 1$ kHz | −85.8 | | dB |
| | | $f_{IN} = 20$ kHz | −79.9 | | dB |
| | | $f_{IN} = 40$ kHz | −72.9 | | dB |
| SETHD | Single-Ended Total Harmonic Distortion | $V_{IN} = 5\ V_{p\text{-}p}$ | | | |
| | | $f_{IN} = 1$ kHz | −80.3 | | dB |
| | | $f_{IN} = 20$ kHz | −75.6 | | dB |
| | | $f_{IN} = 40$ kHz | −72.8 | | dB |
| DENOB | Differential Effective Number of Bits | $V_{IN} = \pm 5V$ | | | |
| | | $f_{IN} = 1$ kHz | 12.6 | | Bits |
| | | $f_{IN} = 20$ kHz | 12.2 | | Bits |
| | | $f_{IN} = 40$ kHz | 12.1 | | Bits |
| SEENOB | Single-Ended Effective Number of Bits | $V_{IN} = 5\ V_{p\text{-}p}$ | | | |
| | | $f_{IN} = 1$ kHz | 11.3 | | Bits |
| | | $f_{IN} = 20$ kHz | 11.2 | | Bits |
| | | $f_{IN} = 40$ kHz | 10.8 | | Bits |
| DSFDR | Differential Spurious Free Dynamic Range | $V_{IN} = \pm 5V$ | | | |
| | | $f_{IN} = 1$ kHz | 87.2 | | dB |
| | | $f_{IN} = 20$ kHz | 78.9 | | dB |
| | | $f_{IN} = 40$ kHz | 72.8 | | dB |
| | Multiplexer Channel-to-Channel Crosstalk | $V_{IN} = 5\ V_{PP}$, $f_{IN} = 40$ kHz | | | |
| | | LM12(H)454 MUXOUT Only | −76 | | dB |
| | | LM12(H)458 MUX plus Converter | −78 | | dB |
| $t_{PU}$ | Power-Up Time | | 10 | | ms |
| $t_{WU}$ | Wake-Up Time | | 10 | | ms |

**DC Characteristics** The following specifications apply to the LM12454, LM12H454, LM12458, and LM12H458 for $V_A^+ = V_D^+ = 5V$, $V_{REF+} = 5V$, $V_{REF-} = 0V$, $f_{CLK} = 8.0$ MHz (LM12H454/8) or $f_{CLK} = 5.0$ MHz (LM12454/8), and minimum acquisition time unless otherwise specified. **Boldface limits apply for $T_A = T_J = T_{MIN}$ to $T_{MAX}$**; all other limits $T_A = T_J = 25°C$. (Notes 6, 7, 8, and 19)

| Symbol | Parameter | Conditions | Typical (Note 10) | Limits (Note 11) | Unit (Limit) |
|---|---|---|---|---|---|
| $I_D^+$ | $V_D^+$ Supply Current | $\overline{CS}$ = "1" | | | |
| | | LM12454/8 | 0.55 | **1.0** | mA (max) |
| | | LM12H454/8 | 0.55 | **1.2** | |
| $I_A^+$ | $V_A^+$ Supply Current | $\overline{CS}$ = "1" | | | |
| | | LM12454/8 | 3.1 | **5.0** | mA (max) |
| | | LM12H454/8 | 3.1 | **5.5** | |
| $I_{ST}$ | Stand-By Supply Current ($I_D^+ + I_A^+$) | Power-Down Mode Selected | | | |
| | | Clock Stopped | 10 | | μA (max) |
| | | 8 MHz Clock | 40 | | μA (max) |
| | Multiplexer ON-Channel Leakage Current | $V_A^+ = 5.5V$ | | | |
| | | ON-Channel = 5.5V | | **0.3** | |
| | | OFF-Channel = 0V | 0.1 | | μA (max) |
| | | LM12(H)458CMEL | | **0.5** | |
| | | ON-Channel = 0V | | **0.3** | |
| | | OFF-Channel = 5.5V | 0.1 | | μA (max) |
| | | LM12(H)458CMEL | | **0.5** | |
| | Multiplexer OFF-Channel Leakage Current | $V_A^+ = 5.5V$ | | | |
| | | ON-Channel = 5.5V | | **0.3** | |
| | | OFF-Channel = 0V | 0.1 | | μA (max) |
| | | LM12(H)458CMEL | | **0.5** | |
| | | ON-Channel = 0V | | **0.3** | |
| | | OFF-Channel = 5.5V | 0.1 | | μA (max) |
| | | LM12(H)458CMEL | | **0.5** | |
| $R_{ON}$ | Multiplexer ON-Resistance | LM12(H)454 | | | |
| | | $V_{IN} = 5V$ | 800 | **1500** | Ω(max) |
| | | $V_{IN} = 2.5V$ | 850 | **1500** | Ω(max) |
| | | $V_{IN} = 0V$ | 760 | **1500** | Ω(max) |
| | Multiplexer Channel-to-Channel $R_{ON}$ matching | LM12(H)454 | | | |
| | | $V_{IN} = 5V$ | ±1.0% | **±3.0%** | (max) |
| | | $V_{IN} = 2.5V$ | ±1.0% | **±3.0%** | (max) |
| | | $V_{IN} = 0V$ | ±1.0% | **±3.0%** | (max) |

**Internal Reference Characteristics** The following specifications apply to the LM12454, LM12H454, LM12458, and LM12H458 for $V_A^+ = V_D^+ = 5V$ unless otherwise specified. **Boldface limits apply for $T_A = T_J = T_{MIN}$ to $T_{MAX}$**; all other limits $T_A = T_J = 25°C$. (Notes 6, 7, and 19)

| Symbol | Parameter | Conditions | Typical (Note 10) | Limits (Note 11) | Unit (Limit) |
|---|---|---|---|---|---|
| $V_{REFOUT}$ | Internal Reference Output Voltage | | 2.5 | **2.5 ±4%** | V (max) |
| | | LM12(H)458CMEL | | **2.5 ±6%** | |
| $\Delta V_{REF}/\Delta T$ | Internal Reference Temperature Coefficient | | 40 | | ppm/°C |
| $\Delta_{REF}/\Delta I_L$ | Internal Reference Load Regulation | Sourcing ($0 < I_L \le +4$ mA) | | **0.2** | %/mA (max) |
| | | Sinking ($-1 \le I_{IL} < 0$ mA) | | **1.2** | %/mA (max) |
| $\Delta V_{REF}$ | Line Regulation | $4.5V \le V_A^+ \le 5.5V$ | 3 | **20** | mV (max) |
| $I_{SC}$ | Internal Reference Short Circuit Current | $V_{REFOUT} = 0V$ | 13 | **25** | mA (max) |
| $\Delta V_{REF}/\Delta t$ | Long Term Stability | | 200 | | ppm/kHr |
| $t_{SU}$ | Internal Reference Start-Up Time | $V_A^+ = V_D^+ = 0V \rightarrow 5V$, $C_L = 100\ \mu F$ | 10 | | ms |

## Digital Characteristics

The following specifications apply to the LM12454, LM12H454, LM12458, and LM12H458 for $V_A^+ = V_D^+ = 5V$, unless otherwise specified. **Boldface limits apply for $T_A = T_J = T_{MIN}$ to $T_{MAX}$**; all other limits $T_A = T_J = 25°C$. (Notes 6, 7, 8, and 19)

| Symbol | Parameter | Conditions | Typical (Note 10) | Limits (Note 11) | Unit (Limit) |
|---|---|---|---|---|---|
| $V_{IN(1)}$ | Logical "1" Input Voltage | $V_A^+ = V_D^+ = 5.5V$ | | **2.0** | V (min) |
| $V_{IN(0)}$ | Logical "0" Input Voltage | $V_A^+ = V_D^+ = 4.5V$ | | **0.8** | V (max) |
| $I_{IN(1)}$ | Logical "1" Input Current | $V_{IN} = 5V$<br>LM12(H)458CMEL | 0.005 | **1.0**<br>**2.0** | μA (max) |
| $I_{IN(0)}$ | Logical "0" Input Current | $V_{IN} = 0V$<br>LM12(H)458CMEL | −0.005 | **−1.0**<br>**−2.0** | μA (max) |
| $C_{IN}$ | D0–D15 Input Capacitance | | 6 | | pF |
| $V_{OUT(1)}$ | Logical "1" Output Voltage | $V_A^+ = V_D^+ = 4.5V$<br>$I_{OUT} = -360\ \mu A$<br>$I_{OUT} = -10\ \mu A$ | | <br>**2.4**<br>**4.25** | <br>V (min)<br>V (min) |
| $V_{OUT(0)}$ | Logical "0" Output Voltage | $V_A^+ = V_D^+ = 4.5V$<br>$I_{OUT} = 1.6\ mA$ | | **0.4** | V (max) |
| $I_{OUT}$ | TRI-STATE® Output Leakage Current | $V_{OUT} = 0V$<br>$V_{OUT} = 5V$ | −0.01<br>0.01 | **−3.0**<br>**3.0** | μA (max)<br>μA (max) |

## Digital Timing Characteristics

The following specifications apply to the LM12454, LM12H454, LM12458, and LM12H458 for $V_A^+ = V_D^+ = 5V$, $t_r = t_f = 3$ ns, and $C_L = 100$ pF on data I/O, $\overline{INT}$ and DMARQ lines unless otherwise specified. **Boldface limits apply for $T_A = T_J = T_{MIN}$ to $T_{MAX}$**; all other limits $T_A = T_J = 25°C$. (Notes 6, 7, 8, and 19)

| Symbol (See *Figures 8a, 8b,* and *8c*) | Parameter | Conditions | Typical (Note 10) | Limits (Note 11) | Unit (Limit) |
|---|---|---|---|---|---|
| 1, 3 | $\overline{CS}$ or Address Valid to ALE Low Set-Up Time | | | **40** | ns (min) |
| 2, 4 | $\overline{CS}$ or Address Valid to ALE Low Hold Time | | | **20** | ns (min) |
| 5 | ALE Pulse Width | | | **45** | ns (min) |
| 6 | $\overline{RD}$ High to Next ALE High | | | **35** | ns (min) |
| 7 | ALE Low to $\overline{RD}$ Low | | | **20** | ns (min) |
| 8 | $\overline{RD}$ Pulse Width | | | **100** | ns (min) |
| 9 | $\overline{RD}$ High to Next $\overline{RD}$ or $\overline{WR}$ Low | | | **100** | ns (min) |
| 10 | ALE Low to $\overline{WR}$ Low | | | **20** | ns (min) |
| 11 | $\overline{WR}$ Pulse Width | | | **60** | ns (min) |
| 12 | $\overline{WR}$ High to Next ALE High | | | **75** | ns (min) |
| 13 | $\overline{WR}$ High to Next $\overline{RD}$ or $\overline{WR}$ Low | | | **140** | ns (min) |
| 14 | Data Valid to $\overline{WR}$ High Set-Up Time | | | **40** | ns (min) |
| 15 | Data Valid to $\overline{WR}$ High Hold Time | | | **30** | ns (min) |
| 16 | $\overline{RD}$ Low to Data Bus Out of TRI-STATE | | 40 | **10**<br>**70** | ns (min)<br>ns (max) |
| 17 | $\overline{RD}$ High to TRI-STATE | $R_L = 1\ k\Omega$ | 30 | **10**<br>**110** | ns (min)<br>ns (max) |
| 18 | $\overline{RD}$ Low to Data Valid (Access Time) | | 30 | **10**<br>**80** | ns (min)<br>ns (max) |

## Digital Timing Characteristics

The following specifications apply to the LM12454, LM12H454, LM12458, and LM12H458 for $V_A+ = V_D+ = 5V$, $t_r = t_f = 3$ ns, and $C_L = 100$ pF on data I/O, $\overline{INT}$ and DMARQ lines unless otherwise specified. **Boldface limits apply for $T_A = T_J = T_{MIN}$ to $T_{MAX}$**; all other limits $T_A = T_J = 25°C$. (Notes 6, 7, 8, and 19) (Continued)

| Symbol (See *Figures 8a, 8b,* and *8c*) | Parameter | Conditions | Typical (Note 10) | Limits (Note 11) | Unit (Limit) |
|---|---|---|---|---|---|
| 20 | Address Valid or $\overline{CS}$ Low to $\overline{RD}$ Low | | | **20** | ns (min) |
| 21 | Address Valid or $\overline{CS}$ Low to $\overline{WR}$ Low | | | **20** | ns (min) |
| 19 | Address Invalid from $\overline{RD}$ or $\overline{WR}$ High | | | **10** | ns (min) |
| 22 | $\overline{INT}$ High from $\overline{RD}$ Low | | 30 | **10**<br>**60** | ns (min)<br>ns (max) |
| 23 | DMARQ Low from $\overline{RD}$ Low | | 30 | **10**<br>**60** | ns (min)<br>ns (max) |

**Note 1:** Absolute Maximum Ratings indicate limits beyond which damage to the device may occur. Operating Ratings indicate conditions for which the device is functional, but do not guarantee specific performance limits. For guaranteed specifications and test conditions, see the Electrical Characteristics. The guaranteed specifications apply only for the test conditions listed. Some performance characteristics may degrade when the device is not operated under the listed test conditions.

**Note 2:** All voltages are measured with respect to GND, unless otherwise specified.

**Note 3:** When the input voltage ($V_{IN}$) at any pin exceeds the power supply rails ($V_{IN} <$ GND or $V_{IN} > (V_A+$ or $V_D+)$), the current at that pin should be limited to 5 mA. The 20 mA maximum package input current rating allows the voltage at any four pins, with an input current of 5 mA, to simultaneously exceed the power supply voltages.

**Note 4:** The maximum power dissipation must be derated at elevated temperatures and is dictated by $T_{Jmax}$ (maximum junction temperature), $\Theta_{JA}$ (package junction to ambient thermal resistance), and $T_A$ (ambient temperature). The maximum allowable power dissipation at any temperature is $PD_{max} = (T_{Jmax} - T_A)/\Theta_{JA}$ or the number given in the Absolute Maximum Ratings, whichever is lower. For this device, $T_{Jmax} = 150°C$, and the typical thermal resistance ($\Theta_{JA}$) of the LM12(H)454 and LM12(H)458 in the V package, when board mounted, is 47°C/W and in the EL package, when board mounted, is 70°C/W.

**Note 5:** Human body model, 100 pF discharged through a 1.5 kΩ resistor.

**Note 6:** Two on-chip diodes are tied to each analog input through a series resistor, as shown below. Input voltage magnitude up to 5V above $V_A+$ or 5V below GND will not damage the LM12(H)454 or the LM12(H)458. However, errors in the A/D conversion can occur if these diodes are forward biased by more than 100 mV. As an example, if $V_A+$ is 4.5 $V_{DC}$, full-scale input voltage must be $\leq 4.6$ $V_{DC}$ to ensure accurate conversions.

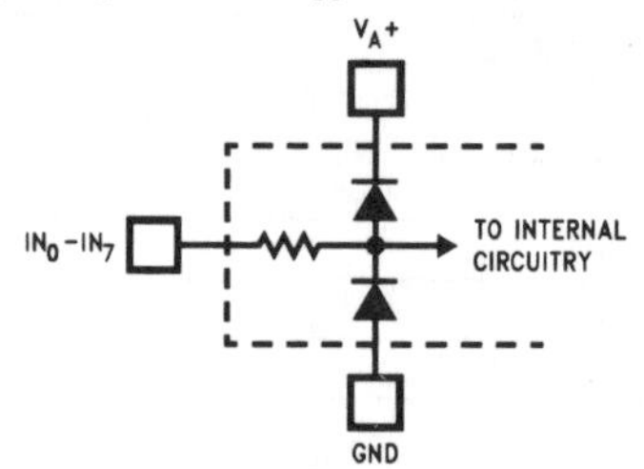


TL/H/11264-3

**Note 7:** $V_A+$ and $V_D+$ must be connected together to the same power supply voltage and bypassed with separate capacitors at each V+ pin to assure conversion/comparison accuracy.

**Note 8:** Accuracy is guaranteed when operating at $f_{CLK} = 5$ MHz for the LM12454/8 and $f_{CLK} = 8$ MHz for the LM12H454/8.

**Note 9:** With the test condition for $V_{REF}$ ($V_{REF+} - V_{REF-}$) given as +5V, the 12-bit LSB is 1.22 mV and the 8-bit/"Watchdog" LSB is 19.53 mV.

**Note 10:** Typicals are at $T_A = 25°C$ and represent most likely parametric norm.

**Note 11:** Limits are guaranteed to National's AOQL (Average Output Quality Level).

**Note 12:** Positive integral linearity error is defined as the deviation of the analog value, expressed in LSBs, from the straight line that passes through positive full-scale and zero. For negative integral linearity error the straight line passes through negative full-scale and zero. (See *Figures 5b* and *5c*).

**Note 13:** Zero error is a measure of the deviation from the mid-scale voltage (a code of zero), expressed in LSB. It is the worst-case value of the code transitions between −1 to 0 and 0 to +1 (see *Figure 6*).

**Note 14:** The DC common-mode error is measured with both inputs shorted together and driven from 0V to 5V. The measured value is referred to the resulting output value when the inputs are driven with a 2.5V signal.

**Note 15:** Power Supply Sensitivity is measured after Auto-Zero and/or Auto-Calibration cycle has been completed with $V_A+$ and $V_D+$ at the specified extremes.

**Note 16:** $V_{REFCM}$ (Reference Voltage Common Mode Range) is defined as $(V_{REF+} + V_{REF-})/2$.

**Note 17:** The LM12(H)454/8's self-calibration technique ensures linearity and offset errors as specified, but noise inherent in the self-calibration process will result in a repeatability uncertainty of ±0.10 LSB.

**Note 18:** The Throughput Rate is for a single instruction repeated continuously. Sequencer states 0 (1 clock cycle), 1 (1 clock cycle), 7 (9 clock cycles) and 5 (44 clock cycles) are used (see *Figure 11*). One additional clock cycle is used to read the conversion result stored in the FIFO, for a total of 56 clock cycles per conversion. The Throughput Rate is $f_{CLK}$ (MHz)/N, where N is the number of clock cycles/conversion.

**Note 19:** A military RETS specification is available upon request. At the time of printing, the LM12(H)458CMEL/883 RETS specification complied with the **boldface** values in the Limits column.

## Electrical Characteristics

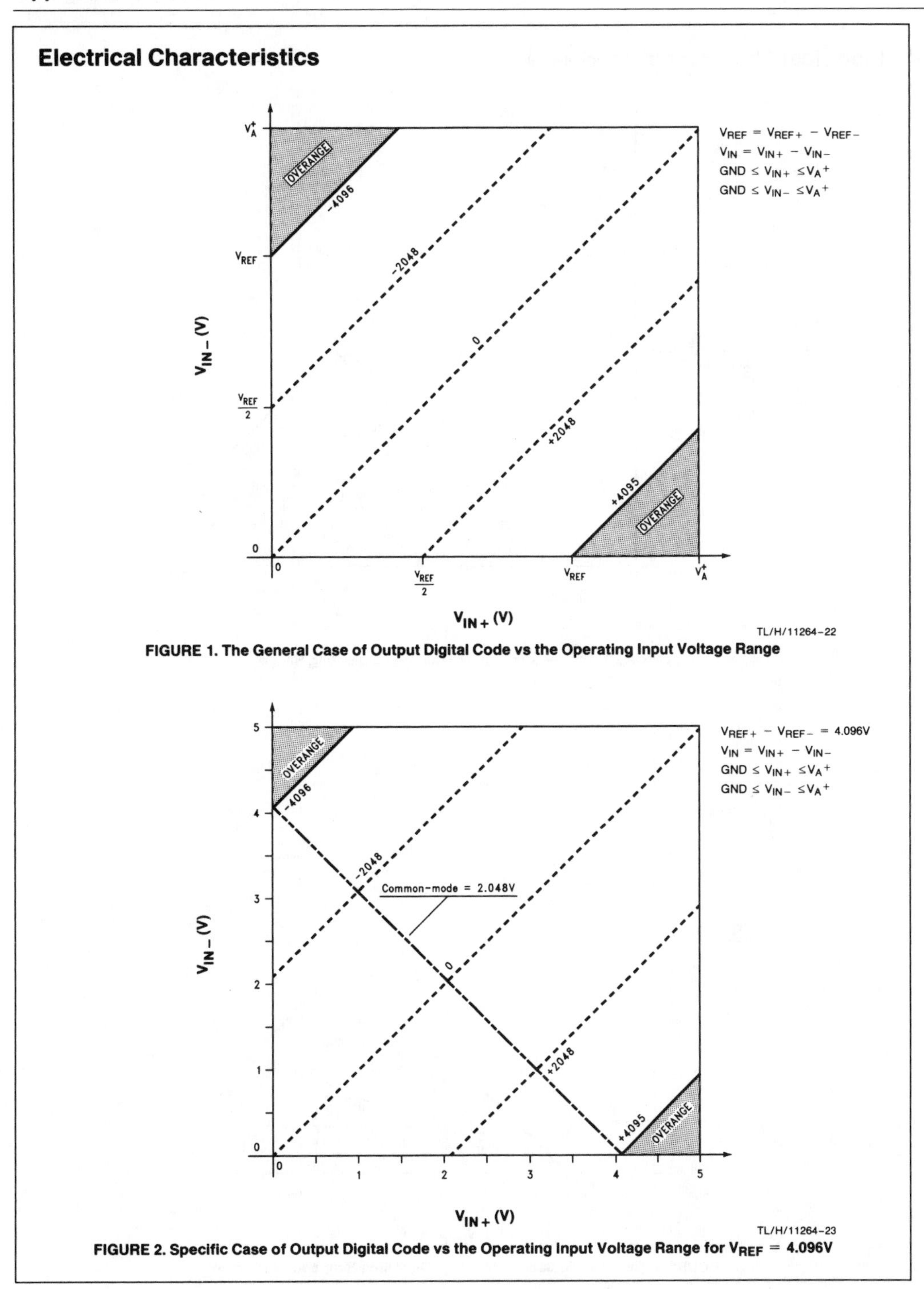


**FIGURE 1. The General Case of Output Digital Code vs the Operating Input Voltage Range**



**FIGURE 2. Specific Case of Output Digital Code vs the Operating Input Voltage Range for $V_{REF}$ = 4.096V**

## Electrical Characteristics (Continued)

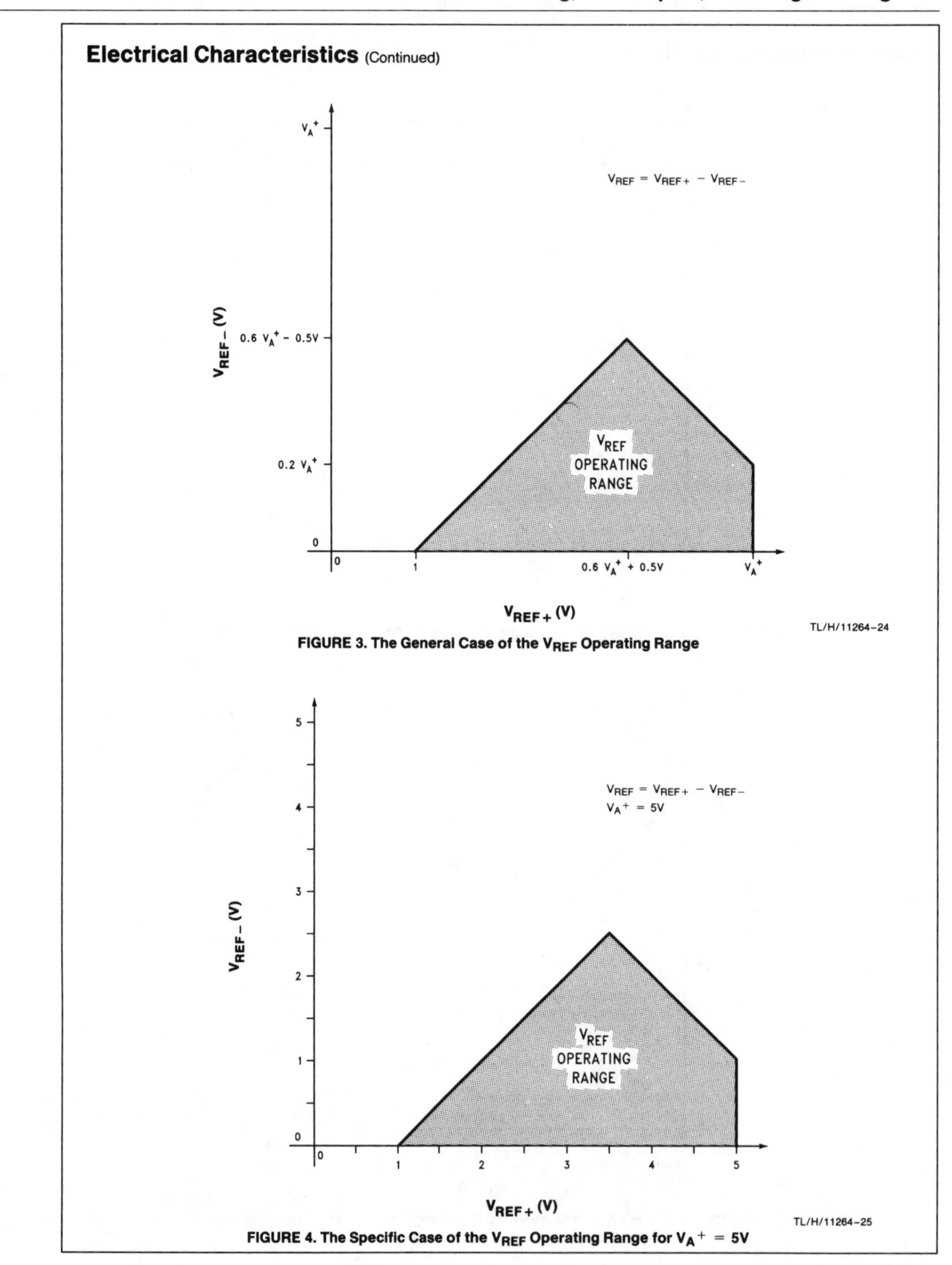


**FIGURE 3. The General Case of the $V_{REF}$ Operating Range**



**FIGURE 4. The Specific Case of the $V_{REF}$ Operating Range for $V_A^+ = 5V$**

## Electrical Characteristics (Continued)

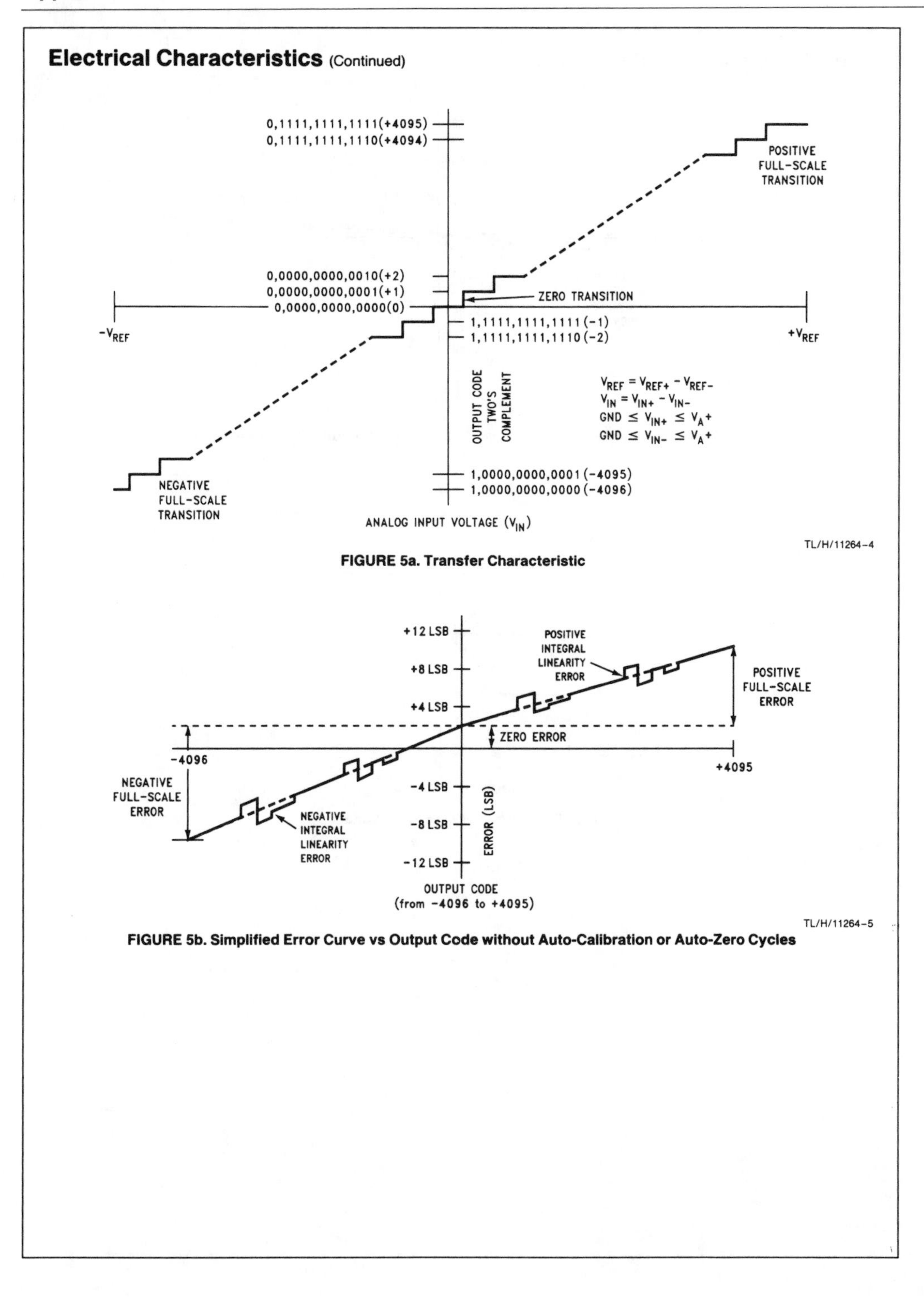


**FIGURE 5a. Transfer Characteristic**



**FIGURE 5b. Simplified Error Curve vs Output Code without Auto-Calibration or Auto-Zero Cycles**

## Electrical Characteristics (Continued)

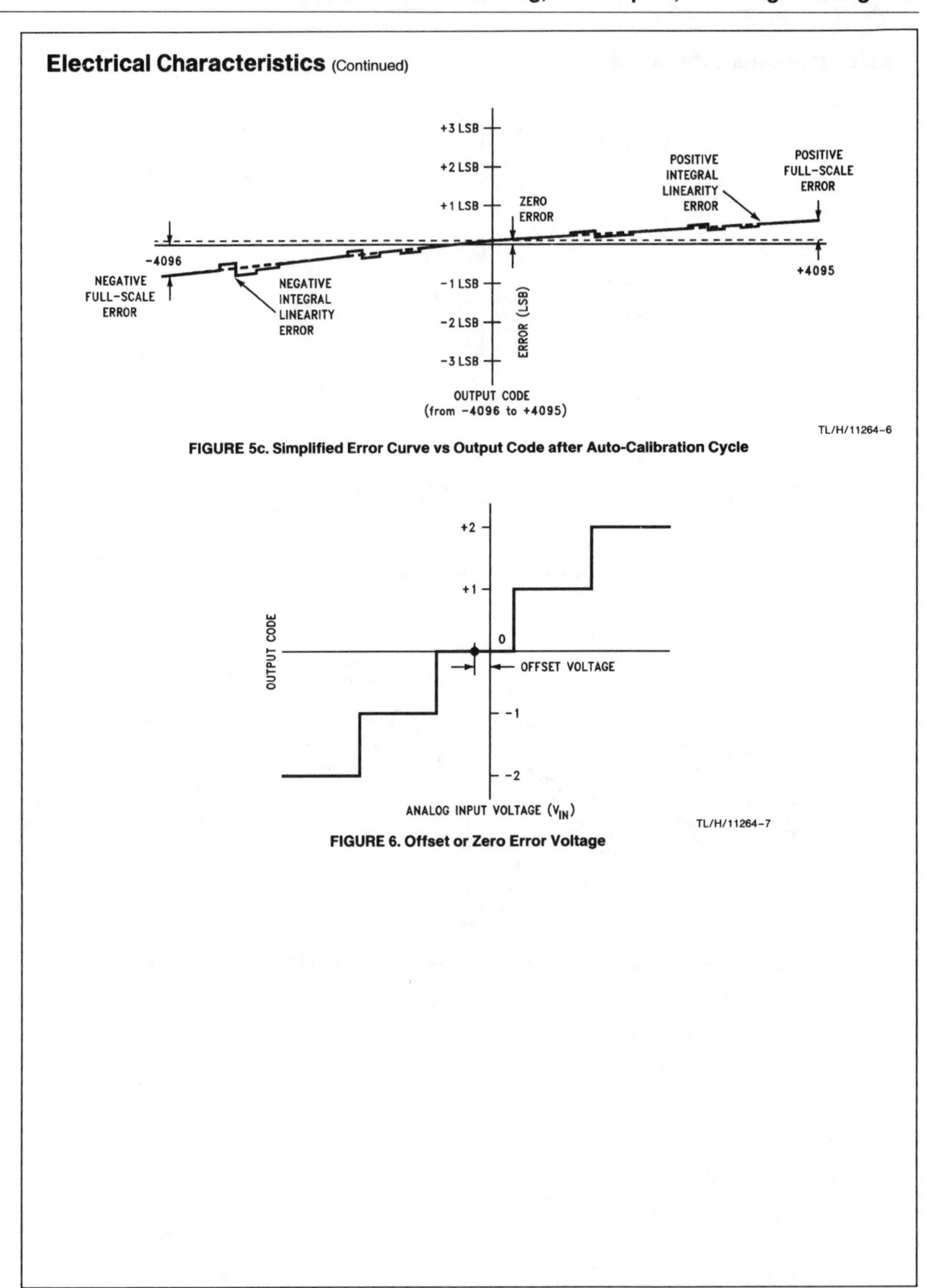


FIGURE 5c. Simplified Error Curve vs Output Code after Auto-Calibration Cycle



FIGURE 6. Offset or Zero Error Voltage

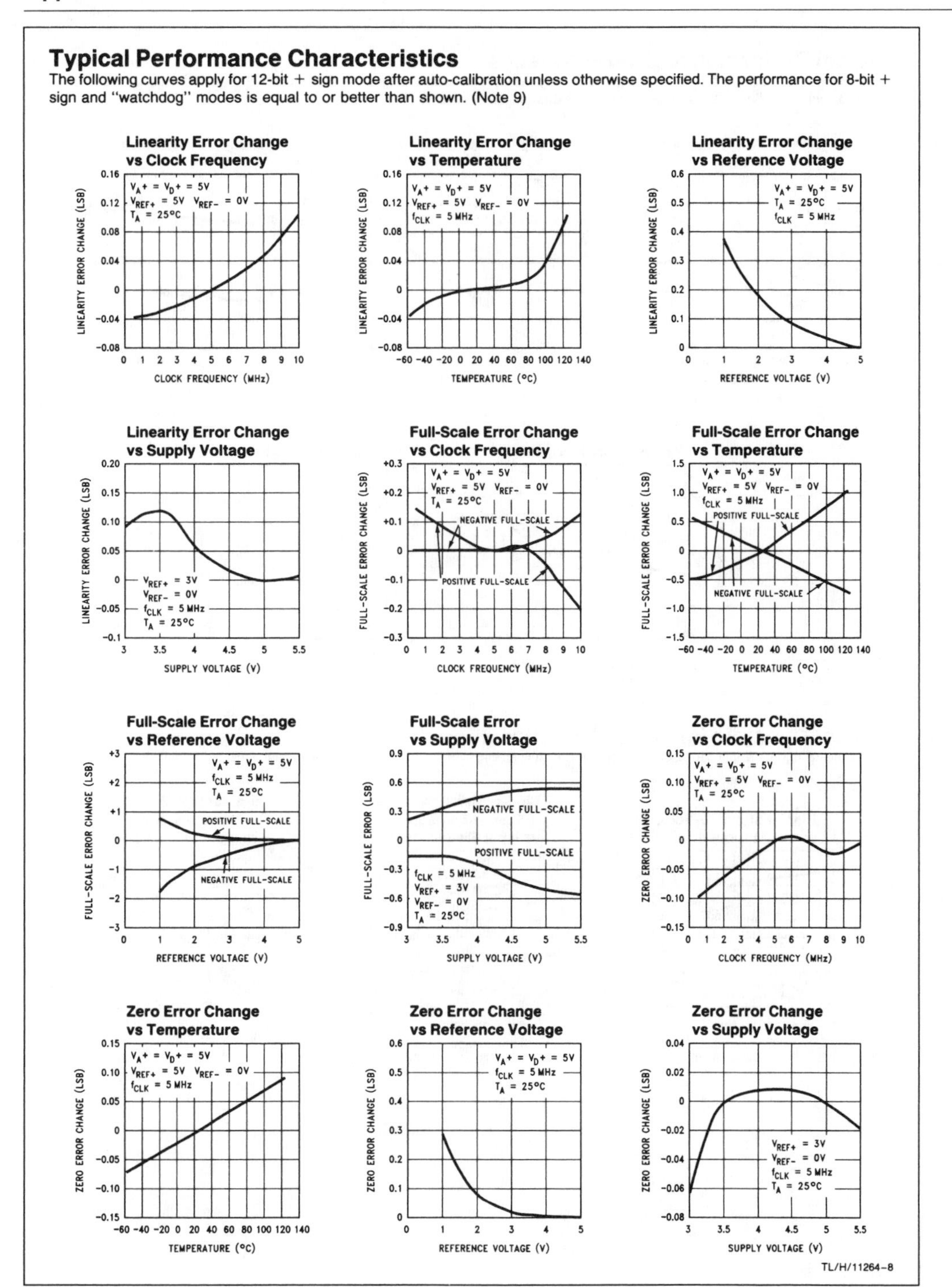
Typical Performance Characteristics
The following curves apply for 12-bit + sign mode after auto-calibration unless otherwise specified. The performance for 8-bit + sign and "watchdog" modes is equal to or better than shown. (Note 9)
Linearity Error Change vs Clock Frequency
Linearity Error Change vs Temperature
Linearity Error Change vs Reference Voltage
Linearity Error Change vs Supply Voltage
Full-Scale Error Change vs Clock Frequency
Full-Scale Error Change vs Temperature
Full-Scale Error Change vs Reference Voltage
Full-Scale Error vs Supply Voltage
Zero Error Change vs Clock Frequency
Zero Error Change vs Temperature
Zero Error Change vs Reference Voltage
Zero Error Change vs Supply Voltage
LINEARITY ERROR CHANGE (LSB)
FULL-SCALE ERROR CHANGE (LSB)
FULL-SCALE ERROR (LSB)
ZERO ERROR CHANGE (LSB)
CLOCK FREQUENCY (MHz)
TEMPERATURE (°C)
REFERENCE VOLTAGE (V)
SUPPLY VOLTAGE (V)
POSITIVE FULL-SCALE
NEGATIVE FULL-SCALE
TL/H/11264-8

## Typical Performance Characteristics

The following curves apply for 12-bit + sign mode after auto-calibration unless otherwise specified. The performance for 8-bit + sign and "watchdog" modes is equal to or better than shown. (Note 9) (Continued)

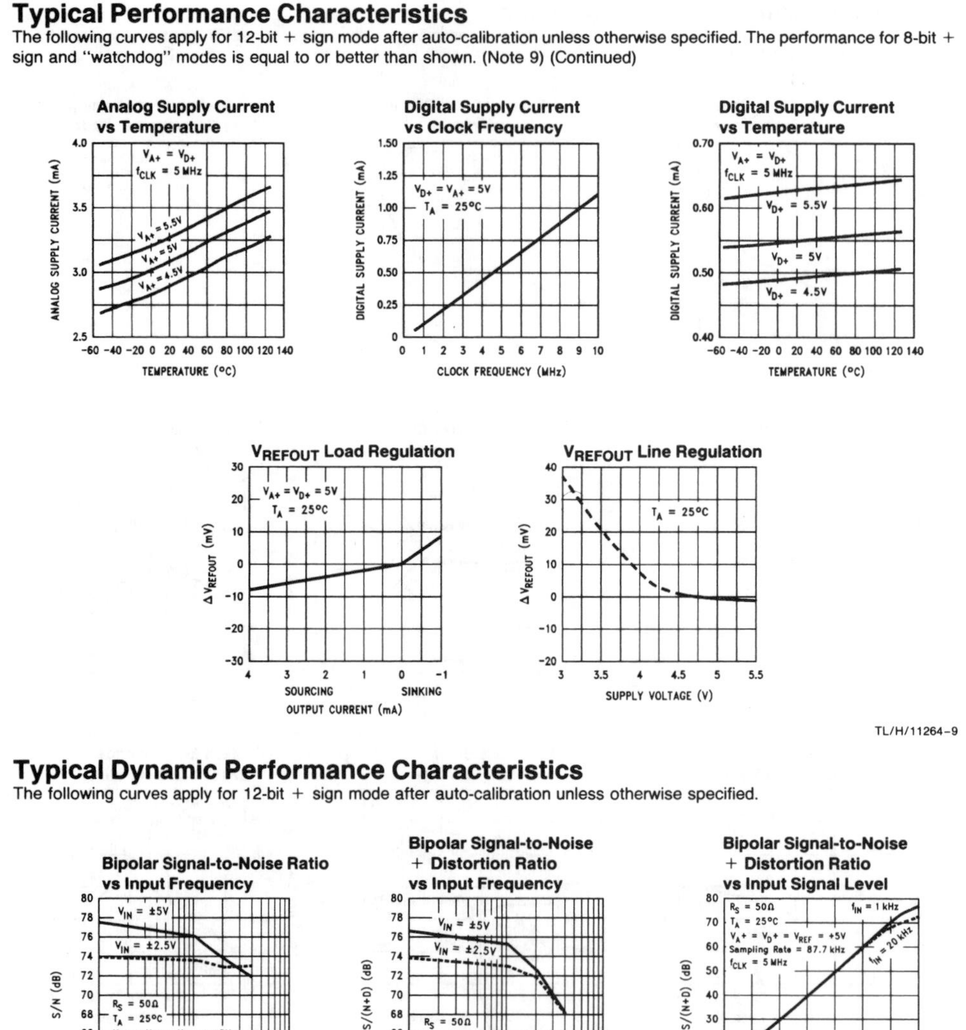


## Typical Dynamic Performance Characteristics

The following curves apply for 12-bit + sign mode after auto-calibration unless otherwise specified.

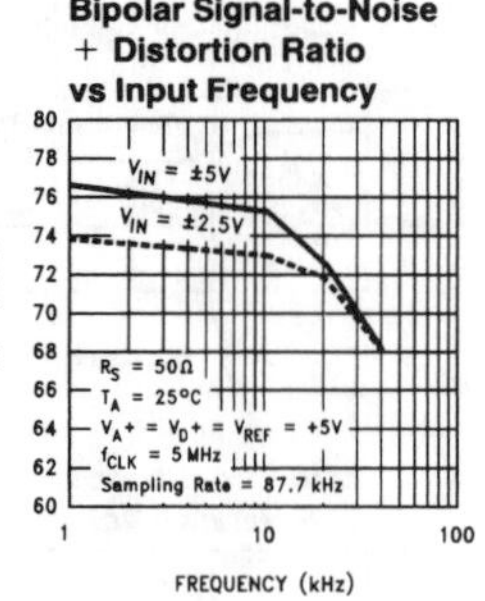

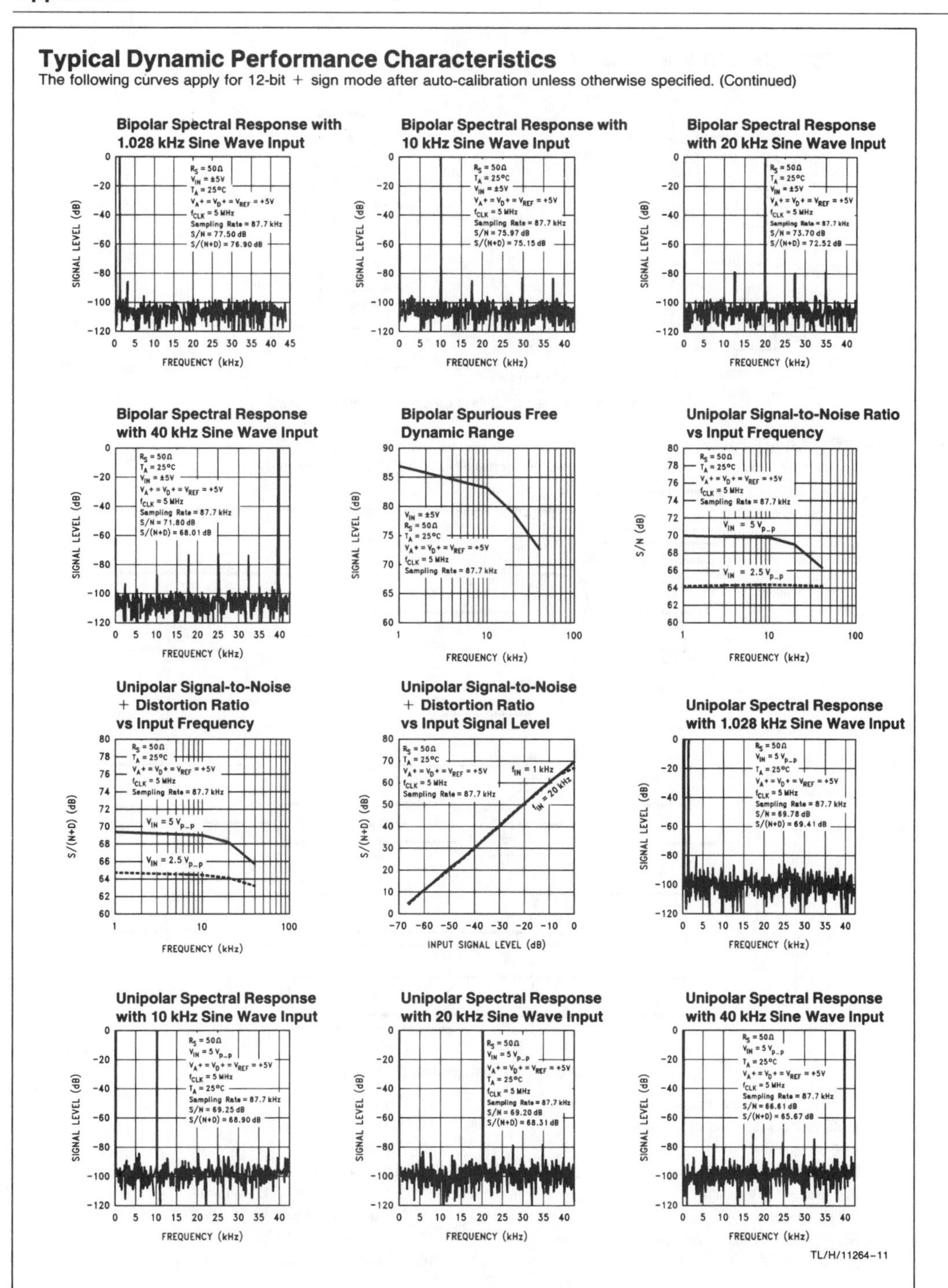
Typical Dynamic Performance Characteristics
The following curves apply for 12-bit + sign mode after auto-calibration unless otherwise specified. (Continued)
Bipolar Spectral Response with 1.028 kHz Sine Wave Input
R_S = 50Ω
V_IN = ±5V
T_A = 25°C
V_A+ = V_D+ = V_REF = +5V
f_CLK = 5 MHz
Sampling Rate = 87.7 kHz
S/N = 77.50 dB
S/(N+D) = 76.90 dB
Bipolar Spectral Response with 10 kHz Sine Wave Input
S/N = 75.97 dB
S/(N+D) = 75.15 dB
Bipolar Spectral Response with 20 kHz Sine Wave Input
S/N = 73.70 dB
S/(N+D) = 72.52 dB
Bipolar Spectral Response with 40 kHz Sine Wave Input
S/N = 71.80 dB
S/(N+D) = 68.01 dB
Bipolar Spurious Free Dynamic Range
Unipolar Signal-to-Noise Ratio vs Input Frequency
V_IN = 5 V_P-P
V_IN = 2.5 V_P-P
Unipolar Signal-to-Noise + Distortion Ratio vs Input Frequency
Unipolar Signal-to-Noise + Distortion Ratio vs Input Signal Level
f_IN = 1 kHz
f_IN = 20 kHz
Unipolar Spectral Response with 1.028 kHz Sine Wave Input
S/N = 69.78 dB
S/(N+D) = 69.41 dB
Unipolar Spectral Response with 10 kHz Sine Wave Input
S/N = 69.25 dB
S/(N+D) = 68.90 dB
Unipolar Spectral Response with 20 kHz Sine Wave Input
S/N = 69.20 dB
S/(N+D) = 68.31 dB
Unipolar Spectral Response with 40 kHz Sine Wave Input
S/N = 66.61 dB
S/(N+D) = 65.67 dB
SIGNAL LEVEL (dB)
S/N (dB)
S/(N+D) (dB)
FREQUENCY (kHz)
INPUT SIGNAL LEVEL (dB)
TL/H/11264-11

## Test Circuits and Waveforms

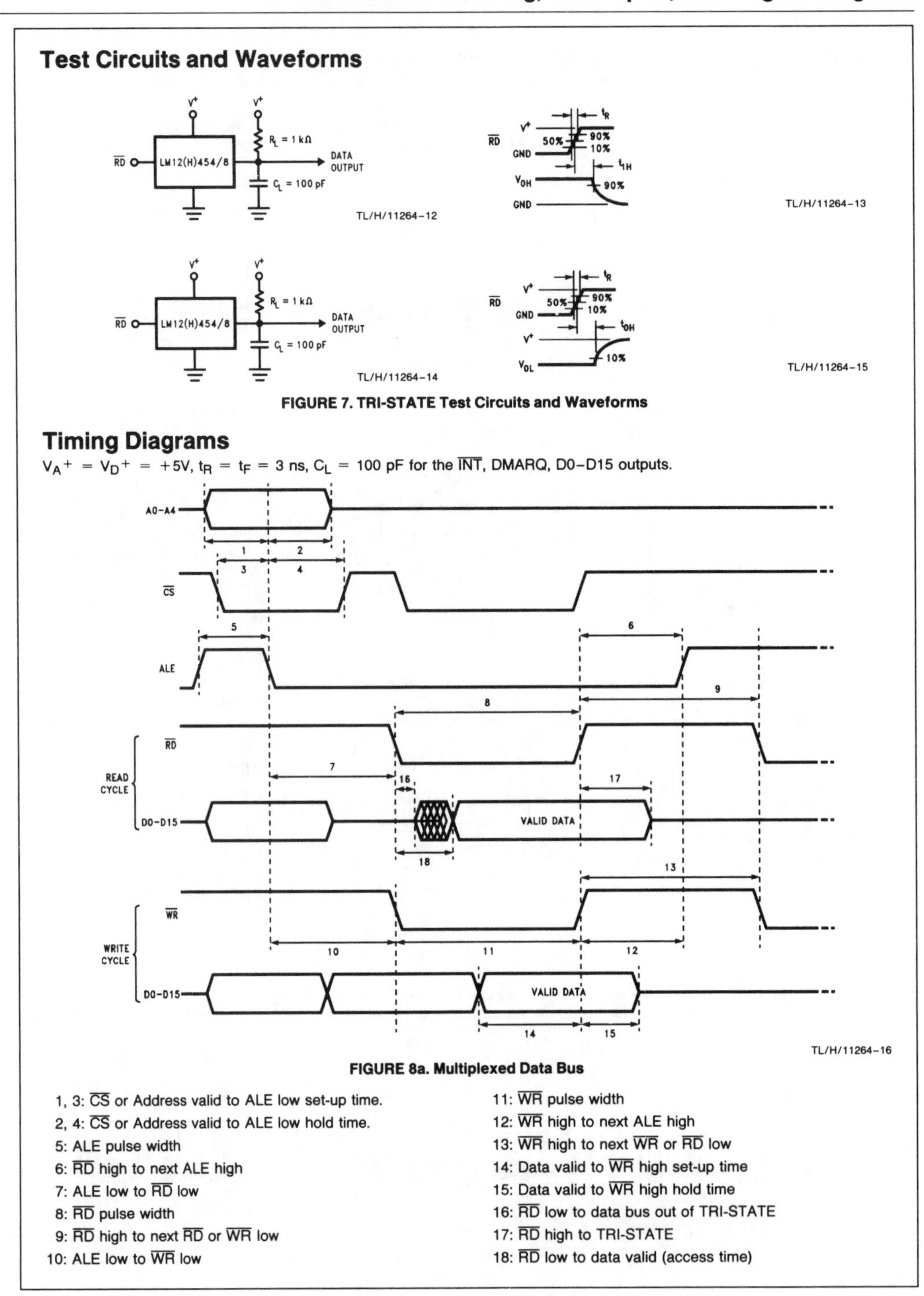


**FIGURE 7. TRI-STATE Test Circuits and Waveforms**

## Timing Diagrams

$V_A{}^+ = V_D{}^+ = +5V$, $t_R = t_F = 3$ ns, $C_L = 100$ pF for the $\overline{INT}$, DMARQ, D0–D15 outputs.

**FIGURE 8a. Multiplexed Data Bus**

1, 3: $\overline{CS}$ or Address valid to ALE low set-up time.
2, 4: $\overline{CS}$ or Address valid to ALE low hold time.
5: ALE pulse width
6: $\overline{RD}$ high to next ALE high
7: ALE low to $\overline{RD}$ low
8: $\overline{RD}$ pulse width
9: $\overline{RD}$ high to next $\overline{RD}$ or $\overline{WR}$ low
10: ALE low to $\overline{WR}$ low
11: $\overline{WR}$ pulse width
12: $\overline{WR}$ high to next ALE high
13: $\overline{WR}$ high to next $\overline{WR}$ or $\overline{RD}$ low
14: Data valid to $\overline{WR}$ high set-up time
15: Data valid to $\overline{WR}$ high hold time
16: $\overline{RD}$ low to data bus out of TRI-STATE
17: $\overline{RD}$ high to TRI-STATE
18: $\overline{RD}$ low to data valid (access time)

## Timing Diagrams

$V_A^+ = V_D^+ = +5V$, $t_R = t_F = 3$ ns, $C_L = 100$ pF for the $\overline{INT}$, DMARQ, D0–D15 outputs. (Continued)

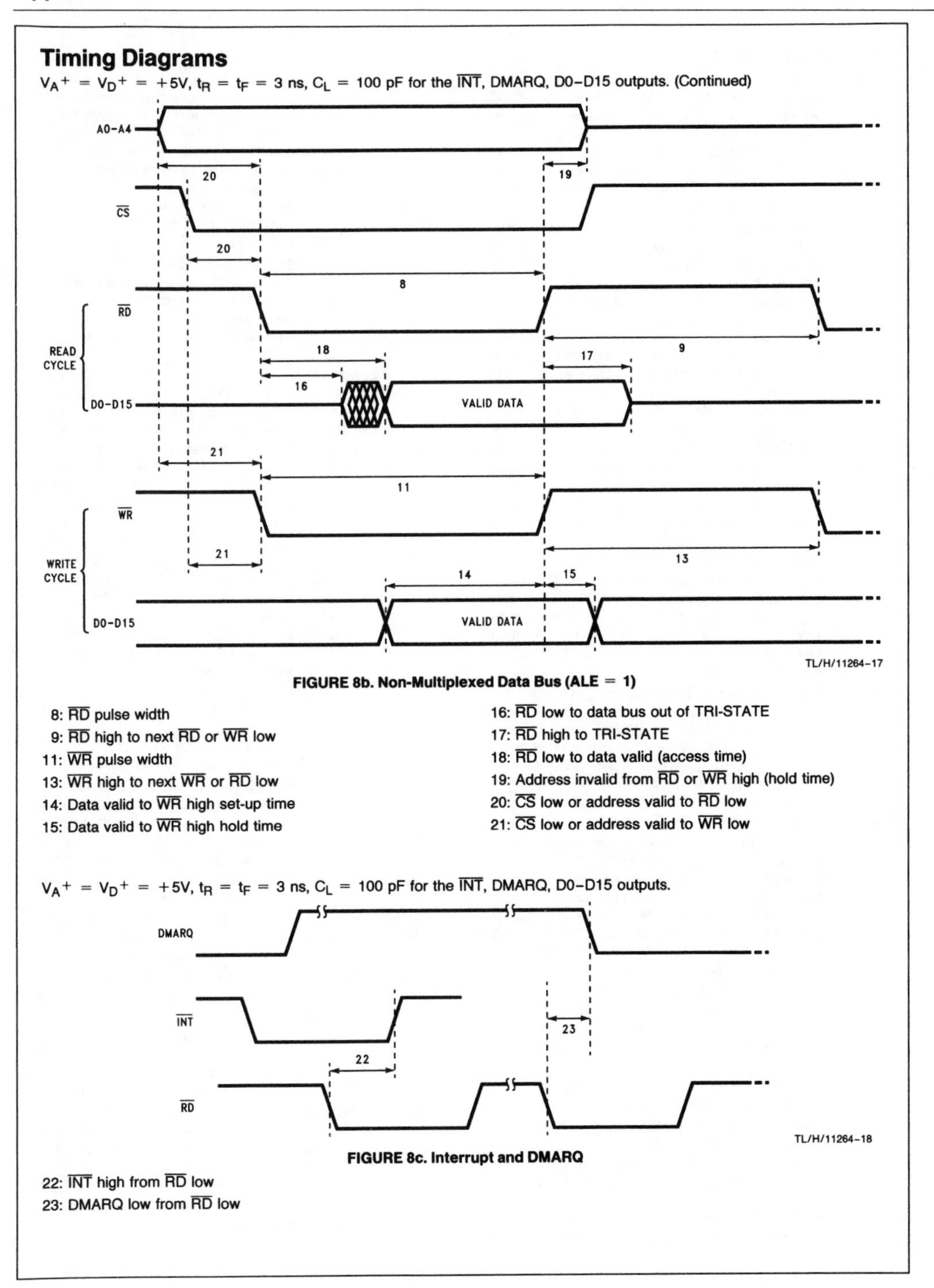


**FIGURE 8b. Non-Multiplexed Data Bus (ALE = 1)**

8: $\overline{RD}$ pulse width
9: $\overline{RD}$ high to next $\overline{RD}$ or $\overline{WR}$ low
11: $\overline{WR}$ pulse width
13: $\overline{WR}$ high to next $\overline{WR}$ or $\overline{RD}$ low
14: Data valid to $\overline{WR}$ high set-up time
15: Data valid to $\overline{WR}$ high hold time
16: $\overline{RD}$ low to data bus out of TRI-STATE
17: $\overline{RD}$ high to TRI-STATE
18: $\overline{RD}$ low to data valid (access time)
19: Address invalid from $\overline{RD}$ or $\overline{WR}$ high (hold time)
20: $\overline{CS}$ low or address valid to $\overline{RD}$ low
21: $\overline{CS}$ low or address valid to $\overline{WR}$ low

$V_A^+ = V_D^+ = +5V$, $t_R = t_F = 3$ ns, $C_L = 100$ pF for the $\overline{INT}$, DMARQ, D0–D15 outputs.



**FIGURE 8c. Interrupt and DMARQ**

22: $\overline{INT}$ high from $\overline{RD}$ low
23: DMARQ low from $\overline{RD}$ low

## Pin Description

$V_A+$
$V_D+$ These are the analog and digital supply voltage pins. The LM12(H)454/8's supply voltage operating range is +3.0V to +5.5V. Accuracy is guaranteed only if $V_A+$ and $V_D+$ are connected to the same power supply. Each pin should have a parallel combination of 10 $\mu$F (electrolytic or tantalum) and 0.1 $\mu$F (ceramic) bypass capacitors connected between it and ground.

D0–D15 The internal data input/output TRI-STATE buffers are connected to these pins. These buffers are designed to drive capacitive loads of 100 pF or less. External buffers are necessary for driving higher load capacitances. These pins allows the user a means of instruction input and data output. With a logic **high** applied to the **BW** pin, data lines D8–D15 are placed in a high impedance state and data lines D0–D7 are used for instruction input and data output when the LM12(H)454/8 is connected to an 8-bit wide data bus. A logic **low** on the **BW** pin allows the LM12(H)454/8 to exchange information over a 16-bit wide data bus.

$\overline{\text{RD}}$ This is the input for the active low READ bus control signal. The data input/output TRI-STATE buffers, as selected by the logic signal applied to the **BW** pin, are enabled when $\overline{\text{RD}}$ and $\overline{\text{CS}}$ are both low. This allows the LM12(H)454/8 to transmit information onto the databus.

$\overline{\text{WR}}$ This is the input for the active low WRITE bus control signal. The data input/output TRI-STATE buffers, as selected by the logic signal applied to the **BW** pin, are enabled when $\overline{\text{WR}}$ and $\overline{\text{CS}}$ are both low. This allows the LM12(H)454/8 to receive information from the databus.

$\overline{\text{CS}}$ This is the input for the active low Chip Select control signal. A logic low should be applied to this pin only during a READ or WRITE access to the LM12(H)454/8. The internal clocking is halted and conversion stops while Chip Select is low. Conversion resumes when the Chip Select input signal returns high.

ALE This is the Address Latch Enable input. It is used in systems containing a multiplexed databus. When ALE is asserted **high**, the LM12(H)454/8 accepts information on the databus as a valid address. A high-to-low transition will latch the address data on A0–A4 and the logic state on the $\overline{\text{CS}}$ input. Any changes on A0–A4 and $\overline{\text{CS}}$ while ALE is low will not affect the LM12(H)454/8. See *Figure 8a*. When a non-multiplexed bus is used, ALE is continuously asserted **high**. See *Figure 8b*.

CLK This is the external clock input pin. The LM12(H)454/8 operates with an input clock frequency in the range of 0.05 MHz to 10.0 MHz.

A0–A4 These are the LM12(H)454/8's address lines. They are used to access all internal registers, Conversion FIFO, and Instruction **RAM**.

SYNC This is the synchronization input/output. When used as an output, it is designed to drive capacitive loads of 100 pF or less. External buffers are necessary for driving higher load capacitances. SYNC is an **input** if the Configuration register's "I/O Select" bit is **low**. A rising edge on this pin causes the internal S/H to hold the input signal. The next rising clock edge either starts a conversion or makes a comparison to a programmable limit depending on which function is requested by a programming instruction. This pin will be an **output** if "I/O Select" is set **high**. The SYNC output goes high when a conversion or a comparison is started and low when completed. (See Section 2.2). An internal reset after power is first applied to the LM12(H)454/8 automatically sets this pin as an input.

BW This is the Bus Width input pin. This input allows the LM12(H)454/8 to interface directly with either an 8- or 16-bit databus. A logic high sets the width to 8 bits and places D8–D15 in a high impedance state. A logic low sets the width to 16 bits.

$\overline{\text{INT}}$ This is the active low interrupt output. This output is designed to drive capacitive loads of 100 pF or less. External buffers are necessary for driving higher load capacitances. An interrupt signal is generated any time a non-masked interrupt condition takes place. There are eight different conditions that can cause an interrupt. Any interrupt is reset by reading the Interrupt Status register. (See Section 2.3.)

DMARQ This is the active high Direct Memory Access Request output. This output is designed to drive capacitive loads of 100 pF or less. External buffers are necessary for driving higher load capacitances. It goes high whenever the number of conversion results in the conversion FIFO equals a programmable value stored in the Interrupt Enable register. It returns to a logic low when the FIFO is empty.

GND This is the LM12(H)454/8 ground connection. It should be connected to a low resistance and inductance analog ground return that connects directly to the system power supply ground.

IN0–IN7 (IN0–IN3 LM12H454 LM12454) These are the eight (LM12(H)458) or four (LM12(H)454) analog inputs. A given channel is selected through the instruction RAM. Any of the channels can be configured as an independent single-ended input. Any pair of channels, whether adjacent or non-adjacent, can operate as a fully differential pair.

S/H IN+
S/H IN− These are the LM12(H)454's non-inverting and inverting inputs to the internal S/H.

MUXOUT+
MUXOUT− These are the LM12(H)454's non-inverting and inverting outputs from the internal multiplexer.

$V_{REF-}$ This is the negative reference input. The LM12(H)454/8 operate with $0V \leq V_{REF-} \leq V_{REF+}$. This pin should be bypassed to ground with a parallel combination of 10 $\mu$F and 0.1 $\mu$F (ceramic) capacitors.

$V_{REF+}$ This is the positive reference input. The LM12(H)454/8 operate with $0V \leq V_{REF+} \leq V_A+$. This pin should be bypassed to ground with a parallel combination of 10 $\mu$F and 0.1 $\mu$F (ceramic) capacitors.

$V_{REFOUT}$ This is the internal 2.5V bandgap's output pin. This pin should be bypassed to ground with a 100 $\mu$F capacitor.

# Application Information

## 1.0 Functional Description

The LM12(H)454 and LM12(H)458 are multi-functional Data Acquisition Systems that include a fully differential 12-bit-plus-sign self-calibrating analog-to-digital converter (ADC) with a two's-complement output format, an 8-channel (LM12(H)458) or a 4-channel (LM12(H)454) analog multiplexer, an internal 2.5V reference, a first-in-first-out (FIFO) register that can store 32 conversion results, and an Instruction RAM that can store as many as eight instructions to be sequentially executed. The LM12(H)454 also has a differential multiplexer output and a differential S/H input. All of this circuitry operates on only a single +5V power supply.

The LM12(H)454/8 have three modes of operation:

- 12-bit + sign with correction
- 8-bit + sign without correction
- 8-bit + sign comparison mode ("watchdog" mode)

The fully differential 12-bit-plus-sign ADC uses a charge re-distribution topology that includes calibration capabilities. Charge re-distribution ADCs use a capacitor ladder in place of a resistor ladder to form an internal DAC. The DAC is used by a successive approximation register to generate intermediate voltages between the voltages applied to $V_{REF-}$ and $V_{REF+}$. These intermediate voltages are compared against the sampled analog input voltage as each bit is generated. The number of intermediate voltages and comparisons equals the ADC's resolution. The correction of each bit's accuracy is accomplished by calibrating the capacitor ladder used in the ADC.

Two different calibration modes are available; one compensates for offset voltage, or zero error, while the other corrects both offset error and the ADC's linearity error.

When correcting offset only, the offset error is measured once and a correction coefficient is created. During the full calibration, the offset error is measured eight times, averaged, and a correction coefficient is created. After completion of either calibration mode, the offset correction coefficient is stored in an internal offset correction register.

The LM12(H)454/8's overall linearity correction is achieved by correcting the internal DAC's capacitor mismatch. Each capacitor is compared eight times against all remaining smaller value capacitors and any errors are averaged. A correction coefficient is then created and stored in one of the thirteen internal linearity correction registers. An internal state machine, using patterns stored in an internal 16 x 8-bit ROM, executes each calibration algorithm.

Once calibrated, an internal arithmetic logic unit (ALU) uses the offset correction coefficient and the 13 linearity correction coefficients to reduce the conversion's offset error and linearity error, in the background, during the 12-bit + sign conversion. The 8-bit + sign conversion and comparison modes use only the offset coefficient. The 8-bit + sign mode performs a conversion in less than half the time used by the 12-bit + sign conversion mode.

The LM12(H)454/8's "watchdog" mode is used to monitor a single-ended or differential signal's amplitude. Each sampled signal has two limits. An interrupt can be generated if the input signal is above or below either of the two limits. This allows interrupts to be generated when analog voltage inputs are "inside the window" or, alternatively, "outside the window". After a "watchdog" mode interrupt, the processor can then request a conversion on the input signal and read the signal's magnitude.

The analog input multiplexer can be configured for any combination of single-ended or fully differential operation. Each input is referenced to ground when a multiplexer channel operates in the single-ended mode. Fully differential analog input channels are formed by pairing any two channels together.

The LM12(H)454's multiplexer outputs and S/H inputs (MUXOUT+, MUXOUT− and S/H IN+, S/H IN−) provide the option for additional analog signal processing. Fixed-gain amplifiers, programmable-gain amplifiers, filters, and other processing circuits can operate on the signal applied to the selected multiplexer channel(s). If external processing is not used, connect MUXOUT+ to S/H IN+ and MUXOUT− to S/H IN−.

The LM12(H)454/8's internal S/H is designed to operate at its minimum acquisition time (1.13 μs, 12 bits) when the source impedance, $R_S$, is $\leq 60\Omega$ ($f_{CLK} \leq 8$ MHz). When $60\Omega < R_S \leq 4.17$ kΩ, the internal S/H's acquisition time can be increased to a maximum of 4.88 μs (12 bits, $f_{CLK} =$ 8 MHz). See Section 2.1 (Instruction RAM "00") Bits 12–15 for more information.

An internal 2.5V bandgap reference output is available at pin 44. This voltage can be used as the ADC reference for ratiometric conversion or as a virtual ground for front-end analog conditioning circuits. The $V_{REFOUT}$ pin should be bypassed to ground with a 100 μF capacitor.

Microprocessor overhead is reduced through the use of the internal conversion FIFO. Thirty-two consecutive conversions can be completed and stored in the FIFO without any microprocessor intervention. The microprocessor can, at any time, interrogate the FIFO and retrieve its contents. It can also wait for the LM12(H)454/8 to issue an interrupt when the FIFO is full or after any number (≤32) of conversions have been stored.

Conversion sequencing, internal timer interval, multiplexer configuration, and many other operations are programmed and set in the Instruction RAM.

A diagnostic mode is available that allows verification of the LM12(H)458's operation. The diagnostic mode is disabled in the LM12(H)454. This mode internally connects the voltages present at the $V_{REFOUT}$, $V_{REF+}$, $V_{REF-}$, and GND pins to the internal $V_{IN+}$ and $V_{IN-}$ S/H inputs. This mode is activated by setting the Diagnostic bit (Bit 11) in the Configuration register to a "1". More information concerning this mode of operation can be found in Section 2.2.

## 2.0 Internal User-Programmable Registers

### 2.1 INSTRUCTION RAM

The instruction RAM holds up to eight sequentially executable instructions. Each 48-bit long instruction is divided into three 16-bit sections. READ and WRITE operations can be issued to each 16-bit section using the instruction's address and the 2-bit "RAM pointer" in the Configuration register. The eight instructions are located at addresses 0000 through 0111 (A4–A1, BW = 0) when using a 16-bit wide data bus or at addresses 00000 through 01111 (A4–A0, BW = 1) when using an 8-bit wide data bus. They can be accessed and programmed in random order.

Any Instruction RAM READ or WRITE can affect the sequencer's operation:

> The Sequencer should be stopped by setting the RESET bit to a "1" or by resetting the START bit in the Configuration Register and waiting for the current instruction to finish execution before any Instruction RAM READ or WRITE is initiated.
>
> A soft RESET should be issued by writing a "1" to the Configuration Register's RESET bit after any READ or WRITE to the Instruction RAM.

The three sections in the Instruction RAM are selected by the Configuration Register's 2-bit "RAM Pointer", bits D8 and D9. The first 16-bit Instruction RAM section is selected with the RAM Pointer equal to "00". This section provides multiplexer channel selection, as well as resolution, acquisition time, etc. The second 16-bit section holds "watchdog" limit #1, its sign, and an indicator that shows that an interrupt can be generated if the input signal is greater or less than the programmed limit. The third 16-bit section holds "watchdog" limit #2, its sign, and an indicator that shows that an interrupt can be generated if the input signal is greater or less than the programmed limit.

**Instruction RAM "00"**

**Bit 0** is the LOOP bit. It indicates the last instruction to be executed in any instruction sequence when it is set to a "1". The next instruction to be executed will be instruction 0.

**Bit 1** is the PAUSE bit. This controls the Sequencer's operation. When the PAUSE bit is set ("1"), the Sequencer will stop after reading the current instruction, but before executing it and the start bit, in the Configuration register, is automatically reset to a "0". Setting the PAUSE also causes an interrupt to be issued. The Sequencer is restarted by placing a "1" in the Configuration register's Bit 0 (Start bit).

After the Instruction RAM has been programmed and the RESET bit is set to "1", the Sequencer retrieves Instruction 000, decodes it, and waits for a "1" to be placed in the Configuration's START bit. The START bit value of "0" "overrides" the action of Instruction 000's PAUSE bit when the Sequencer is started. Once started, the Sequencer executes Instruction 000 and retrieves, decodes, and executes each of the remaining instructions. No PAUSE Interrupt (INT 5) is generated the first time the Sequencer executes Instruction 000 having a PAUSE bit set to "1". When the Sequencer encounters a LOOP bit or completes all eight instructions, Instruction 000 is retrieved and decoded. A set PAUSE bit in Instruction 000 now halts the Sequencer before the instruction is executed.

**Bits 2–4** select which of the eight input channels ("000" to "111" for IN0–IN7) will be configured as non-inverting inputs to the LM12(H)458's ADC. (See Page 25, Table I.) They select which of the four input channels ("000" to "011" for IN0–IN4) will be configured as non-inverting inputs to the LM12(H)454's ADC. (See Page 25, Table II.)

**Bits 5–7** select which of the seven input channels ("001" to "111" for IN1 to IN7) will be configured as inverting inputs to the LM12(H)458's ADC. (See Page 25, Table I.) They select which of the three input channels ("001" to "011" for IN1–IN4) will be configured as inverting inputs to the LM12(H)454's ADC. (See Page 25, Table II.) Fully differential operation is created by selecting two multiplexer channels, one operating in the non-inverting mode and the other operating in the inverting mode. A code of "000" selects ground as the inverting input for single ended operation.

**Bit 8** is the SYNC bit. Setting Bit 8 to "1" causes the Sequencer to suspend operation at the end of the internal S/H's acquisition cycle and to wait until a rising edge appears at the SYNC pin. When a rising edge appears, the S/H acquires the input signal magnitude and the ADC performs a conversion on the clock's next rising edge. When the SYNC pin is used as an input, the Configuration register's "I/O Select" bit (Bit 7) must be set to a "0". With SYNC configured as an input, it is possible to synchronize the start of a conversion to an external event. This is useful in applications such as digital signal processing (DSP) where the exact timing of conversions is important.

When the LM12(H)454/8 are used in the "watchdog" mode with external synchronization, two rising edges on the SYNC input are required to initiate two comparisons. The first rising edge initiates the comparison of the selected analog input signal with Limit #1 (found in Instruction RAM "01") and the second rising edge initiates the comparison of the same analog input signal with Limit #2 (found in Instruction RAM "10").

**Bit 9** is the TIMER bit. When Bit 9 is set to "1", the Sequencer will halt until the internal 16-bit Timer counts down to zero. During this time interval, no "watchdog" comparisons or analog-to-digital conversions will be performed.

**Bit 10** selects the ADC conversion resolution. Setting Bit 10 to "1" selects 8-bit + sign and when reset to "0" selects 12-bit + sign.

**Bit 11** is the "watchdog" comparison mode enable bit. When operating in the "watchdog" comparison mode, the selected analog input signal is compared with the programmable values stored in Limit #1 and Limit #2 (see Instruction RAM "01" and Instruction RAM "10"). Setting Bit 11 to "1" causes two comparisons of the selected analog input signal with the two stored limits. When Bit 11 is reset to "0", an 8-bit + sign or 12-bit + sign (depending on the state of Bit 10 of Instruction RAM "00") conversion of the input signal can take place.

## 2.0 Internal User-Programmable Registers (Continued)

<table>
<tr><th>A4 A3 A2 A1</th><th>Purpose</th><th>Type</th><th>D15</th><th>D14</th><th>D13</th><th>D12</th><th>D11</th><th>D10</th><th>D9</th><th>D8</th><th>D7</th><th>D6</th><th>D5</th><th>D4</th><th>D3</th><th>D2</th><th>D1</th><th>D0</th></tr>
<tr><td>0 000 to 111</td><td>Instruction RAM (RAM Pointer = 00)</td><td>R/W</td><td colspan="4">Acquisition Time</td><td>Watch-dog</td><td>8/12</td><td>Timer</td><td>Sync</td><td colspan="3">$V_{IN-}$ (MUXOUT−)*</td><td colspan="3">$V_{IN+}$ (MUXOUT+)*</td><td>Pause</td><td>Loop</td></tr>
<tr><td>0 000 to 111</td><td>Instruction RAM (RAM Pointer = 01)</td><td>R/W</td><td colspan="6">Don't Care</td><td>>/<</td><td>Sign</td><td colspan="8">Limit #1</td></tr>
<tr><td>0 000 to 111</td><td>Instruction RAM (RAM Pointer = 10)</td><td>R/W</td><td colspan="6">Don't Care</td><td>>/<</td><td>Sign</td><td colspan="8">Limit #2</td></tr>
<tr><td>1 0 0 0</td><td>Configuration Register</td><td>R/W</td><td colspan="4">Don't Care</td><td>DIAG†</td><td>Test = 0</td><td colspan="2">RAM Pointer</td><td>I/O Sel</td><td>Auto $Zero_{ec}$</td><td>Chan Mask</td><td>Stand-by</td><td>Full CAL</td><td>Auto-Zero</td><td>Reset</td><td>Start</td></tr>
<tr><td>1 0 0 1</td><td>Interrupt Enable Register</td><td>R/W</td><td colspan="5">Number of Conversions in Conversion FIFO to Generate INT2</td><td colspan="3">Sequencer Address to Generate INT1</td><td>INT7</td><td>INT6</td><td>INT5</td><td>INT4</td><td>INT3</td><td>INT2</td><td>INT1</td><td>INT0</td></tr>
<tr><td>1 0 1 0</td><td>Interrupt Status Register</td><td>R</td><td colspan="5">Actual Number of Conversion Results in Conversion FIFO</td><td colspan="3">Address of Sequencer Instruction being Executed</td><td>INST7</td><td>INST6</td><td>INST5</td><td>INST4</td><td>INST3</td><td>INST2</td><td>INST1</td><td>INST0</td></tr>
<tr><td>1 0 1 1</td><td>Timer Register</td><td>R/W</td><td colspan="8">Timer Preset High Byte</td><td colspan="8">Timer Preset Low Byte</td></tr>
<tr><td>1 1 0 0</td><td>Conversion FIFO</td><td>R</td><td colspan="3">Address or Sign</td><td>Sign</td><td colspan="4">Conversion Data: MSBs</td><td colspan="8">Conversion Data: LSBs</td></tr>
<tr><td>1 1 0 1</td><td>Limit Status Register</td><td>R</td><td colspan="8">Limit #2: Status</td><td colspan="8">Limit #1: Status</td></tr>
</table>

*LM12(H)454 (Refer to Table II).
†LM12(H)458 only. Must be set to "0" for the LM12(H)454.

**FIGURE 9. LM12(H)454/8 Memory Map for 16-Bit Wide Databus (BW = "0", Test Bit = "0" and A0 = Don't Care)**

## 2.0 Internal User-Programmable Registers (Continued)

| A4 | A3 | A2 | A1 | A0 | Purpose | Type | D7 | D6 | D5 | D4 | D3 | D2 | D1 | D0 |
|---|---|---|---|---|---|---|---|---|---|---|---|---|---|---|
| 0 | 0<br>1 | 0<br>to<br>1 | 0<br>1 | 0 | Instruction RAM (RAM Pointer = 00) | R/W | $V_{IN}-$ (MUXOUT−)* | | | $V_{IN}+$ (MUXOUT+)* | | | Pause | Loop |
| 0 | 0<br>1 | 0<br>to<br>1 | 0<br>1 | 1 | | R/W | Acquisition Time | | | | Watch-dog | 8/12 | Timer | Sync |
| 0 | 0<br>1 | 0<br>to<br>1 | 0<br>1 | 0 | Instruction RAM (RAM Pointer = 01) | R/W | Comparison Limit #1 | | | | | | | |
| 0 | 0<br>1 | 0<br>to<br>1 | 0<br>1 | 1 | | R/W | Don't Care | | | | | | >/< | Sign |
| 0 | 0<br>1 | 0<br>to<br>1 | 0<br>1 | 0 | Instruction RAM (RAM Pointer = 10) | R/W | Comparison Limit #2 | | | | | | | |
| 0 | 0<br>1 | 0<br>to<br>1 | 0<br>1 | 1 | | R/W | Don't Care | | | | | | >/< | Sign |
| 1 | 0 | 0 | 0 | 0 | Configuration Register | R/W | I/O Sel | Auto $Zero_{ec}$ | Chan Mask | Stand-by | Full Cal | Auto-Zero | Reset | Start |
| 1 | 0 | 0 | 0 | 1 | | R/W | Don't Care | | | | DIAG† | Test = 0 | RAM Pointer | |
| 1 | 0 | 0 | 1 | 0 | Interrupt Enable Register | R/W | INT7 | INT6 | INT5 | INT4 | INT3 | INT2 | INT1 | INT0 |
| 1 | 0 | 0 | 1 | 1 | | R/W | Number of Conversions in Conversion FIFO to Generate INT2 | | | | | Sequencer Address to Generate INT1 | | |
| 1 | 0 | 1 | 0 | 0 | Interrupt Status Register | R | INST7 | INST6 | INST5 | INST4 | INST3 | INST2 | INST1 | INST0 |
| 1 | 0 | 1 | 0 | 1 | | R | Actual Number of Conversions Results in Conversion FIFO | | | | | Address of Sequencer Instruction being Executed | | |
| 1 | 0 | 1 | 1 | 0 | Timer Register | R/W | Timer Preset: Low Byte | | | | | | | |
| 1 | 0 | 1 | 1 | 1 | | R/W | Timer Preset: High Byte | | | | | | | |
| 1 | 1 | 0 | 0 | 0 | Conversion FIFO | R | Conversion Data: LSBs | | | | | | | |
| 1 | 1 | 0 | 0 | 1 | | R | Address or Sign | | | Sign | Conversion Data: MSBs | | | |
| 1 | 1 | 0 | 1 | 0 | Limit Status Register | R | Limit #1 Status | | | | | | | |
| 1 | 1 | 0 | 1 | 1 | | R | Limit #2 Status | | | | | | | |

*LM12(H)454 (Refer to Table II).

†LM12(H)458 only. Must be set to "0" for the LM12(H)454.

**FIGURE 10. LM12(H)454/8 Memory Map for 8-Bit Wide Databus (BW = "1" and Test Bit = "0")**

## 2.0 Internal User-Programmable Registers (Continued)

**Bits 12–15** are used to store the user-programmable acquisition time. The Sequencer keeps the internal S/H in the acquisition mode for a fixed number of clock cycles (nine clock cycles, for 12-bit + sign conversions and two clock cycles for 8-bit + sign conversions or "watchdog" comparisons) plus a variable number of clock cycles equal to twice the value stored in Bits 12–15. Thus, the S/H's acquisition time is (9 + 2D) clock cycles for 12-bit + sign conversions and (2 + 2D) clock cycles for 8-bit + sign conversions or "watchdog" comparisons, where D is the value stored in Bits 12–15. The minimum acquisition time compensates for the typical internal multiplexer series resistance of 2 kΩ, and any additional delay created by Bits 12–15 compensates for source resistances greater than 60Ω (100Ω). (For this acquisition time discussion, numbers in ( ) are shown for the LM12(H)454/8 operating at 5 MHz.) The necessary acquisition time is determined by the source impedance at the multiplexer input. If the source resistance ($R_S$) < 60Ω (100Ω) and the clock frequency is 8 MHz, the value stored in bits 12–15 (D) can be 0000. If $R_S$ > 60Ω (100Ω), the following equations determine the value that should be stored in bits 12–15.

$$D = 0.45 \times R_S \times f_{CLK}$$

for 12-bits + sign

$$D = 0.36 \times R_S \times f_{CLK}$$

for 8-bits + sign and "watchdog"

$R_S$ is in kΩ and $f_{CLK}$ is in MHz. Round the result to the next higher integer value. If D is greater than 15, it is advisable to lower the source impedance by using an analog buffer between the signal source and the LM12(H)458's multiplexer inputs. The value of D can also be used to compensate for the settling or response time of external processing circuits connected between the LM12(H)454's MUXOUT and S/H IN pins.

**Instruction RAM "01"**

The second Instruction RAM section is selected by placing a "01" in Bits 8 and 9 of the Configuration register.

**Bits 0–7** hold "watchdog" **limit #1**. When Bit 11 of Instruction RAM "00" is set to a "1", the LM12(H)454/8 performs a "watchdog" comparison of the sampled analog input signal with the limit #1 value first, followed by a comparison of the same sampled analog input signal with the value found in limit #2 (Instruction RAM "10").

**Bit 8** holds limit #1's sign.

**Bit 9**'s state determines the limit condition that generates a "watchdog" interrupt. A "1" causes a voltage greater than limit #1 to generate an interrupt, while a "0" causes a voltage less than limit #1 to generate an interrupt.

**Bits 10–15** are not used.

**Instruction RAM "10"**

The third Instruction RAM section is selected by placing a "10" in Bits 8 and 9 of the Configuration register.

**Bits 0–7** hold "watchdog" **limit #2**. When Bit 11 of Instruction RAM "00" is set to a "1", the LM12(H)454/8 performs a "watchdog" comparison of the sampled analog input signal with the limit #1 value first (Instruction RAM "01"), followed by a comparison of the same sampled analog input signal with the value found in limit #2.

**Bit 8** holds limit #2's sign.

**Bit 9**'s state determines the limit condition that generates a "watchdog" interrupt. A "1" causes a voltage greater than limit #2 to generate an interrupt, while a "0" causes a voltage less than limit #2 to generate an interrupt.

**Bits 10–15** are not used.

**2.2 CONFIGURATION REGISTER**

The Configuration register, 1000 (A4–A1, BW = 0) or 1000x (A4–A0, BW = 1) is a 16-bit control register with read/write capability. It acts as the LM12(H)454's and LM12(H)458's "control panel" holding global information as well as start/stop, reset, self-calibration, and stand-by commands.

**Bit 0** is the START/STOP bit. Reading Bit 0 returns an indication of the Sequencer's status. A "0" indicates that the Sequencer is stopped and waiting to execute the next instruction. A "1" shows that the Sequencer is running. Writing a "0" halts the Sequencer when the current instruction has finished execution. The next instruction to be executed is pointed to by the instruction pointer found in the status register. A "1" restarts the Sequencer with the instruction currently pointed to by the instruction pointer. (See Bits 8–10 in the Interrupt Status register.)

**Bit 1** is the LM12(H)454/8's system RESET bit. Writing a "1" to Bit 1 stops the Sequencer (resetting the Configuration register's START/STOP bit), resets the Instruction pointer to "000" (found in the Interrupt Status register), clears the Conversion FIFO, and resets all interrupt flags. The RESET bit will return to "0" after two clock cycles unless it is forced high by writing a "1" into the Configuration register's Standby bit. A reset signal is internally generated when power is first applied to the part. No operation should be started until the RESET bit is "0".

Writing a "1" to **Bit 2** initiates an auto-zero offset voltage calibration. Unlike the eight-sample auto-zero calibration performed during the full calibration procedure, Bit 2 initiates a "short" auto-zero by sampling the offset once and creating a correction coefficient (full calibration averages eight samples of the converter offset voltage when creating a correction coefficient). If the Sequencer is running when Bit 2 is set to "1", an auto-zero starts immediately after the conclusion of the currently running instruction. Bit 2 is reset automatically to a "0" and an interrupt flag (Bit 3, in the Interrupt Status register) is set at the end of the auto-zero (76 clock cycles). After completion of an auto-zero calibration, the Sequencer fetches the next instruction as pointed to by the Instruction RAM's pointer and resumes execution. If the Sequencer is stopped, an auto-zero is performed immediately at the time requested.

Writing a "1" to **Bit 3** initiates a complete calibration process that includes a "long" auto-zero offset voltage correction (this calibration averages eight samples of the comparator offset voltage when creating a correction coefficient) followed by an ADC linearity calibration. This complete calibration is started after the currently running instruction is completed if the Sequencer is running when Bit 3 is set to "1". Bit 3 is reset automatically to a "0" and an interrupt flag (Bit 4, in the Interrupt Status register) will be generated at the end of the calibration procedure (4944 clock cycles). After completion of a full auto-zero and linearity calibration, the Sequencer fetches the next instruction as pointed to by the Instruction RAM's pointer and resumes execution. If the Sequencer is stopped, a full calibration is performed immediately at the time requested.

## 2.0 Internal User-Programmable Registers (Continued)

**Bit 4** is the Standby bit. Writing a "1" to Bit 4 immediately places the LM12(H)454/8 in Standby mode. Normal operation returns when Bit 4 is reset to a "0". The Standby command ("1") disconnects the external clock from the internal circuitry, decreases the LM12(H)454/8's internal analog circuitry power supply current, and preserves all internal RAM contents. After writing a "0" to the Standby bit, the LM12(H)454/8 returns to an operating state identical to that caused by exercising the RESET bit. A Standby completion interrupt is issued after a power-up completion delay that allows the analog circuitry to settle. The Sequencer should be restarted only after the Standby completion is issued. The Instruction RAM can still be accessed through read and write operations while the LM12(H)454/8 are in Standby Mode.

**Bit 5** is the Channel Address Mask. If Bit 5 is set to a "1", Bits 13–15 in the conversion FIFO will be equal to the sign bit (Bit 12) of the conversion data. Resetting Bit 5 to a "0" causes conversion data Bits 13 through 15 to hold the instruction pointer value of the instruction to which the conversion data belongs.

**Bit 6** is used to select a "short" auto-zero correction for every conversion. The Sequencer automatically inserts an auto-zero before every conversion or "watchdog" comparison if Bit 6 is set to "1". No automatic correction will be performed if Bit 6 is reset to "0".

The LM12(H)454/8's offset voltage, after calibration, has a typical drift of 0.1 LSB over a temperature range of −40°C to +85°C. This small drift is less than the variability of the change in offset that can occur when using the auto-zero correction with each conversion. This variability is the result of using only one sample of the offset voltage to create a correction value. This variability decreases when using the full calibration mode because eight samples of the offset voltage are taken, averaged, and used to create a correction value.

**Bit 7** is used to program the SYNC pin (29) to operate as either an input or an output. The SYNC pin becomes an output when Bit 7 is a "1" and an input when Bit 7 is a "0". With SYNC programmed as an input, the rising edge of any logic signal applied to pin 29 will start a conversion or "watchdog" comparison. Programmed as an output, the logic level at pin 29 will go high at the start of a conversion or "watchdog" comparison and remain high until either have finished. See Instruction RAM "00", Bit 8.

**Bits 8** and **9** form the RAM Pointer that is used to select each of a 48-bit instruction's three 16-bit sections during read or write actions. A "00" selects Instruction RAM section one, "01" selects section two, and "10" selects section three.

**Bit 10** activates the Test mode that is used only during production testing. Leave this bit reset to "0".

**Bit 11** is the Diagnostic bit and is available only in the LM12(H)458. It can be activated by setting it to a "1" (the Test bit must be reset to a "0"). The Diagnostic mode, along with a correctly chosen instruction, allows verification that the LM12(H)458's ADC is performing correctly. When activated, the inverting and non-inverting inputs are connected as shown in Table I. As an example, an instruction with "001" for both $V_{IN+}$ and $V_{IN-}$ while using the Diagnostic mode typically results in a full-scale output.

**2.3 INTERRUPTS**

The LM12(H)454 and LM12(H)458 have eight possible interrupts, all with the same priority. Any of these interrupts will cause a hardware interrupt to appear on the $\overline{INT}$ pin (31) if they are not masked (by the Interrupt Enable register). The Interrupt Status register is then read to determine which of the eight interrupts has been issued.

**TABLE I. LM12(H)458 Input Multiplexer Channel Configuration Showing Normal Mode and Diagnostic Mode**

| Channel Selection Data | Normal Mode | | Diagnostic Mode | |
|---|---|---|---|---|
| | $V_{IN+}$ | $V_{IN-}$ | $V_{IN+}$ | $V_{IN-}$ |
| 000 | IN0 | GND | $V_{REFOUT}$ | GND |
| 001 | IN1 | IN1 | $V_{REF+}$ | $V_{REF-}$ |
| 010 | IN2 | IN2 | IN2 | IN2 |
| 011 | IN3 | IN3 | IN3 | IN3 |
| 100 | IN4 | IN4 | IN4 | IN4 |
| 101 | IN5 | IN5 | IN5 | IN5 |
| 110 | IN6 | IN6 | IN6 | IN6 |
| 111 | IN7 | IN7 | IN7 | IN7 |

**TABLE II. LM12(H)454 Input Multiplexer Channel Configuration**

| Channel Selection Data | MUX+ | MUX− |
|---|---|---|
| 000 | IN0 | GND |
| 001 | IN1 | IN1 |
| 010 | IN2 | IN2 |
| 011 | IN3 | IN3 |
| 1XX | OPEN | OPEN |

The Interrupt Status register, 1010 (A4–A1, BW = 0) or 1010x (A4–A0, BW = 1) must be cleared by reading it after writing to the Interrupt Enable register. This removes any spurious interrupts on the $\overline{INT}$ pin generated during an Interrupt Enable register access.

**Interrupt 0** is generated whenever the analog input voltage on a selected multiplexer channel crosses a limit while the LM12(H)454/8 are operating in the "watchdog" comparison mode. Two sequential comparisons are made when the LM12(H)454/8 are executing a "watchdog" instruction. Depending on the logic state of Bit 9 in the Instruction RAM's second and third sections, an interrupt will be generated either when the input signal's magnitude is greater than or less than the programmable limits. (See the Instruction RAM, Bit 9 description.) The Limit Status register will indicate which preprogrammed limit, #1 or #2 and which instruction was executing when the limit was crossed.

**Interrupt 1** is generated when the Sequencer reaches the instruction counter value specified in the Interrupt Enable register's bits 8–10. This flag appears before the instruction's execution.

**Interrupt 2** is activated when the Conversion FIFO holds a number of conversions equal to the programmable value

## 2.0 Internal User-Programmable Registers (Continued)

stored in the Interrupt Enable register's Bits 11–15. This value ranges from 0001 to 1111, representing 1 to 31 conversions stored in the FIFO. A user-programmed value of 0000 has no meaning. See Section 3.0 for more FIFO information.

The completion of the short, single-sampled auto-zero calibration generates **Interrupt 3**.

The completion of a full auto-zero and linearity self-calibration generates **Interrupt 4**.

**Interrupt 5** is generated when the Sequencer encounters an instruction that has its Pause bit (Bit 1 in Instruction RAM "00") set to "1".

The LM12(H)454/8 issues **Interrupt 6** whenever it senses that its power supply voltage is dropping below 4V (typ). This interrupt indicates the potential corruption of data returned by the LM12(H)454/8.

**Interrupt 7** is issued after a short delay (10 ms typ) while the LM12(H)454/8 returns from Standby mode to active operation using the Configuration register's Bit 4. This short delay allows the internal analog circuitry to settle sufficiently, ensuring accurate conversion results.

### 2.4 INTERRUPT ENABLE REGISTER

The Interrupt Enable register at address location 1001 (A4–A1, BW = 0) or 1001x (A4–A0, BW = 1) has READ/WRITE capability. An individual interrupt's ability to produce an external interrupt at pin 31 ($\overline{\text{INT}}$) is accomplished by placing a "1" in the appropriate bit location. Any of the internal interrupt-producing operations will set their corresponding bits to "1" in the Interrupt Status register regardless of the state of the associated bit in the Interrupt Enable register. See Section 2.3 for more information about each of the eight internal interrupts.

**Bit 0** enables an external interrupt when an internal "watchdog" comparison limit interrupt has taken place.

**Bit 1** enables an external interrupt when the Sequencer has reached the address stored in Bits 8–10 of the Interrupt Enable register.

**Bit 2** enables an external interrupt when the Conversion FIFO's limit, stored in Bits 11–15 of the Interrupt Enable register, has been reached.

**Bit 3** enables an external interrupt when the single-sampled auto-zero calibration has been completed.

**Bit 4** enables an external interrupt when a full auto-zero and linearity self-calibration has been completed.

**Bit 5** enables an external interrupt when an internal Pause interrupt has been generated.

**Bit 6** enables an external interrupt when a low power supply condition ($V_A^+ < 4V$) has generated an internal interrupt.

**Bit 7** enables an external interrupt when the LM12(H)454/8 return from power-down to active mode.

**Bits 8–10** form the storage location of the user-programmable value against which the Sequencer's address is compared. When the Sequencer reaches an address that is equal to the value stored in Bits 8–10, an internal interrupt is generated and appears in Bit 1 of the Interrupt Status register. If Bit 1 of the Interrupt Enable register is set to "1", an external interrupt will appear at pin 31 ($\overline{\text{INT}}$).

The value stored in bits 8–10 ranges from 000 to 111, representing 0 to 7 instructions stored in the Instruction RAM. After the Instruction RAM has been programmed and the RESET bit is set to "1", the Sequencer is started by placing a "1" in the Configuration register's START bit. Setting the INT 1 trigger value to 000 **does not generate** an INT 1 the **first** time the Sequencer retrieves and decodes Instruction 000. The Sequencer **generates** INT 1 (by placing a "1" in the Interrupt Status register's Bit 1) the **second time and after** the Sequencer encounters Instruction 000. It is important to remember that the Sequencer continues to operate even if an Instruction interrupt (INT 1) is internally or externally generated. The only mechanisms that stop the Sequencer are an instruction with the PAUSE bit set to "1" (halts before instruction execution), placing a "0" in the Configuration register's START bit, or placing a "1" in the Configuration register's RESET bit.

**Bits 11–15** hold the number of conversions that must be stored in the Conversion FIFO in order to generate an internal interrupt. This internal interrupt appears in Bit 2 of the Interrupt Status register. If Bit 2 of the Interrupt Enable register is set to "1", an external interrupt will appear at pin 31 ($\overline{\text{INT}}$).

### 2.5 INTERRUPT STATUS REGISTER

This read-only register is located at address 1010 (A4–A1, BW = 0) or 1010x (A4–A0, BW = 1). The corresponding flag in the Interrupt Status register goes high ("1") any time that an interrupt condition takes place, whether an interrupt is enabled or disabled in the Interrupt Enable register. Any of the active ("1") Interrupt Status register flags are reset to "0" whenever this register is read or a device reset is issued (see Bit 1 in the Configuration Register).

**Bit 0** is set to "1" when a "watchdog" comparison limit interrupt has taken place.

**Bit 1** is set to "1" when the Sequencer has reached the address stored in Bits 8–10 of the Interrupt Enable register.

**Bit 2** is set to "1" when the Conversion FIFO's limit, stored in Bits 11–15 of the Interrupt Enable register, has been reached.

**Bit 3** is set to "1" when the single-sampled auto-zero has been completed.

**Bit 4** is set to "1" when an auto-zero and full linearity self-calibration has been completed.

**Bit 5** is set to "1" when a Pause interrupt has been generated.

**Bit 6** is set to "1" when a low-supply voltage condition ($V_A^+ < 4V$) has taken place.

**Bit 7** is set to "1" when the LM12(H)454/8 return from power-down to active mode.

**Bits 8–10** hold the Sequencer's actual instruction address while it is running.

**Bits 11–15** hold the actual number of conversions stored in the Conversion FIFO while the Sequencer is running.

### 2.6 LIMIT STATUS REGISTER

The read-only register is located at address 1101 (A4–A1, BW = 0) or 1101x (A4–A0, BW = 1). This register is used in tandem with the Limit #1 and Limit #2 registers in the Instruction RAM. Whenever a given instruction's input voltage exceeds the limit set in its corresponding Limit register (#1 or #2), a bit, corresponding to the instruction number, is set in the Limit Status register. Any of the active ("1") Limit Status flags are reset to "0" whenever this register is

## 2.0 Internal User-Programmable Registers (Continued)

read or a device reset is issued (see Bit 1 in the Configuration register). This register holds the status of limits #1 and #2 for each of the eight instructions.

**Bits 0–7** show the Limit #1 status. Each bit will be set high ("1") when the corresponding instruction's input voltage exceeds the threshold stored in the instruction's Limit #1 register. When, for example, instruction 3 is a "watchdog" operation (Bit 11 is set high) and the input for instruction 3 meets the magnitude and/or polarity data stored in instruction 3's Limit #1 register, Bit 3 in the Limit Status register will be set to a "1".

**Bits 8–15** show the Limit #2 status. Each bit will be set high ("1") when the corresponding instruction's input voltage exceeds the threshold stored in the instruction's Limit #2 register. When, for example, the input to instruction 6 meets the value stored in instruction 6's Limit #2 register, Bit 14 in the Limit Status register will be set to a "1".

### 2.7 TIMER

The LM12(H)454/8 have an on-board 16-bit timer that includes a 5-bit pre-scaler. It uses the clock signal applied to pin 23 as its input. It can generate time intervals of 0 through $2^{21}$ clock cycles in steps of $2^5$. This time interval can be used to delay the execution of instructions. It can also be used to slow the conversion rate when converting slowly changing signals. This can reduce the amount of redundant data stored in the FIFO and retrieved by the controller.

The user-defined timing value used by the Timer is stored in the 16-bit READ/WRITE Timer register at location 1011 (A4–A1, BW = 0) or 1011x (A4–A0, BW = 1) and is preloaded automatically. Bits 0–7 hold the preset value's low byte and Bits 8–15 hold the high byte. The Timer is activated by the Sequencer only if the current instruction's Bit 9 is set ("1"). If the equivalent decimal value "N" $(0 \leq N \leq 2^{16} - 1)$ is written inside the 16-bit Timer register and the Timer is enabled by setting an instruction's bit 9 to a "1", the Sequencer will delay the same instruction's execution by halting at state 3 (S3), as shown in *Figure 11,* for $32 \times N + 2$ clock cycles.

### 2.8 DMA

The DMA works in tandem with Interrupt 2. An active DMA Request on pin 32 (DMARQ) requires that the FIFO interrupt be enabled. The voltage on the DMARQ pin goes high when the number of conversions in the FIFO equals the 5-bit value stored in the Interrupt Enable register (bits 11–15). The voltage on the $\overline{\text{INT}}$ pin goes low at the same time as the voltage on the DMARQ pin goes high. The voltage on the DMARQ pin goes low when the FIFO is emptied. The Interrupt Status register must be read to clear the FIFO interrupt flag in order to enable the next DMA request.

DMA operation is optimized through the use of the 16-bit databus connection (a logic "0" applied to the BW pin). Using this bus width allows DMA controllers that have single address Read/Write capability to easily unload the FIFO. Using DMA on an 8-bit databus is more difficult. Two read operations (low byte, high byte) are needed to retrieve each conversion result from the FIFO. Therefore, the DMA controller must be able to repeatedly access two constant addresses when transferring data from the LM12(H)454/8 to the host system.

## 3.0 FIFO

The result of each conversion stored in an internal read-only FIFO (First-In, First-Out) register. It is located at 1100 (A4–A1, BW = 0) or 1100x (A4–A0, BW = 1). This register has 32 16-bit wide locations. Each location holds 13-bit data. Bits 0–3 hold the four LSB's in the 12 bits + sign mode or "1110" in the 8 bits + sign mode. Bits 4–11 hold the eight MSB's and Bit 12 holds the sign bit. Bits 13–15 can hold either the sign bit, extending the register's two's complement data format to a full sixteen bits or the instruction address that generated the conversion and the resulting data. These modes are selected according to the logic state of the Configuration register's Bit 5.

The FIFO status should be read in the Interrupt Status register (Bits 11–15) to determine the number of conversion results that are held in the FIFO before retrieving them. This will help prevent conversion data corruption that may take place if the number of reads are greater than the number of conversion results contained in the FIFO. Trying to read the FIFO when it is empty may corrupt new data being written into the FIFO. Writing more than 32 conversion data into the FIFO by the ADC results in loss of the first conversion data. Therefore, to prevent data loss, it is recommended that the LM12(H)454/8's interrupt capability be used to inform the system controller that the FIFO is full.

The lower portion (A0 = 0) of the data word (Bits 0–7) should be read first followed by a read of the upper portion (A0 = 1) when using the 8-bit bus width (BW = 1). Reading the upper portion first causes the data to shift down, which results in loss of the lower byte.

**Bits 0–12** hold 12-bit + sign conversion data. **Bits 0–3** will be 1110 (LSB) when using 8-bit plus sign resolution.

**Bits 13–15** hold either the instruction responsible for the associated conversion data or the sign bit. Either mode is selected with Bit 5 in the Configuration register.

Using the FIFO's full depth is achieved as follows. Set the value of the Interrupt Enable registers's Bits 11–15 to 1111 and the Interrupt Enable register's Bit 2 to a "1". This generates an external interrupt when the 31st conversion is stored in the FIFO. This gives the host processor a chance to send a "0" to the LM12(H)454/8's Start bit (Configuration register) and halt the ADC before it completes the 32nd conversion. The Sequencer halts after the current (32) conversion is completed. The conversion data is then transferred to the FIFO and occupies the 32nd location. FIFO overflow is avoided if the Sequencer is halted before the start of the 32nd conversion by placing a "0" in the Start bit (Configuration register). It is important to remember that the Sequencer **continues to operate even if a FIFO interrupt (INT 2) is internally or externally generated**. The only mechanisms that stop the Sequencer are an instruction with the PAUSE bit set to "1" (halts before instruction execution), placing a "0" in the Configuration register's START bit, or placing a "1" in the Configuration register's RESET bit.

## 4.0 Sequencer

The Sequencer uses a 3-bit counter (Instruction Pointer, or IP, in *Figure 7)* to retrieve the programmable conversion instructions stored in the Instruction RAM. The 3-bit counter is reset to 000 during chip reset or if the current executed instruction has its Loop bit (Bit 1 in any Instruction RAM "00") set high ("1"). It increments at the end of the currently executed instruction and points to the next instruction. It will continue to increment up to 111 unless an instruction's Loop bit is set. If this bit is set, the counter resets to "000" and execution begins again with the first instruction. If all instructions have their Loop bit reset to "0", the Sequencer will execute all eight instructions continuously. Therefore, it is important to realize that if less than eight instructions are programmed, the Loop bit on the last instruction must be set. Leaving this bit reset to "0" allows the Sequencer to execute "unprogrammed" instructions, the results of which may be unpredictable.

The Sequencer's Instruction Pointer value is readable at any time and is found in the Status register at Bits 8–10. The Sequencer can go through eight states during instruction execution:

**State 0:** The current instruction's first 16 bits are read from the Instruction RAM "00". This state is one clock cycle long.

**State 1:** Checks the state of the Calibration and Start bits. This is the "rest" state whenever the Sequencer is stopped using the reset, a Pause command, or the Start bit is reset low ("0"). When the Start bit is set to a "1", this state is one clock cycle long.

**State 2:** Perform calibration. If bit 2 or bit 6 of the Configuration register is set to a "1", state 2 is 76 clock cycles long. If the Configuration register's bit 3 is set to a "1", state 2 is 4944 clock cycles long.

**State 3:** Run the internal 16-bit Timer. The number of clock cycles for this state varies according to the value stored in the Timer register. The number of clock cycles is found by using the expression below

$$32T + 2$$

where $0 \leq T \leq 2^{16} - 1$.

**State 7:** Run the acquisition delay and read Limit #1's value if needed. The number of clock cycles for 12-bit + sign mode varies according to

$$9 + 2D$$

where D is the user-programmable 4-bit value stored in bits 12–15 of Instruction RAM "00" and is limited to $0 \leq D \leq 15$.

The number of clock cycles for 8-bit + sign or "watchdog" mode varies according to

$$2 + 2D$$

where D is the user-programmable 4-bit value stored in bits 12–15 of Instruction RAM "00" and is limited to $0 \leq D \leq 15$.

**State 6:** Perform first comparison. This state is 5 clock cycles long.

**State 4:** Read Limit #2. This state is 1 clock cycle long.

**State 5:** Perform a conversion or second comparison. This state takes 44 clock cycles when using the 12-bit + sign mode or 21 clock cycles when using the 8-bit + sign mode. The "watchdog" mode takes 5 clock cycles.

## 4.0 Sequencer (Continued)

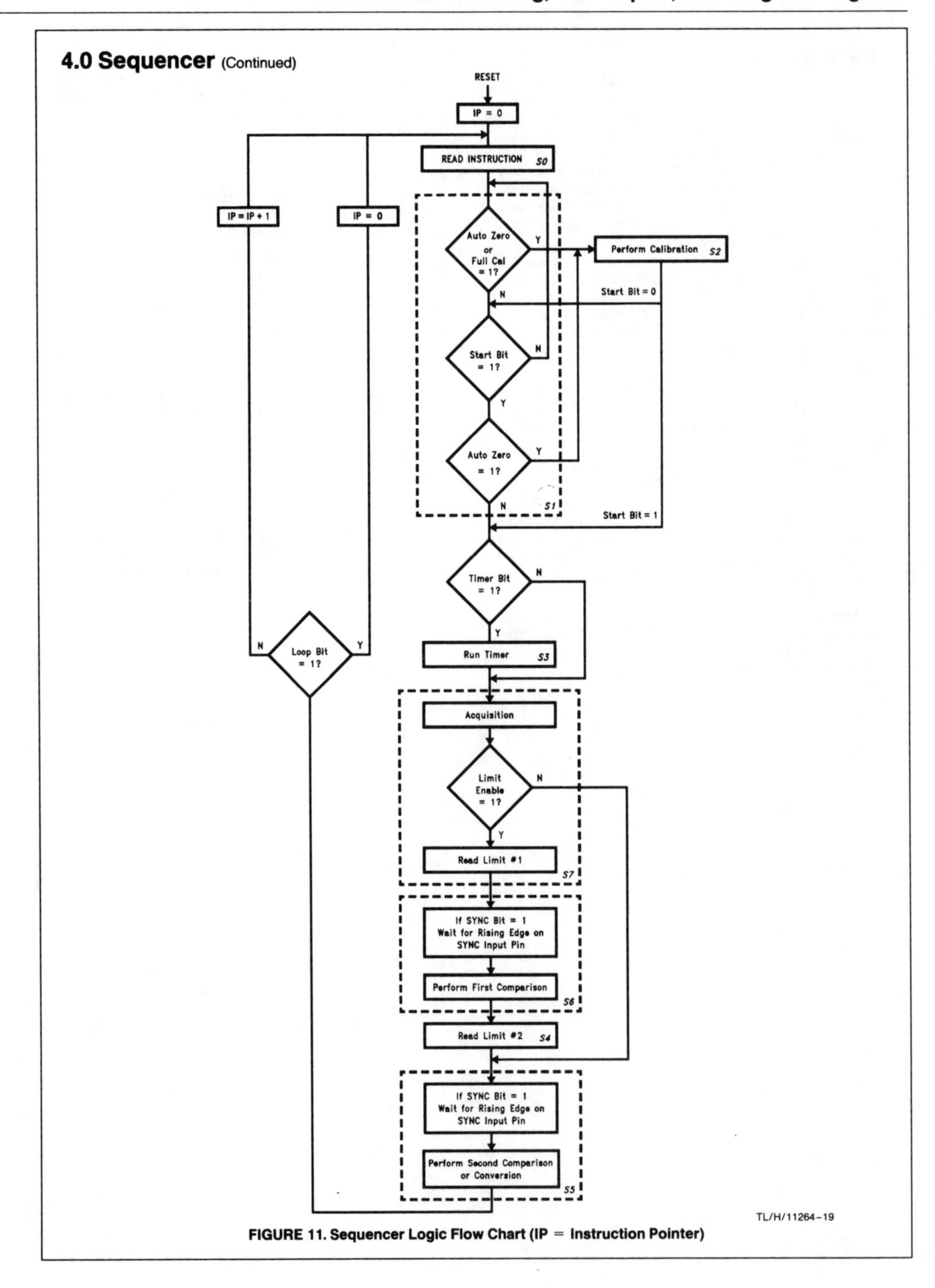


**FIGURE 11. Sequencer Logic Flow Chart (IP = Instruction Pointer)**

# 5.0 Analog Considerations

### 5.1 REFERENCE VOLTAGE

The difference in the voltages applied to the $V_{REF+}$ and $V_{REF-}$ defines the analog input voltage span (the difference between the voltages applied between two multiplexer inputs or the voltage applied to one of the multiplexer inputs and analog ground), over which 4095 positive and 4096 negative codes exist. The voltage sources driving $V_{REF+}$ or $V_{REF-}$ must have very low output impedance and noise. The circuit in *Figure 12* is an example of a very stable reference appropriate for use with the LM12(H)454/8.

The ADC can be used in either ratiometric or absolute reference applications. In ratiometric systems, the analog input voltage is proportional to the voltage used for the ADC's reference voltage. When this voltage is the system power supply, the $V_{REF+}$ pin is connected to $V_A{}^+$ and $V_{REF-}$ is connected to GND. This technique relaxes the system reference stability requirements because the analog input voltage and the ADC reference voltage move together. This maintains the same output code for given input conditions.

For absolute accuracy, where the analog input voltage varies between very specific voltage limits, a time and temperature stable voltage source can be connected to the reference inputs. Typically, the reference voltage's magnitude will require an initial adjustment to null reference voltage induced full-scale errors.

When using the LM12(H)454/8's internal 2.5V bandgap reference, a parallel combination of a 100 μF capacitor and a 0.1 μF capacitor connected to the $V_{REFOUT}$ pin is recommended for low noise operation. When left unconnected, the reference remains stable without a bypass capacitor. However, ensure that stray capacitance at the $V_{REFOUT}$ pin remains below 50 pF.

### 5.2 INPUT RANGE

The LM12(H)454/8's fully differential ADC and reference voltage inputs generate a two's-complement output that is found by using the equation below.

$$\text{output code} = \frac{V_{IN+} - V_{IN-}}{V_{REF+} - V_{REF-}}(4096) - \tfrac{1}{2} \qquad \text{(12-bit)}$$

$$\text{output code} = \frac{V_{IN+} - V_{IN-}}{V_{REF+} - V_{REF-}}(256) - \tfrac{1}{2} \qquad \text{(8-bit)}$$

Round up to the next integer value between −4096 to 4095 for 12-bit resolution and between −256 to 255 for 8-bit resolution if the result of the above equation is not a whole number. As an example, $V_{REF+} = 2.5V$, $V_{REF-} = 1V$, $V_{IN+} = 1.5V$ and $V_{IN-}$ = GND. The 12-bit + sign output code is positive full-scale, or 0,1111,1111,1111. If $V_{REF+} = 5V$, $V_{REF-} = 1V$, $V_{IN+} = 3V$, and $V_{IN-}$ = GND, the 12-bit + sign output code is 0,1100,0000,0000.

### 5.3 INPUT CURRENT

A charging current flows into or out of (depending on the input voltage polarity) the analog input pins, IN0–IN7 at the start of the analog input acquisition time ($t_{ACQ}$). This current's peak value will depend on the actual input voltage applied.

### 5.4 INPUT SOURCE RESISTANCE

For low impedance voltage sources (<100Ω for 5 MHz operation and <60Ω for 8 MHz operation), the input charging current will decay, before the end of the S/H's acquisition time, to a value that will not introduce any conversion errors. For higher source impedances, the S/H's acquisition time can be increased. As an example, operating with a 5 MHz clock frequency and maximum acquisition time, the LM12(H)454/8's analog inputs can handle source impedance as high as 6.67 kΩ. When operating at 8 MHz and maximum acquisition time, the LM12H454/8's analog inputs can handle source impedance as high as 4.17 kΩ. Refer to Section 2.1, Instruction RAM "00", Bits 12–15 for further information.

### 5.5 INPUT BYPASS CAPACITANCE

External capacitors (0.01 μF–0.1 μF) can be connected between the analog input pins, IN0–IN7, and analog ground to filter any noise caused by inductive pickup associated with long input leads. It will not degrade the conversion accuracy.

### 5.6 NOISE

The leads to each of the analog multiplexer input pins should be kept as short as possible. This will minimize input noise and clock frequency coupling that can cause conversion errors. Input filtering can be used to reduce the effects of the noise sources.

### 5.7 POWER SUPPLIES

Noise spikes on the $V_A{}^+$ and $V_D{}^+$ supply lines can cause conversion errors; the comparator will respond to the noise. The ADC is especially sensitive to any power supply spikes that occur during the auto-zero or linearity correction. Low inductance tantalum capacitors of 10 μF or greater paralleled with 0.1 μF monolithic ceramic capacitors are recom-

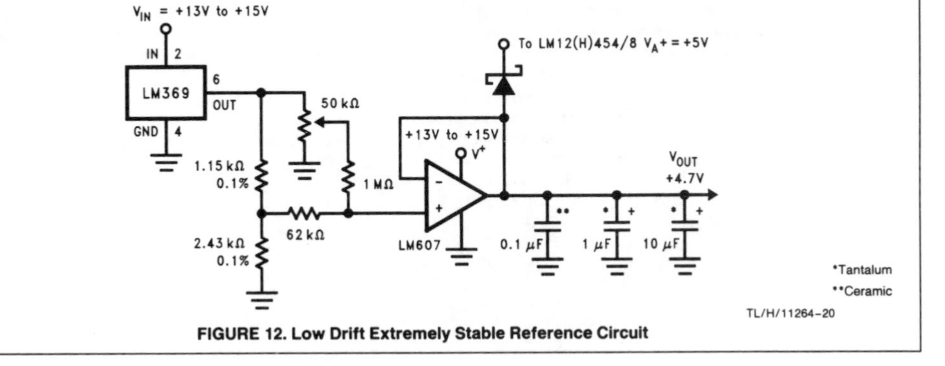


**FIGURE 12. Low Drift Extremely Stable Reference Circuit**

# 5.0 Analog Considerations (Continued)

mended for supply bypassing. Separate bypass capacitors should be used for the $V_A+$ and $V_D+$ supplies and placed as close as possible to these pins.

**5.8 GROUNDING**

The LM12(H)454/8's nominal high resolution performance can be maximized through proper grounding techniques. These include the use of separate analog and digital ground planes. The digital ground plane is placed under all components that handle digital signals, while the analog ground plane is placed under all analog signal handling circuitry. The digital and analog ground planes are connected at only one point, the power supply ground. This greatly reduces the occurrence of ground loops and noise.

It is recommended that stray capacitance between the analog inputs or outputs (LM12(H)454: IN0–IN3, MUXOUT+, MUXOUT−, S/H IN+, S/H IN−; LM12(H)458: IN0–IN7, $V_{REF+}$, and $V_{REF-}$) be reduced by increasing the clearance (+1/16th inch) between the analog signal and reference pins and the ground plane.

**5.9 CLOCK SIGNAL LINE ISOLATION**

The LM12(H)454/8's performance is optimized by routing the analog input/output and reference signal conductors (pins 34–44) as far as possible from the conductor that carries the clock signal to pin 23. Ground traces parallel to the clock signal trace can be used on printed circuit boards to reduce clock signal interference on the analog input/output pins.

# 6.0 Application Circuits

**6.1 PC EVALUATION/INTERFACE BOARD**

*Figure 13* is the schematic of an evaluation/interface board designed to interface the LM12(H)454 or LM12(H)458 with an XT or AT style computer. The board can be used to develop both software and hardware. The board hardwires the BW (Bus Width) pin to a logic high, selecting an 8-bit wide databus. Therefore, it is designed for an 8-bit expansion slot on the computer's motherboard.

The circuit operates on a single +5V supply derived from the computer's +12V supply using an LM340 regulator. This greatly attenuates noise that may be present on the computer's power supply lines. However, your application may only need an LC filter.

*Figure 13* also shows the recommended supply ($V_A+$ and $V_D+$) and reference input ($V_{REF+}$ and $V_{REF-}$) bypassing. The digital and analog supply pins can be connected together to the same supply voltage. However, they need separate, multiple bypass capacitors. Multiple capacitors on the supply pins and the reference inputs ensures a low impedance bypass path over a wide frequency range.

All digital interface control signals (IOR, IOW, and AEN), data lines (DB0–DB7), address lines (A0–A9), and IRQ (interrupt request) lines (IRQ2, IRQ3, and IRQ5) connections are made through the motherboard slot connector. All analog signals applied to, or received by, the input multiplexer (IN0–IN7 for the LM12(H)458 and IN0–IN3, MUXOUT+, MUXOUT−, S/H IN+ and S/H IN− for the LM12(H)454), $V_{REF+}$, $V_{REF-}$, $V_{REFOUT}$, and the SYNC signal input/output are applied through a DB-37 connector on the rear side of the board. *Figure 13* shows that there are numerous analog ground connections available on the DB-37 connector.

The voltage applied to $V_{REF-}$ and $V_{REF+}$ is selected using two jumpers, JP1 and JP2. JP1 selects between the voltage applied to the DB-37's pin 24 or GND and applies it to the LM12(H)454/8's $V_{REF-}$ input. JP2 selects between the LM12(H)454/8's internal reference output, $V_{REFOUT}$, and the voltage applied to the DB-37's pin 22 and applies it to the LM12(H)454/8's $V_{REF+}$ input.

**TABLE III. LM12(H)454/8 Evaluation/Interface Board SW DIP-8 Switch Settings for Available I/O Memory Locations**

| Hexidecimal I/O Memory Base Address | SW DIP-8 | | | |
|---|---|---|---|---|
| | SW1 (SEL0) | SW2 (SEL1) | SW3 (SEL2) | SW4 (SEL3) |
| 100 | ON | ON | ON | ON |
| 120 | OFF | ON | ON | ON |
| 140 | ON | OFF | ON | ON |
| 160 | OFF | OFF | ON | ON |
| 180 | ON | ON | OFF | ON |
| 1A0 | OFF | ON | OFF | ON |
| 1C0 | ON | OFF | OFF | ON |
| 300 | OFF | OFF | OFF | ON |
| 340 | ON | ON | ON | OFF |
| 280 | OFF | ON | ON | OFF |
| 2A0 | ON | OFF | ON | OFF |

The board allows the use of one of three Interrupt Request (IRQ) lines IRQ2, IRQ3, and IRQ5. The individual IRQ line can be selected using switches 5, 6, and 7 of SW DIP-8. When using any of these three IRQs, the user needs to ensure that there are no conflicts between the evaluation board and any other boards attached to the computer's motherboard.

Switches 1–4, along with address lines A5–A9 are used as inputs to GAL16V8 Programmable Gate Array (U2). This device forms the interface between the computer's control and address lines and generates the control signals used by the LM12(H)454/8 for $\overline{CS}$, $\overline{WR}$, and $\overline{RD}$. It also generates the signal that controls the data buffers. Several address ranges within the computer's I/O memory map are available. Refer to Table III for the switch settings that gives the desired I/O memory address range. Selection of an address range must be done so that there are no conflicts between the evaluation board and any other boards attached to the computer's motherboard. The GAL equations are shown in *Figure 14*. The GAL functional block diagram is shown in *Figure 15*.

*Figures 16–19* show the layout of each layer in the 3-layer evaluation/interface board plus the silk-screen layout showing parts placement. *Figure 17* is the top or component side, *Figure 18* is the middle or ground plane layer, *Figure 19* is the circuit side, and *Figure 16* is the parts layout.

## 6.0 Application Circuits (Continued)

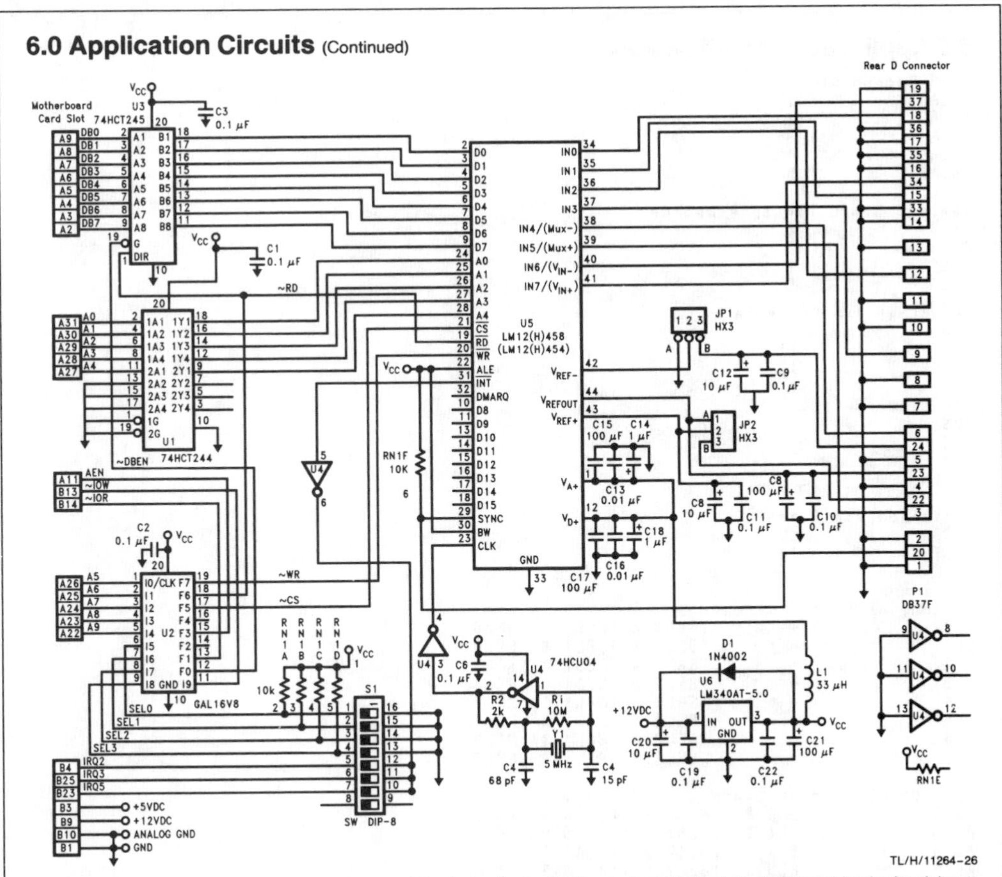


**Note:** The layout utilizes a split ground plane. The analog ground plane is placed under all analog signals and U5 pins 1, 34–44. The remaining signals and pins are placed over the digital ground. The single point ground connection is at U6, pin 2, and this is connected to the motherboard pin B1.

## Parts List:

| Part | Description | Part | Description |
|---|---|---|---|
| Y1 | HC49U, 8 MHz crystal | C7, C21 | 100 µF, 25V, electrolytic |
| D1 | 1N4002 | C8, C12, C20 | 10 µF, 35V, electrolytic |
| L1 | 33 µH | C13, C16 | 0.01 µF, 50V, monolithic ceramic |
| P1 | DB37F; parallel connector | C14, C18 | 1 µF, 35V, tantalum |
| R1 | 10 MΩ, 5%, ¼W | C15, C17 | 100 µF, 50V, ceramic disk |
| R2 | 2 kΩ, 5%, ¼W | U1 | MM74HCT244N |
| RN1 | 10 kΩ, 6 resistor SIP, 5%, ⅛W | U2 | GAL16V8-20LNC |
| JP1, JP2 | HX3, 3-pin jumper | U3 | MM74HCT245N |
| S1 | SW DIP-8; 8 SPST switches | U4 | MM74HCU04N |
| C1–3, C6, C9–11, C19, C22 | 0.1 µF, 50V, monolithic ceramic | U5 | LM12(H)458CIV or LM12(H)454CIV |
| | | U6 | LM340AT-5.0 |
| C4 | 68 pF, 50V, ceramic disk | SK1 | 44-pin PLCC socket |
| C5 | 15 pF, 50V, ceramic disk | A1 | LM12(H)458/4 Rev. D PC Board |

**FIGURE 13. Schematic and Parts List for the LM12(H)454/8 Evaluation/Interface Board for XT and AT Style Computers, Order Number LM12458EVAL**

## 6.0 Application Circuits (Continued)

```
;I/O Decode Lines
io_A5           1
io_A6           2
io_A7           3
io_A8           4
io_A9           5

;Select Lines for Zone Decode
SEL0            6
SEL1            7
SEL2            8
SEL3            9
```

TL/H/11264-33

```
;Physical I/O Controls
AEN                     15
!io_WR          11
!io_RD          13

;Physical Outputs
!CS                     17
!WR                     19
!RD                     18
!DBEN           12

;Intermediate Terms:
DEC0            16
FILT            14

Equations
;Decode of Select Lines:
SL0             =       !SEL2 & !SEL1 & !SEL0;
SL1             =       !SEL2 & !SEL1 &  SEL0;
SL2             =       !SEL2 &  SEL1 & !SEL0;
SL3             =       !SEL2 &  SEL1 &  SEL0;
SL4             =        SEL2 & !SEL1 & !SEL0;
SL5             =        SEL2 & !SEL1 &  SEL0;
SL6             =        SEL2 &  SEL1 & !SEL0;
SL7             =        SEL2 &  SEL1 &  SEL0;

;Decode of Address Lines:
AL00    =       SL0 & !io_A7 & !io_A6 & !io_A5;
AL20    =       SL1 & !io_A7 & !io_A6 &  io_A5;
AL40    =       SL2 & !io_A7 &  io_A6 & !io_A5;
AL60    =       SL3 & !io_A7 &  io_A6 &  io_A5;
AL80    =       SL4 &  io_A7 & !io_A6 & !io_A5;
ALA0    =       SL5 &  io_A7 & !io_A6 &  io_A5;
ALC0    =       SL6 &  io_A7 &  io_A6 & !io_A5;

AH01    =       !SEL3 & !io_A9 &  io_A8;
AH02    =        SEL3 &  io_A9 & !io_A8 &  io_A7 & !io_A6;
AH03    =        SEL3 &  io_A9 &  io_A8 & !io_A7 & !io_A5;

;Intermediate Address Groups:
DEC0    =       !AEN & (AL00 + AL20 + AL40 + AL60 + AL80 + ALA0 + ALC0);

;DAS Chip Select Decode:
FILT    =       CS & ( io_WR + io_RD);
CS              =       ( io_WR  + io_RD) & DEC0 & ( AH01 + AH02 + AH03);
DBEN    =       CS & DEC0 & ( io_WR + io_RD);

;Delayed Read/ Write Decodes:
WR              =       io_WR & FILT;
RD              =       io_RD & FILT;
```

TL/H/11264-32

**FIGURE 14. Logic Equations Used to Program the GAL16V8**

## 6.4 Application Circuits (Continued)

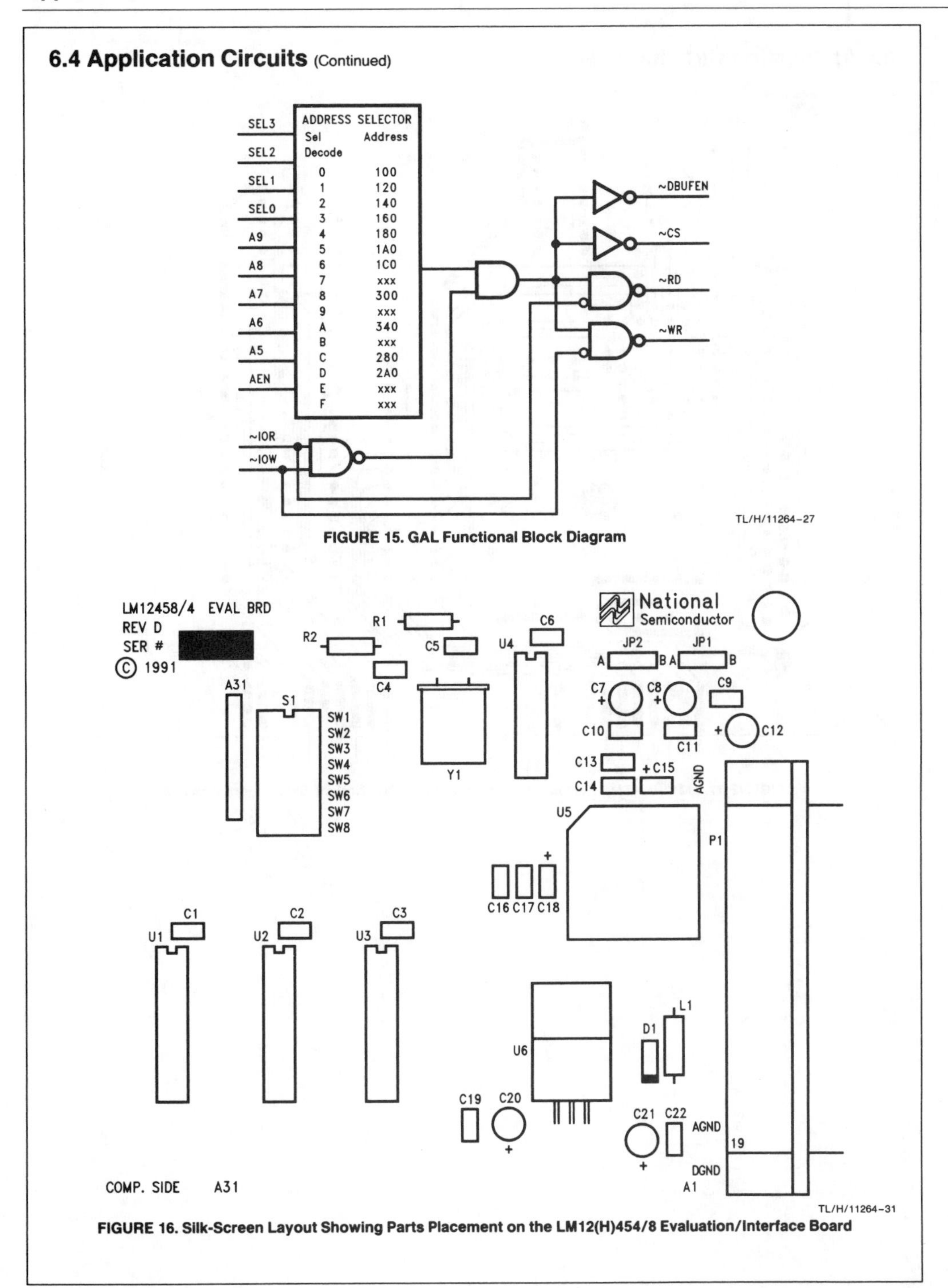


**FIGURE 15. GAL Functional Block Diagram**



**FIGURE 16. Silk-Screen Layout Showing Parts Placement on the LM12(H)454/8 Evaluation/Interface Board**

## 6.4 Application Circuits (Continued)

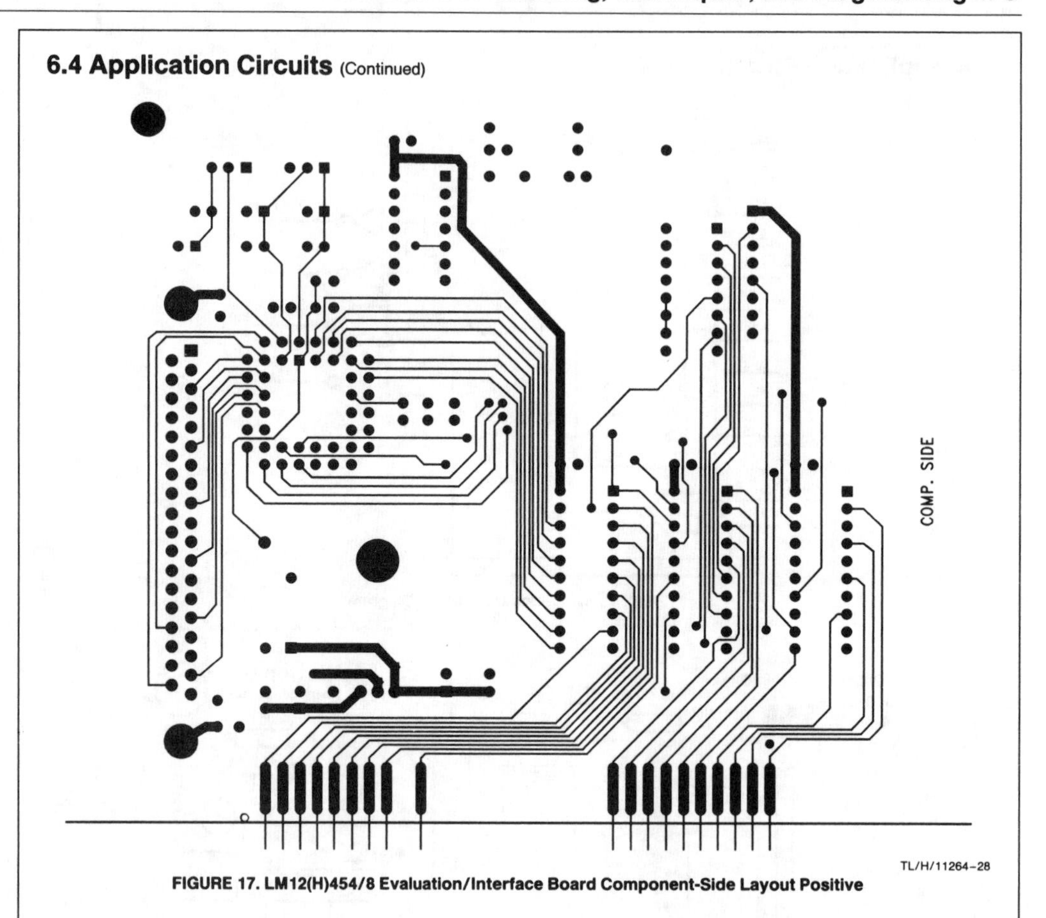


TL/H/11264-28

**FIGURE 17. LM12(H)454/8 Evaluation/Interface Board Component-Side Layout Positive**

## 6.4 Application Circuits (Continued)

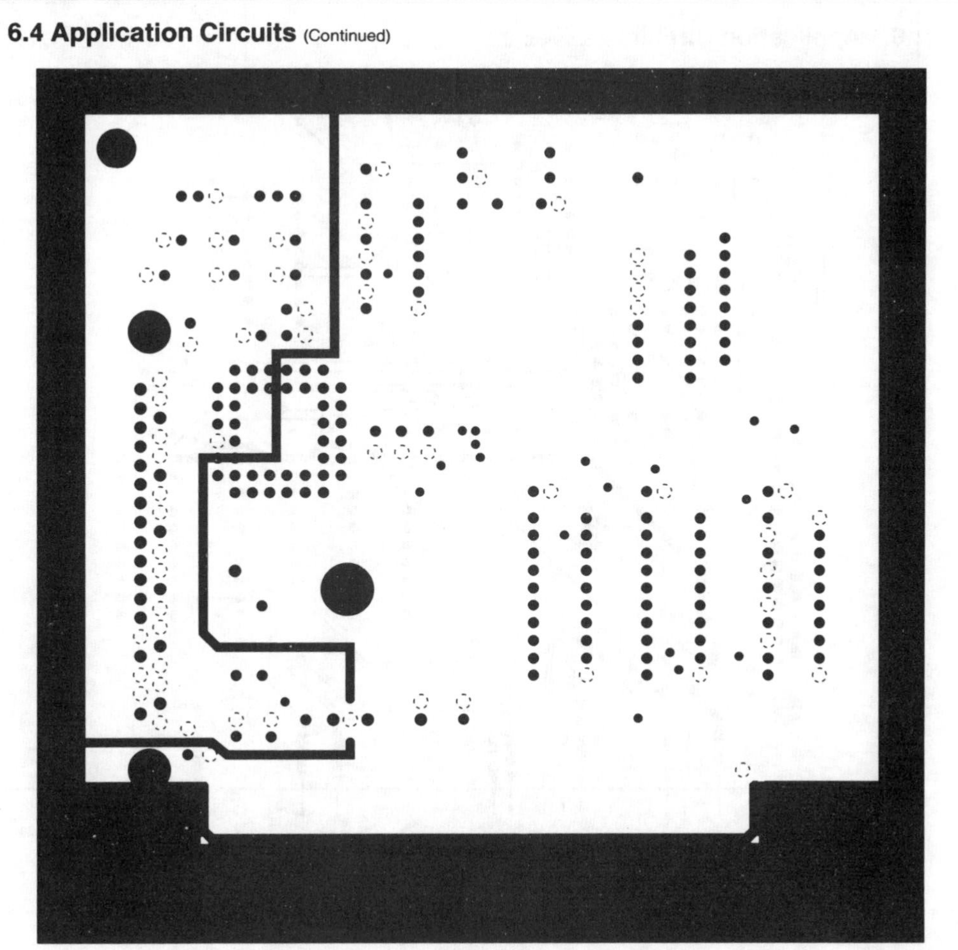

TL/H/11264-29

**FIGURE 18. LM12(H)454/8 Evaluation/Interface Board Ground-Plane Layout Negative**

## 6.4 Application Circuits (Continued)

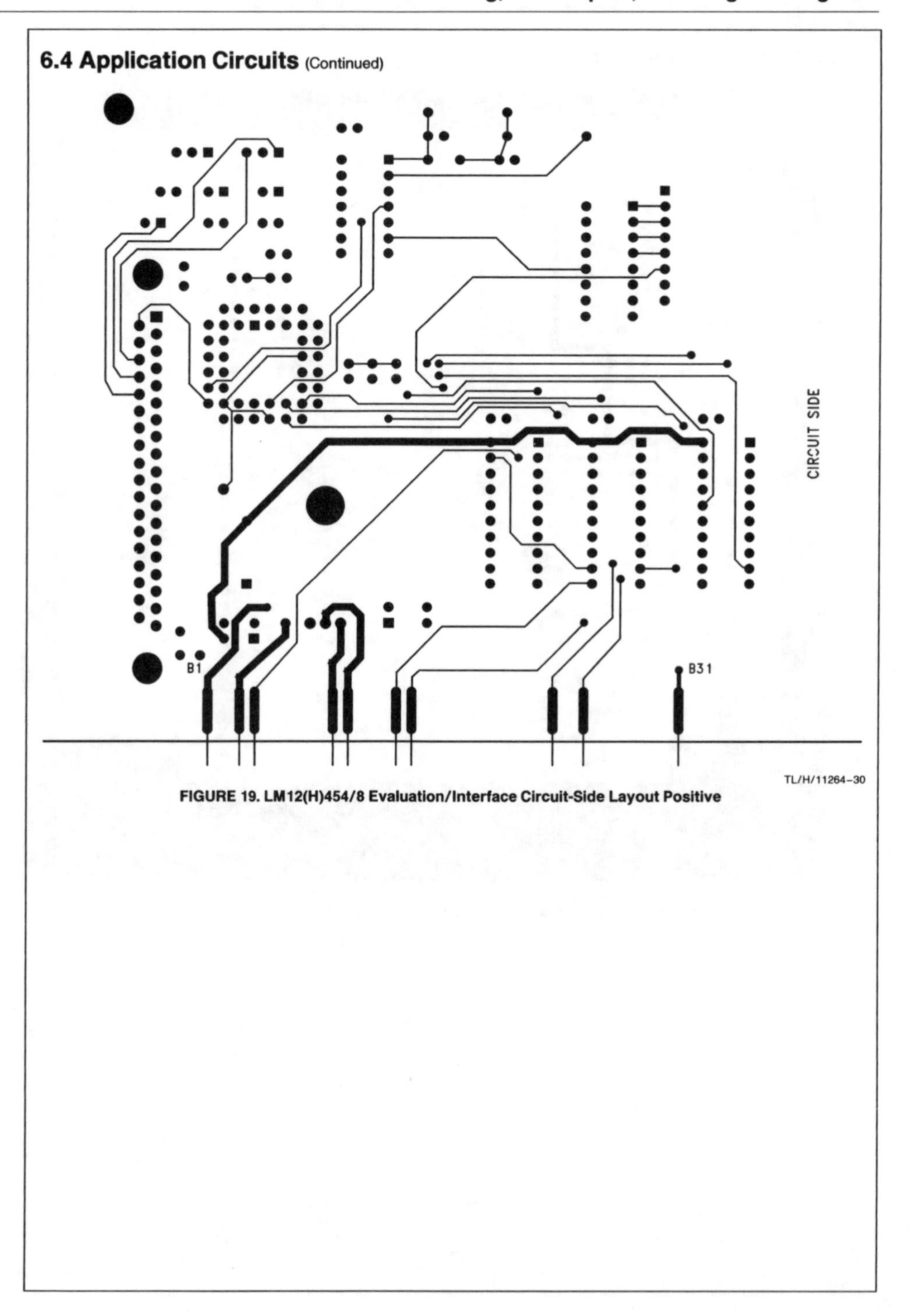


**FIGURE 19. LM12(H)454/8 Evaluation/Interface Circuit-Side Layout Positive**

## Physical Dimensions inches (millimeters)

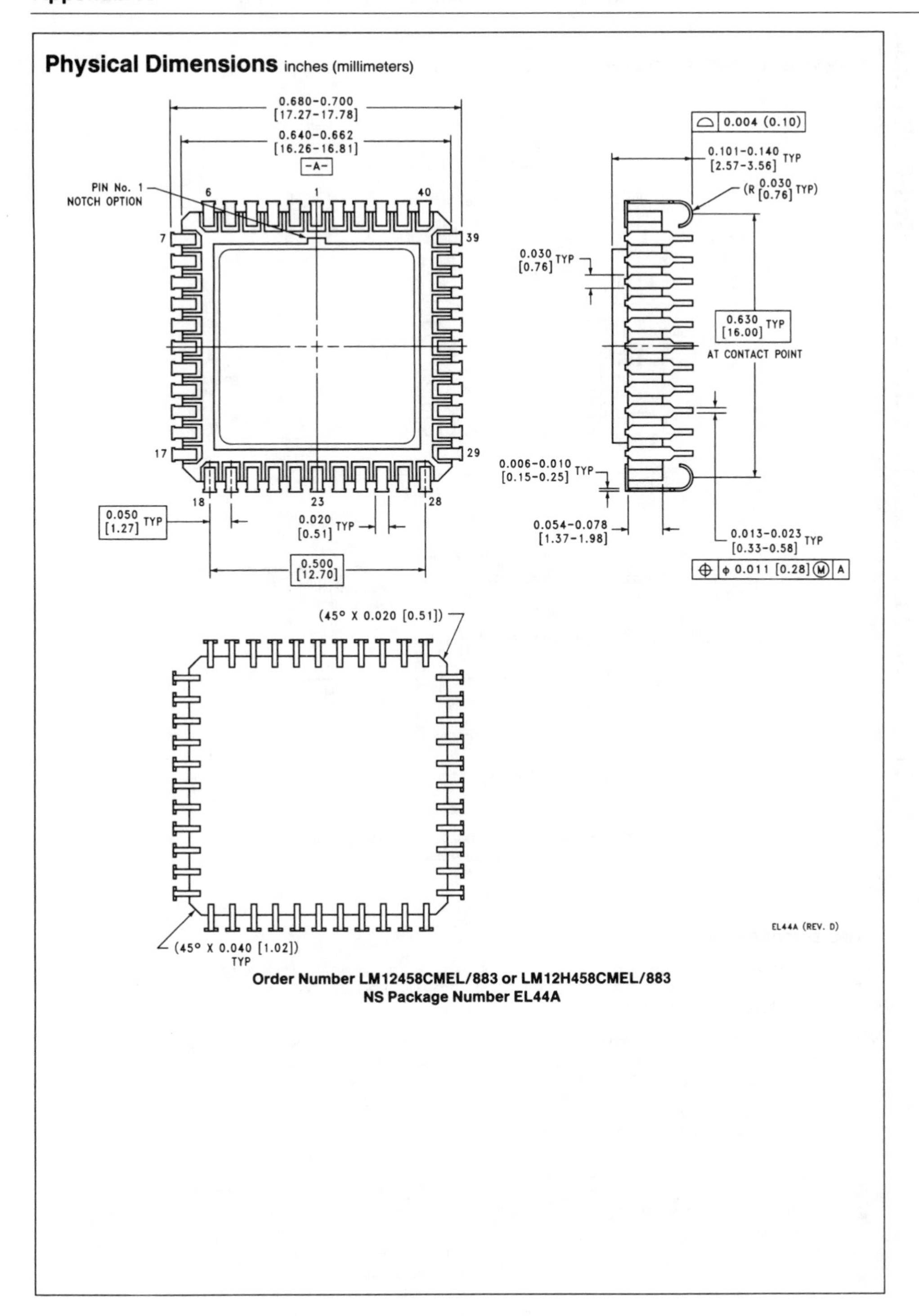


**Order Number LM12458CMEL/883 or LM12H458CMEL/883**
**NS Package Number EL44A**

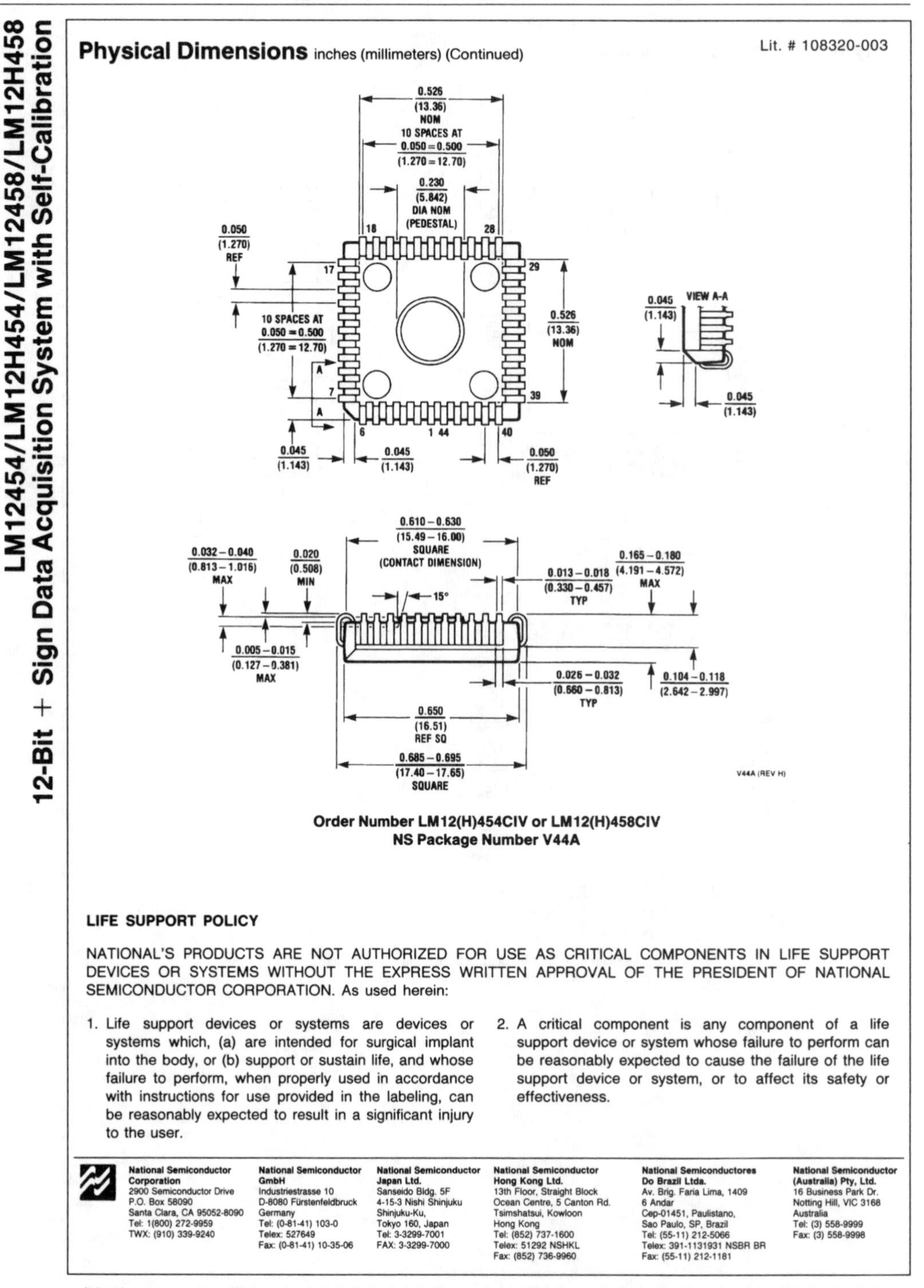

LM12454/LM12H454/LM12458/LM12H458 12-Bit + Sign Data Acquisition System with Self-Calibration

## Physical Dimensions inches (millimeters) (Continued)

Lit. # 108320-003



**Order Number LM12(H)454CIV or LM12(H)458CIV**
**NS Package Number V44A**

### LIFE SUPPORT POLICY

NATIONAL'S PRODUCTS ARE NOT AUTHORIZED FOR USE AS CRITICAL COMPONENTS IN LIFE SUPPORT DEVICES OR SYSTEMS WITHOUT THE EXPRESS WRITTEN APPROVAL OF THE PRESIDENT OF NATIONAL SEMICONDUCTOR CORPORATION. As used herein:

1. Life support devices or systems are devices or systems which, (a) are intended for surgical implant into the body, or (b) support or sustain life, and whose failure to perform, when properly used in accordance with instructions for use provided in the labeling, can be reasonably expected to result in a significant injury to the user.
2. A critical component is any component of a life support device or system whose failure to perform can be reasonably expected to cause the failure of the life support device or system, or to affect its safety or effectiveness.


**National Semiconductor Corporation**
2900 Semiconductor Drive
P.O. Box 58090
Santa Clara, CA 95052-8090
Tel: 1(800) 272-9959
TWX: (910) 339-9240

**National Semiconductor GmbH**
Industriestrasse 10
D-8080 Fürstenfeldbruck
Germany
Tel: (0-81-41) 103-0
Telex: 527649
Fax: (0-81-41) 10-35-06

**National Semiconductor Japan Ltd.**
Sanseido Bldg. 5F
4-15-3 Nishi Shinjuku
Shinjuku-Ku,
Tokyo 160, Japan
Tel: 3-3299-7001
FAX: 3-3299-7000

**National Semiconductor Hong Kong Ltd.**
13th Floor, Straight Block
Ocean Centre, 5 Canton Rd.
Tsimshatsui, Kowloon
Hong Kong
Tel: (852) 737-1600
Telex: 51292 NSHKL
Fax: (852) 736-9960

**National Semiconductores Do Brazil Ltda.**
Av. Brig. Faria Lima, 1409
6 Andar
Cep-01451, Paulistano,
Sao Paulo, SP, Brazil
Tel: (55-11) 212-5066
Telex: 391-1131931 NSBR BR
Fax: (55-11) 212-1181

**National Semiconductor (Australia) Pty, Ltd.**
16 Business Park Dr.
Notting Hill, VIC 3168
Australia
Tel: (3) 558-9999
Fax: (3) 558-9998

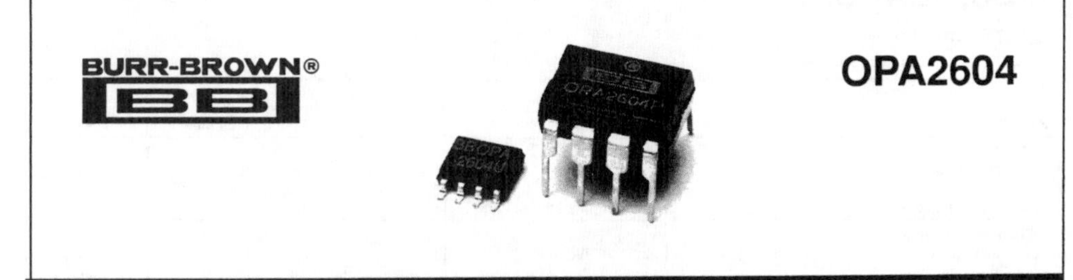


# OPA2604

# Dual FET-Input, Low Distortion OPERATIONAL AMPLIFIER

## FEATURES

- LOW DISTORTION: 0.0003% at 1kHz
- LOW NOISE: 10nV/√Hz
- HIGH SLEW RATE: 25V/μs
- WIDE GAIN-BANDWIDTH: 20MHz
- UNITY-GAIN STABLE
- WIDE SUPPLY RANGE: $V_S$ = ±4.5 to ±24V
- DRIVES 600Ω LOADS

## APPLICATIONS

- PROFESSIONAL AUDIO EQUIPMENT
- PCM DAC I/V CONVERTER
- SPECTRAL ANALYSIS EQUIPMENT
- ACTIVE FILTERS
- TRANSDUCER AMPLIFIER
- DATA ACQUISITION

## DESCRIPTION

The OPA2604 is a dual, FET-input operational amplifier designed for enhanced AC performance. Very low distortion, low noise and wide bandwidth provide superior performance in high quality audio and other applications requiring excellent dynamic performance.

New circuit techniques and special laser trimming of dynamic circuit performance yield very low harmonic distortion. The result is an op amp with exceptional sound quality. The low-noise FET input of the OPA2604 provides wide dynamic range, even with high source impedance. Offset voltage is laser-trimmed to minimize the need for interstage coupling capacitors.

The OPA2604 is available in 8-pin plastic mini-DIP and SO-8 surface-mount packages, specified for the –25°C to +85°C temperature range.

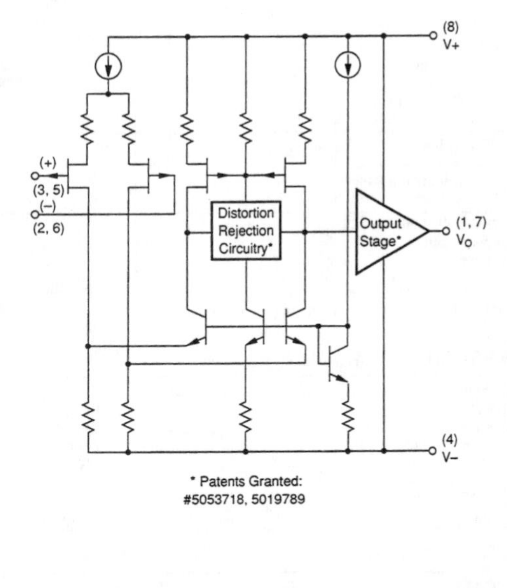


* Patents Granted: #5053718, 5019789

International Airport Industrial Park • Mailing Address: PO Box 11400 • Tucson, AZ 85734 • Street Address: 6730 S. Tucson Blvd. • Tucson, AZ 85706
Tel: (602) 746-1111 • Twx: 910-952-1111 • Cable: BBRCORP • Telex: 066-6491 • FAX: (602) 889-1510 • Immediate Product Info: (800) 548-6132


PDS-1069C Printed in U.S.A. September, 1993

# SPECIFICATIONS

## ELECTRICAL

$T_A$ = +25°C, $V_S$ = ±15V unless otherwise noted.

| PARAMETER | CONDITION | OPA2604AP, AU MIN | TYP | MAX | UNITS |
|---|---|---|---|---|---|
| **OFFSET VOLTAGE** | | | | | |
| Input Offset Voltage | | | ±1 | ±3 | mV |
| Average Drift | | | ±8 | | μV/°C |
| Power Supply Rejection | $V_S$ = ±5 to ±24V | 80 | 100 | | dB |
| **INPUT BIAS CURRENT**[1] | | | | | |
| Input Bias Current | $V_{CM}$ = 0V | | 100 | | pA |
| Input Offset Current | $V_{CM}$ = 0V | | ±4 | | pA |
| **NOISE** | | | | | |
| Input Voltage Noise | | | | | |
| Noise Density: f = 10Hz | | | 25 | | nV/ $\sqrt{Hz}$ |
| f = 100Hz | | | 15 | | nV/ $\sqrt{Hz}$ |
| f = 1kHz | | | 11 | | nV/ $\sqrt{Hz}$ |
| f = 10kHz | | | 10 | | nV/ $\sqrt{Hz}$ |
| Voltage Noise, BW = 20Hz to 20kHz | | | 1.5 | | μVp-p |
| Input Bias Current Noise | | | | | |
| Current Noise Density, f = 0.1Hz to 20kHz | | | 6 | | fA/ $\sqrt{Hz}$ |
| **INPUT VOLTAGE RANGE** | | | | | |
| Common-Mode Input Range | | ±12 | ±13 | | V |
| Common-Mode Rejection | $V_{CM}$ = ±12V | 80 | 100 | | dB |
| **INPUT IMPEDANCE** | | | | | |
| Differential | | | $10^{12}$ ‖ 8 | | Ω ‖ pF |
| Common-Mode | | | $10^{12}$ ‖ 10 | | Ω ‖ pF |
| **OPEN-LOOP GAIN** | | | | | |
| Open-loop Voltage Gain | $V_O$ = ±10V, $R_L$ = 1kΩ | 80 | 100 | | dB |
| **FREQUENCY RESPONSE** | | | | | |
| Gain-Bandwidth Product | G = 100 | | 20 | | MHz |
| Slew Rate | 20Vp-p, $R_L$ = 1kΩ | 15 | 25 | | V/μs |
| Settling Time: 0.01% | G = –1, 10V Step | | 1.5 | | μs |
| 0.1% | | | 1 | | μs |
| Total Harmonic Distortion + Noise (THD + N) | G = 1, f = 1kHz $V_O$ = 3.5Vrms, $R_L$ = 1kΩ | | 0.0003 | | % |
| Channel Separation | f = 1kHz, $R_L$ = 1kΩ | | 142 | | dB |
| **OUTPUT** | | | | | |
| Voltage Output | $R_L$ = 600Ω | ±11 | ±12 | | V |
| Current Output | $V_O$ = ±12V | | ±35 | | mA |
| Short Circuit Current | | | ±40 | | mA |
| Output Resistance, Open-Loop | | | 25 | | Ω |
| **POWER SUPPLY** | | | | | |
| Specified Operating Voltage | | | ±15 | | V |
| Operating Voltage Range | | ±4.5 | | ±24 | V |
| Current, Total Both Amplifiers | | | ±10.5 | ±12 | mA |
| **TEMPERATURE RANGE** | | | | | |
| Specification | | –25 | | +85 | °C |
| Storage | | –40 | | +125 | °C |
| Thermal Resistance[2], $\theta_{JA}$ | | | 90 | | °C/W |

NOTES: (1) Typical performance, measured fully warmed-up. (2) Soldered to circuit board—see text.

## ABSOLUTE MAXIMUM RATINGS

| | |
|---|---|
| Power Supply Voltage | ±25V |
| Input Voltage | (V–)–1V to (V+)+1V |
| Output Short Circuit to Ground | Continuous |
| Operating Temperature | –40°C to +100°C |
| Storage Temperature | –40°C to +125°C |
| Junction Temperature | +150°C |
| Lead Temperature (soldering, 10s) AP | +300°C |
| Lead Temperature (soldering, 3s) AU | +260°C |

## ORDERING INFORMATION

| MODEL | PACKAGE | TEMP. RANGE |
|---|---|---|
| OPA2604AP | 8-Pin Plastic DIP | –25°C to +85°C |
| OPA2604AU | SO-8 Surface-Mount | –25°C to +85°C |

## PACKAGING INFORMATION[1]

| MODEL | PACKAGE | PACKAGE DRAWING NUMBER |
|---|---|---|
| OPA2604AP | 8-Pin Plastic DIP | 006 |
| OPA2604AU | SO-8 Surface-Mount | 182 |

NOTE: (1) For detailed drawing and dimension table, please see end of data sheet, or Appendix D of Burr-Brown IC Data Book.

OPA2604

**PIN CONFIGURATION**

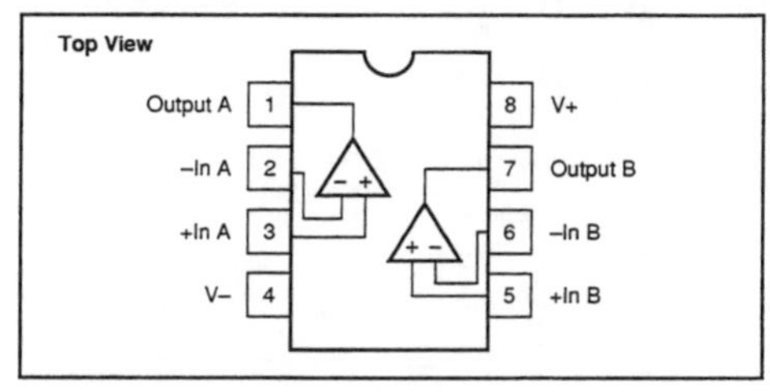


## ELECTROSTATIC DISCHARGE SENSITIVITY

Any integrated circuit can be damaged by ESD. Burr-Brown recommends that all integrated circuits be handled with appropriate precautions. Failure to observe proper handling and installation procedures can cause damage.

ESD damage can range from subtle performance degradation to complete device failure. Precision integrated circuits may be more susceptible to damage because very small parametric changes could cause the device not to meet published specifications.

## TYPICAL PERFORMANCE CURVES

$T_A$ = +25°C, $V_S$ = ±15V unless otherwise noted.

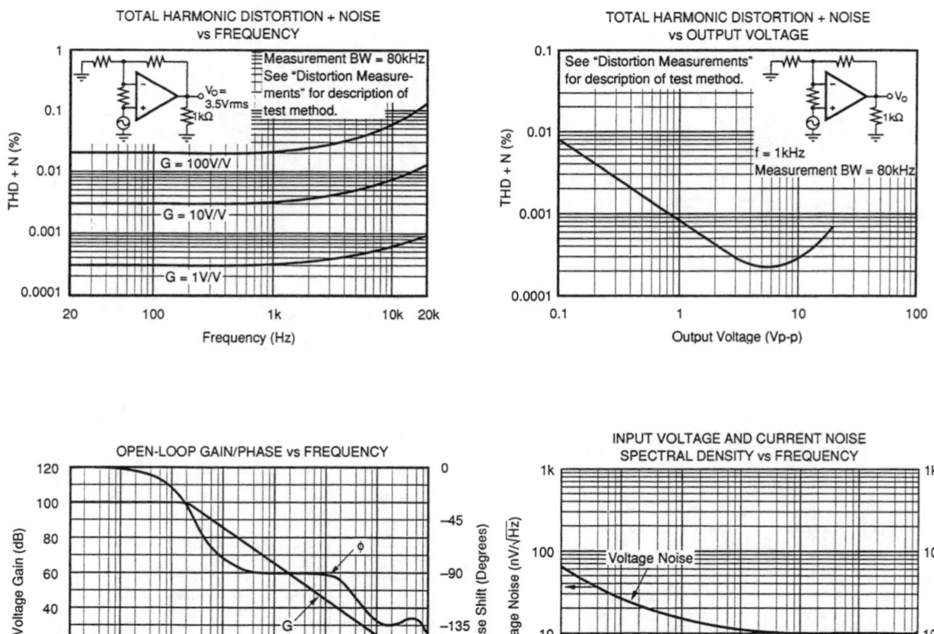


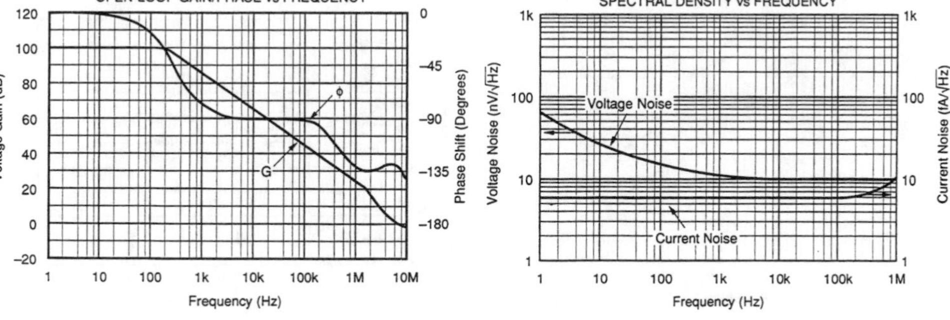

# TYPICAL PERFORMANCE CURVES (CONT)

$T_A = +25°C$, $V_S = ±15V$ unless otherwise noted.

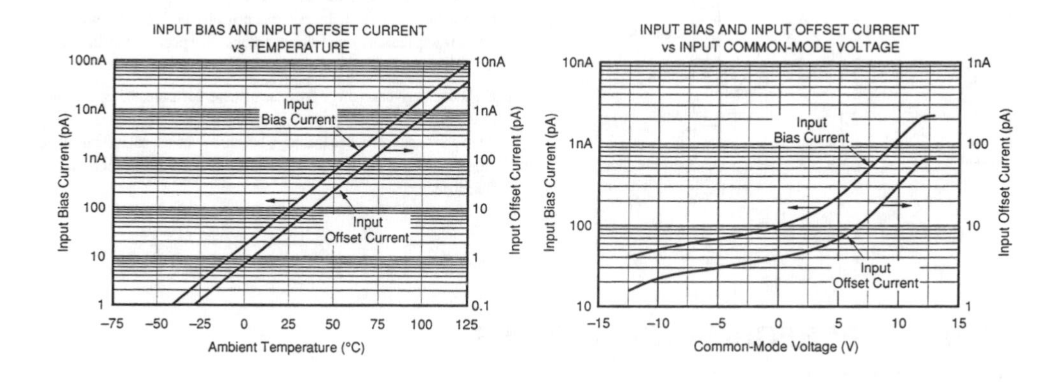


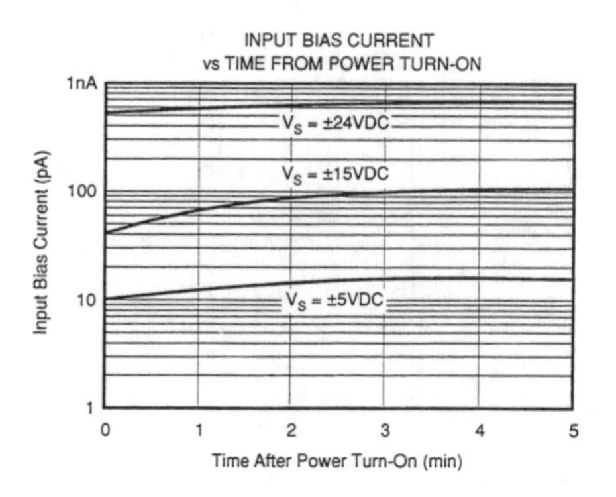


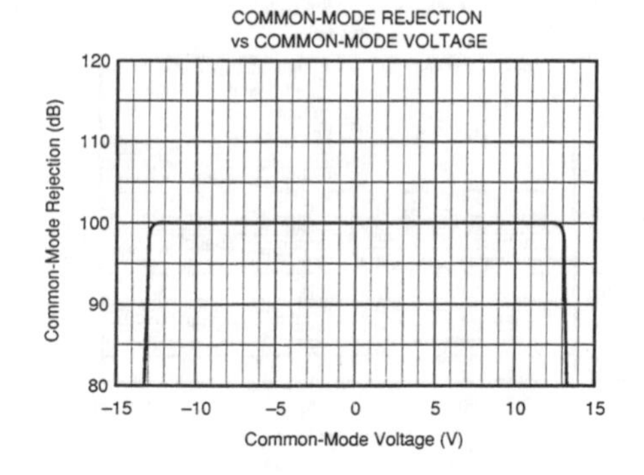


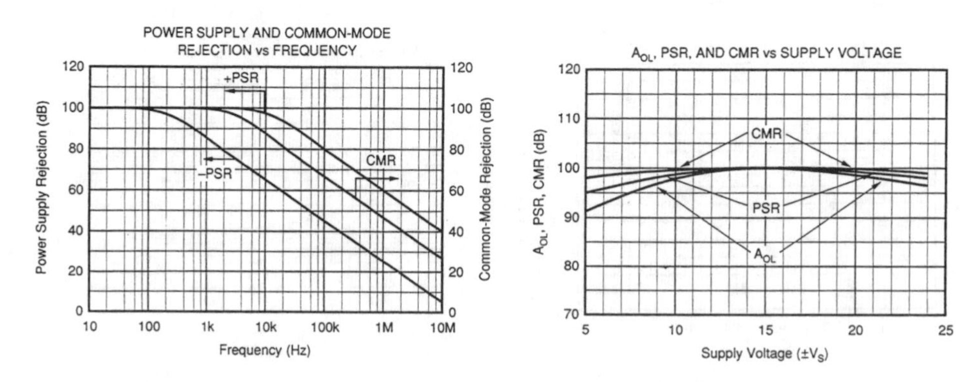

# TYPICAL PERFORMANCE CURVES (CONT)

$T_A$ = +25°C, $V_S$ = ±15V unless otherwise noted.

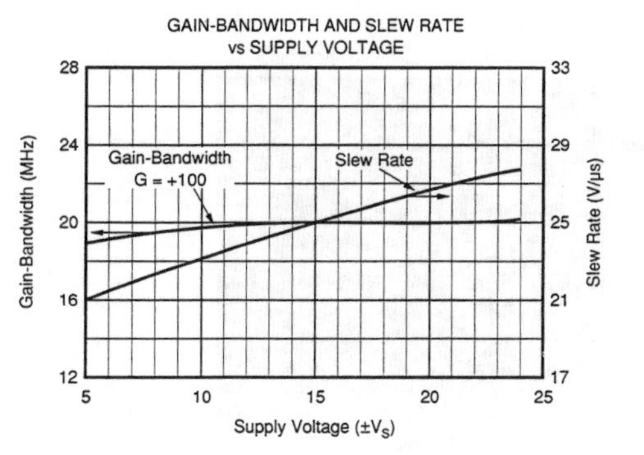


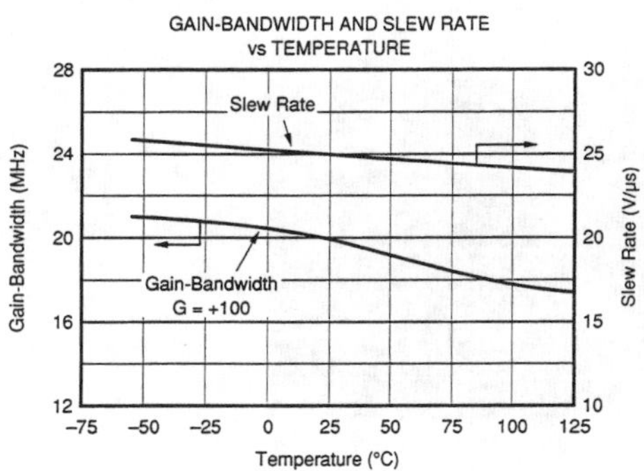


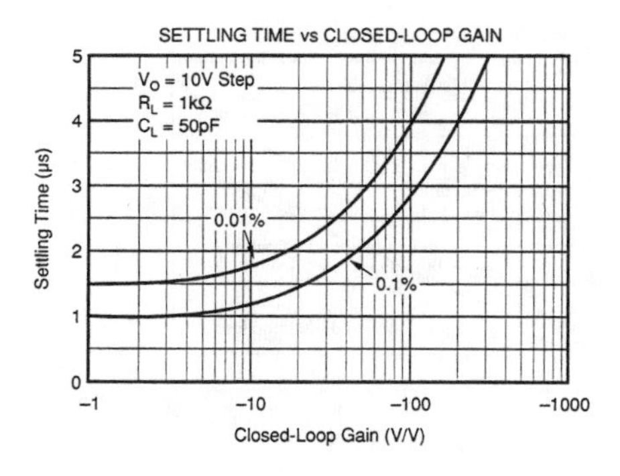


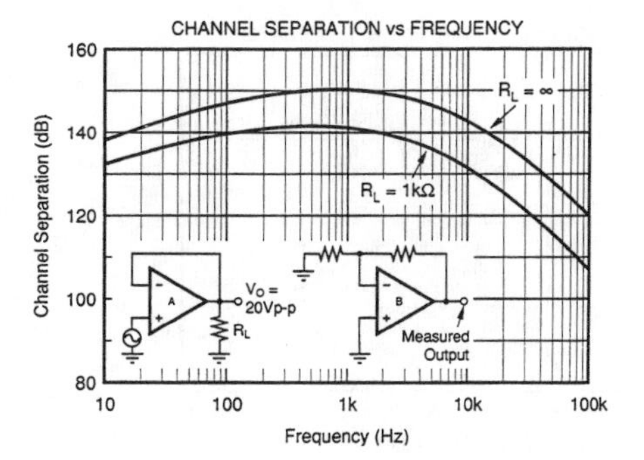


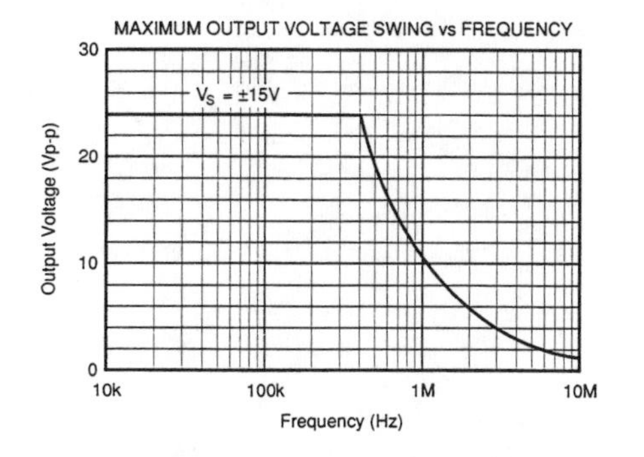


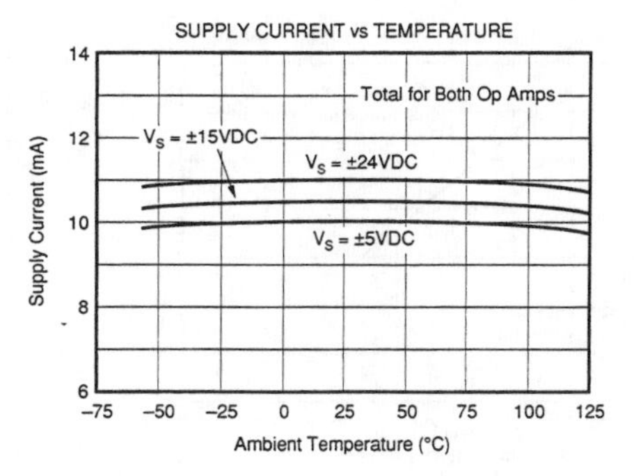

# TYPICAL PERFORMANCE CURVES (CONT)

$T_A$ = +25°C, $V_S$ = ±15V unless otherwise noted.

LARGE-SIGNAL TRANSIENT RESPONSE

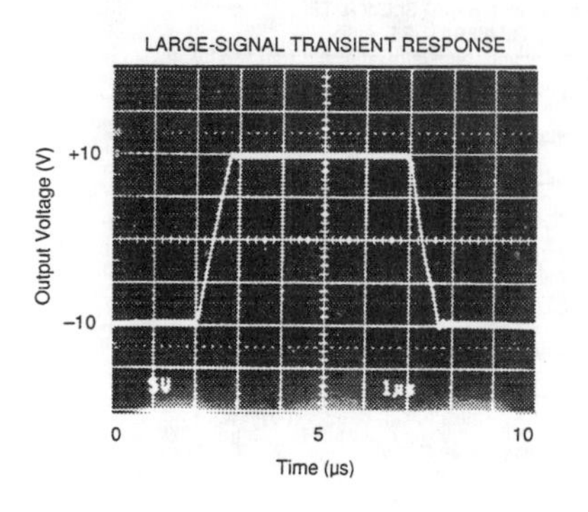


SMALL-SIGNAL TRANSIENT RESPONSE

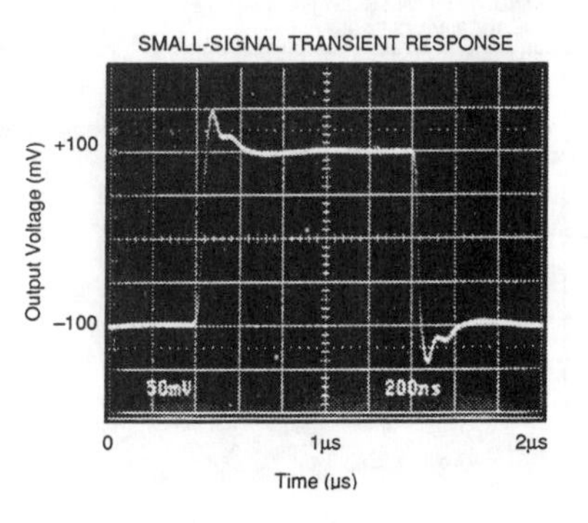


SHORT-CIRCUIT CURRENT vs TEMPERATURE

Short-Circuit Current (mA): 20, 30, 40, 50, 60

$I_{SC+}$ and $I_{SC-}$

Ambient Temperature (°C): –75, –50, –25, 0, 25, 50, 75, 100, 125

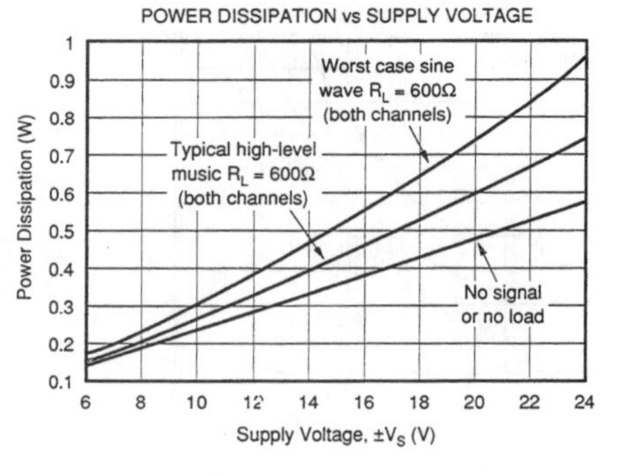

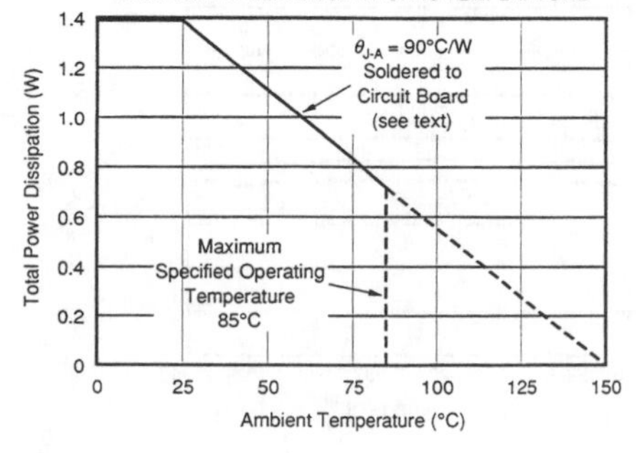

# APPLICATIONS INFORMATION

The OPA2604 is unity-gain stable, making it easy to use in a wide range of circuitry. Applications with noisy or high impedance power supply lines may require decoupling capacitors close to the device pins. In most cases 1μF tantalum capacitors are adequate.

## DISTORTION MEASUREMENTS

The distortion produced by the OPA2604 is below the measurement limit of virtually all commercially available equipment. A special test circuit, however, can be used to extend the measurement capabilities.

Op amp distortion can be considered an internal error source which can be referred to the input. Figure 1 shows a circuit which causes the op amp distortion to be 101 times greater than normally produced by the op amp. The addition of $R_3$ to the otherwise standard non-inverting amplifier configuration alters the feedback factor or noise gain of the circuit. The closed-loop gain is unchanged, but the feedback available for error correction is reduced by a factor of 101. This extends the measurement limit, including the effects of the signal-source purity, by a factor of 101. Note that the input signal and load applied to the op amp are the same as with conventional feedback without $R_3$.

Validity of this technique can be verified by duplicating measurements at high gain and/or high frequency where the distortion is within the measurement capability of the test equipment. Measurements for this data sheet were made with the Audio Precision, System One which greatly simplifies such repetitive measurements. The measurement technique can, however, be performed with manual distortion measurement instruments.

## CAPACITIVE LOADS

The dynamic characteristics of the OPA2604 have been optimized for commonly encountered gains, loads and operating conditions. The combination of low closed-loop gain and capacitive load will decrease the phase margin and may lead to gain peaking or oscillations. Load capacitance reacts with the op amp's open-loop output resistance to form an additional pole in the feedback loop. Figure 2 shows various circuits which preserve phase margin with capacitive load. Request Application Bulletin AB-028 for details of analysis techniques and applications circuits.

For the unity-gain buffer, Figure 2a, stability is preserved by adding a phase-lead network, $R_C$ and $C_C$. Voltage drop across $R_C$ will reduce output voltage swing with heavy loads. An alternate circuit, Figure 2b, does not limit the output with low load impedance. It provides a small amount of positive feedback to reduce the net feedback factor. Input impedance of this circuit falls at high frequency as op amp gain rolloff reduces the bootstrap action on the compensation network.

Figures 2c and 2d show compensation techniques for noninverting amplifiers. Like the follower circuits, the circuit in Figure 2d eliminates voltage drop due to load current, but at the penalty of somewhat reduced input impedance at high frequency.

Figures 2e and 2f show input lead compensation networks for inverting and difference amplifier configurations.

## NOISE PERFORMANCE

Op amp noise is described by two parameters—noise voltage and noise current. The voltage noise determines the noise performance with low source impedance. Low noise bipolar-input op amps such as the OPA27 and OPA37 provide very low voltage noise. But if source impedance is greater than a few thousand ohms, the current noise of bipolar-input op amps react with the source impedance and will dominate. At a few thousand ohms source impedance and above, the OPA2604 will generally provide lower noise.

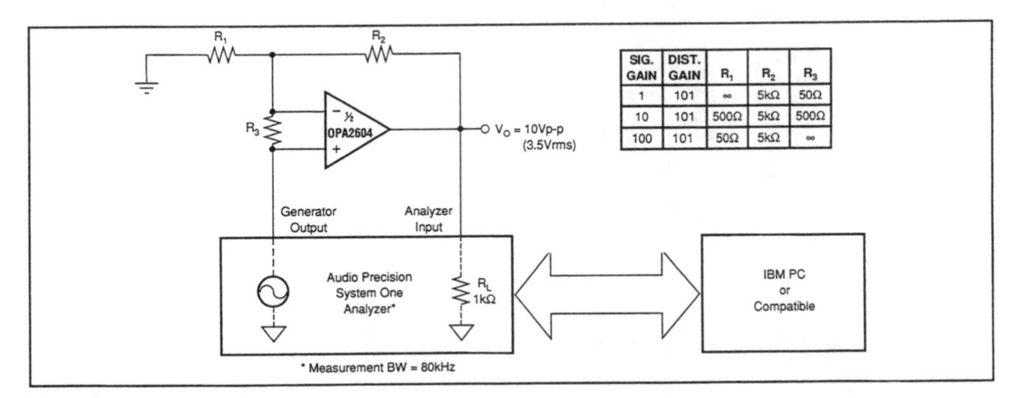


| SIG. GAIN | DIST. GAIN | $R_1$ | $R_2$ | $R_3$ |
|---|---|---|---|---|
| 1 | 101 | ∞ | 5kΩ | 50Ω |
| 10 | 101 | 500Ω | 5kΩ | 500Ω |
| 100 | 101 | 50Ω | 5kΩ | ∞ |

FIGURE 1. Distortion Test Circuit.

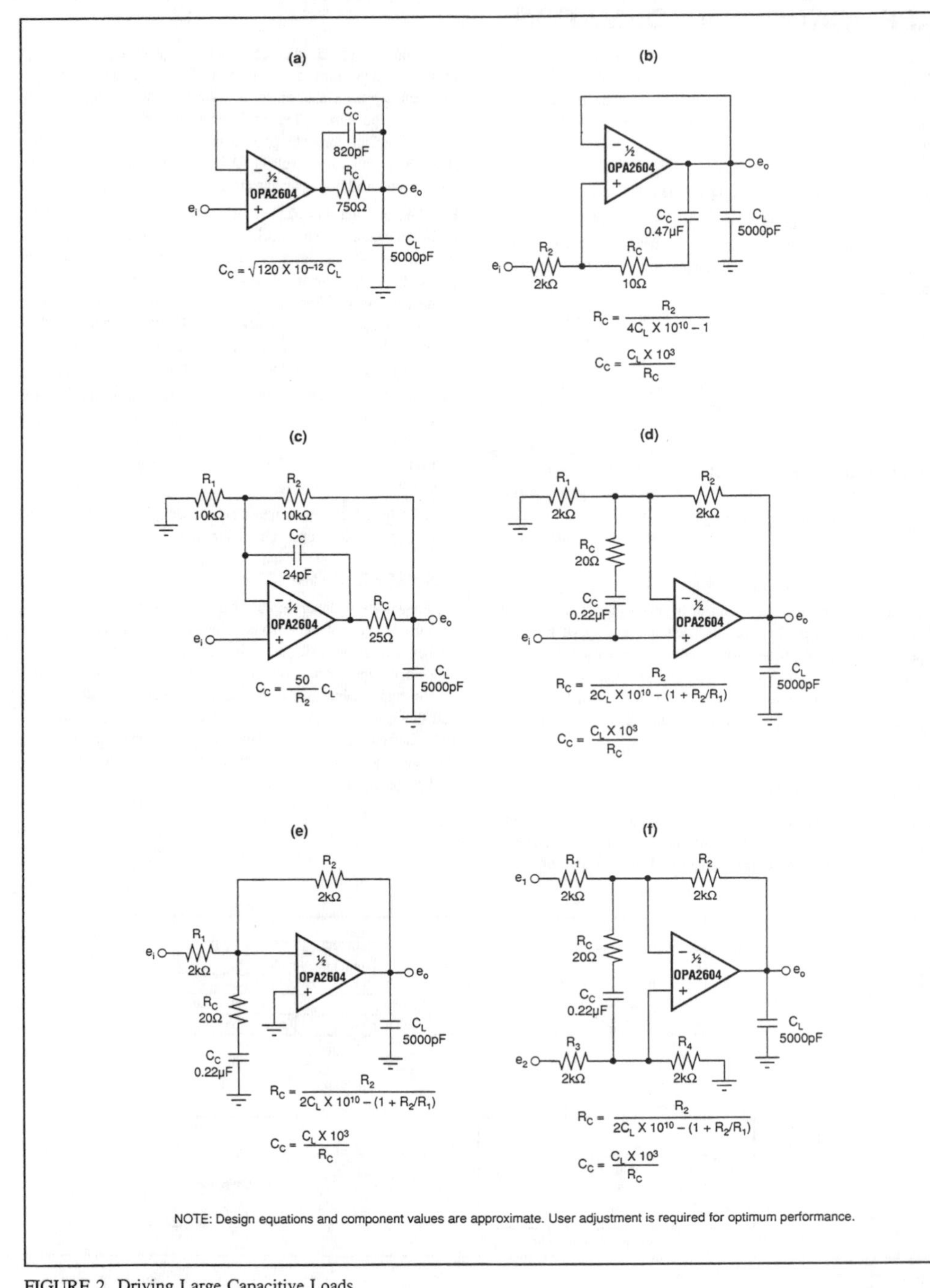


FIGURE 2. Driving Large Capacitive Loads.

## POWER DISSIPATION

The OPA2604 is capable of driving 600Ω loads with power supply voltages up to ±24V. Internal power dissipation is increased when operating at high power supply voltage. The typical performance curve, Power Dissipation vs Power Supply Voltage, shows quiescent dissipation (no signal or no load) as well as dissipation with a worst case continuous sine wave. Continuous high-level music signals typically produce dissipation significantly less than worst case sine waves.

Copper leadframe construction used in the OPA2604 improves heat dissipation compared to conventional plastic packages. To achieve best heat dissipation, solder the device directly to the circuit board and use wide circuit board traces.

## OUTPUT CURRENT LIMIT

Output current is limited by internal circuitry to approximately ±40mA at 25°C. The limit current decreases with increasing temperature as shown in the typical curves.

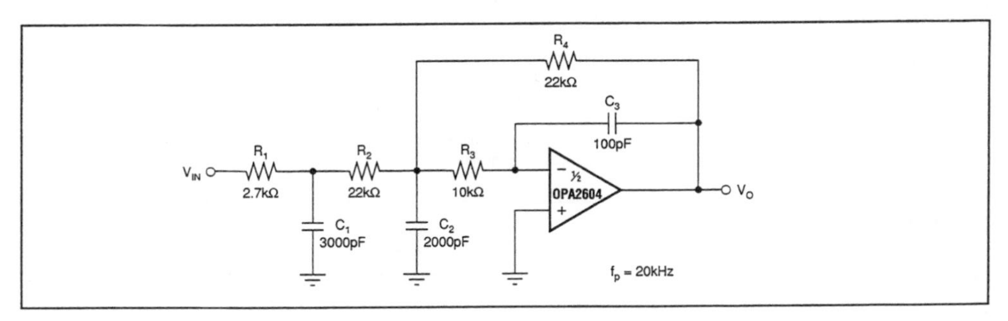


FIGURE 3. Three-Pole Low-Pass Filter.

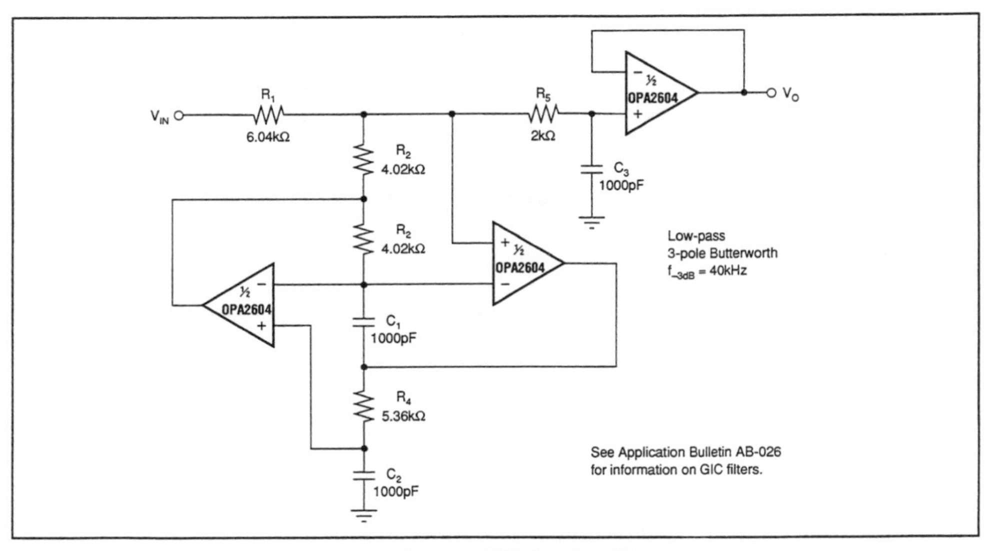


FIGURE 4. Three-Pole Generalized Immittance Converter (GIC) Low-Pass Filter.

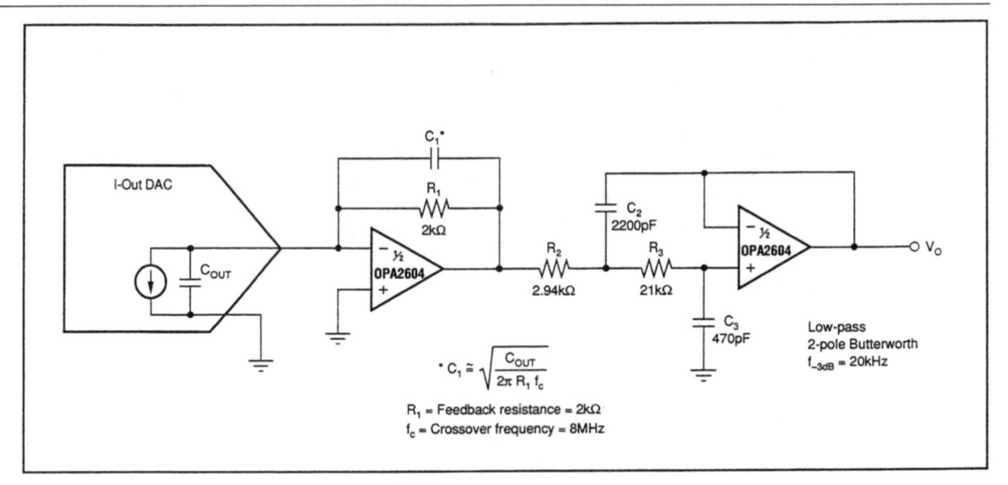


FIGURE 5. DAC I/V Amplifier and Low-Pass Filter.

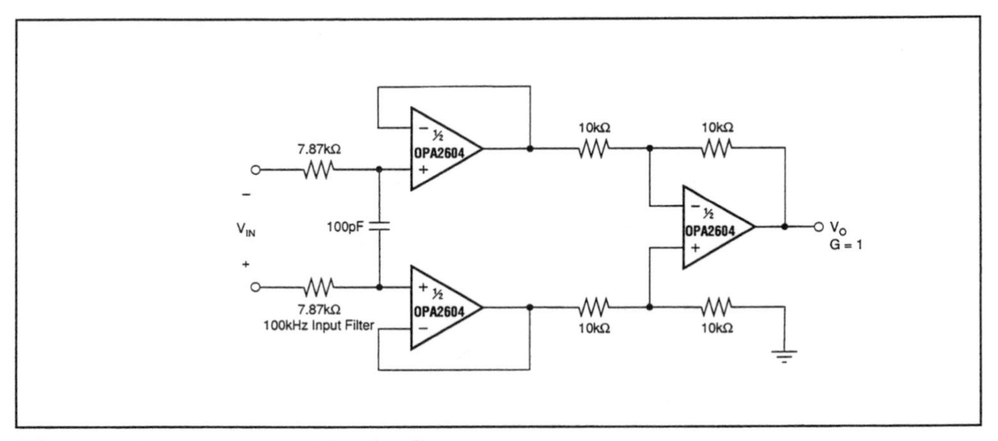


FIGURE 6. Differential Amplifier with Low-Pass Filter.

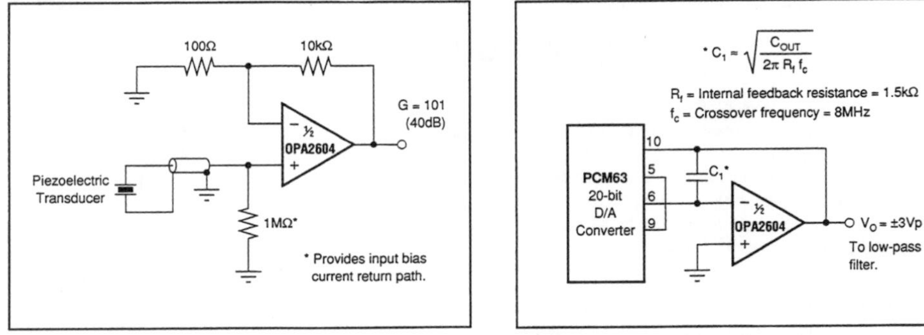


$$^{*}C_1 = \sqrt{\frac{C_{OUT}}{2\pi R_f f_c}}$$

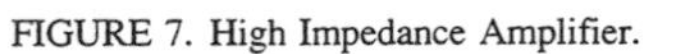

FIGURE 7. High Impedance Amplifier.

FIGURE 8. Digital Audio DAC I-V Amplifier.

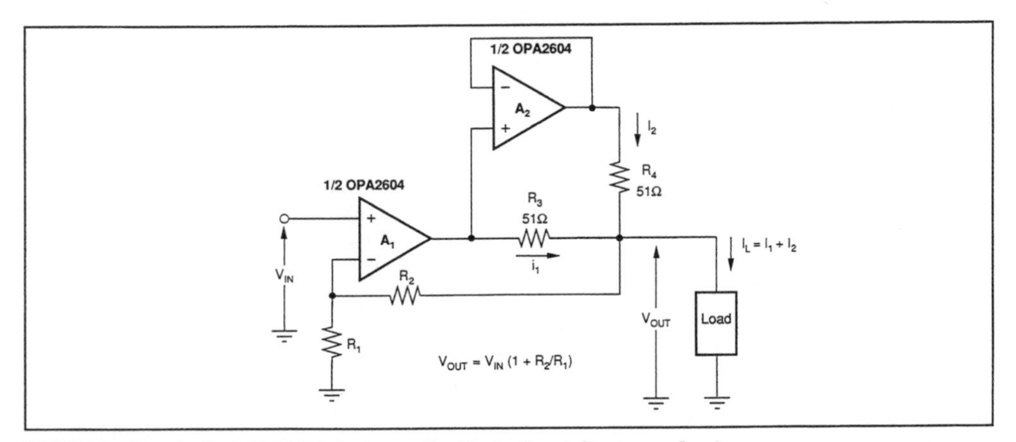


FIGURE 9. Using the Dual OPA2604 Op Amp to Double the Output Current to a Load.

# SOUND QUALITY

The following discussion is provided, recognizing that not all measured performance behavior explains or correlates with listening tests by audio experts. The design of the OPA2604 included consideration of both objective performance measurements, as well as an awareness of widely held theory on the success and failure of previous op amp designs.

## SOUND QUALITY

The sound quality of an op amp is often the crucial selection criteria—even when a data sheet claims exceptional distortion performance. By its nature, sound quality is subjective. Furthermore, results of listening tests can vary depending on application and circuit configuration. Even experienced listeners in controlled tests often reach different conclusions.

Many audio experts believe that the sound quality of a high performance FET op amp is superior to that of bipolar op amps. A possible reason for this is that bipolar designs generate greater odd-order harmonics than FETs. To the human ear, odd-order harmonics have long been identified as sounding more unpleasant than even-order harmonics. FETs, like vacuum tubes, have a square-law I-V transfer function which is more linear than the exponential transfer function of a bipolar transistor. As a direct result of this square-law characteristic, FETs produce predominantly even-order harmonics. Figure 10 shows the transfer function of a bipolar transistor and FET. Fourier transformation of both transfer functions reveals the lower odd-order harmonics of the FET amplifier stage.

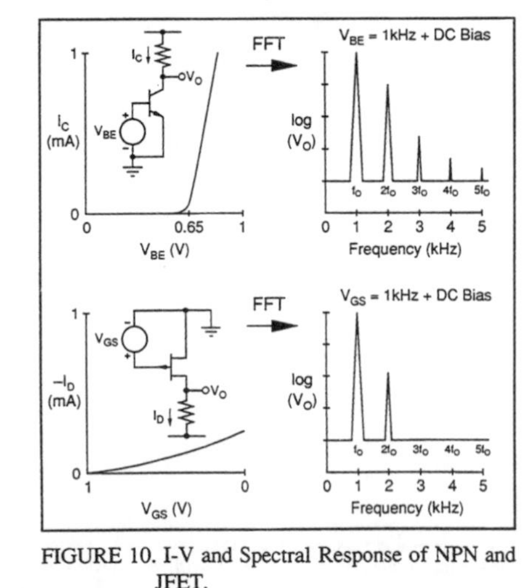


FIGURE 10. I-V and Spectral Response of NPN and JFET.

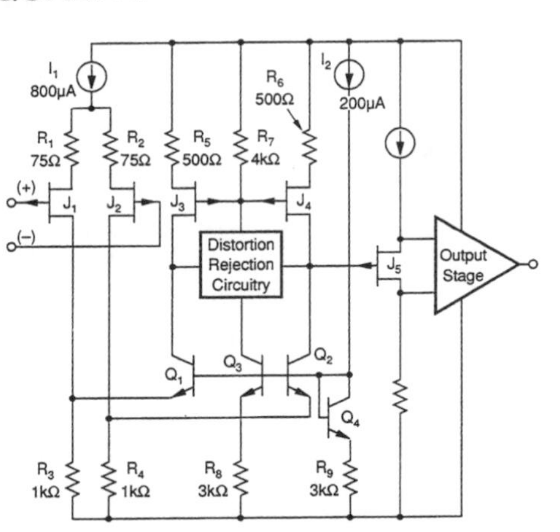


## THE OPA2604 DESIGN

The OPA2604 uses FETs throughout the signal path, including the input stage, input-stage load, and the important phase-splitting section of the output stage. Bipolar transistors are used where their attributes, such as current capability are important and where their transfer characteristics have minimal impact.

The topology consists of a single folded-cascode gain stage followed by a unity-gain output stage. Differential input transistors $J_1$ and $J_2$ are special large-geometry, P-channel JFETs. Input stage current is a relatively high 800μA, providing high transconductance and reducing voltage noise. Laser trimming of stage currents and careful attention to symmetry yields a nearly symmetrical slew rate of ±25V/μs.

The JFET input stage holds input bias current to approximately 100pA, or roughly 3000 times lower than common bipolar-input audio op amps. This dramatically reduces noise with high-impedance circuitry.

The drains of $J_1$ and $J_2$ are cascoded by $Q_1$ and $Q_2$, driving the input stage loads, FETs $J_3$ and $J_4$. Distortion reduction circuitry (patent pending) linearizes the open-loop response and increases voltage gain. The 20MHz bandwidth of the OPA2604 further reduces distortion through the user-connected feedback loop.

The output stage consists of a JFET phase-splitter loaded into high speed all-NPN output drivers. Output transistors are biased by a special circuit to prevent cutoff, even with full output swing into 600Ω loads.

The two channels of the OPA2604 are completely independent, including all bias circuitry. This eliminates any possibility of crosstalk through shared circuits—even when one channel is overdriven.

## MECHANICALS

### Package Number 006 — 8-Pin Plastic Single-Wide DIP

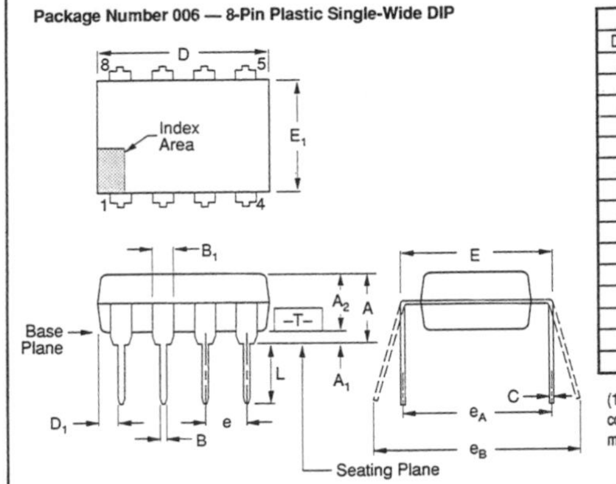


| DIM | INCHES MIN | INCHES MAX | MILLIMETERS MIN | MILLIMETERS MAX |
|---|---|---|---|---|
| A [3] | — | .210 | — | 5.33 |
| $A_1$ [3] | .015 | — | 0.38 | — |
| $A_2$ | .115 | .195 | 2.92 | 4.95 |
| B | .014 | .022 | 0.36 | 0.56 |
| $B_1$ | .045 | .070 | 1.14 | 1.78 |
| C | .008 | .015 | 0.20 | 0.38 |
| D [4] | .348 | .430 | 8.84 | 10.92 |
| $D_1$ | .005 | — | 0.13 | — |
| E [5] | .300 | .325 | 7.62 | 8.26 |
| $E_1$ [4] | .240 | .280 | 6.10 | 7.11 |
| e | .100 BASIC | | 2.54 BASIC | |
| eA [5] | .300 BASIC | | 7.63 BASIC | |
| eB [6] | — | .430 | — | 10.92 |
| L [3] | .115 | .160 | 2.92 | 4.06 |
| N [7] | 8 | | 8 | |

(1) Controlling dimension: Inch. In case of conflict between the English and metric dimensions, the inch dimensions control.

(2) Dimensioning and tolerancing per ANSI Y14.5M-1982.

(3) Dimensions A, $A_1$, and L are measured with the package seated in JEDEC seating plane gauge GS-3.

(4) D and $E_1$ dimensions for plastic packages do not include mold flash or protrusions. Mold flash or protrusions shall not exceed .010 inch (0.25mm).

(5) E and eA measured with the leads constrained to be perpendicular to plane T.

(6) eB is measured at the lead tips with the leads unconstrained.

(7) N is the maximum number of terminal positions.

(8) Corner leads (1, 4, 5, and 8) may be configured as shown in Figure 2.

(9) For automatic insertion, any raised irregularity on the top surface (step, mesa, etc.) shall be symmetrical about the lateral and longitudinal package center-lines.

### Package Number 182 — 8-Pin SO-8 Surface Mount

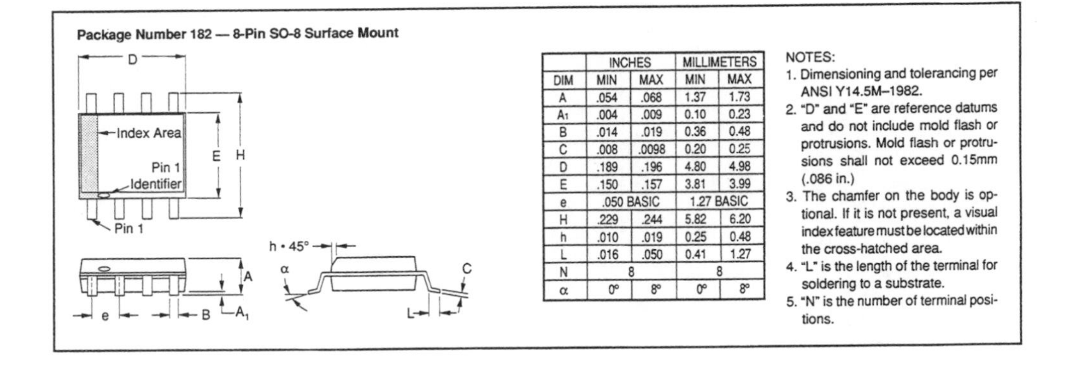


| DIM | INCHES MIN | INCHES MAX | MILLIMETERS MIN | MILLIMETERS MAX |
|---|---|---|---|---|
| A | .054 | .068 | 1.37 | 1.73 |
| $A_1$ | .004 | .009 | 0.10 | 0.23 |
| B | .014 | .019 | 0.36 | 0.48 |
| C | .008 | .0098 | 0.20 | 0.25 |
| D | .189 | .196 | 4.80 | 4.98 |
| E | .150 | .157 | 3.81 | 3.99 |
| e | .050 BASIC | | 1.27 BASIC | |
| H | .229 | .244 | 5.82 | 6.20 |
| h | .010 | .019 | 0.25 | 0.48 |
| L | .016 | .050 | 0.41 | 1.27 |
| N | 8 | | 8 | |
| α | 0° | 8° | 0° | 8° |

NOTES:

1. Dimensioning and tolerancing per ANSI Y14.5M–1982.
2. "D" and "E" are reference datums and do not include mold flash or protrusions. Mold flash or protrusions shall not exceed 0.15mm (.086 in.)
3. The chamfer on the body is optional. If it is not present, a visual index feature must be located within the cross-hatched area.
4. "L" is the length of the terminal for soldering to a substrate.
5. "N" is the number of terminal positions.

19-4191; Rev. 0; 1/92

MAXIM

# 8th-Order Continuous-Time Active Filter

MAX274

## General Description

The MAX274 is an 8th-order, continuous-time active filter consisting of four independent cascadable 2nd-order sections. Each filter section can implement any all-pole bandpass or lowpass filter response, such as Butterworth, Bessel, and Chebyshev, and is programmed by four external resistors. The MAX274 provides lower noise than switched-capacitor filters, as well as superior dynamic performance - both due to its continuous-time design. Since continuous-time filters do not require a clock signal, clock noise and aliased signals are eliminated in the MAX274.

Center frequencies range from 100Hz to 150kHz. Center-frequency accuracy is within ±1% over the full operating temperature range. For center frequencies up to 300kHz, the MAX275 4th-order continuous-time filter is recommended.

Low noise and total harmonic distortion (THD) typically better than -86dB make the MAX274 ideal for lowpass anti-aliasing and DAC output smoothing in high-resolution data-acquisition applications.

## Applications

Low-Distortion Anti-Aliasing Filters
DAC Output Smoothing Filters
Modems
Audio/Sonar/Avionics Frequency Filtering
Vibration Analysis

## Features

- ♦ **Continuous-Time Filter - No Clock, No Clock Noise**
- ♦ **8th-Order - Four 2nd-Order Sections**
- ♦ **150kHz Center-Frequency Range**
- ♦ **Implements Butterworth, Chebyshev, Bessel and Other Filter Responses**
- ♦ **Lowpass, Bandpass Outputs**
- ♦ **Low Noise: -86dB THD Typical**
- ♦ **Center-Frequency Accuracy within ±1% Guaranteed Over Temperature**
- ♦ **Operates from a Single +5V Supply or Dual ±5V Supplies**
- ♦ **Cascadable for Higher Order**

## Ordering Information

| PART | TEMP. RANGE | PIN-PACKAGE |
|---|---|---|
| MAX274ACNG | 0°C to +70°C | 24 Narrow Plastic DIP |
| MAX274BCNG | 0°C to +70°C | 24 Narrow Plastic DIP |
| MAX274ACWI | 0°C to +70°C | 28 Wide SO |
| MAX274BCWI | 0°C to +70°C | 28 Wide SO |
| MAX274BC/D | 0°C to +70°C | Dice* |

*Ordering Information continued on last page*

* *Contact factory for dice specifications.*

## Typical Operating Circuit

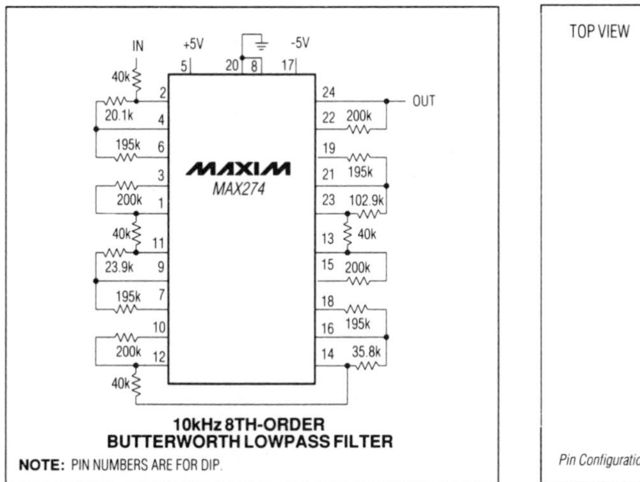


**10kHz 8TH-ORDER BUTTERWORTH LOWPASS FILTER**

**NOTE:** PIN NUMBERS ARE FOR DIP.

## Pin Configurations

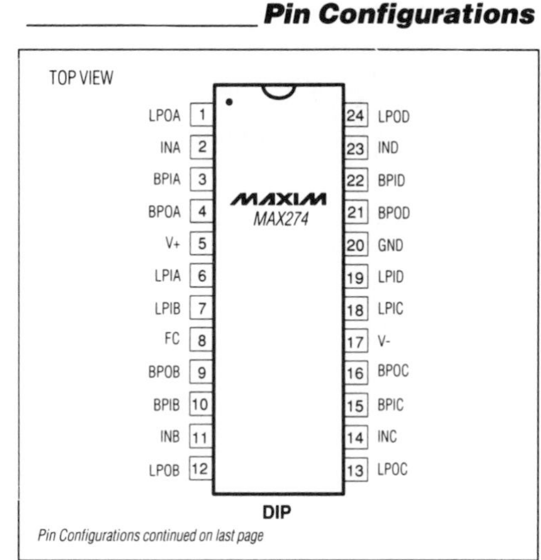


**DIP**

*Pin Configurations continued on last page*

# 8th-Order Continuous-Time Active Filter



## ABSOLUTE MAXIMUM RATINGS

V+ to V- . . . . . . . . -0.3V, 12V
Input Voltage to GND (any input) . . . . . . . . V- - 0.3V, V+ + 0.3V
Continuous Power Dissipation ($T_A$ = +70°C)
  Narrow Plastic DIP (derate 8mW/°C above +70°C) . . 696mW
  Wide SO (derate 10mW/°C above +70°) . . . . . . . . . . 1000mW
  CERDIP (derate 11.1mW/°C above +70°C) . . . . . . . 1000mW

Operating Temperature Ranges:
  MAX274_C_ _ . . . . . . . . 0°C to +70°C
  MAX274_E_ _ . . . . . . . . -40°C to +85°C
  MAX274_MRG . . . . . . . . -55°C to +125°C
Storage Temperature Range . . . . . . . . -65°C to +165°C
Lead Temperature (soldering, 10 sec) . . . . . . . . +300°C

*Stresses beyond those listed under "Absolute Maximum Ratings" may cause permanent damage to the device. These are stress ratings only, and functional operation of the device at these or any other conditions beyond those indicated in the operational sections of the specifications is not implied. Exposure to absolute maximum rating conditions for extended periods may affect device reliability.*

## ELECTRICAL CHARACTERISTICS

(V+ = 5V, V- = -5V, Test Circuit A of Figure 1, $T_A$ = $T_{MIN}$ to $T_{MAX}$, unless otherwise noted.)

| PARAMETER | SYMBOL | CONDITIONS | | MIN | TYP | MAX | UNITS |
|---|---|---|---|---|---|---|---|
| **FILTER CHARACTERISTICS** | | | | | | | |
| Maximum Operating Frequency | | | | | 10 | | MHz |
| Center-Frequency Range | $F_O$ | (Note 1) | | | 100 to 150k | | Hz |
| Center-Frequency Accuracy | $F_O$ | | MAX274A | -1.0 | | 1.0 | % |
| | | | MAX274B | -1.4 | | 1.4 | |
| Q Accuracy - Unadjusted | | | MAX274A | -10 | | 10 | % |
| | | | MAX274B | -15 | | 15 | |
| Q Accuracy - Adjusted | | Scaled for bandwidth compensation | | | ±2.8 | | % |
| $F_O$ Temperature Coefficient | $\Delta F_O/\Delta T$ | (Note 2) | | | -28 | | ppm/°C |
| Q Temperature Coefficient | $\Delta Q/\Delta T$ | (Note 2) | | | 160 | | ppm/°C |
| Wideband Noise | $V_{NOISE}$ | LPO_, Figure 1, test circuit B | 1Hz to 10Hz | | 23 | | $\mu V_{RMS}$ |
| | | | 10Hz to 10kHz | | 120 | | |
| **DC CHARACTERISTICS** | | | | | | | |
| DC Lowpass Gain Accuracy | $H_{OLP}$ | Assume ideal resistors | MAX274A | -2 | | 2 | % |
| | | | MAX274B | -3 | | 3 | |
| Offset Voltage at Outputs | $V_{OS}$ | LPO_ | MAX274A | -200 | | 200 | mV |
| | | | MAX274B | -300 | | 300 | |
| | | BPO_ | MAX274A | -40 | | 40 | |
| | | | MAX274B | -80 | | 80 | |
| Offset Voltage Drift | $\Delta V_{OS}/\Delta T$ | | | | 20 | | µV/°C |
| Leakage Current at FC Pin | $I_{FC}$ | | | -10 | | 10 | µA |
| **DYNAMIC FILTER CHARACTERISTICS** | | | | | | | |
| Signal-to-Noise plus Distortion | SINAD | $F_{TEST}$ = 1kHz, Figure 1, test circuit B | LPO_, $V_{LPO}$ = 8Vp-p | | -86 | | dB |
| | | $F_{TEST}$ = 10kHz, Figure 1, test circuit C | | | -82 | | |

# 8th-Order Continuous-Time Active Filter

## ELECTRICAL CHARACTERISTICS (continued)

(V+ = 5V, V- = -5V, Test Circuit A of Figure 1, $T_A$ = $T_{MIN}$ to $T_{MAX}$, unless otherwise noted.)

| PARAMETER | SYMBOL | CONDITIONS | MIN | TYP | MAX | UNITS |
|---|---|---|---|---|---|---|
| Output Voltage Swing | $V_{OUT}$ | LPO_ , BPO_ , $R_{LOAD}$ = 5kΩ | ±3.25 | ±4.5 | | V |
| Slew Rate | SR | | | 10 | | V/µs |
| Gain Bandwidth Product | GBW | | | 7.5 | | MHz |
| **POWER REQUIREMENTS** | | | | | | |
| Supply Voltage Range | $V_{SUPP}$ | (Note 3) | ±2.37 | | ±5.5 | V |
| Supply Current | $I_C$ | For V+, V- | | 20 | 30 | mA |
| Power-Supply Rejection Ratio | PSRR | V+ = 5V + 100mVp-p at 1kHz, V- = -5V | | -30 | | dB |

**Note 1:** Center frequencies ($F_O$s) below 100Hz are possible at reduced dynamic range.
**Note 2:** Assume no drift for external resistors.
**Note 3:** See Figure 9 for single-supply operation.

## Typical Operating Characteristics

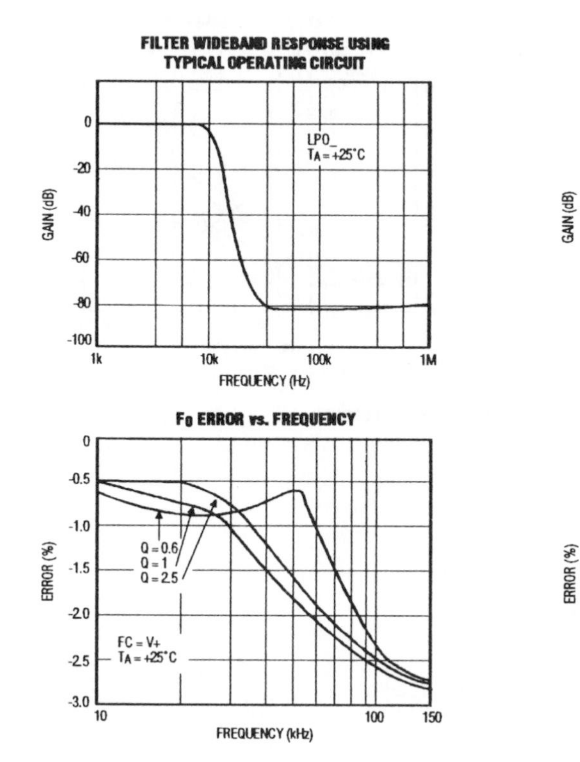


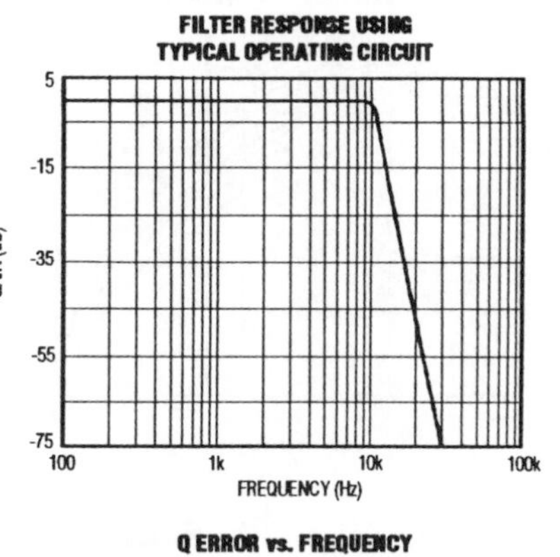


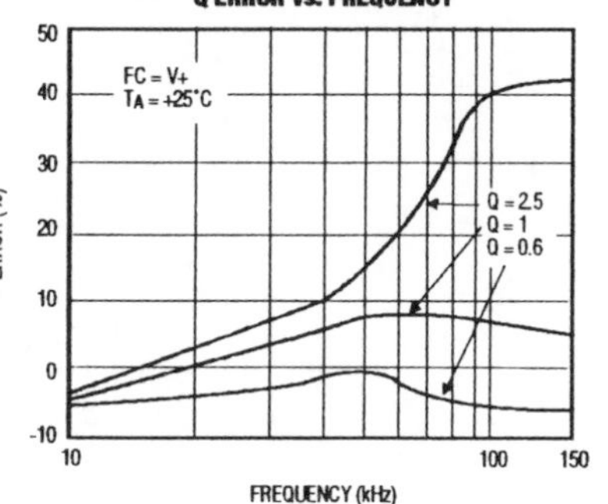

# 8th-Order Continuous-Time Active Filter

MAX274

## Typical Operating Characteristics (continued)

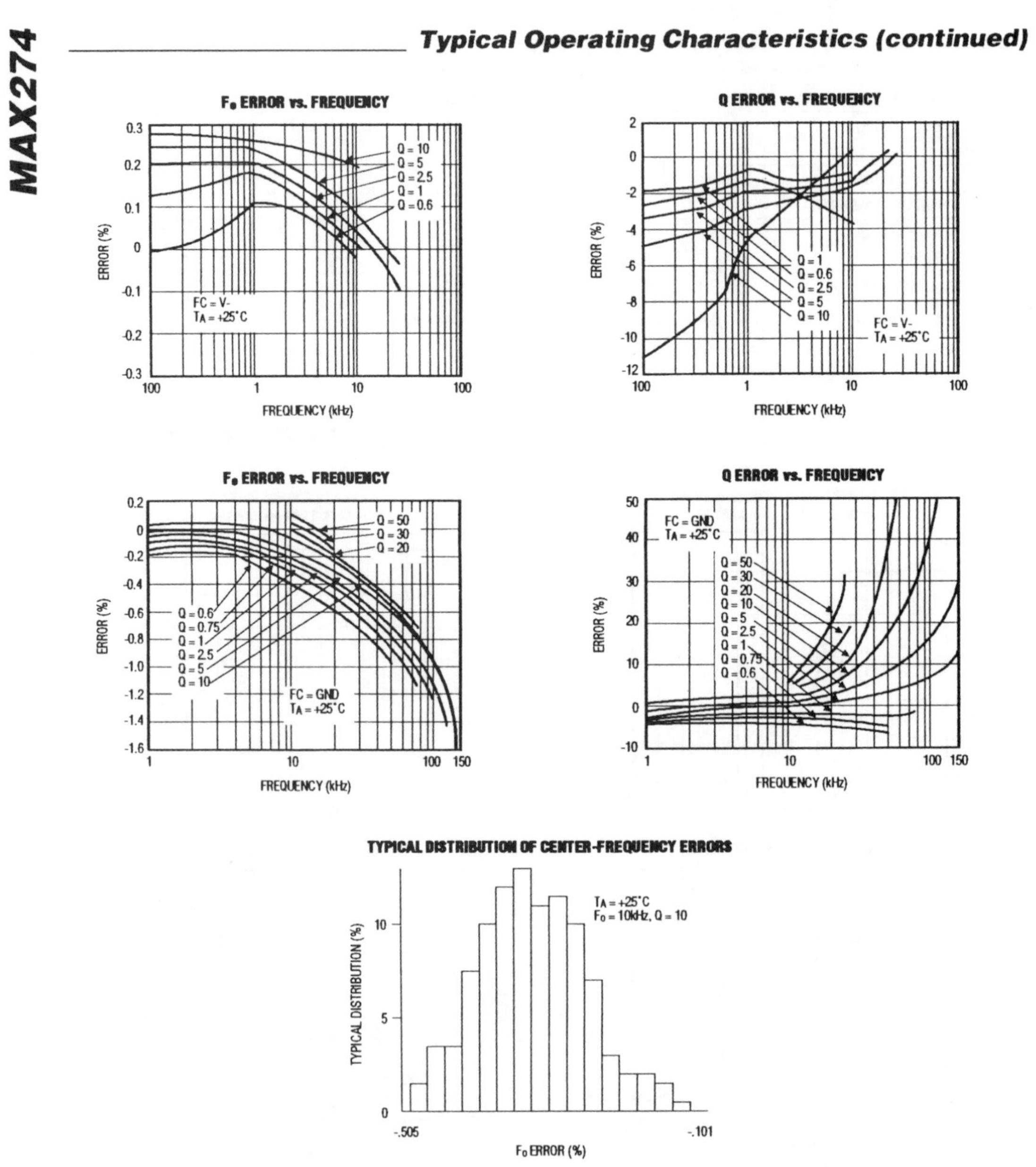

# 8th-Order Continuous-Time Active Filter

## Typical Operating Characteristics (continued)

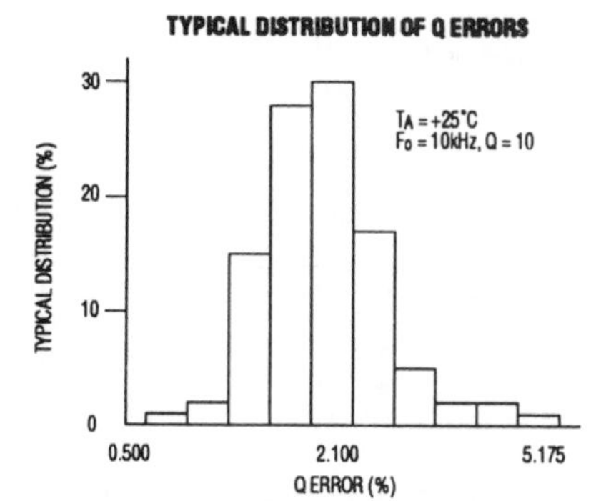


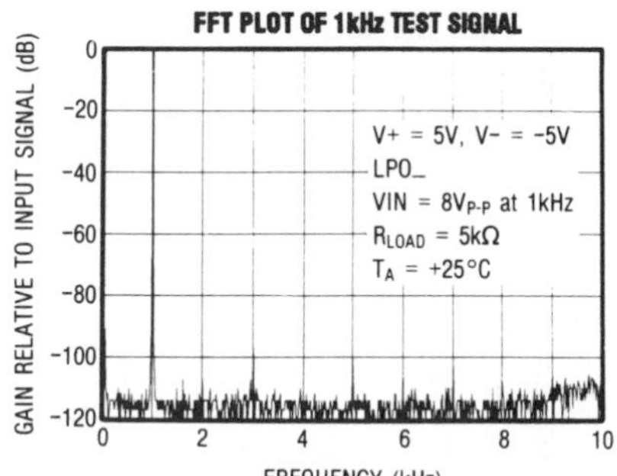


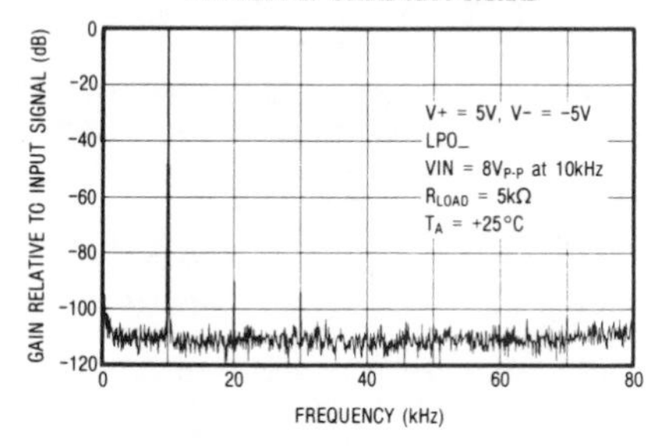


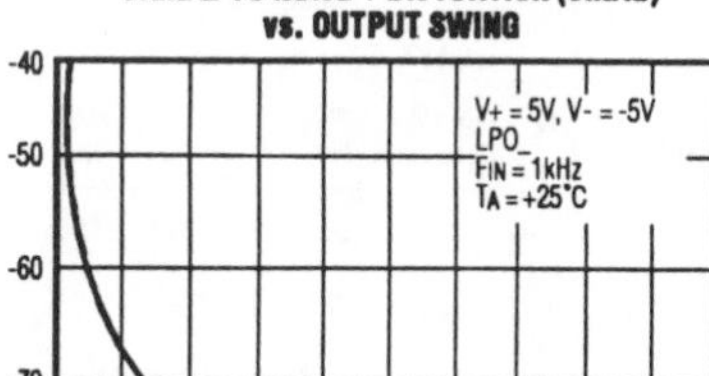
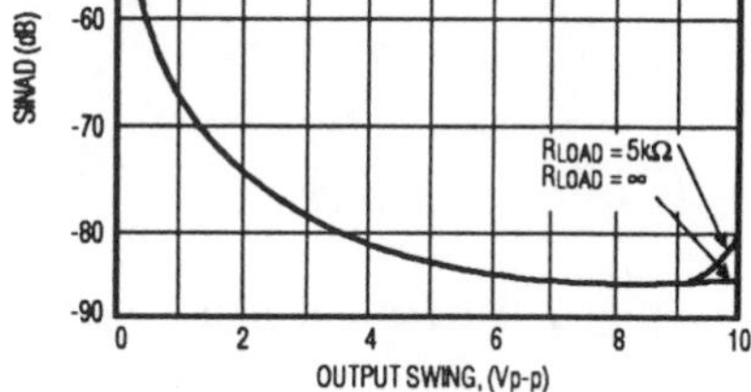

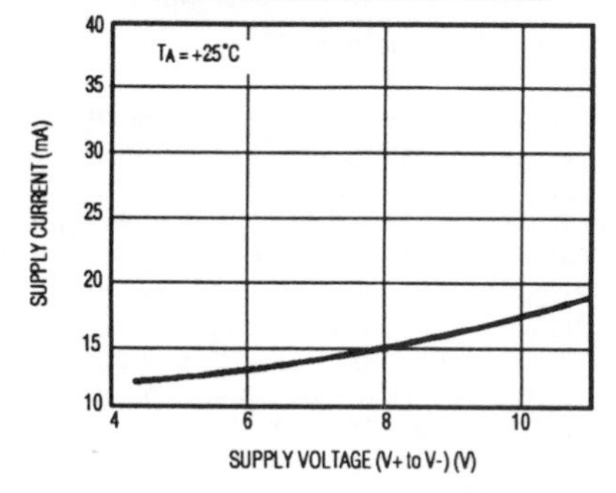


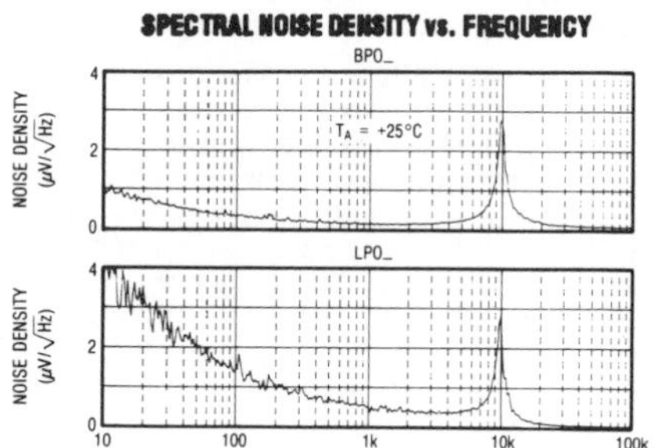

# *8th-Order Continuous-Time Active Filter*

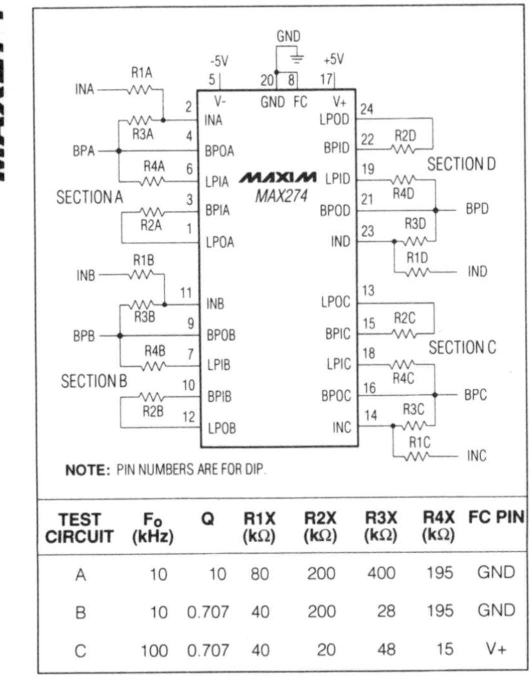


| TEST CIRCUIT | $F_O$ (kHz) | Q | R1X (kΩ) | R2X (kΩ) | R3X (kΩ) | R4X (kΩ) | FC PIN |
|---|---|---|---|---|---|---|---|
| A | 10 | 10 | 80 | 200 | 400 | 195 | GND |
| B | 10 | 0.707 | 40 | 200 | 28 | 195 | GND |
| C | 100 | 0.707 | 40 | 20 | 48 | 15 | V+ |

*Figure 1. Connection Diagram and Test Circuit*

## *Detailed Description*

The MAX274 contains four identical 2nd-order filter sections. Figure 2 shows the state-variable topography employed in each filter section. This topography allows simultaneous lowpass and bandpass functions at separate outputs.

The MAX274 employs a four-amplifier design, chosen for its relative insensitivity to parasitic capacitances and high bandwidth. The built-in capacitors and amplifiers, together with external resistors, form cascaded integrators with feedback to provide simultaneous lowpass and bandpass filtered outputs. To maximize bandwidth, the highpass (HP) node is not accessible. A 5kΩ resistor is connected in series with the input of the last stage amplifier to isolate the integration capacitor from external parasitic capacitances that could alter the filter's pole accuracy.

Although a notch output pin is not available, a notch can be created at the pole frequency by summing the input and bandpass output.

## *Filter Design Procedure*

Figure 3 outlines the overall filter design procedure. Maxim's Filter Design Software, available in the MAX274EVKIT, is highly recommended. The software automatically calculates filter order, poles, and Qs based on the required filter shape, so no manual calculations are necessary. Menu-driven commands and on-screen filter response graphs take the user through the complete design process, including the selection of resistor values for implementing a filter with the MAX274.

If designing without the software, see the filter design references listed at the end of this data sheet. These references provide numerical tables and equations needed to translate a desired filter response into order, poles, and Q. Once these three parameters have been calculated, see *Translating $F_O$/Q Pairs into MAX274 Hardware (Resistor Selection)* below.

### *Translating Calculated $F_O$/Q Pairs into MAX274 Hardware (Resistor Selection)*

If the filter design procedure has been completed as outlined in Figure 3, with the exception of external resistor selection, follow these four steps:

**1. Check all $F_O$/Q pairs for realizability.** The MAX274 has limits on which $F_O$/Q values can be implemented. These limits are bound by finite amplifier gain bandwidth and load-drive capability (which limit the highest frequency $F_O$/highest Qs), and by amplifier noise pickup and susceptibility to errors caused by stray capacitance (which sets a low frequency limit on the poles). Refer to Figure 4 to ensure each $F_O$/Q pair is within the "realizable" portion of the graph. If filter Qs are too high, reduce them by increasing the filter order (that is, increase the number of poles in the overall filter).

High-frequency $F_O$s (up to 400kHz) and high Qs outside of Figure 4's limits are also realizable, but $F_O$ and Q will deviate significantly from the ideal. Adjust resistor values by prototyping.

To implement $F_O$s less than 100Hz, see *High-Value Resistor Transformation* Section.

**2. Calculate resistor values for each section ($F_O$/Q pair).** Calculate resistor values using the graphs and equations in steps A through D of this section. Begin by estimating required values according to the graphs, then use the given equations to derive a precise value.

Resistor values should not exceed 4MΩ because parasitic capacitances shunting such high values cause excessive $F_O$/Q errors. Values lower than 5kΩ for R2 and R3 are not recommended due to limited amplifier output drive capability. For cases where higher values are unavoidable (as in low-frequency sections) refer to the *High-Value Resistor Transformation* section.

# 8th-Order Continuous-Time Active Filter

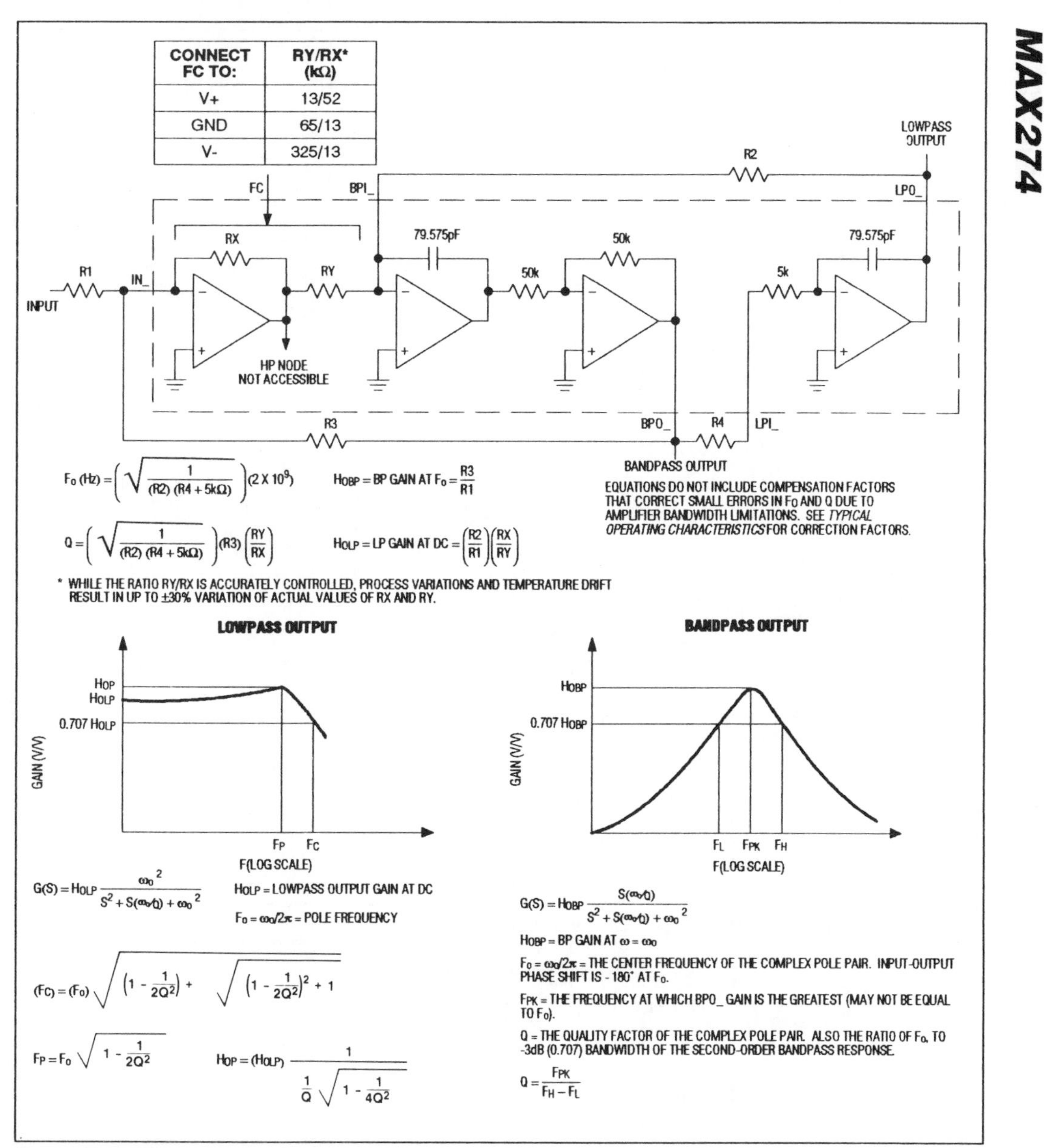

| CONNECT FC TO: | RY/RX* (kΩ) |
|---|---|
| V+ | 13/52 |
| GND | 65/13 |
| V- | 325/13 |

LOWPASS OUTPUT
R2
FC
BPI_
LPO_
RX
79.575pF
50k
79.575pF
R1
IN_
RY
50k
5k
INPUT
HP NODE
NOT ACCESSIBLE
R3
BPO_
R4
LPI_
BANDPASS OUTPUT
$F_0\,(Hz) = \left(\sqrt{\frac{1}{(R2)\,(R4+5k\Omega)}}\right)(2 \times 10^9)$
$H_{OBP}$ = BP GAIN AT $F_0 = \frac{R3}{R1}$
$Q = \left(\sqrt{\frac{1}{(R2)\,(R4+5k\Omega)}}\right)(R3)\left(\frac{RY}{RX}\right)$
$H_{OLP}$ = LP GAIN AT DC = $\left(\frac{R2}{R1}\right)\left(\frac{RX}{RY}\right)$
EQUATIONS DO NOT INCLUDE COMPENSATION FACTORS THAT CORRECT SMALL ERRORS IN $F_0$ AND Q DUE TO AMPLIFIER BANDWIDTH LIMITATIONS. SEE *TYPICAL OPERATING CHARACTERISTICS* FOR CORRECTION FACTORS.
* WHILE THE RATIO RY/RX IS ACCURATELY CONTROLLED, PROCESS VARIATIONS AND TEMPERATURE DRIFT RESULT IN UP TO ±30% VARIATION OF ACTUAL VALUES OF RX AND RY.
LOWPASS OUTPUT
$H_{OP}$
$H_{OLP}$
0.707 $H_{OLP}$
GAIN (V/V)
$F_P$ $F_C$
F(LOG SCALE)
BANDPASS OUTPUT
$H_{OBP}$
0.707 $H_{OBP}$
GAIN (V/V)
$F_L$ $F_{PK}$ $F_H$
F(LOG SCALE)
$G(S) = H_{OLP}\frac{\omega_0^2}{S^2 + S(\omega_0/Q) + \omega_0^2}$
$H_{OLP}$ = LOWPASS OUTPUT GAIN AT DC
$F_0 = \omega_0/2\pi$ = POLE FREQUENCY
$(F_C) = (F_0)\sqrt{\left(1 - \frac{1}{2Q^2}\right) + \sqrt{\left(1 - \frac{1}{2Q^2}\right)^2 + 1}}$
$F_P = F_0\sqrt{1 - \frac{1}{2Q^2}}$
$H_{OP} = (H_{OLP})\frac{1}{\frac{1}{Q}\sqrt{1 - \frac{1}{4Q^2}}}$
$G(S) = H_{OBP}\frac{S(\omega_0/Q)}{S^2 + S(\omega_0/Q) + \omega_0^2}$
$H_{OBP}$ = BP GAIN AT $\omega = \omega_0$
$F_0 = \omega_0/2\pi$ = THE CENTER FREQUENCY OF THE COMPLEX POLE PAIR. INPUT-OUTPUT PHASE SHIFT IS - 180° AT $F_0$.
$F_{PK}$ = THE FREQUENCY AT WHICH BPO_ GAIN IS THE GREATEST (MAY NOT BE EQUAL TO $F_0$).
Q = THE QUALITY FACTOR OF THE COMPLEX POLE PAIR. ALSO THE RATIO OF $F_0$ TO -3dB (0.707) BANDWIDTH OF THE SECOND-ORDER BANDPASS RESPONSE.
$Q = \frac{F_{PK}}{F_H - F_L}$

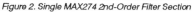

*Figure 2. Single MAX274 2nd-Order Filter Section*

# 8th-Order Continuous-Time Active Filter

MAX274

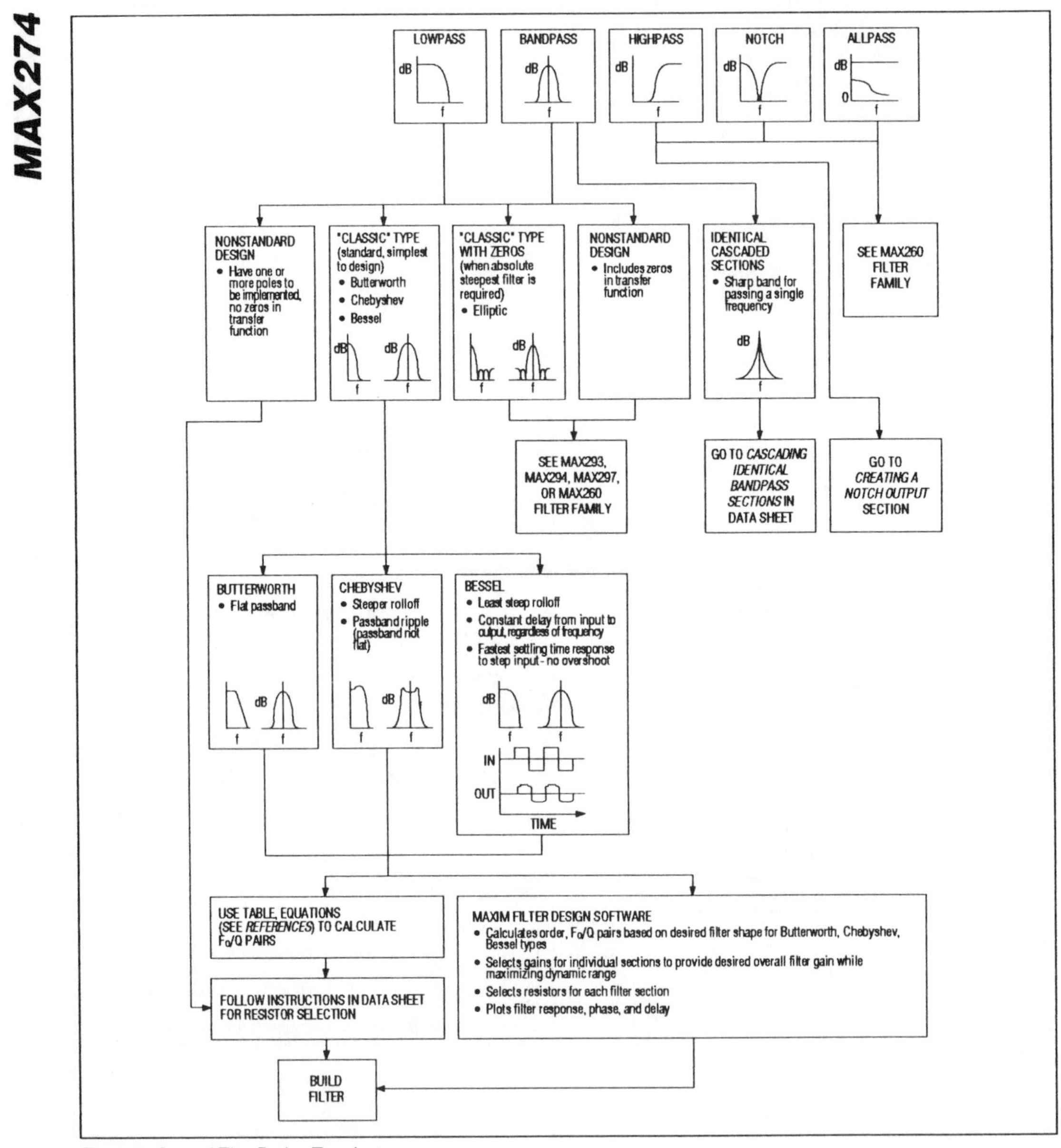


*Figure 3. General Filter Design Flowchart*

# 8th-Order Continuous-Time Active Filter



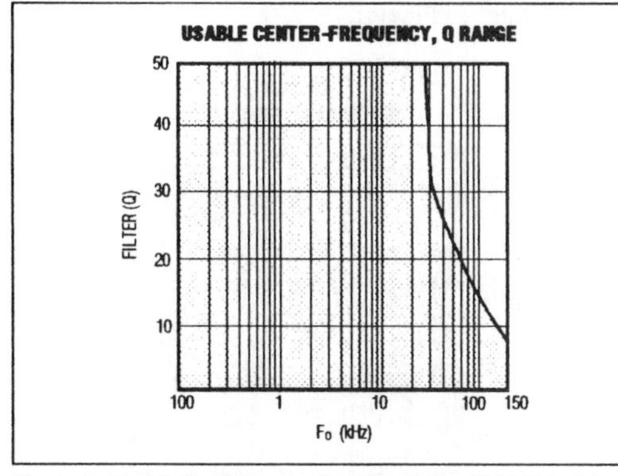


*Figure 4. Usable $F_o$, Q Range. See Translating Fo/Q Pairs into MAX274 Hardwork (Resistor Selection).*

The Frequency Control (FC) pin is connected to V+, GND, or V- and scales R3 and R1 to accomodate a wide range of gains and Q values. Note that the FC scaling factor is common to all four filter sections; the chosen setting must permit reasonable resistor values for all filter sections. Refer to the *FC Pin Connection* section.

The steps for calculating resistor values are given below.

**STEP A. CALCULATE R2.**

$$R2 = \frac{(2X10^9)}{F_O}$$

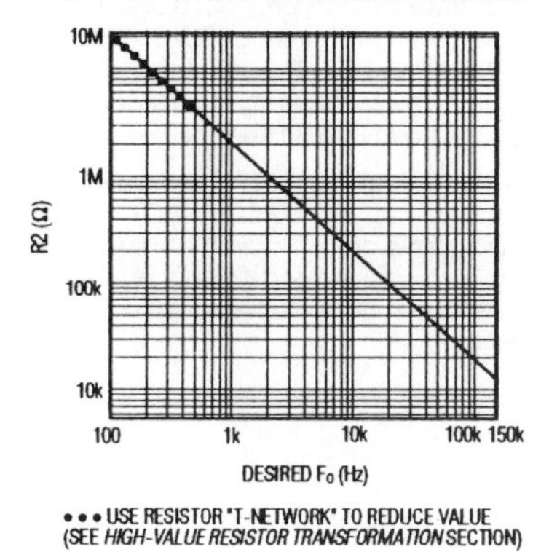


Resistors R2 and R4 set the center frequency.

**STEP B. CALCULATE R4.**

$$R4 = R2 - 5k\Omega$$

R4 may be less than 5kΩ because an internal series 5kΩ resisitor limits BPO-loading.

**STEP C. CALCULATE R3.**

R3 sets the Q for the section. R3 values are plotted assuming Q = 1; since R3 is proportional to Q, multiply the graph's value by the desired Q.

Given Q, three choices exist for R3, depending on the FC pin setting. Choose a setting which provides a reasonable resistor value (5kΩ < R3 < 4MΩ). R3 > 4MΩ may be used if unavoidable - refer to the *High-Value Resistor Transformation* section for an explanation of resistor "Ts."

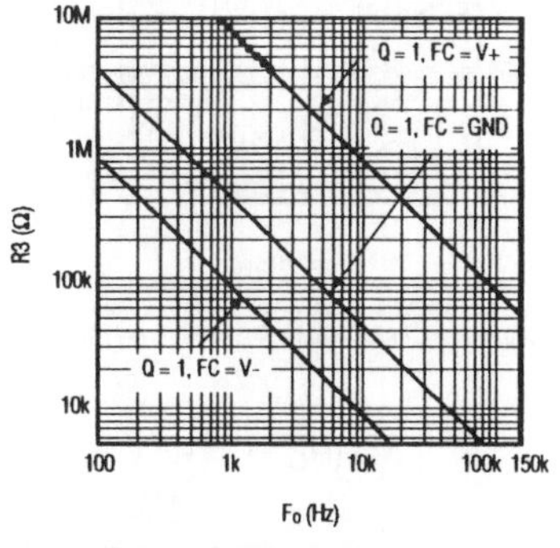


Scale R3 to desired Q

$$R3 = \frac{(Q)\,(2X10^9)}{F_O} \times \left(\frac{RX}{RY}\right)$$

| CONNECT FC TO: | RX/RY |
| --- | --- |
| V+ | 4/1 |
| GND | 1/5 |
| V- | 1/25 |

# 8th-Order Continuous-Time Active Filter



**STEP D. CALCULATE R1.**

R1 sets the gain. If individual section gains have not yet been calculated, refer to *Cascaded Filter Gain Optimization, Ordering of Sections.*

R1 is inversely proportional to LP gain. R1 values for gains of 1 and 10 are plotted; scale R1 according to the desired gain.

**Lowpass Filters:**

The FC pin setting was chosen in Step C (or from previous section calculations).

LP GAIN
RESISTOR R1 vs. FREQUENCY

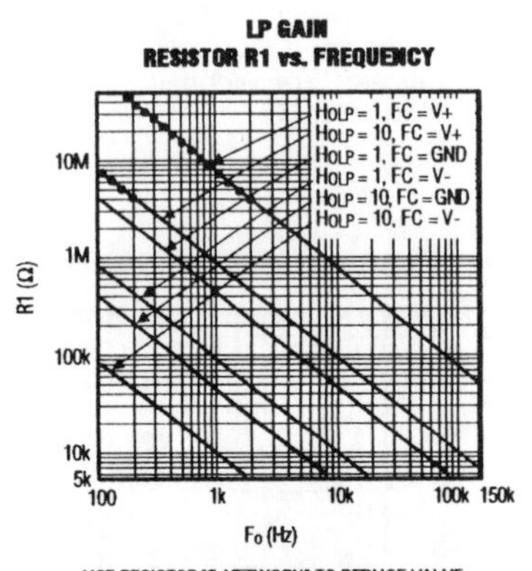

• • • USE RESISTOR "T-NETWORK" TO REDUCE VALUE (SEE *HIGH-VALUE RESISTOR TRANSFORMATION* SECTION)

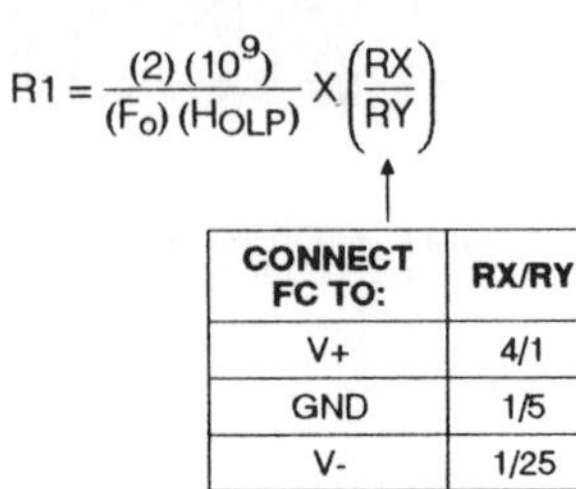

$$R1 = \frac{(2)(10^9)}{(F_O)(H_{OLP})} \times \left(\frac{RX}{RY}\right)$$

| CONNECT FC TO: | RX/RY |
|---|---|
| V+ | 4/1 |
| GND | 1/5 |
| V- | 1/25 |

where $H_{OLP}$ is the gain at LP at DC.

**Bandpass Filters:**

BP GAIN
RESISTOR R1 vs. RESISTOR R3

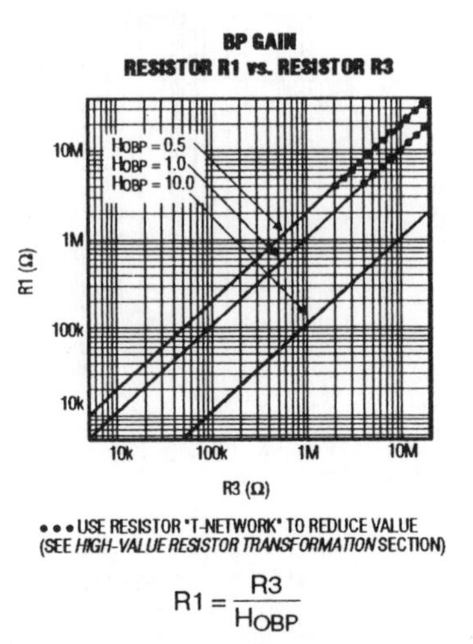


• • • USE RESISTOR "T-NETWORK" TO REDUCE VALUE (SEE *HIGH-VALUE RESISTOR TRANSFORMATION* SECTION)

$$R1 = \frac{R3}{H_{OBP}}$$

where $H_{OBP}$ is the gain at BP at $F_O$.

**3. Recalculate resistor values to compensate for filter amplifier bandwidth errors.** Some of the *Typical Operating Characteristics* graphs show deviations in $F_O$ and Q compared with expected values, due to gain rolloff of the internal amplifiers. If desired, correct these deviations by recalculating values R1-R4.

**4. Build a filter prototype.** Build and test all filter designs! Refer to the *Prototyping, PC-Board Layout* section of this data sheet. For faster prototyping, the MAX274EVKIT includes a PC-board circuit and design software to build and test filter designs instantly.

For applications that require high accuracy (for example, those with filter sections containing Qs greater than 10), or those that use a ground plane, a final prototype tuning procedure is recommended. Build a prototype filter; then adjust resistor values of each section until desired accuracy is achieved.

# 8th-Order Continuous-Time Active Filter



### High-Value Resistor Transformation

High-value resistors (greater than about 4MΩ) used in the MAX274 filter circuit introduce excessive $F_O$ and Q errors due to PC-board parasitics. To reduce the impedance of these feedback paths while maintaining equivalent feedback current, use the resistor "T" method shown in Figure 5.

$F_O$s less than 100Hz can be realized using T-networks. T-networks provide the equivalent of large resistor values for R2, R3, and R4, necessary for low-frequency filters; however, T-networks reduce dynamic range by attenuating the input signal level. Note that parasitic capacitances across these high resistor values affect the filter response at high frequencies. For best results, build a prototype and check its performance thoroughly.

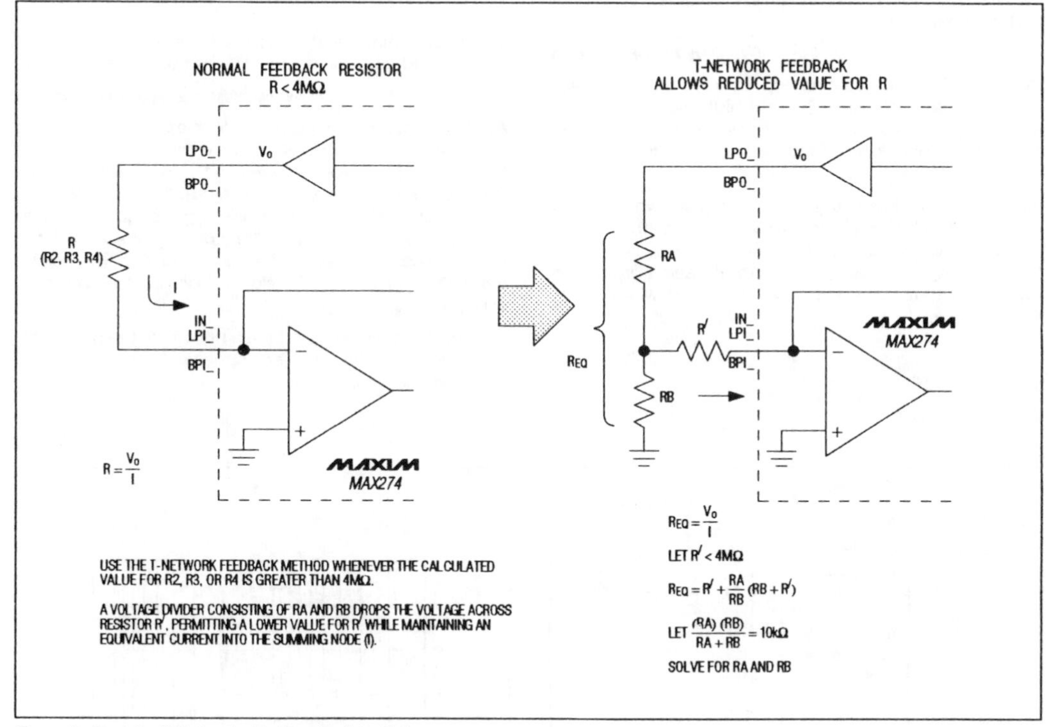


*Figure 5. Resistor T-Networks Reduce Resistor Values*

# 8th-Order Continuous-Time Active Filter



### Odd Number of Poles

For lowpass designs containing an odd number of poles, add an RC lowpass filter after the final filter section. The value of RC should be:

$$RC = 1/2\pi F_O$$

where $F_O$ is the desired real pole frequency. If required, buffer the RC with an op amp.

In many cases it may be advantageous to simply increase the filter order by 1, and implement it with an additional 2nd-order section.

### FC Pin Connection

Connect FC to GND for all applications, except where resistor values fall below 5kΩ (at high $F_O$s, low Qs). In these cases connect FC to V+. For low $F_O$s and high Qs, connect FC to V- to keep the value of R1 and R3 below 4MΩ.

$F_O$ and Q errors are significantly higher when FC is connected to V+ or V- (see *Typical Operating Characteristics*). Adjusting resistor values compensates for these errors, since the errors are repeatable from part to part. Note that noise increases threefold when FC is connected to V+.

### Cascading Identical Sections for Simplest Bandpass

If designing a bandpass filter where a single frequency (or a very narrow band of frequencies) must be passed, several 2nd-order sections with identical $F_O$s and Qs may be cascaded. The resulting Q (selectivity) of the filter is a function of the individual sections' Qs and the number of sections cascaded:

$$Q_t = \frac{Q}{\sqrt{2^{1/n} - 1}}$$

where $Q_t$ is the overall cascaded filter Q, Q is the Q of each individual section, and N is the number of sections.

### Creating a Notch Output

A notch (zero) can be created in the filter response by summing the input signal with BPO_ using an external op amp (Figure 6a). The notch will have the poles and Q characteristics of the 2nd-order section, as well as a zero at the pole frequency (transfer function given in Figure 6a). $H_{OBP}$ (BP gain at $F_O$) must be accurately set to unity so the input signal summed with BPO_ cancels precisely at the pole frequency. The notch's maximum attenuation is therefore a function of the accuracy of R1, R3, $R_{IN}$, and $R_{BP}$.

A notch can be used to create a null within the passband of a lowpass filter to reject specific frequencies (see *Applications* section).

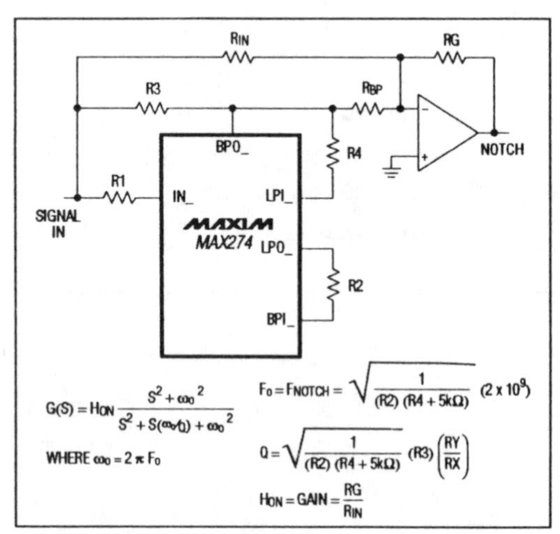

*Figure 6a. Creating a Notch Output*

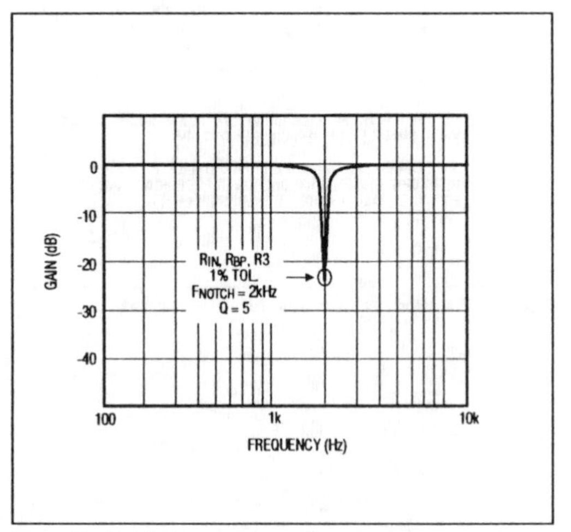


*Figure 6b. Notch Response*

# 8th-Order Continuous-Time Active Filter



### Cascaded Filter Gain Optimization, Ordering of Sections

Gains across the individual sections in a filter may be set an infinite number of ways, as long as the total gain from filter input to output is correct. Often, gains cannot be equally divided among sections, since different $F_O$s and Qs create gain peaks and valleys at different frequencies for each section.

The goal in choosing gains is to prevent section outputs from swinging beyond the ±3.25V limit (using ±5V supplies) while the full input signal is applied. On the other hand, if section gains are set too low and only a small proportion of output range is used, the noise factor increases. An optimal gain distribution between sections allows each section to swing as close to ±3.25V as possible in a wide range of frequencies.

Check the unused output (BPO_ or LPO_), and the internal HP node for overvoltage, since clipping at any node will cause distortion at the outputs. The HP node is not available for probing (Figure 2); however, its gain may approach RX / R1. Low R1 values and connection of FC to V+ (which sets RX as high as 64kΩ) may cause this node to clip.

Maxim's Filter Design Software allows optimum gain by plotting output gains of each successive cascaded filter section, including the internal node. Gains may be adjusted manually and sections reordered for the best overall dynamic range.

To optimize gain without the help of software, begin by ordering the sections from lowest Q to highest Q. Divide gains equally between sections, setting each section gain to:

$$H_O = A^{(1/N)}$$

where A = overall filter gain

$H_O = H_{OBP}$ for bandpass designs (gain at $F_O$)

$H_O = H_{OLP}$ for lowpass designs (gain at DC)

N = total number of sections

This approach offers a good first-pass solution to clipping problems in the high Q sections by keeping gains low in the first (low Q) sections. The gains may then be adjusted in hardware to maximize overall dynamic range.

### Resistors

Aside from accuracy, the most important criterion for resistor selection is parasitic capacitance across the resistor. Typical capacitance should be less than 1pF. Precision wire-wound resistors exhibit several picofarads, as well as unacceptable inductance – DO NOT USE THESE. Capacitance effectively reduces the resistance at high frequencies (especially when using high-value resistors), and causes phase shifts in feedback loops. Do not mount resistors in sockets. Socket capacitance appearing across resistors is often several picofarads, and will cause significant errors in $F_O$ and Q. Metal-film resistors minimize noise better than carbon types.

### Prototyping, PC-Board Layout

For highest accuracy filters, build the filter prototype on a PC board with a layout as similar as possible to the final production circuit. If a ground plane will be used in production, build prototype filters on a copper board. Do not use push-in type breadboards for prototyping – pin-to-pin capacitance is too high. For faster prototyping, the MAX274EVKIT includes a PC-board circuit to test designs.

Layout-sensitive errors, though repeatable from part to part, vary according to resistor placement, trace routing, and ground-plane layout. For highest accuracy, use the recommended layout provided in Figure 7. Keep all traces, especially LPI_ and BPI_ , as short as possible. LPI_ and BPI_ are particularly sensitive to ground capacitance, and may cause errors in Q. If a ground plane is used, tune the prototype filter by adjusting resistor values to cancel errors caused by ground capacitance.

Prevent capacitive coupling between pins. Coupling between BPI_ and BPO_ can cause $F_O$ errors; capacitance across resistors connecting IN and BPO_ (R3), BPI_ and LPO_ (R2), and BPO_ and LPI_ (R4) cause $F_O$ and Q errors. Minimize these errors with "tight" (shortest trace) layout practices.

### Measuring $F_o$ and Q

For multiple-order filters, measure each section individually, before cascading, to verify correct $F_O$ and Q. For best results, measure BPO_ with a spectrum analyzer. $F_O$ is the frequency at which the input and BPO_ are 180° out of phase. Q is the ratio of $F_{PK}$ to BPO_'s - 3dB bandwidth (Figure 2), where $F_{PK}$ is the frequency at which BPO_ gain is the greatest (which may not be equal to $F_O$).

### Filter $F_o$ and Q Accuracy

$F_O$ sensitivity to external resistor tolerance is 1:1 – for example, use of 1% tolerant resistors for R2 and R4 adds ±1% error to $F_O$ (which should be added to the ±1% tolerance of the MAX274 itself, guaranteed over temperature). Q errors are of greater magnitude, since they are a function of the internal resistor divider (controlled by the FC pin) and also involve R3. Typical Q error distributions are given in the *Typical Operating Characteristics*; additional Q errors associated with resistor tolerances are a function of R2, R3, and R4, and must be calculated according to the values used.

# 8th-Order Continuous-Time Active Filter

MAX274

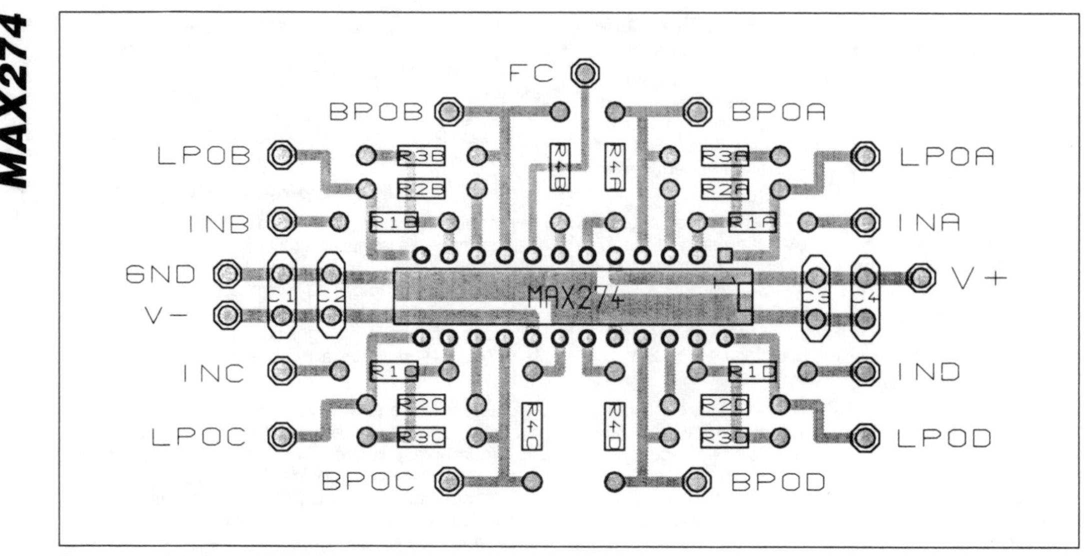


*Figure 7. MAX274 Suggested PC-Board Layout for DIP*

### *DC Offset Removal*

Figures 8a and 8b show methods for removing the DC offset voltage at LPO_. The first method shows adjustable DC nulling signals injected into either BPI_ or the filter input. $R_{TRIM}$ must be adjusted until DC offset is nulled at the LPO_ (Figure 8a). Figure 8b shows a trimless solution for lowpass filters that removes DC offset by AC coupling the LPO_ output, while allowing a DC path through R from the input. At DC and low frequencies, the output is equal to the prefiltered signal input (across R); at higher frequencies, C conducts and the output equals the signal at LPO_. The external RC pole should be set at least one frequency decade lower than the overall filter $F_O$. A low offset amplifier can buffer the output signal, if desired. For bandpass filters, a simple buffered RC highpass filter at the output removes DC offset.

### *Noise and Distortion*

Noise spectral density is shown in the *Typical Operating Characteristics*. The noise frequency distribution is shaped by the filter gain and response (higher Q section will have a proportionally higher noise peak around the pole frequency), as well as by amplifier 1/f noise. With FC set to V+, noise is 3 times greater than if set to GND or V-; therefore, avoid this setting for noise-sensitive applications. The noise density graphs from the *Typical Operating Characteristics* can be scaled to any gain or Q for an accurate noise estimation.

The MAX274 can drive 5kΩ loads to typically within ±500mV of the supply rails with negligible distortion. The outputs can drive up to 100pF; however, filters with high $F_O$s and Qs will undergo some phase shift (1° at 100kHz driving 130pF, $F_O$ = 100kHz, Q = 10 section).

# 8th-Order Continuous-Time Active Filter



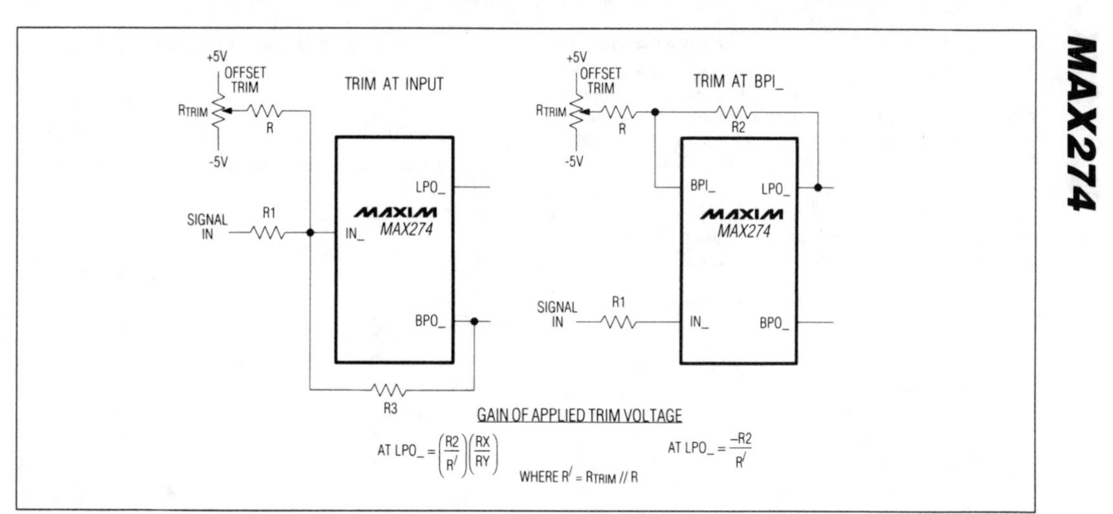


*Figure 8a. Trimmed Offset Removal*

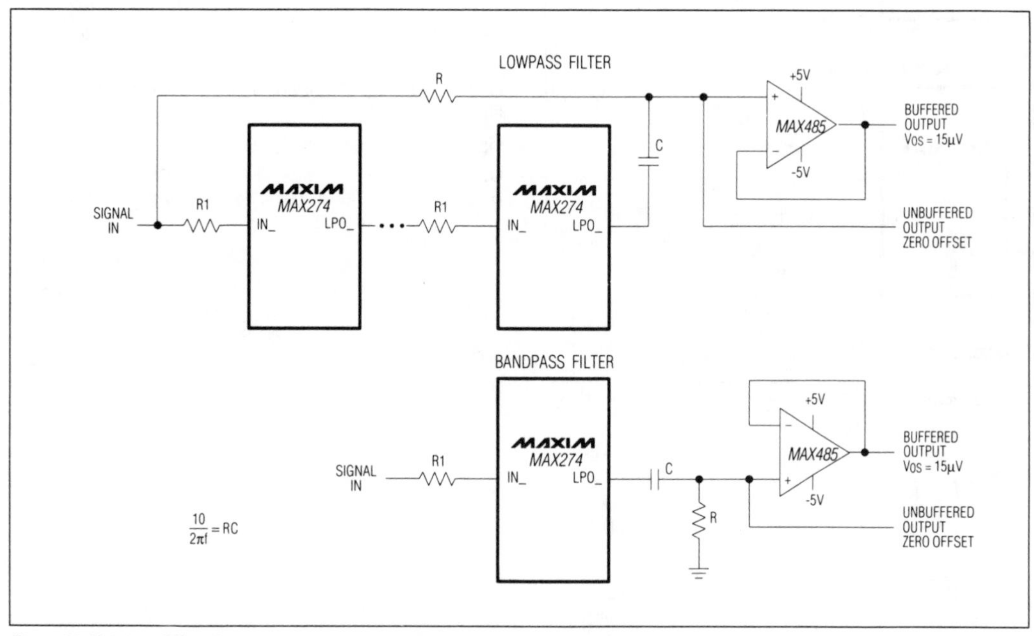


*Figure 8b. Trimless Offset Removal*

# 8th-Order Continuous-Time Active Filter



### Power Supplies

The MAX274 can be operated from a single power supply or dual supplies (Figure 9). V+ and V- pins must be properly bypassed to GND with 4.7μF electrolytic (tantalum preferred) and 0.1μF ceramic capacitors in parallel. These should be as close as possible to the chip supply pins.

For single-supply applications, GND must be centered between V+ and V- voltages so signals remain in the common-mode range of the internal amplifiers.

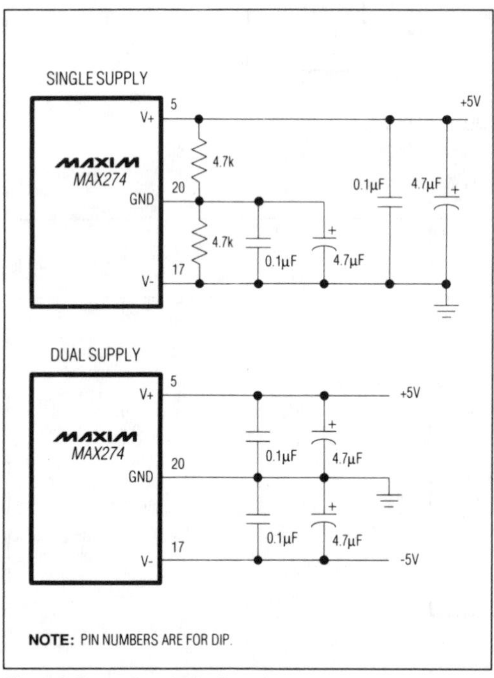


*Figure 9. Power-Supply Configurations*

### Maxim Filter Design Software

Maxim's new Filter Design Software aids in the filter design procedure by performing the following tasks:

* Designs Butterworth, Chebyshev, and Bessel filters
* Calculates order, poles, and Qs from user-defined filter requirements
* Plots filter gain, phase, and delay for inspection
* Calculates required resistor values for each MAX274 filter section, compensating for filter gain-bandwidth errors
* Allows gain optimization for each cascaded filter section
* Determines filter response based on actual resistor values

In the USA and Canada, the software may be ordered directly from Maxim. In other countries, call your local Maxim representative.

### References

The following references contain information and tables to aid in filter designs:

Carson, Chen. *Active Filter Design*, Hayden, 1982.

Tedeschi, Franck. *Active Filter Cookbook*, Tab Books No 1133, 1979.

Hilburn, Johnson. *Manual of Active Filter Design*, McGraw Hill, 1973.

German Language:

U. Tietze; Ch. Schenk. *Halbleiter-Schaltungstecknik Springer-Verlag*, Berlin Heidelberg, New York/Tokyo 1991.

# 8th-Order Continuous-Time Active Filter



## Applications

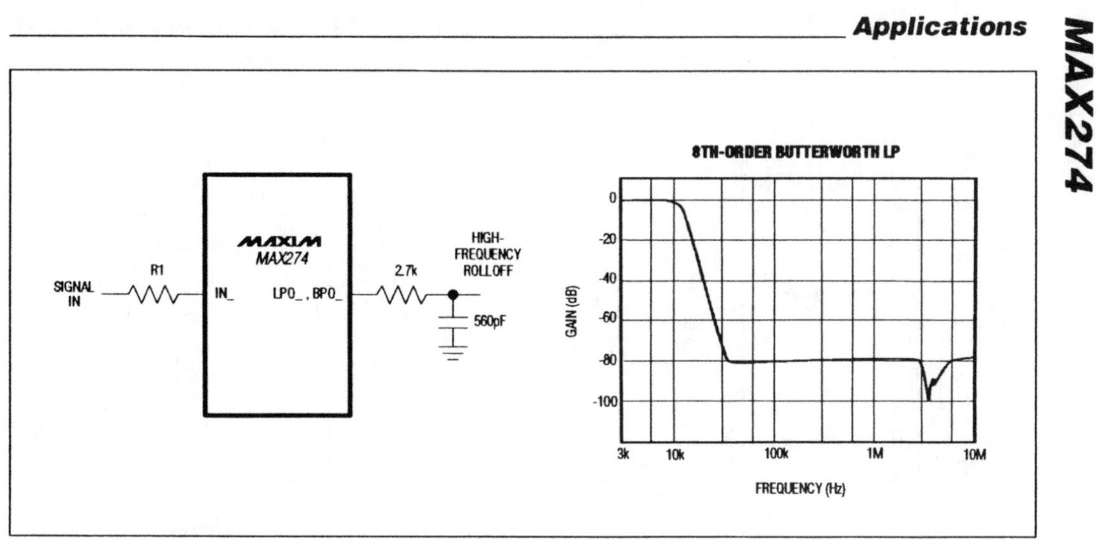


*Figure 10. External RC Lowpass for High-Frequency Rolloff*

# 8th-Order Continuous-Time Active Filter

MAX274

## Applications (continued)

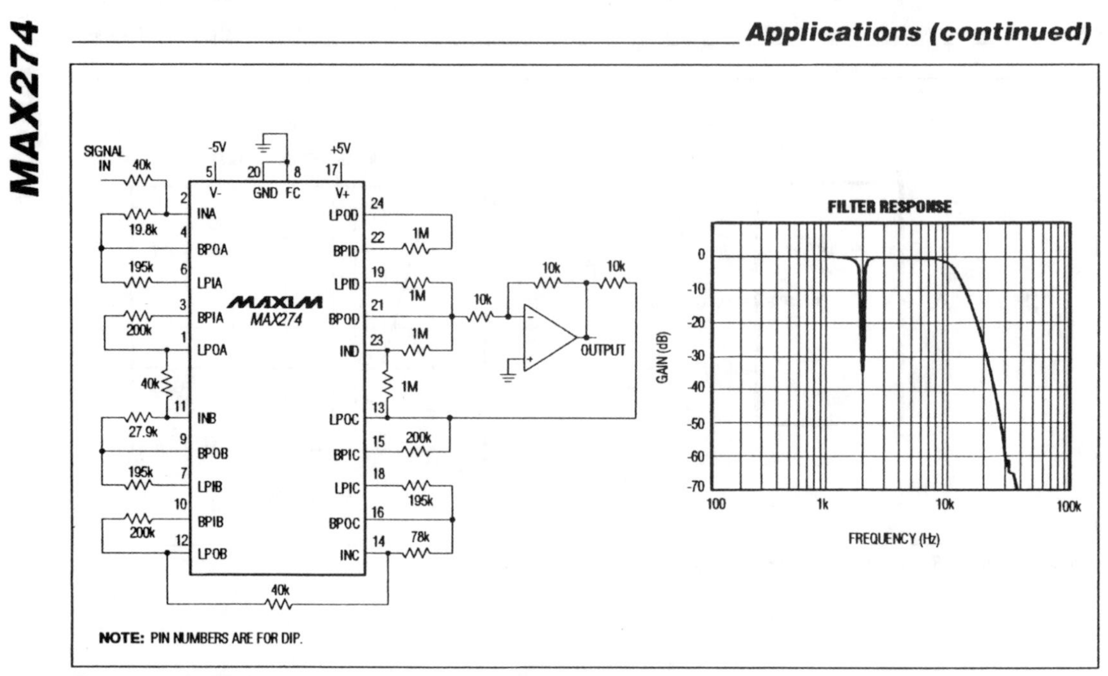


*Figure 11. 10kHz 6th-Order Butterworth Lowpass Filter with 2kHz Notch*

# 8th-Order Continuous-Time Active Filter

## Applications (continued)



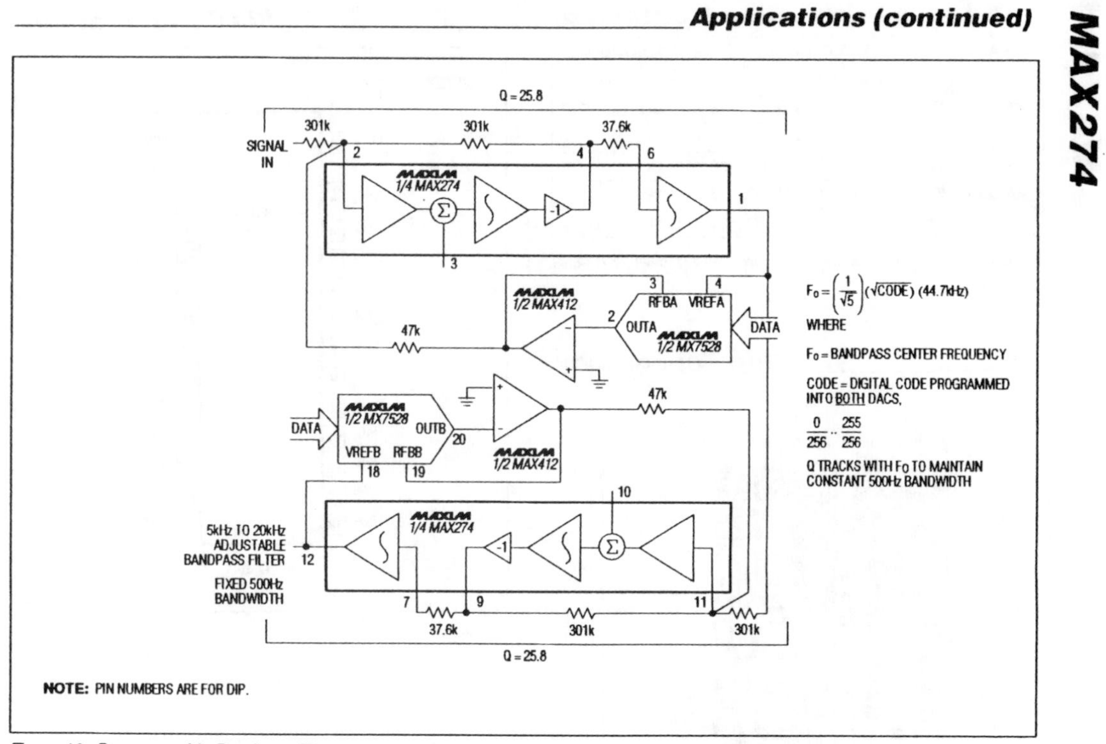


*Figure 12. Programmable Bandpass Filter*

# 8th-Order Continuous-Time Active Filter

MAX274

## Ordering Information (continued)

| PART | TEMP. RANGE | PIN-PACKAGE |
|---|---|---|
| MAX274AENG | -40°C to +85°C | 24 Narrow Plastic DIP |
| MAX274BENG | -40°C to +85°C | 24 Narrow Plastic DIP |
| MAX274AEWI | -40°C to +85°C | 28 Wide SO |
| MAX274BEWI | -40°C to +85°C | 28 Wide SO |
| MAX274AMRG | -55°C to +125°C | 24 CERDIP** |
| MAX274BMRG | -55°C to +125°C | 24 CERDIP** |

** Contact factory for availability and processing to MIL-STD-883.

## Chip Topography

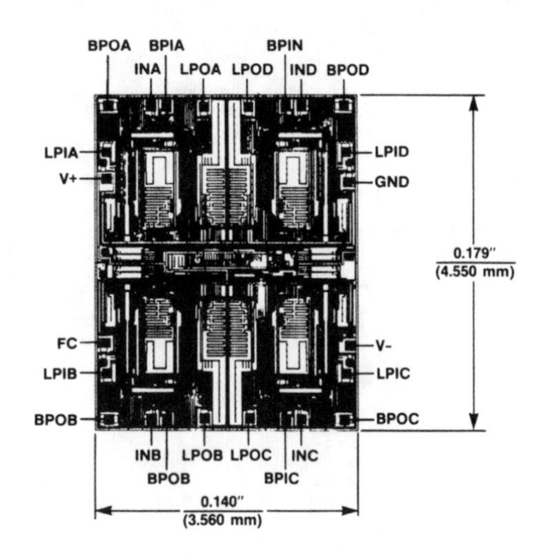


## Pin Configurations

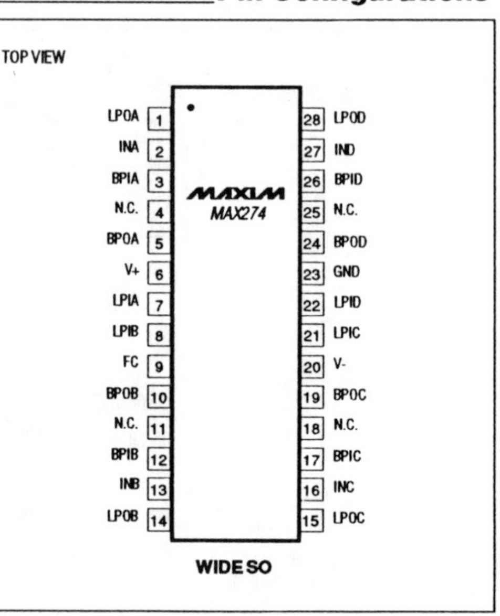

MAXIM

# Voltage-Output, 12-Bit DACs with Internal Reference

MAX507/MAX508

## General Description

The MAX507/MAX508 are complete 12-bit, voltage-output digital-to-analog converters (DACs). The DAC output voltage and the reference have the same polarity, allowing single-supply operation. Both DACs include an internal buried-zener reference. Integrating a DAC, voltage-output amplifier, and reference on one monolithic device greatly enhances reliability over multi-chip circuits.

Double-buffered logic inputs interface easily to microprocessors (μPs). Data is transferred into the input register either from a 12-bit-wide data bus (MAX507) for 16-bit μPs, or in a right-justified (8+4)-bit format (MAX508) for 8- or 16-bit μPs. All logic signals are level triggered and are TTL and CMOS compatible. Interface timing specifications insure compatibility with all common μPs.

The DACs are specified and tested for both dual- and single-supply operation. Usable supplies range from single +12V to dual ±15V.

On-board gain-setting resistors allow three output-voltage ranges: 0V to +5V and 0V to +10V can be generated when using either single or dual supplies. With dual supplies, ±5V is also available. The output amplifier can drive a 2kΩ load to +10V.

## Applications

- Digital Offset and Gain Adjustment
- Industrial Controls
- Arbitrary Function Waveform Generators
- Automatic Test Equipment
- Automated Calibration
- Machine and Motion Control

## Features

- ♦ 12-Bit Voltage Output
- ♦ Internal Voltage Reference
- ♦ Fast μP Interface
- ♦ 12 (MAX507) and 8+4 (MAX508) Data-Bus Widths
- ♦ Single +12V to Dual ±15V Supply Operation
- ♦ 20- and 24-Pin DIP and Wide SO Packages

## Ordering Information

| PART | TEMP. RANGE | PIN-PACKAGE | ERROR (LSBs) |
|---|---|---|---|
| MAX507ACNG | 0°C to +70°C | 24 Narrow Plastic DIP | ±1/2 |
| MAX507BCNG | 0°C to +70°C | 24 Narrow Plastic DIP | ±3/4 |
| MAX507ACWG | 0°C to +70°C | 24 Wide SO | ±1/2 |
| MAX507BCWG | 0°C to +70°C | 24 Wide SO | ±3/4 |
| MAX507BC/D | 0°C to +70°C | Dice* | ±3/4 |
| MAX507AENG | -40°C to +85°C | 24 Narrow Plastic DIP | ±1/2 |
| MAX507BENG | -40°C to +85°C | 24 Narrow Plastic DIP | ±3/4 |
| MAX507AEWG | -40°C to +85°C | 24 Wide SO | ±1/2 |
| MAX507BEWG | -40°C to +85°C | 24 Wide SO | ±3/4 |
| MAX507AMRG | -55°C to +125°C | 24 Narrow CERDIP** | ±1/2 |
| MAX507BMRG | -55°C to +125°C | 24 Narrow CERDIP** | ±3/4 |

**Ordering Information continued on page 12.**

* *Contact factory for dice specifications.*
** *Contact factory for availability and processing to MIL-STD-883.*

## Functional Diagram

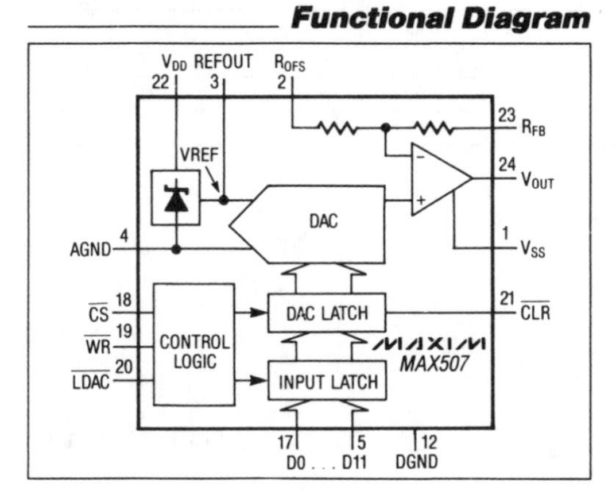


## Pin Configurations

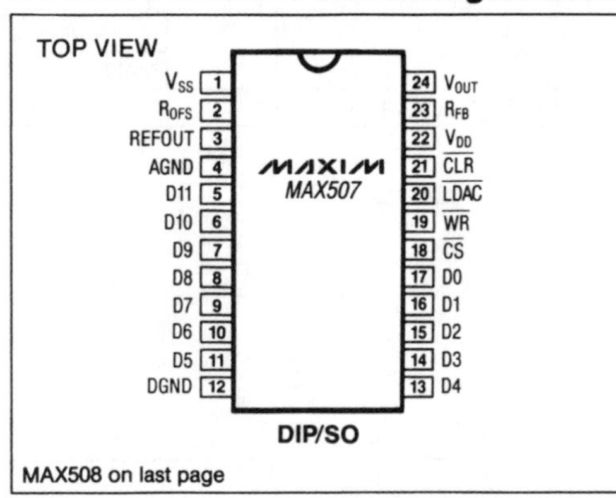


MAX508 on last page

# *Voltage-Output, 12-Bit DACs with Internal Reference*

**MAX507/MAX508**

## ABSOLUTE MAXIMUM RATINGS

$V_{DD}$ to AGND ........................ -0.3V, +17V
$V_{DD}$ to DGND ........................ -0.3V, +17V
$V_{DD}$ to $V_{SS}$ ........................ -0.3V, +34V
AGND to DGND ........................ -0.3V, $V_{DD}$
Digital Input Voltage to GND .......... -0.3V, $V_{DD}$ +0.3V
$V_{OUT}$ to AGND (Note 1) .................. $V_{SS}$, $V_{DD}$
$V_{OUT}$ to $V_{SS}$ (Note 1) .................... 0V, +34V
$V_{OUT}$ to $V_{DD}$ (Note 1) .................... -34V, 0V
REFOUT to AGND (Note 1) ........... -0.3V, $V_{DD}$ +0.3V

Continuous Power Dissipation (any package)
to +75°C ................................ 450mW
derate above +75°C .................... 6mW/°C
Operating Temperature Ranges:
MAX507_C__, MAX508_C__ ............ 0°C to +70°C
MAX507_E__, MAX508_E__ ........... -40°C to +85°C
MAX507_M__, MAX508_M__ ......... -55°C to +125°C
Storage Temperature Range ........... -65°C to +150°C
Lead Temperature (soldering, 10 sec) ........... +300°C

**Note 1:** The output can be shorted to either supply rail if the package power dissipation is not exceeded. Typical short-circuit current to AGND is 25mA.

*Stresses beyond those listed under "Absolute Maximum Ratings" may cause permanent damage to the device. These are stress ratings only, and functional operation of the device at these or any other conditions beyond those indicated in the operational sections of the specification is not implied. Exposure to absolute maximum rating conditions for extended periods may affect device reliability.*

## ELECTRICAL CHARACTERISTICS

Single Supply ($V_{DD}$ = +11.4V to +15.75V, $V_{SS}$ = AGND = DGND = 0V, $R_L$ = 2kΩ, $C_L$ = 100pF, REFOUT unloaded, all grades, $T_A$ = $T_{MIN}$ to $T_{MAX}$, unless otherwise noted.)

| PARAMETER | SYMBOL | CONDITIONS | | MIN | TYP | MAX | UNITS |
|---|---|---|---|---|---|---|---|
| **STATIC PERFORMANCE** | | | | | | | |
| Resolution | N | | | 12 | | | Bits |
| Relative Accuracy | INL | $T_A$ = +25°C | MAX507/508A | | | ±1/2 | LSB |
| | | | MAX507/508B | | | ±3/4 | |
| | | $T_A$ = $T_{MIN}$ to $T_{MAX}$ | MAX507/508A | | | ±3/4 | |
| | | | MAX507/508B | | | ±1 | |
| Differential Nonlinearity | DNL | | | | | ±1 | LSB |
| Unipolar Offset Error | | $T_A$ = +25°C | | | | ±3 | LSB |
| | | $T_A$ = $T_{MIN}$ to $T_{MAX}$ | | | | ±5 | |
| DAC Gain Error | | | | | | ±2 | LSB |
| Full-Scale Output Voltage Error | | $V_{DD}$ = +12V or +15V | $T_A$ = +25°C | | | ±0.2 | %FSR |
| | | | $T_A$ = $T_{MIN}$ to $T_{MAX}$ | | | ±0.6 | |
| Full-Scale Output Voltage Change | | $V_{DD}$ over full range | $T_A$ = +25°C | | | ±0.12 | %FSR/V |
| | | | $T_A$ = $T_{MIN}$ to $T_{MAX}$ | | | ±0.2 | |
| Full-Scale Tempco | | MAX507/508_C/E | | | | ±30 | ppm FSR/°C |
| | | MAX507/508_M | | | | ±40 | |
| Unipolar Offset Error Change | | $V_{DD}$ = +12V ± 5% or +15V ± 5% | | | | ±1 | mV |

# Voltage-Output, 12-Bit DACs with Internal Reference



## ELECTRICAL CHARACTERISTICS (continued)

Single Supply ($V_{DD}$ = +11.4V to +15.75V, $V_{SS}$ = AGND = DGND = 0V, $R_L$ = 2kΩ, $C_L$ = 100pF, REFOUT unloaded, all grades, $T_A$ = $T_{MIN}$ to $T_{MAX}$, unless otherwise noted.)

| PARAMETER | SYMBOL | CONDITIONS | | MIN | TYP | MAX | UNITS |
|---|---|---|---|---|---|---|---|
| **REFERENCE** | | | | | | | |
| Reference Output | | $V_{DD}$ = +12V or +15V | $T_A$ = +25°C | 4.99 | | 5.01 | V |
| Reference Voltage Change | | $V_{DD}$ = +12V ± 5% or +15V ± 5% | $T_A$ = +25°C | | | 2 | mV/V |
| | | | $T_A$ = $T_{MIN}$ to $T_{MAX}$ | | | 6 | |
| Reference Temperature Coefficient | | MAX507/508_C/E | | | ±30 | | ppm/°C |
| | | MAX507/508_M | | | ±40 | | |
| Reference Load Sensitivity | | $I_{LOAD}$ = 0μA to 100μA | | | | ±1 | mV |
| **ANALOG OUTPUT** | | | | | | | |
| Ranges (Note 2) | | | | | | 0 to 5 | V |
| | | | | | | 0 to 10 | |
| Output Range Resistors | | | | 15 | | 30 | kΩ |
| DC Output Impedance | | | | | 0.5 | | Ω |
| Short-Circuit Current | | | | | 40 | | mA |
| **DYNAMIC PERFORMANCE** (Note 3) | | | | | | | |
| Voltage-Output Slew Rate | | | | 2 | | | V/μs |
| $V_{OUT}$ Settling Time | | To ±1/2 LSB for full-scale change | | | | 5 | μs |
| Digital Feedthrough | | | | | 10 | | nV-s |
| Digtal-to-Analog Glitch Impulse | | Major carry transition | | | 30 | | nV-s |
| Output Load Resistance (Note 2) | | $V_{OUT}$ = 0V to +10V | | 2 | | | kΩ |
| **POWER SUPPLIES** | | | | | | | |
| $V_{DD}$ Range | | For specified performance | | 11.4 | | 15.75 | V |
| $I_{DD}$ | | Outputs unloaded | $T_A$ = +25°C | | | 9 | mA |
| | | | $T_A$ = $T_{MIN}$ to $T_{MAX}$ | | | 12 | |

# *Voltage-Output, 12-Bit DACs with Internal Reference*



## ELECTRICAL CHARACTERISTICS

Dual Supply ($V_{DD}$ = +11.4V to +15.75V, $V_{SS}$ = −11.4V to −15.75V, DGND = AGND = 0V, $R_L$ = 2kΩ, $C_L$ = 100pF, REFOUT unloaded, all grades, $T_A$ = $T_{MIN}$ to $T_{MAX}$, unless otherwise noted.)

| PARAMETER | SYMBOL | CONDITIONS | | MIN | TYP | MAX | UNITS |
|---|---|---|---|---|---|---|---|
| **STATIC PERFORMANCE** | | | | | | | |
| Resolution | N | | | 12 | | | Bits |
| Relative Accuracy | INL | $T_A$ = +25°C | MAX507/508A | | | ±1/2 | LSB |
| | | | MAX507/508B | | | ±3/4 | |
| | | $T_A$ = $T_{MIN}$ to $T_{MAX}$ | MAX507/508A | | | ±3/4 | |
| | | | MAX507/508B | | | ±1 | |
| Differential Nonlinearity | DNL | | | | | ±1 | LSB |
| Bipolar Zero Offset Error | BZOE | MAX507/508A | $T_A$ = +25°C | | | ±2 | LSB |
| | | | $T_A$ = $T_{MIN}$ to $T_{MAX}$ | | | ±4 | |
| | | MAX507/508B | $T_A$ = +25°C | | | ±3 | |
| | | | $T_A$ = $T_{MIN}$ to $T_{MAX}$ | | | ±5 | |
| DAC Gain Error | | | | | | ±2 | LSB |
| Full-Scale Output Voltage Error | | $V_{DD}$ = +15V, $V_{SS}$ = −15V | $T_A$ = +25°C | | | ±0.2 | %FSR |
| | | | $T_A$ = $T_{MIN}$ to $T_{MAX}$ | | | ±0.6 | |
| | | $V_{DD}$ = +12V, $V_{SS}$ = −12V | $T_A$ = +25°C | | | ±0.2 | |
| | | | $T_A$ = $T_{MIN}$ to $T_{MAX}$ | | | ±0.6 | |
| Full-Scale Output Change with $V_{DD}$ | | $V_{DD}$ = +12V ± 5% or +15V ± 5% $V_{SS}$ = −12V or −15V | $T_A$ = +25°C | | | ±0.12 | %FSR/V |
| | | | $T_A$ = $T_{MIN}$ to $T_{MAX}$ | | | ±0.2 | |
| Full-Scale Output Change with $V_{SS}$ | $V_{SS}$ | $V_{SS}$ = −12V ± 5% or −15V ± 5% $V_{DD}$ = +12V or +5V | | | | 0.01 | %FSR/V |
| Full-Scale Tempco | | MAX507/508_C/E | | | | ±30 | ppm FSR/°C |
| | | MAX507/508_M | | | | ±40 | |
| Bipolar Zero Offset Change | | $V_{DD}$ = +12V ± 5% or +15V ± 5% $V_{SS}$ = −12V or −15V | | | | ±1 | mV |
| | | $V_{SS}$ = −12V ± 5% or −15V ± 5% $V_{DD}$ = +12V or +15V | | | | ±1 | |
| **REFERENCE** | | | | | | | |
| Reference Output | | $V_{DD}$ = +12V or +15V | $T_A$ = +25°C | 4.99 | | 5.01 | V |
| Reference Output Change | | $V_{DD}$ over full range | $T_A$ = +25°C | | | 2 | mV/V |
| | | | $T_A$ = $T_{MIN}$ to $T_{MAX}$ | | | 6 | |
| Reference Temperature Coefficient | | MAX507/508_C/E | | | ±30 | | ppm/°C |
| | | MAX507/508_M | | | ±40 | | |
| Reference Load Sensitivity | | $I_{LOAD}$ = 0μA to 100μA | | | | ±1 | mV |

# Voltage-Output, 12-Bit DACs with Internal Reference



## ELECTRICAL CHARACTERISTICS (continued)

Dual Supply ($V_{DD}$ = +11.4V to +15.75V, $V_{SS}$ = −11.4V to −15.75V, DGND = AGND = 0V, $R_L$ = 2kΩ, $C_L$ = 100pF, REFOUT unloaded, all grades, $T_A$ = $T_{MIN}$ to $T_{MAX}$, unless otherwise noted.)

| PARAMETER | SYMBOL | CONDITIONS | | MIN | TYP | MAX | UNITS |
|---|---|---|---|---|---|---|---|
| **ANALOG OUTPUT** | | | | | | | |
| Ranges (Notes 2, 4) | | | | | 0 to +5 or +10, -5 to +5 | | V |
| Output Range Resistors | | | | 15 | | 30 | kΩ |
| DC Output Impedance | | | | | 0.5 | | Ω |
| Short-Circuit Current | | | | | 40 | | mA |
| **DYNAMIC PERFORMANCE** (Note 3) | | | | | | | |
| Voltage-Output Slew Rate | | | | 2 | | | V/μs |
| $V_{OUT}$ Settling Time | | to ±1/2 LSB | | | | 5 | μs |
| Digital Feedthrough | | | | | 10 | | nV-s |
| Digtal-to-Analog Glitch Impulse | | Major carry transition | | | 30 | | nV-s |
| Output Load Resistance | | $V_{OUT}$ = -5V to +10V | | 2 | | | kΩ |
| **POWER SUPPLIES** | | | | | | | |
| $V_{DD}$ Range | | For specified performance | | 11.4 | | 15.75 | V |
| $V_{SS}$ Range | | For specified performance | | -11.4 | | -15.75 | V |
| $I_{DD}$ | | Outputs unloaded | $T_A$ = +25°C | | | 9 | mA |
| | | | $T_A$ = $T_{MIN}$ to $T_{MAX}$ | | | 12 | |
| $I_{SS}$ | | Outputs unloaded | $T_A$ = +25°C | | | 3 | mA |
| | | | $T_A$ = $T_{MIN}$ to $T_{MAX}$ | | | 5 | |

# Voltage-Output, 12-Bit DACs with Internal Reference

MAX507/MAX508

## ELECTRICAL CHARACTERISTICS

Single or Dual Supply ($V_{DD}$ = +11.4V to +15.75V, $V_{SS}$ = 0V to −15.75V, DGND = AGND = 0V, REFOUT unloaded, $R_L$ = 2kΩ, $C_L$ = 100pF, all grades, $T_A$ = $T_{MIN}$ to $T_{MAX}$, unless otherwise noted.)

| PARAMETER | SYMBOL | CONDITIONS | | MIN | TYP | MAX | UNITS |
|---|---|---|---|---|---|---|---|
| **DIGITAL INPUTS** | | | | | | | |
| $V_{INH}$ | | | | 2.4 | | | V |
| $V_{INL}$ | | | | | | 0.8 | V |
| Input Current | $I_{IN}$ | D0–D11 | $T_A$ = +25°C | | | ±1 | μA |
| | | | $T_A$ = $T_{MIN}$ to $T_{MAX}$ | | | ±10 | |
| $I_{INH}$ | | $\overline{CS}$, $\overline{WR}$, $\overline{LDAC}$, $\overline{CLR}$ | $T_A$ = +25°C | | | ±1 | μA |
| | | | $T_A$ = $T_{MIN}$ to $T_{MAX}$ | | | ±10 | |
| $I_{INL}$ | | $\overline{CS}$, $\overline{WR}$, $\overline{LDAC}$, $\overline{CLR}$ | $T_A$ = +25°C | | | ±150 | μA |
| | | | $T_A$ = $T_{MIN}$ to $T_{MAX}$ | | | ±200 | |
| Digital Input Capacitance | | | | | 8 | | pF |

## TIMING CHARACTERISTICS

(All grades, $T_A$ = $T_{MIN}$ to $T_{MAX}$, unless otherwise noted.)

| PARAMETER | SYMBOL | CONDITIONS | MIN | TYP | MAX | UNITS |
|---|---|---|---|---|---|---|
| $\overline{CS}$ Pulse Width (Note 5) | $t_1$ | $T_A$ = +25°C | 80 | | | ns |
| | | $T_A$ = $T_{MIN}$ to $T_{MAX}$ | 100 | | | |
| $\overline{WR}$ Pulse Width | $t_2$ | $T_A$ = +25°C | 80 | | | ns |
| | | $T_A$ = $T_{MIN}$ to $T_{MAX}$ | 100 | | | |
| $\overline{CS}$ to $\overline{WR}$ Setup Time (Note 5) | $t_3$ | | 0 | | | ns |
| $\overline{CS}$ to $\overline{WR}$ Hold Time (Note 5) | $t_4$ | | 0 | | | ns |
| Data to $\overline{WR}$ Setup Time | $t_5$ | $T_A$ = +25°C | 100 | | | ns |
| | | $T_A$ = $T_{MIN}$ to $T_{MAX}$ | 110 | | | |
| Data to $\overline{WR}$ Hold Time | $t_6$ | | 10 | | | ns |
| $\overline{LDAC}$ Pulse Width | $t_7$ | $T_A$ = +25°C | 80 | | | ns |
| | | $T_A$ = $T_{MIN}$ to $T_{MAX}$ | 100 | | | |
| $\overline{CLR}$ Pulse Width (MAX507) | $t_8$ | $T_A$ = +25°C | 80 | | | ns |
| | | $T_A$ = $T_{MIN}$ to $T_{MAX}$ | 100 | | | |

**Note 2:** $V_{OUT}$ must be less than ($V_{DD}$ - 2.5V).
**Note 3:** Dynamic performance is included for design guidance, not subject to test.
**Note 4:** The 0V to +5V or +10V ranges can be used with $V_{SS}$ = −5V with no degradation.
**Note 5:** $\overline{CS}$ = $\overline{CSLSB}$ and $\overline{CSMSB}$ for MAX508.

# Voltage-Output, 12-Bit DACs with Internal Reference



## Detailed Description

### Digital-to-Analog Converters

The MAX507/MAX508 are 12-bit voltage-output DACs. The DAC output voltage has the same polarity as the reference, allowing single-supply operation.

The basic DAC circuit consists of a laser-trimmed, thin-film, R-2R resistor array with NMOS voltage switches (Figure 1).

### Output-Buffer Amplifier

The output amplifier is noninverting and configurable for a gain of 1 or 2. Three output voltage ranges can be configured for: 0V to +5V, 0V to +10V, and -5V to +5V. The output amplifier can drive 2kΩ in parallel with 100pF connected to GND.

The MAX507/MAX508 can operate from a single supply with a 0V to +5V or a 0V to +10V output range by tying $V_{SS}$ to 0V. However, the speed and current-sinking capability of the amplifier decreases as the output falls within 0.5V of $V_{SS}$. Speed and current-sinking capability can be maintained by including a negative supply. Table 1 lists the allowable single and dual supplies for each range.

The output amplifier's small-signal bandwidth is typically 2MHz. Output noise is approximately 25nV/$\sqrt{Hz}$ at 1kHz, and output broadband noise is approximately 25$\mu V_{RMS}$.

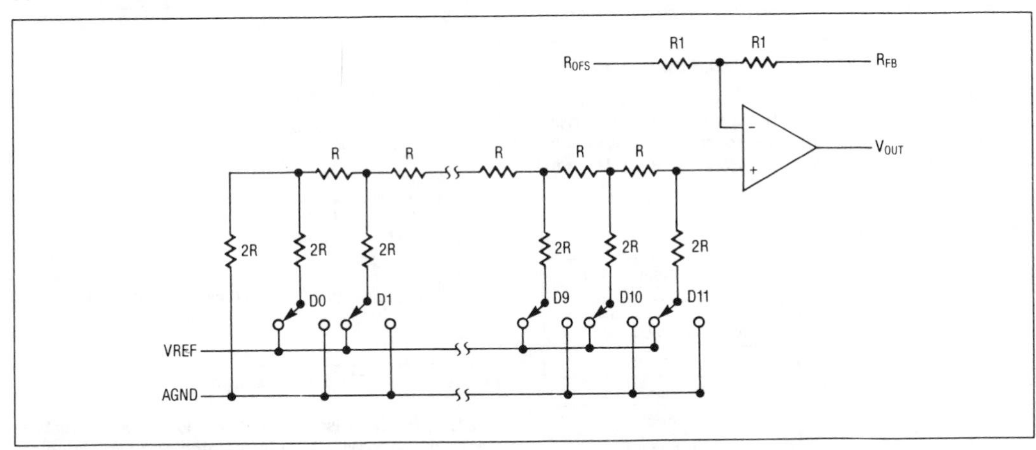


*Figure 1. Simplified MAX507 DAC Circuit*

**Table 1. Output Voltage Range vs. Supply Voltage**

| Range | Single Supply $V_{DD}$ | Dual Supply $V_{DD}$ | Dual Supply $V_{SS}$ |
|---|---|---|---|
| 0V to +5V | +11.4V to +15.75V | +11.4V to +15.75V | -4.5V to -15.75V |
| 0V to +10V | +14.25V to +15.75V | +14.25V to +15.75V | -4.5V to -15.75V |
| -5V to +5V | | +11.4V to +15.75V | -11.4V to -15.75V |

# Voltage-Output, 12-Bit DACs with Internal Reference

### Voltage Reference

The voltage at REFOUT is 5V ± 10mV at +25°C. The reference is internally connected to the DAC and is buffered to accommodate the DAC's variable impedance. This buffer is capable of driving the DAC, the $R_{OFS}$ resistor, and up to 500µA of external current. MAX507/MAX508 specifications are determined with the internal reference. The reference should be decoupled at REFOUT with 10Ω in series with the recommended decoupling capacitors, 10µF in parallel with 0.1µF.

### Digital Inputs and Interface Logic

All logic inputs are compatible with both TTL and 5V CMOS logic. Supply current is specified for TTL input levels, but is reduced by about 450µA when the data inputs are driven near DGND or $V_{DD}$. The control inputs ($\overline{CLR}$, $\overline{LDAC}$, $\overline{WR}$, $\overline{CS}$, $\overline{CSMSB}$, and $\overline{CSLSB}$) each draw 100µA from $I_{DD}$ when low.

### MAX507 Interface

Table 2 is the MAX507 truth table. The MAX507 accepts a 12-bit input word that can be latched or transferred directly to the DAC. $\overline{CS}$ and $\overline{WR}$ control the input latch, and $\overline{LDAC}$ transfers information from the input latch to the DAC latch.

**Table 2. MAX507 Truth Table**

| $\overline{CLR}$ | $\overline{LDAC}$ | $\overline{WR}$ | $\overline{CS}$ | Function |
|---|---|---|---|---|
| 1 | 0 | 0 | 0 | Both latches transparent |
| 1 | 1 | 1 | X | Both latches latched |
| 1 | 1 | X | 1 | Both latches latched |
| 1 | 1 | 0 | 0 | Input latch transparent |
| 1 | 1 | ↑ | 0 | Input latch latched |
| 1 | 0 | 1 | 1 | DAC latch transparent |
| 1 | ↑ | 1 | 1 | DAC latch latched |
| 0 | X | X | X | DAC latch all 0s |
| ↑ | 1 | 1 | 1 | DAC latch latched with 0s; output at 0V or -5V |
| ↑ | 0 | 0 | 0 | Both latches transparent; output follows input data |

1 = High State  X = Don't Care
0 = Low State  ↑ = Rising Edge

The input latch is transparent when $\overline{CS}$ and $\overline{WR}$ are low; the DAC latch is transparent when $\overline{LDAC}$ is low. Data is latched within the input latch on the rising edge of $\overline{WR}$ when $\overline{CS}$ is low. The rising edge of $\overline{LDAC}$ latches data into the DAC when $\overline{CS}$ and $\overline{WR}$ are low. After $\overline{CS}$ and $\overline{WR}$ are high, $\overline{LDAC}$ must be held low for $t_7$ or longer (Figure 2).

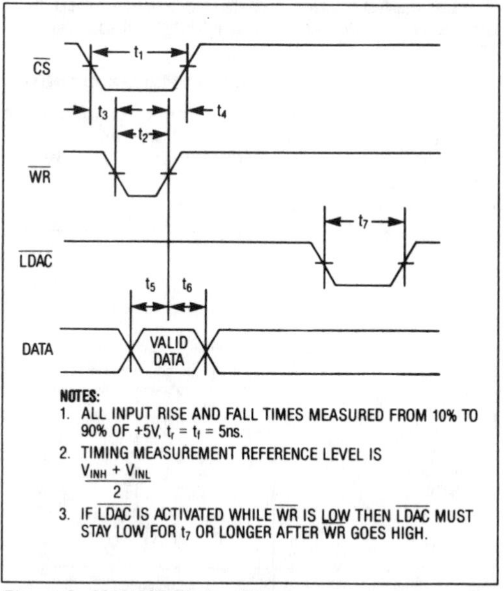

CS
WR
LDAC
DATA
VALID DATA
t1
t2
t3
t4
t5
t6
t7
NOTES:
1. ALL INPUT RISE AND FALL TIMES MEASURED FROM 10% TO 90% OF +5V, tr = tf = 5ns.
2. TIMING MEASUREMENT REFERENCE LEVEL IS (VINH + VINL)/2
3. IF LDAC IS ACTIVATED WHILE WR IS LOW THEN LDAC MUST STAY LOW FOR t7 OR LONGER AFTER WR GOES HIGH.

*Figure 2. MAX507 Timing Diagram*

The DAC latch is reset to zeros with $\overline{CLR}$ low. $\overline{CLR}$ acts as a zero override when the input latch and DAC latch are transparent. Then, a low-to-high $\overline{CLR}$ transition loads all zeros into the DAC latch, and the output remains low (0V to -5V).

### MAX508 Interface

The MAX508's 8-bit-wide data bus interfaces with 8-bit µPs. The MAX508 contains an input latch and a DAC latch. The data held in the DAC latch determines the output of the DAC. Table 3 is the MAX508 truth table, Figure 3 shows the input control logic, and Figure 4 shows the write-cycle timing.

# Voltage-Output, 12-Bit DACs with Internal Reference



**Table 3. MAX508 Truth Table**

| $\overline{CSLSB}$ | $\overline{CSMSB}$ | $\overline{WR}$ | $\overline{LDAC}$ | Function |
|---|---|---|---|---|
| 0 | 1 | 0 | 1 | Loads LSBs to input latches |
| 0 | 1 | ↑ | 1 | Locks LSBs in input latches |
| ↑ | 1 | 0 | 1 | Locks LSBs in input latches |
| 1 | 0 | 0 | 1 | Loads MSBs to input latches |
| 1 | 0 | ↑ | 1 | Locks MSBs in input latches |
| 1 | ↑ | 0 | 1 | Locks MSBs in input latches |
| 1 | 1 | 1 | 0 | Loads input into DAC latch |
| 1 | 1 | 1 | ↑ | Locks input into DAC latch |
| 1 | 0 | 0 | 0 | Loads MSBs to input latches and loads input into DAC latch |
| 1 | 1 | 1 | 1 | No data transfer |

1 = High State  0 = Low State  ↑ = Rising Edge

Right-justified data is loaded into the MAX508 using $\overline{CSMSB}$, $\overline{CSLSB}$, and $\overline{WR}$. Data can be latched into the input latch on the rising edge of $\overline{WR}$ for the most significant bit (MSB) and least significant bit (LSB), or on the rising edge of $\overline{CSMSB}$ for the MSB and $\overline{CSLSB}$ for the LSB. Either the MSB or the LSB can be loaded first.

The complete, 12-bit word loads into the DAC register when $\overline{LDAC}$ is low, and latches on $\overline{LDAC}$'s rising edge. $\overline{LDAC}$ is asynchronous and independent of $\overline{WR}$, so it is ideal for simultaneously updating multiple MAX508 outputs. Because $\overline{LDAC}$ can occur during a write cycle, it must stay low for $t_7$ (or longer) after $\overline{WR}$ goes high to ensure correct data is latched to the output.

The MAX508 output can be updated in two write cycles by tying $\overline{CSMSB}$ and $\overline{LDAC}$. In this automatic transfer mode, $\overline{CSLSB}$ and $\overline{WR}$ latch the lower 8 bits into the input latch; then $\overline{CSMSB}$, $\overline{WR}$, and $\overline{LDAC}$ load the upper 4 bits into the input latch and transfer the 12-bit word into the DAC latch. Alternatively, the MAX507 can be updated in two writes by tying $\overline{CSLSB}$ to $\overline{LDAC}$ if the upper 4 bits are input first, followed by the lower 8 bits.

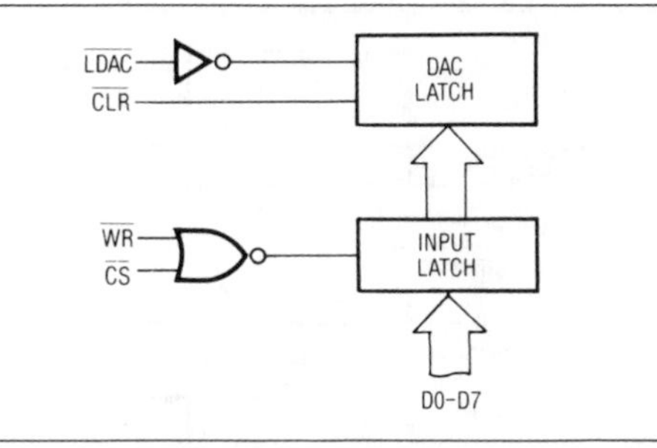


*Figure 3a. MAX507 Input Control Logic*

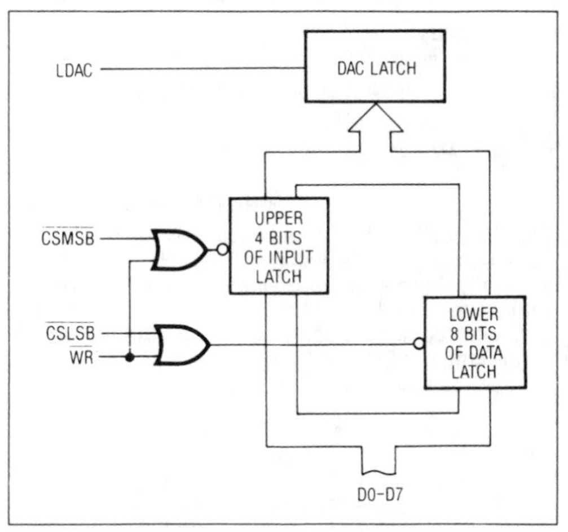


*Figure 3b. MAX508 Input Control Logic*

# Voltage-Output, 12-Bit DACs with Internal Reference



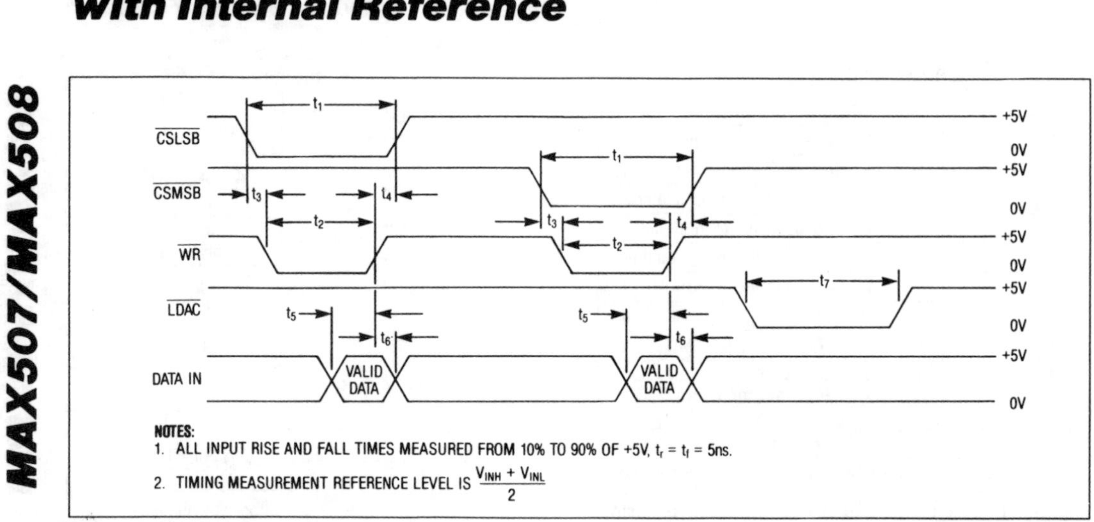


*Figure 4. MAX508 Timing Diagram*

### *Unipolar Configuration*

The MAX507/MAX508 are set up for a 0V to +5V unipolar output range by connecting $R_{OFS}$, $R_{FB}$, and $V_{OUT}$ (Figure 5). The converters operate from either a single or a dual supply in this configuration. See Table 4 for the DAC-latch contents (input) vs. analog output (output). In this range, 1LSB = VREF ($2^{-12}$).

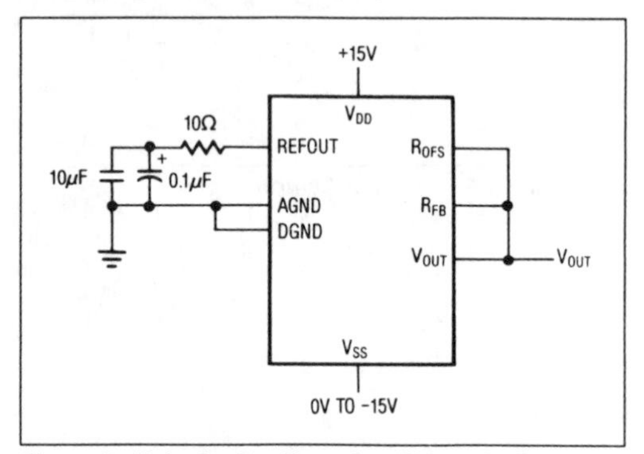

*Figure 5. Unipolar Configuration (0V to +5V Output)*

**Table 4. Unipolar-Code Table (0V to +5V Output)**

| INPUT | OUTPUT |
|---|---|
| 1111 1111 1111 | (VREF) $\frac{4095}{4096}$ |
| 1000 0000 0001 | (VREF) $\frac{2049}{4096}$ |
| 1000 0000 0000 | (VREF) $\frac{2048}{4096}$ =+VREF/2 |
| 0111 1111 1111 | (VREF) $\frac{2047}{4096}$ |
| 0000 0000 0001 | (VREF) $\frac{1}{4096}$ |
| 0000 0000 0000 | 0V |

A 0V to +10V unipolar output range is set up by connecting $R_{OFS}$ to AGND and $R_{FB}$ to $V_{OUT}$ (Figure 6). See Table 5 for the DAC-latch contents (input) vs. analog output (output). The MAX507/MAX508 operate from either a single or a dual supply in this configuration. In this range, 1LSB = VREF ($2^{-11}$).

# Voltage-Output, 12-Bit DACs with Internal Reference



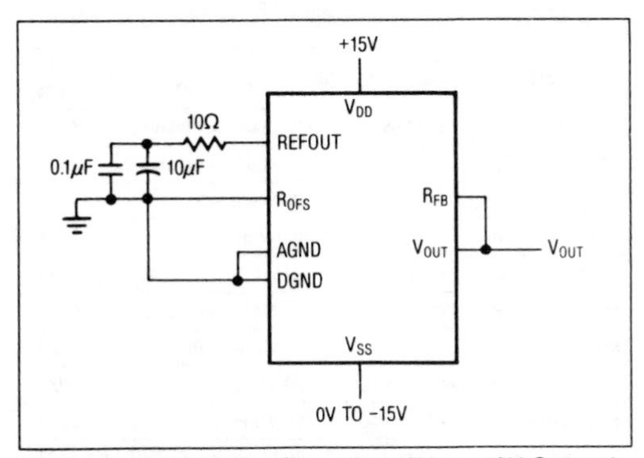


*Figure 6. Unipolar Configuration (0V to +10V Output)*

**Table 5. Unipolar-Code Table (0V to +10V Output)**

| INPUT | | | OUTPUT |
|---|---|---|---|
| 1111 | 1111 | 1111 | $+2\ (VREF)\ \frac{4095}{4096}$ |
| 1000 | 0000 | 0001 | $+2\ (VREF)\ \frac{2049}{4096}$ |
| 1000 | 0000 | 0000 | $+2\ (VREF)\ \frac{2048}{4096} = +VREF$ |
| 0111 | 1111 | 1111 | $+2\ (VREF)\ \frac{2047}{4096}$ |
| 0000 | 0000 | 0001 | $+2\ (VREF)\ \frac{1}{4096}$ |
| 0000 | 0000 | 0000 | 0V |

### Bipolar Configuration

A -5V to +5V bipolar range is set up by connecting $R_{OFS}$ to REFOUT and $R_{FB}$ to $V_{OUT}$, and operating from dual power supplies (Table 1). See Table 6 for the DAC-latch contents (input) vs. analog output (output). In this range, 1LSB = (2) VREF ($2^{-11}$) = (VREF) 1/2048.

**Table 6. Bipolar-Code Table (-5V to +5V Output)**

| INPUT | | | OUTPUT |
|---|---|---|---|
| 1111 | 1111 | 1111 | $(+VREF)\ \frac{2047}{2048}$ |
| 1000 | 0000 | 0001 | $(+VREF)\ \frac{1}{2048}$ |
| 1000 | 0000 | 0000 | 0V |
| 0111 | 1111 | 1111 | $(-VREF)\ \frac{1}{2048}$ |
| 0000 | 0000 | 0001 | $(-VREF)\ \frac{2047}{2048}$ |
| 0000 | 0000 | 0000 | $(-VREF)\ \frac{2048}{2048} = -VREF$ |

# Voltage-Output, 12-Bit DACs with Internal Reference

MAX507/MAX508

## Pin Configurations (continued)

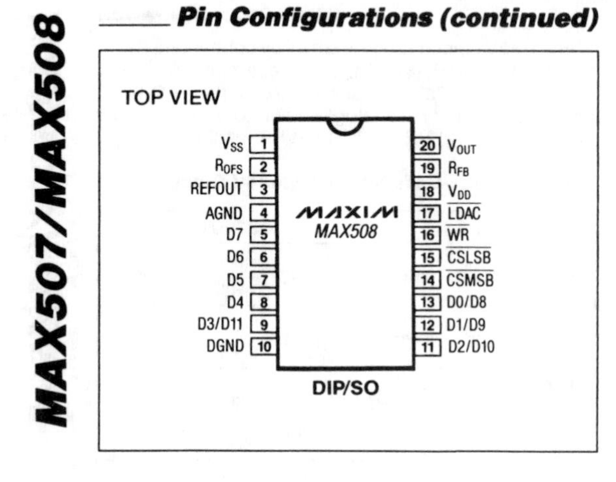


## Ordering Information (continued)

| PART | TEMP. RANGE | PIN-PACKAGE | ERROR (LSBs) |
|---|---|---|---|
| MAX508ACPP | 0°C to +70°C | 20 Narrow Plastic DIP | ±1/2 |
| MAX508BCPP | 0°C to +70°C | 20 Narrow Plastic DIP | ±3/4 |
| MAX508ACWP | 0°C to +70°C | 20 Wide SO | ±1/2 |
| MAX508BCWP | 0°C to +70°C | 20 Wide SO | ±3/4 |
| MAX508BC/D | 0°C to +70°C | Dice* | ±3/4 |
| MAX508AEPP | -40°C to +85°C | 20 Narrow Plastic DIP | ±1/2 |
| MAX508BEPP | -40°C to +85°C | 20 Narrow Plastic DIP | ±3/4 |
| MAX508AEWP | -40°C to +85°C | 20 Wide SO | ±1/2 |
| MAX508BEWP | -40°C to +85°C | 20 Wide SO | ±3/4 |
| MAX508AMJP | -55°C to +125°C | 20 Narrow CERDIP** | ±1/2 |
| MAX508BMJP | -55°C to +125°C | 20 Narrow CERDIP** | ±3/4 |

* *Contact factory for dice specifications.*
** *Contact factory for availability and processing to MIL-STD-883.*

# Index


## A

A/D conversion, 84
accuracy, 263
active filters, 95
adaptive filtering, 331
additivity, 13
aliasing, 147
all pole, 65
allpass filter, 62, 79
amplitude response, 221
analog, 4
analytic, 134
angular frequency, 44
antisymmetrical, 141
aperture jitter, 85
aperture time, 84
approximating functions, 17
Argand diagram, 28
autocorrelation, 332

## B

bandpass filter, 61, 84
bandstop, 85
bandwidth, 84, 117
barrel shifter, 301, 312
basis functions, 17
Bessel-like polynomials, 78
bi-quad filter, 108
bi-quad transfer function, 108
bilateral Z transform, 203
bilinear transform, 202, 241
bit reversal, 166, 170, 282, 308
Blackman, 178
block floating point, 301
BLT, 241
butterfly, 168
Butterworth, 65, 99, 104

## C

capacitor, 25, 50
cascade structures, 248–249
cascading lower-order filters, 95
causal, 8
center frequency, 61
Chebyshev, 68, 100

Chebyshev polynomials, 68
chopping, 272
circular buffer, 260, 281, 314
classic FFT, 179
combine, 191
complex frequency, 44
continuous spectra, 41, 136
continuous time signals, 4
continuous-time filter, 115, 123
convergence, 36
convolution, 11, 135, 156, 172, 231, 329, 332
convolution in frequency domain, 255
convolution in the frequency domain, 231–235
convolution in the time domain, 235
convolution in time domain, 258
convolution sum, 159
convolution with a rectangular pulse, 173
correlation, 158, 332
cosine bell, 177
critical resistance, 27
critically damped, 29, 91
cross-correlation, 332
current, 25
cutoff frequency, 59
cyclic convolution, 161

## D

D/A, 86
DACQ, 120, 181
damping, 90
damping constant, 27
damping ratio, 27
data acquisition, 54
decimation, 58, 146, 152
decimation in frequency, 169
decimation in time, 168
decimation in time, 179
deconvolution in frequency domain, 236
deconvolution in time domain, 238
delay equalizers, 79
delta function, 51
derivatives, 47
Dirac delta function, 10, 146
Dirac delta function, 51
direct form I, 249, 261
direct form II, 250
Dirichlet conditions, 36
discontinuity, 37
discrete Fourier transform, 159
discrete spectrum, 41
division, 274
domain, 13
dot product, 14
downsampling, 152
drive capability (op amp), 119
dynamic range of A/D, 57, 86

## E

elliptic, 72
ENOB, 58
equal-ripple approximation, 69
Euler's equation, 39
even and odd, 171–172
even function, 37

## F

fast Fourier transform, 166
fast Hartley transform, 189
FDNR, 111–112
FFT, 290, 308, 328
filter response, 89
filter transformations, 83
filters, 56
finite duration sequence, 207
finite impulse response, 202
finite length arithmetic, 263
FIR, 201, 305, 315
fixed point, 264
floating point, 264
forced oscillators, 23
Fourier coefficients, 138
Fourier integral, 41

Fourier series, 33, 137
Fourier series method, 202, 226, 256
Fourier transform, 143, 153–159, 172, 204, 217
fractional binary form, 267
frequency domain, 4
frequency domain response, 135
frequency sampling method, 202
frequency sampling method, 224
frequency shift, 47, 156
frequency transformation, 246
frequency warping, 244
frequency-dependent negative resistor, 111
frequency-dependent voltage divider, 87
friction, 21

## G

general immitance networks, 86
generalized immitance converter, 111
generalized linear phase system, 221
Gibbs phenomenon, 38, 176
GIC, 96, 111, 121
GIN, 112
Goertzel, 328
group delay, 59, 64, 75
group delay calculations, 77

## H

half-power frequencies, 33
half-power points, 32, 60
Hamming, 178
harmonic analysis, 36
harmonic signal, 19
harmonics, 33, 137, 154
Hartley transform, 171, 187
harvard architecture, 300
hidden bit, 265
highpass filter, 62, 88
homogenous differential equation, 33
homogenous linear differential equation, 19
Horner's Rule, 276

## I

ideal brickwall filter, 219
ideal transmission, 74
IIR, 201, 307, 315
imaginary axis, 29
impulse function, 5, 159
impulse invariant method, 239
impulse response, 9, 134
inductor, 25
infinite gain filter, 109
infinite impulse response, 202
infinite impulse response filter, 212
infinite sum, 138
initial conditions, 22, 42
input currents (op amp), 118
interpolation, 146, 153, 276
inverse Z transform, 208

## K

Kaiser window, 196
Kaiser/Bessel, 178

## L

Laguerre root finding, 78
Lanczo's $\sigma$ factors, 38
Laplace, 203
Laplace transform, 44, 159
lattice, 252
leakage, 173
limit cycle, 216, 271
line spectrum, 41, 136, 144
linear constant coefficient difference equation, 212
linear independence, 13
linear phase, 73, 79, 82, 220
linear phase filter types, 221–224
linear phase filters, 73
linear time invariant filters, 213
linearity, 8, 156

LM12H458, 123
LMS, 334
lowpass, 61, 83, 87
lowpass filter transfer function, 94
LRC network, 92
LRC to FDNR, 113

## M

magnitude, 81, 135
magnitude squared function, 64
magnitude squared response, 51
maximally flat system response, 64
modified Bessel function, 200
monotonic, 59
moving average filter, 213
MSE, 333
multiple feedback networks, 109
multiplicative, 8, 13, 156

## N

natural frequency, 22
neper, 44
network analysis, 49
Newton's method, 275
noncyclical convolution, 161
nonrecursive difference equation, 213
nonrecursive filter, 220
norm, 15
normalization, 312
notch filter, 62, 85
numerical methods, 254
Nyquist, 55, 147, 161

## O

odd and even, 140–143, 145
odd function, 37
one-sided Laplace transform, 44
operational transconductance amplifier, 111
order of a filter, 81
orthogonal, 14
orthogonality, 5
orthonormal, 15
oscill.cpp, 181
oscill.exe, 126
oscillate, 19
OTA, 111
output voltage swing, 116
overdamped, 91
overflow, 270
oversampling, 58
overshoot, 135, 176

## P

Parseval's theorem, 156, 158, 332
passband, 58–59
period, 19, 41
periodic, 35, 48
periodic function, 6
periodic signal, 137
periodicity, 163
permutation, 170
permute, 190
phase, 59, 64, 135
phase characteristics, 74
phase delay, 76, 220, 274
phase distortion, 56
phase factors, 162
phase response, 77
phase shift, 59
piecewise continuous, 37
pipelined, 302
poles, 31
poles and zeros, 29
poles and zeros of allpass, 81
power factor, 93
precision, 263
properties of the Laplace transform, 46
prototype filter, 83
pulse train, 147

## Q

Q15, 31, 33, 66, 90, 94, 95, 264
quadratic equation, 26
quantization, 271

## R

radian frequency, 44
radix, 4
decimation in frequency FFT, 317
ramp function, 11
RC filter, 90
reactance, 91
real FFT, 186
real transform, 171
rectangular function, 152, 177
recursive difference equations, 212
region of convergence, 206
Remez exchange method, 225
resistor, 25, 50
resolution of A/D, 86
resonance, 31, 32, 60
resonant circuit, 90, 93
resonant frequency, 90
ringing, 176
rise time, 60
rolloff, 60
rms quantization noise level, 57
roll off, 56, 57, 89
ROM tables, 283
root locus, 29
rounding, 272

## S

s to z plane mapping, 215, 245
s-plane, 45, 204
Sallen and Key, 98, 123
Sallen-Key, 98
sample frequency, 4
sample function, 10
sample interval, 4
sample-and-hold, 55, 85
sampling, 140, 146–152
sampling theorem, 4
saturation, 273
scaled data, 94
scaling, 9
second-order filter, 90
settling time, 60, 117, 135
Shannon, 147
shift theorem, 166, 173
signal, 1, 19
signed magnitude, 265
similarity theorem, 156, 158, 166
simple harmonic oscillator, 19
sinc, 143
singular functions, 50
slew rate, 116
SNR, 58, 116
sources of noise, 120
spectra, 136
spectral leakage, 176
spring constant, 21
state variable filter, 106
steady state, 42
steady state error, 135
steady state response, 23, 42
step function, 11
step response, 135
stopband, 59
sum of products, 258, 313
superposition, 9
switched capacitor, 86, 114
symmetrical, 141
symmetry, 37
synthesis, 134
system identification, 335

## T

tables, 278
tapped delay filter, 220
tee circuit, 119

third-order Sallen and Key, 102
Thomson, 73, 101
time domain, 4
time domain response, 134
time invariance, 8
time shifting, 156, 159
transfer function, 7
transforms in the limit, 137
transients, 42
transition band, 56, 64
transversal filter, 220
truncation, 272
turn-off transient, 42
turn-on transient, 42
twiddle factors, 162
two-side Laplace transform, 44

## U

unbiased rounding, 272
undamped natural angular frequency, 27
underdamped, 91
unilateral Z transform, 203
unit circle, 204
unit impulse, 146, 151
unit step function, 152

## V

VCIS, 111
VCVS, 59, 96, 98
voltage follower, 96
voltage-controlled voltage source, 96
Von Neumann architecture, 300
Vos, 117

## W

white noise, 274
width of main lobe, 174
window weighting, 173
windowing, 173
windows, 173, 196, 229–231

## Z

Z tranform, 203
zero phase frequency response, 221
zeros, 29, 31

POSTAGE WILL BE PAID BY ADDRESSEE

**JOHN WILEY & SONS, INC.**
**Attn: Continuation Department**
**P.O. Box 6792**
**Somerset, N.J. 08875-9976**

**PRACTICAL DSP MODELING, TECHNIQUES, AND PROGRAMMING IN C**

Programs are available on a 3½ inch high-density disk.

The companion diskette contains the code examples from the book, a software digital oscilloscope, and Mathcad documents that model some of the analysis tools described in the book.

---

*DISKETTE:* Please send me_______ copy(ies) of the diskette for use with the book **PRACTICAL DSP MODELING, TECHNIQUES, AND PROGRAMMING IN C** at $13.00 each. Morgan/Practical DSP Modeling, Techniques, and Programming in C diskette, ISBN: 0-471-00613-0.

*BOOK:* Please send me _______ copy(ies) of the book **MORGAN/PRACTICAL DSP MODELING, TECHNIQUES, AND PROGRAMMING IN C**, at $36.95, ISBN: 0-471-00606-8.

*BOOK/DISK SET*: Please send me _______ copy(ies) of the book/disk set for **MORGAN/PRACTICAL DSP MODELING, TECHNIQUES, AND PROGRAMMING IN C**, at $49.95, ISBN: 0-471-00434-0.

☐ Payment enclosed ☐ Visa ☐ MasterCard ☐ American Express

Card Number____________________________ Expiration Date___________

Signature_____________________________________________________

NAME_________________________________________________________

COMPANY NAME_______________________________________________

ADDRESS_____________________________________________________

CITY/STATE_________________________________ ZIP CODE___________